Elementary Algebra

Fourth Edition

Vivian Shaw Groza
Sacramento City College

BROOKS/COLE
THOMSON LEARNING ™

Printed in the United States of America

Brooks.Cole
511 Forest Lodge Road
Pacific Grove, CA 93950
USA

For information about our products, contact us:
Thomson Learning Academic Resource Center
1-800-423-0563
http://www.brookcole.com

International Headquarters
Thomson Learning
International Division
290 Harbor Drive, 2nd Floor
Stamford, CT 06902-7477
USA

UK/Europe/Middle East/South Africa
Thomson Learning
Berkshire House
168-173 High Holborn
London WCIV 7AA

Asia
Thomson Learning
60 Albert Street, #15-01
Albert Complex
Singapore 189969

Canada
Nelson Thomson Learning
1120 Birchmount Road
Toronto, Ontario MIK 5G4
Canada
United Kingdom

ISBN 0-15-504431-1

The Adaptable Courseware Program consists of products and additions to existing Brooks/Cole products that are produced from camera-ready copy. Peer review, class testing, and accuracy are primarily the responsibility of the author(s).

PREFACE

Revisions for this 4th edition of ELEMENTARY ALGEBRA are based on instructors' suggestions from extensive classroom use.

Word problems no longer appear as a separate chapter but have been integrated throughout the text. New applications have been inserted and old problems have been simplified and modernized.

The student gains an early introduction to verbal problems in Chapters 1 and 2, where there is preparation by translation and evaluation of formulas. The formulas now include percentage applications. Then, in Chapter 3, Linear Equations in One Variable, formula and percentage word problems are presented. Chapter 3 also includes number and age problems.

The solution of quadratic equations by factoring, including problems based on the Theorem of Pythagoras, have been moved to Chapter 4, Operations on Polynomials. In Chapter 5, Algebraic Fractions, a new section on decimals has been added. The ratio and proportion problems also appear in this chapter.

The mixture and work problems now occur in Chapter 6, Fractional Equations. The mixture problems are preceded by a section on equations involving decimal and rational coefficients. The work problems are preceded by a section dealing with equations typical of work problems. The units on mixture and work problems are each divided into sections dealing with the various types of these problems.

New slope applications appear in Chapter 7, Graphing Points and Lines, and motion problems are placed in Chapter 8, Linear Systems. New space applications have been added to the word problems of Chapter 9, Radicals and Quadratic Equations.

Another new feature of this edition is the use of Marginal Notes. These are used to highlight important ideas, to provide further explanation, and to caution the student against common errors.

Many of the discussions in the exposition have been simplified, augmented, or modified. Also some new sections have been added. The Exercise Sets have been improved, modified, and brought up to date. There are also some new exercise sets.

In Chapter 1, the opening section on the operations of algebra has been rewritten with a special section on the numbers 0 and 1. There is a new section on Formulas, introducing percentage and concepts used in later word problems.

The absolute value discussion has been changed in Chapter 2, and subtraction of signed numbers now includes number line examples. There is also a new section here on reciprocals, and a new problem set on evaluations and substitution of signed numbers.

The axioms that were formerly introduced in Chapter 1 have now been moved to Chapter 3 where their application is more immediate. New and revised word problems also occur in Chapter 3.

In Chapter 5, the exposition has been simplified and the section on Equal Fractions and Simplification has been rewritten. The sections on Multiplication and Division have been moved to precede those on Addition and Subtraction. This seems to be a more natural and logical order. A section on LCM has been inserted for Addition and Subtraction. The Sample Problems have been improved and more explanations have been provided.

Major revisions have been made in Chapter 6, Fractional Equations. The Sample Problems have been redone, simplified, and improved with better explanations. The method of solution has been changed to that of multiplying both sides by the LCD. The various types of equations have been divided into sections with appropriate examples for each type.

Chapter 7 on Graphing contains a new optional section on Families of Lines and Equations of Lines, a request of several reviewers. Also new slope applications have been added.

There are some minor revisions in Chapter 8 on Linear Systems. A new feature is the presence of motion problems, now moved to this chapter.

The applications in Chapter 9, Radicals and Quadratic Equations, have been improved and modernized. In the section on solution by completing the square, the method for solving $ax^2 + bx + x = 0$ has been changed to "first divide by a". This technique is then used in the next section to derive the quadratic formula.

Finally, a human touch has been added by the artwork placed at the beginning of each chapter. Here the student sees the relationship of algebra to sports, recreation and travel, the home, careers, consumer matters, business, transportation, and space. Applications in the word problems for each chapter are directly related to the opening art for the chapter.

There is an Instructor's Manual to accompany the text. It contains 4 tests for each chapter. It also contains 6 Final Examinations, 3 of the traditional type, and 3 multiple choice. In addition, the manual provides chapter comments and suggestions for the use of the text. This includes Sample Lecture Outlines.

ACKNOWLEDGEMENTS

I would like to thank the following people for their many helpful and constructive suggestions:

Nancy Calland, Clark Technical College
Troy Chaffin, Gaston College
Chris Kardaras, Wright College
Jack Lobb, Cumberland County College
Jean Moran, Donnelly College
Donald Poulson, Mesa Community College
Larry Smith, Midplains Community College
Shirley Sorenson, University of Maryland
Ann Tiber, Washington Technical College
Doug Vollom, Vermilion Community College
Sharon Wellman, Everett Community College

Vivian Shaw Groza

CONTENTS

Numbers and Operations

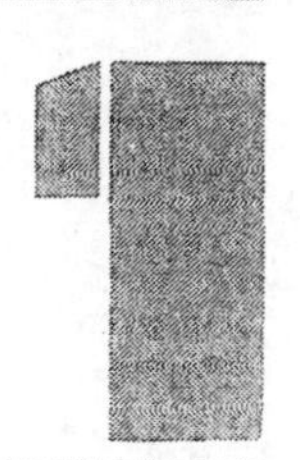

ALGEBRA IN ACTION: SPORTS

By using formulas, you can take raw sports statistics and compute standings, special averages, and other useful information.

Percentage of Games Won

$$PCT = \frac{100\,W}{W + L}$$

Games Behind Leader

$$GBL = \frac{W - L - (w - l)}{2}$$

Chapter 1 Objectives

Objectives	Typical Problems
Express a sum, difference, product, or quotient in symbols.	**1–10.** Write in symbols. **1.** The sum of 8 and x. **2.** c more than b. **3.** x decreased by 3. **4.** 4 subtracted from y. **5.** k less than n. **6.** The product of 6 and x. **7.** b multiplied by a. **8.** The product of y and 4. **9.** 7 divided by x. **10.** One half of x.
Express a square, cube, or nth power of a number in symbols.	**11–13.** Write in symbols, using exponents. **11.** The square of n. **12.** The cube of x. **13.** The fifth power of y.
Write sums, differences, products, and quotients involving the special numbers 1 and 0.	**14–20.** Write in symbols. **14.** The product of 1 and t. **15.** The product of 0 and t. **16.** The sum of y and 0. **17.** y decreased by 0. **18.** 0 divided by 1. **19.** x divided by 0. **20.** x divided by 1.
Use the listing method to represent a set of digits, natural numbers, or whole numbers.	**21–23.** Use the listing method to represent each set described. **21.** The set of digits greater than 5. **22.** The set of natural numbers greater than 5. **23.** The set of whole numbers x for which $\frac{12}{x}$ is a whole number.
Identify and use symbols to represent the empty set or a universal set.	**24–25.** Represent in symbols. **24.** The set of natural numbers n for which $1 \cdot n = n$. **25.** The set of whole numbers w for which $w + 1 = w$.
List the multiples or factors of a given natural number.	**26–27.** List the elements in each set described. **26.** The multiples of 4. **27.** The factors of 12.
List a subset of the set of even, odd, prime, or composite numbers.	**28–32.** List the members of each set described. **28.** Even numbers less than 15. **29.** Odd numbers less than 15. **30.** Prime numbers less than 20. **31.** Composite numbers less than 20. **32.** Odd composite factors of 90.

Objectives	Typical Problems
Evaluate a numerical expression involving two or more operations or one set of grouping symbols.	**33–34.** Evaluate. **33.** $20 - 2(3 + 4)$ **34.** $\frac{4^2 - 16}{5}$
Translate an expression involving two or more operations into symbols.	**35–37.** Express in symbols. **35.** Twice the sum of x and 5. **36.** The square of the difference of 3 less than y. **37.** The quotient obtained when the product of x and y is divided by the sum of x and y.
Evaluate a numerical expression involving several sets of grouping symbols.	**38–39.** Evaluate. **38.** $8 - [30 - (2 + 3)^2]$ **39.** $2\{15 - [4^2 - (7 - 2)]\}$
Evaluate an algebraic expression by using the substitution principle.	**40–41.** Evaluate. **40.** $(2t^2 - 9)(t + 1)$ for $t = 5$. **41.** $\frac{x^2 - 2xy + y^2}{x - y}$ for $x = 7$ and $y = 2$.
Evaluate a formula.	**42–43.** Evaluate. **42.** Given $C = \frac{5(F - 32)}{9}$ find C for $F = 95$. **43.** Given $A = \frac{h(a + b)}{2}$ find A for $h = 5$, $a = 9$, and $b = 7$.

ANSWERS to Typical Problems

1. $x + 8$ **2.** $b + c$ **3.** $x - 3$ **4.** $y - 4$ **5.** $n - k$ **6.** $6x$ **7.** ab **8.** $4y$ **9.** $\frac{7}{x}$ **10.** $\frac{x}{2}$
11. n^2 **12.** x^3 **13.** y^5 **14.** t **15.** 0 **16.** y **17.** y **18.** 0 **19.** Undefined **20.** x
21. {6,7,8,9} **22.** {6, 7, 8, 9, 10, . . .} **23.** {1, 2, 3, 4, 6, 12} **24.** N or {1, 2, 3, . . .} **25.** ∅ or { }
26. {4, 8, 12, 16, . . .} **27.** {1, 2, 3, 4, 6, 12} **28.** {2, 4, 6, 8, 10, 12, 14} **29.** {1, 3, 5, 7, 9, 11, 13}
30. {2, 3, 5, 7, 11, 13, 17, 19} **31.** {4, 6, 8, 9, 10, 12, 14, 15, 16, 18} **32.** {9, 15, 45} **33.** 6 **34.** 0
35. $2(x + 5)$ **36.** $(y - 3)^2$ **37.** $\frac{xy}{x + y}$ **38.** 3 **39.** 8 **40.** 246 **41.** 5 **42.** 35 **43.** 40

ONE OPERATION 1.1

A. ADDITION, SUBTRACTION, MULTIPLICATION, DIVISION

Algebra is a generalization of arithmetic. It is a study of numbers, operations on numbers, and relations between numbers. The numbers of arithmetic are also some of the numbers of algebra. The term **numbers of arithmetic** includes the **counting numbers,**

The three dots ... mean "and so on in a like manner."

$$1, \; 2, \; 3, \; 4, \; 5, \; 6, \; 7, \; 8, \; 9, \; 10, \; 11, \; 12, \; \ldots$$

the number **zero, 0,** and the quotients of counting numbers, also called **fractions,** such as $\frac{1}{2}$, $\frac{3}{5}$, and $\frac{16}{7}$.

The other numbers of algebra will be introduced gradually as the study progresses. For now, "number" shall mean one of the numbers of arithmetic.

A **numeral** is a symbol that names a number according to a specified system of numeration, such as the decimal system. Examples of numerals are 3, 5, 12, $\frac{1}{2}$, and $10\frac{1}{4}$.

In algebra, a **letter** may also be used as the name of a number.

A **constant** is a numeral or letter that is the name of exactly one number. The letters a, b, and c are commonly used as constants.

A **variable** is a letter that names an unspecified number that belongs to a set of numbers. The letters x, y, and z are commonly used as variables.

The **equal sign, =**, is used to indicate that two expressions represent the same number. For example, when we write $3 + 4 = 7$, we mean that $3 + 4$ is the same number as 7.

Like arithmetic, algebra is concerned with six operations. These are addition, subtraction, multiplication, division, raising to a power, and root extraction. The first five of these operations are discussed in this chapter. Root extraction is discussed in Chapter 9.

Addition

The **sum** of two numbers is the result obtained by adding the two numbers. The numbers that are added are called the **terms** of the sum. For $3 + 4 = 7$, the **terms** are 3 and 4 and the **sum** is 7.

TERM + TERM = SUM

The **terms of a sum can be written in any order.** As an example, $3 + 4$ names the same number as $4 + 3$. Also, $x + 4$ is the same number as $4 + x$ and $y + x$ is the same number as $x + y$.

Conventions are used to make algebraic work easier and to make answers easier to compare.

CONVENTIONS FOR SUMS

1. Write letter terms in alphabetical order, when possible.
2. Write a letter term to the left of a numeral term.

As examples,

$x + 4$ is the conventional form for $4 + x$

$x + y$ is the conventional form for $y + x$

SAMPLE PROBLEMS 1–6

Write each sum in symbols.

1. A number x plus a number y. $x + y$
2. The sum of x and 5. $x + 5$
3. The sum of 6 and y. $y + 6$
4. 5 more than x. $x + 5$
5. y increased by 4. $y + 4$
6. s more than r. $r + s$

NOTE TO STUDENT:
To avoid errors, write neatly and clearly. The letter b is often mistaken for 6 or 10. A printed Y often looks like X. Don't print letters.

PRACTICE EXERCISES 1–6

Write each sum in symbols.

1. The sum of t and 2.
2. The sum of 3 and x.
3. c plus d.
4. b plus a.
5. 7 more than y.
6. y increased by x.

Answers

1. $t + 2$ 2. $x + 3$ 3. $c + d$ 4. $a + b$ 5. $y + 7$ 6. $x + y$

Subtraction

The **difference,** or **remainder,** is the result obtained by subtracting one number from another. The numbers that form the difference are called the terms of the difference. For $10 - 4 = 6$, the terms are 10 and 4. The difference is 6.

The name **subtrahend** is given to the number that is subtracted from the other number, called the **minuend.** For $10 - 4$, the minuend is 10 and the subtrahend is 4.

MINUEND – SUBTRAHEND = DIFFERENCE

While $5 - 4 = 1$, the difference $4 - 5$ is a negative number. Negative numbers will be discussed in Chapter 2.

The order in which the terms of a difference are written is very important.

When a and b are different numbers, the differences $a - b$ and $b - a$ are different numbers.

A special effort must be made to learn how to translate the expressions **a less than b** and **a subtracted from b.** These must be written as **$b - a$.**

SAMPLE PROBLEMS 7–14

Write each difference in symbols.

7. A number x minus a number y. $x - y$
8. x decreased by 6. $x - 6$
9. 5 subtracted from y. $y - 5$
10. t subtracted from 9. $9 - t$
11. 5 less than x. $x - 5$
12. n less than 12. $12 - n$
13. x less than y. $y - x$
14. x subtracted from y. $y - x$

PRACTICE EXERCISES 7–14

Write each difference in symbols.

7. x minus 8.
8. 3 subtracted from x.
9. x subtracted from 10.
10. y decreased by 12.
11. 15 decreased by y.
12. 7 less than t.
13. t less than 2.
14. t subtracted from n.

Answers

7. $x - 8$ 8. $x - 3$ 9. $10 - x$ 10. $y - 12$ 11. $15 - y$
12. $t - 7$ 13. $2 - t$ 14. $n - t$

Multiplication

The **product** is the result obtained by multiplying two numbers. The numbers that are being multiplied are called **factors.** In the arithmetic problem $3 \times 4 = 12$, the **factors** are 3 and 4 and the **product** is 12.

In algebra, the $\times$ symbol is not used for multiplication. A product involving two numerals is written by using a dot between them or by enclosing one or both of the numerals in parentheses. For example, the product of 4 and 5 is written as $4 \cdot 5$ or as 4(5) or as (4)(5). Note that $4 \cdot 5 = 20$, whereas 45 means forty-five.

FACTOR · FACTOR = PRODUCT

A product involving a numeral and a letter, or two letters, is shown by writing the factors next to each other. For example, $4x$ means 4 times x and xy means x times y.

While it is correct to use the dot and write $4 \cdot x$ or $x \cdot y$, the most commonly used forms for these products are $4x$ and xy.

The **factors of a product can be written in any order.** As an example, $5 \cdot 4$ names the same number as $4 \cdot 5$. Also, $5x$ is the same number as $x5$ and yx is the same number as xy.

CONVENTIONS FOR PRODUCTS

1. **Write letter factors in alphabetical order, when possible.**
2. **Write a numeral factor to the left of a letter factor.**

SAMPLE PROBLEMS 15–20

Write each product in symbols.

15. The product of 6 and 8.	$6 \cdot 8$ or $6(8)$ or $(6)(8)$
16. 8 times a number x.	$8x$
17. x multiplied by y.	xy
18. The product of y and 3.	$3y$
19. b multiplied by a.	ab
20. Twice x.	$2x$

PRACTICE EXERCISES 15–20

Write each product in symbols.

15. The product of 6 and 7.
16. 6 times a number y.
17. x multiplied by 2.
18. The product of r and s.
19. The product of y and x.
20. Twice t.

Answers

15. $6 \cdot 7$ 16. $6y$ 17. $2x$ 18. rs 19. xy 20. $2t$

Division

The **quotient** is the result obtained when one number, the **dividend**, is divided by another number, the **divisor**. For the division problem $35 \div 5 = 7$, the dividend is 35, the divisor is 5, and the quotient is 7.

In algebra, the symbol $\div$ is seldom used to indicate division. Instead, the dividend is written over the divisor with a bar between them. As an example, the division of 35 by 5 is written $\frac{35}{5}$.

$$\frac{\text{DIVIDEND}}{\text{DIVISOR}} = \text{QUOTIENT}$$

In a quotient, the dividend must be written on top of the divisor. In general, a different number will be obtained if this order is changed. For example, $\frac{35}{5} = 7$ but $\frac{5}{35} = \frac{1}{7}$. Similarly, $\frac{x}{5}$ and $\frac{5}{x}$ are not the same number.

SAMPLE PROBLEMS 21–26

Write each quotient in symbols.

21. 200 divided by 4. $\frac{200}{4}$

22. x divided by 4. $\frac{x}{4}$

23. 4 divided by x. $\frac{4}{x}$

24. y divided by x. $\frac{y}{x}$

25. One half of x. $\frac{x}{2}$

26. One third of x. $\frac{x}{3}$

Note: The forms $\frac{x}{2}$ and $\frac{x}{3}$ are preferred to $\frac{1}{2}x$ and $\frac{1}{3}x$ to prevent the errors of writing $\frac{1}{2x}$ or $\frac{1}{3x}$.

PRACTICE EXERCISES 21–26

Write each quotient in symbols.

21. 84 divided by 7.
22. y divided by 7.
23. 2 divided by t.
24. x divided by y.
25. One half of t.
26. One third of n.

Answers

21. $\frac{84}{7}$ 22. $\frac{y}{7}$ 23. $\frac{2}{t}$ 24. $\frac{x}{y}$ 25. $\frac{t}{2}$ 26. $\frac{n}{3}$

EXERCISES 1.1A

Write each of the following in symbols.

1. The product of 7 and a number y. 1. ____
2. The sum of a number s and 11. 2. ____
3. The sum of 3 and a number x. 3. ____
4. The difference when a number x is subtracted from 5. 4. ____
5. The difference when 5 is subtracted from a number x. 5. ____
6. The quotient when a number n is divided by 4. 6. ____
7. The quotient when a number x is divided by 8. 7. ____
8. The quotient when 4 is divided by a number n. 8. ____
9. 6 more than a number q. 9. ____
10. The difference when q is subtracted from 6. 10. ____
11. q less than 6. 11. ____
12. The quotient when q is divided into 6. 12. ____
13. The product of z and y. 13. ____
14. The product of y and 9. 14. ____
15. The product of 3 and 25. 15. ____
16. The difference when r is subtracted from 8. 16. ____
17. The quotient when 7 is divided by z. 17. ____
18. The sum of x and 9. 18. ____
19. The product of x and 9. 19. ____
20. The difference when 2 is subtracted from t. 20. ____
21. t decreased by z. 21. ____
22. The product of a number x and a number y. 22. ____
23. The product of 2 and 3. 23. ____
24. The product of 2 and x. 24. ____

25. The quotient when n is divided by 2. 25. ____________

26. The quotient when 2 is divided by n. 26. ____________

27. The difference when x is subtracted from y. 27. ____________

28. The difference when y is subtracted from x. 28. ____________

29. The sum of y and z. 29. ____________

30. The quotient when r is divided by s. 30. ____________

31. Twice y. 31. ____________

32. One half of y. 32. ____________

33. One third of t. 33. ____________

34. One fourth of n. 34. ____________

35. One fifth of x. 35. ____________

B. SQUARING, CUBING, RAISING TO A POWER

The **square** of a number is the product obtained by using the given number as a factor two times.

The squaring operation is indicated in symbols by writing a small 2 to the upper right of the number that is to be squared.

As examples,

x^2 means xx

$$3^2 \text{ means } 3 \cdot 3, \text{ or } 9 \qquad 5^2 \text{ means } 5 \cdot 5, \text{ or } 25$$

The **cube** of a number is the product obtained by using the given number as a factor three times.

The cubing operation is indicated in symbols by writing a small 3 to the upper right of the number that is to be cubed.

As examples,

x^3 means xxx

$$3^3 \text{ means } 3 \cdot 3 \cdot 3, \text{ or } 27 \qquad 5^3 \text{ means } 5 \cdot 5 \cdot 5, \text{ or } 125$$

In general, a counting number, called an **exponent,** written to the upper right of another number, called the **base,** shows how many times the base is to be used as a factor. This special type of multiplication is called **raising to a power.**

$\text{base}^{\text{exponent}} = \text{power}$

As an example, 2^5 (read "two to the fifth" or "the fifth power of two") means $2 \cdot 2 \cdot 2 \cdot 2 \cdot 2$, or 32.

Similarly,

x^n means $xx \cdots x$ (n factors of x)

$$x^4, \quad x \text{ to the fourth, means } \quad xxxx$$

$$x^5, \quad x \text{ to the fifth, means } \quad xxxxx$$

SAMPLE PROBLEMS 1–10

Write each of the following in symbols, using exponents.

WORDS	SYMBOLS
1. The square of 5 (5 times 5).	1. 5^2
2. The square of x (x times x).	2. x^2
3. y squared.	3. y^2
4. The cube of 2 (2 times 2 times 2).	4. 2^3
5. The cube of x (x times x times x).	5. x^3
6. y cubed.	6. y^3
7. The fourth power of 3.	7. 3^4
8. The fourth power of x.	8. x^4
9. The sixth power of y.	9. y^6
10. z to the fifth power.	10. z^5

Note that while x times x can be written as xx, the preferred form for the product is x^2. Similarly, x^3 is preferred to xxx.

EXERCISES 1.1B

Write each of the following in symbols, using exponents.

1. The product of y times itself. 1. ____________
2. The square of n. 2. ____________
3. The product of 4 times 4 times 4. 3. ____________
4. The product of n times n times n. 4. ____________
5. The cube of 6. 5. ____________
6. The cube of t. 6. ____________
7. The square of t. 7. ____________
8. The third power of x. 8. ____________
9. The square of 10. 9. ____________
10. The square of 9. 10. ____________
11. The cube of 9. 11. ____________
12. The cube of y. 12. ____________
13. The fourth power of y. 13. ____________

14. The fourth power of 10. 14. ______

15. The fifth power of 8. 15. ______

16. The fifth power of x. 16. ______

17. 4 to the fifth power. 17. ______

18. 9 to the seventh power. 18. ______

19. The sixth power of n. 19. ______

20. y to the eighth power. 20. ______

21. Six factors of 7. 21. ______

22. Six factors of n. 22. ______

23. Four factors of t. 23. ______

24. Two factors of x. 24. ______

C. THE SPECIAL NUMBERS 1 AND 0

The **number one,** 1, is a very special number for multiplication. It is called the **multiplication identity.** Note that $1 \cdot 5 = 5$, $1 \cdot 2 = 2$, $20 \cdot 1 = 20$, and so on. For all numbers x,

$$1 \cdot x = x \quad \text{and} \quad x \cdot 1 = x$$

While it is correct to write the product of 1 and x as $1 \cdot x$ or as $1x$, the most commonly used form for the product is x.

Although the quotient of x divided by 1 can be written as $\frac{x}{1}$, it is more customary to write $\frac{x}{1}$ as x. Note that $\frac{5}{1} = 5$ because $1 \cdot 5 = 5$ and $\frac{8}{1} = 8$ because $1 \cdot 8 = 8$. In general, since $1 \cdot x = x$ for all x,

$$\frac{x}{1} = x$$

The **number zero,** 0, is a very special number for addition. It is called the **addition identity.** Note that $3 + 0 = 3$, $12 + 0 = 12$, and $0 + 5 = 5$, and so on. For all numbers x,

$$x + 0 = x \quad \text{and} \quad 0 + x = x$$

While it is correct to write the sum of 0 and x as $x + 0$ or as $0 + x$, the most common form for this sum is x.

The value of a number does not change when 0 is subtracted from that number. Note that $5 - 0 = 5$, $7 - 0 = 7$, and so on. Thus it is conventional to write $x - 0$ more simply as x. For all x,

$$x - 0 = x$$

The product of any number x and 0 is 0. Note that $5 \cdot 0 = 0$ and $0 \cdot 8 = 0$. For all x,

$$x \cdot 0 = 0 \quad \text{and} \quad 0 \cdot x = 0$$

When the dividend is 0 and the divisor is not 0, then the quotient is 0. As examples, $\frac{0}{5} = 0$ because $5 \cdot 0 = 0$ and $\frac{0}{3} = 0$ because $3 \cdot 0 = 0$. However, $\frac{5}{0}$ does not represent a number. The expression $\frac{5}{0}$ is undefined because there is no number that can be multiplied by 0 to get a product 5.

Division by zero is undefined

$\frac{x}{0}$ is undefined for all x, including 0

Also note that $\frac{x}{1}$ and $\frac{1}{x}$ are different numbers in general. They are the same number only for the special case that $x = 1$. In other words, $\frac{1}{x}$ must be left as $\frac{1}{x}$.

SAMPLE PROBLEMS 1–10

Express in symbols. If an expression is undefined, write "undefined."

1.	The product of 1 and y.	y
2.	The product of t and 1.	t
3.	The quotient of y divided by 1.	y
4.	The sum of t and 0.	t
5.	The sum of 0 and c.	c
6.	The difference when 0 is subtracted from y.	y
7.	x decreased by 0.	x
8.	The product of y and 0.	0
9.	The quotient when 0 is divided by 6.	0
10.	The quotient when n is divided by 0.	Undefined

EXERCISES 1.1C

Write in symbols. If an expression is undefined, write "undefined."

1. The product of k and 1. **1.** ____________

2. The product of 1 and n. **2.** ____________

3. The quotient when z is divided by 1. **3.** ____________

4. The sum of 0 and y. **4.** ____________

5. The sum of b and 0. **5.** ____________

6. The difference when 0 is subtracted from t. **6.** ____________

7. n decreased by 0. **7.** ____________

8. The product of t and 0. **8.** ____________

9. The quotient when 0 is divided by 9. **9.** ____________

10. The quotient when t is divided by 0. **10.** ____________

11. The sum of x and 0. **11.** ____________

12. The product of x and 0. **12.** ____________

13. The product of 1 and x. **13.** ____________

14. The sum of 1 and x. **14.** ____________

15. The quotient when x is divided by 0. **15.** ____________

16. The quotient when x is divided by 1. **16.** ____________

17. The quotient when 1 is divided by x. **17.** ____________

18. The quotient when 0 is divided by x. **18.** ____________

19. The quotient when 1 is divided by 0. **19.** ____________

20. The quotient when 0 is divided by 1. **20.** ____________

EXERCISES 1.1S

1–50. Write in symbols.

1. The sum of a number n and 8. **1.** ______
2. The product of a number n and 8. **2.** ______
3. The difference when 12 is subtracted from t. **3.** ______
4. The difference when t is subtracted from 12. **4.** ______
5. The quotient when x is divided by 7. **5.** ______
6. The quotient when 7 is divided by x. **6.** ______
7. The product of a and b. **7.** ______
8. The sum of a and b. **8.** ______
9. The quotient when a is divided by b. **9.** ______
10. The difference when a is subtracted from b. **10.** ______
11. The product of 1 and n. **11.** ______
12. 4 less than x. **12.** ______
13. The quotient when 4 is divided into t. **13.** ______
14. The quotient when t is divided into 4. **14.** ______
15. 9 increased by y. **15.** ______
16. 9 decreased by y. **16.** ______
17. Twice t. **17.** ______
18. Three times r. **18.** ______
19. One half of z. **19.** ______
20. One third of n. **20.** ______
21. The square of b. **21.** ______
22. The cube of z. **22.** ______

NAME ______ DATE ______ COURSE ______

23. The cube of 7. 23. ______
24. The square of 18. 24. ______
25. The third power of 5. 25. ______
26. The fourth power of r. 26. ______
27. The third power of s. 27. ______
28. The fifth power of t. 28. ______
29. The sum of x and 0. 29. ______
30. The quotient when y is divided by 1. 30. ______
31. The difference when 0 is subtracted from t. 31. ______
32. 20 more than k. 32. ______
33. The difference when z is decreased by 0. 33. ______
34. The product of x and 1. 34. ______
35. The quotient when 1 is divided by c. 35. ______
36. The product of y and x. 36. ______
37. The sum of y and x. 37. ______
38. The product of 0 and x. 38. ______
39. The quotient when 0 is divided by 6. 39. ______
40. The quotient when 6 is divided by 0. 40. ______
41. The product of 0 and k. 41. ______
42. The product of 1 and k. 42. ______
43. The sum whose terms are 7 and x. 43. ______
44. The product whose factors are 7 and y. 44. ______
45. The difference between z and 10 if the subtrahend is z. 45. ______
46. The quotient whose divisor is 6 and whose dividend is x. 46. ______
47. The power whose base is x and whose exponent is 4. 47. ______
48. The product of 5 factors of x. 48. ______
49. The product of 6 factors of y. 49. ______
50. The product of n factors of t. 50. ______

SETS, MULTIPLES, FACTORS

A. THE SET CONCEPT, THE LISTING METHOD, UNIVERSAL SET, EMPTY SET

A **set** is a well-defined collection of objects which are called elements or members of the set.

"Well-defined" means that it is always possible to determine if a particular object is or is not a member of a given set. A capital letter is often used as the name of a set.

The **listing method** is a commonly used method for defining a particular set. When this method is used, the elements of the set are written inside braces and are separated by commas.

EXAMPLE 1

Represent the **set of digits, *D*,** by the listing method.

Digits

Solution

$D = \{0, 1, 2, 3, 4, 5, 6, 7, 8, 9\}$

EXAMPLE 2

Represent the **set of natural numbers, *N*** (another name for the counting numbers), by the listing method.

Natural numbers

Solution

$N = \{1, 2, 3, 4, 5, \ldots\}$

The three dots, called an ellipsis, mean that this pattern continues without end.

EXAMPLE 3

Represent the **set of whole numbers, *W*,** by the listing method.

Whole numbers

Solution

$W = \{0, 1, 2, 3, 4, 5, \ldots\}$

The set of whole numbers consists of the natural numbers and the number 0.

For a particular discussion, the universal set is a set to which all elements of the discussion must belong.

Universal set

The universal set may differ for different discussions.

Also called the null set, the empty set is the set that has no members. Either $\varnothing$ or $\{\ \}$ is used to indicate the empty set.

Empty set

WARNING:
Do not use {0} or {∅} to name the empty set. Each of these symbols names a set that has one member in it. The empty set has no members.

As an example, the set of digits greater than 10 is the empty set, ∅. This set has **no members,** since no digit is greater than 9. On the other hand, the set of digits less than 1 is the set {0}. This set has **one member,** the number zero, 0. Note that {0} **is not** the empty set. Also note that ∅ and 0 are not the same; 0 is a number, while ∅ is a set that has no member in it.

In Sample Problems 1–7, use the listing method to represent each set that is described.

SAMPLE PROBLEM 1
The set of digits greater than 4.

SOLUTION
{5, 6, 7, 8, 9}

SAMPLE PROBLEM 2
The set of natural numbers greater than 20.

SOLUTION
{21, 22, 23, 24, 25, . . .}

SAMPLE PROBLEM 3
The set of digits x for which $x + 2$ is less than 7.

SOLUTION
Test each digit.

$x = 0$ Is $0 + 2$ less than 7? Yes
$x = 1$ Is $1 + 2$ less than 7? Yes
$x = 2$ Is $2 + 2$ less than 7? Yes
$x = 3$ Is $3 + 2$ less than 7? Yes.
$x = 4$ Is $4 + 2$ less than 7? Yes
$x = 5$ Is $5 + 2$ less than 7? No

Digits larger than 4 do *not* meet the requirement. Thus the set is {0, 1, 2, 3, 4}.

SAMPLE PROBLEM 4
The set of natural numbers x for which $\frac{12}{x}$ is a natural number.

SOLUTION
Test each natural number.

$x = 1$ Is $\frac{12}{1}$ a natural number? Yes
$x = 2$ Is $\frac{12}{2}$ a natural number? Yes
$x = 3$ Is $\frac{12}{3}$ a natural number? Yes
$x = 4$ Is $\frac{12}{4}$ a natural number? Yes
$x = 5$ Is $\frac{12}{5}$ a natural number? No
$x = 6$ Is $\frac{12}{6}$ a natural number? Yes

Continue in this way. The set is {1, 2, 3, 4, 6, 12}.

SAMPLE PROBLEM 5
The set of whole numbers w for which $w + 0 = w$.

SOLUTION
Test $0 + 0 = 0$, $1 + 0 = 1$, $2 + 0 = 2$, $3 + 0 = 3$, and so on. We see that $w + 0 = w$ is true for every whole number.
The solution set may be written as W or {0, 1, 2, 3, 4, 5, . . .}.

SAMPLE PROBLEM 6
The set of whole numbers w for which $w + 1 = 1$.

SOLUTION
Test each whole number: $0 + 1 = 1$ is true, and 0 is in the set.
$1 + 1 = 1$ is false, so 1 is not in the set.
$2 + 1 = 1$ is false, so 2 is not in the set.
It is easy to see that $w + 1 = 1$ is true only for $w = 0$.
The solution is $\{0\}$, the set containing the number 0.

SAMPLE PROBLEM 7
The set of whole numbers w for which $w + 1 = w$.

SOLUTION
Test the whole numbers: $0 + 1 = 0$ is false, $1 + 1 = 1$ is false, $2 + 1 = 2$ is false, and so on. We see that there is no number for which $w + 1 = w$.
The solution is the empty set, written as $\varnothing$ or $\{\ \}$.

EXERCISES 1.2A

Use the listing method to represent each set that is described.

1. The set of digits less than 5. 1. ____________

2. The set of digits greater than 6. 2. ____________

3. The set of natural numbers greater than 6. 3. ____________

4. The set of natural numbers x for which $x + 5$ is less than 10. 4. ____________

5. The set of digits d for which $d + 4$ is greater than 9. 5. ____________

6. The set of natural numbers n for which $n + 4$ is greater than 9. 6. ____________

7. The set of whole numbers w for which $\frac{45}{w}$ is a whole number. 7. ______

8. The set of digits y for which $\frac{45}{y}$ is a natural number. 8. ______

9. The set of digits d for which $\frac{d}{2}$ is a digit. 9. ______

10. The set of natural numbers n for which $\frac{n}{2}$ is a natural number. 10. ______

11. The set of whole numbers w for which $\frac{w}{2}$ is a whole number. 11. ______

12. The set of whole numbers x for which $x \cdot 1 = x$. 12. ______

13. The set of digits d for which $d + 0 = d$. 13. ______

14. The set of digits d for which $d + 2 = d$. 14. ______

15. The set of digits d for which $d \cdot 0 = d$. 15. ______

16. The set of natural numbers n for which $\frac{n}{1} = n$. 16. ______

17. The set of natural numbers n for which $\frac{n}{0}$ is a natural number. 17. ______

18. The set of natural numbers x for which $\frac{1}{x}$ is a natural number. 18. ______

19. The set of whole numbers x for which $x \cdot 0 = 0$. 19. ______

20. The set of whole numbers x for which $x + 5 = x$. 20. ______

B. MULTIPLES, FACTORS

For this section, the universal set is N, the set of natural numbers.

A natural number m is a multiple of another natural number n if it can be obtained by multiplying n by a natural number.

Multiple

For example, 20 is a multiple of 5, since $20 = 5(4)$. We also say that 20 is **divisible** by 5.

A natural number a is a factor of another natural number n if it can be obtained by dividing n by a natural number.

Factor

For example, 6 is a factor of 18 because $\frac{18}{3} = 6$. Note that 3 is also a factor of 18.

SAMPLE PROBLEM 1

List the multiples of 4.

SOLUTION

A multiple of 4 is a number obtained by multiplying 4 by a natural number.

Since $4(1) = \mathbf{4}$, $4(2) = \mathbf{8}$, $4(3) = \mathbf{12}$, $4(4) = \mathbf{16}$, $4(5) = \mathbf{20}$, and so on,

the set of multiples of 4 = {4, 8, 12, 16, 20, . . .}.

SAMPLE PROBLEM 2

List the factors of 12.

SOLUTION

Divide the given number 12 by each natural number.

If the quotient is a natural number, accept both the divisor and the quotient as factors.

$\frac{12}{\mathbf{1}} = \mathbf{12}$ Thus, **1** and **12** are factors.

$\frac{12}{\mathbf{2}} = \mathbf{6}$ Thus **2** and **6** are factors.

$\frac{12}{\mathbf{3}} = \mathbf{4}$ Thus **3** and **4** are factors.

$\frac{12}{4} = 3$ We already have the factors 4 and 3 so no more trials are necessary.

The factors of 12 are {1, 2, 3, 4, 6, 12}.

SAMPLE PROBLEM 3

List the factors of 15.

SOLUTION

$\frac{15}{\mathbf{1}} = \mathbf{15}$ Thus **1** and **15** are factors.

$\frac{15}{2}$ is not a natural number. 2 is not a factor.

$\frac{15}{\mathbf{3}} = \mathbf{5}$ Thus **3** and **5** are factors.

$\frac{15}{4}$ is not a natural number. 4 is not a factor.

$\frac{15}{5} = 3$ The trials are over.
We already have 5 and 3.

The factors of 15 are {1, 3, 5, 15}.

DIVISIBILITY RULES

A number is divisible by

2 if its last digit is 0, 2, 4, 6, or 8.

3 if the sum of its digits is divisible by 3.

5 if its last digit is 0 or 5.

SAMPLE PROBLEM 4

List the natural numbers divisible by 12.

SOLUTION

Numbers divisible by 12 are multiples of 12. Multiply 12 by each natural number. 12(1) = 12, 12(2) = 24, 12(3) = 36, and so on.

{12, 24, 36, 48, 60, . . .} is the answer set.

EXERCISES 1.2B

List the elements in each set described. The universal set is N, the set of natural numbers. This means that each indicated number must be a natural number.

1. The factors of 20.

2. The multiples of 20.

3. The factors of 21.

4. The numbers divisible by 8.

5. The factors of 35 different from 1 and 35.

6. The factors of 17 different from 1 and 17.

7. The multiples of 30.

8. The factors of 30.

9. The numbers divisible by 9.

10. The multiples of 1.

1. ____________________

2. ____________________

3. ____________________

4. ____________________

5. ____________________

6. ____________________

7. ____________________

8. ____________________

9. ____________________

10. ____________________

1–20. Use the listing method to represent each set described.

1. The set of natural numbers less than 7. **1.** ______

2. The set of whole numbers less than 7. **2.** ______

3. The set of digits greater than 4. **3.** ______

4. The set of natural numbers greater than 4. **4.** ______

5. The set of digits x for which $\frac{60}{x}$ is a natural number. **5.** ______

6. The set of natural numbers x for which $\frac{60}{x}$ is a natural number. **6.** ______

7. The set of digits d for which $\frac{d}{3}$ is a digit. **7.** ______

8. The set of natural numbers n for which $\frac{n}{3}$ is a natural number. **8.** ______

9. The set of natural numbers x for which $x + 0 = x$. **9.** ______

10. The set of digits d for which $\frac{0}{d}$ is a digit. **10.** ______

11. The multiples of 45. **11.** ______

NAME ______ DATE ______ COURSE ______

12. The factors of 45. 12. ____________

13. The natural numbers divisible by 30. 13. ____________

14. The factors of 30. 14. ____________

15. The factors of 28 different from 1 and 28. 15. ____________

16. The factors of 43 different from 1 and 43. 16. ____________

17. The factors of 42. 17. ____________

18. The multiples of 10. 18. ____________

19. All whole numbers x for which $x + 5 = 5 + x$. 19. ____________

20. All whole numbers x for which $x + 5 = x$. 20. ____________

EVEN, ODD, PRIME, AND COMPOSITE NUMBERS 1.3

In this section, the universal set is N, the set of natural numbers. When the word "number" is used, number means natural number.

A. EVEN AND ODD NUMBERS

A natural number n is even if and only if 2 is a factor of n.

Even numbers

As examples, 2, 4, 6, 8, and 10 are even numbers.

A natural number n is odd if and only if 2 is not a factor of n.

Odd numbers

As examples, 1, 3, 5, 7, and 9 are odd numbers.

For Sample Problems 1–4, list the members of each set described.

SAMPLE PROBLEM 1
The set of even numbers less than 15.

SOLUTION
{2, 4, 6, 8, 10, 12, 14}

SAMPLE PROBLEM 2
The set of odd numbers less than 15.

SOLUTION
{1, 3, 5, 7, 9, 11, 13}

SAMPLE PROBLEM 3
The set of even multiples of 3.

SOLUTION
The multiples of 3 are {3, 6, 9, 12, 15, 18, . . .}.
The even multiples of 3 are {6, 12, 18, . . .}.

NOTE TO STUDENT:
Read the text aloud. The sense of sound helps the sense of sight.

SAMPLE PROBLEM 4
The set of odd multiples of 3.

SOLUTION
The odd multiples of 3 are {3, 9, 15, . . .}.

SAMPLE PROBLEM 5	SAMPLE PROBLEM 6
The set of even factors of 42.	The set of odd factors of 42.
SOLUTION	**SOLUTION**
The factors of 42 are {1, 2, 3, 6, 7, 14, 21, 42}. The even factors of 42 are {2, 6, 14, 42}.	The odd factors of 42 are {1, 3, 7, 21}.

EXERCISES 1.3A

List the members of each set described.

1. Even numbers greater than 10 and less than 20. **1.** ____________

2. Odd numbers greater than 20 and less than 30. **2.** ____________

3. Even multiples of 5. **3.** ____________

4. Odd multiples of 5. **4.** ____________

5. Even factors of 30. **5.** ____________

6. Odd factors of 30. **6.** ____________

7. Odd factors of 8. **7.** ____________

8. Even factors of 15. **8.** ____________

9. Odd multiples of 6. **9.** ____________

10. Natural numbers that are either even or odd. **10.** ____________

B. PRIME AND COMPOSITE NUMBERS

Let us examine the factors of the natural numbers from 1 to 10.

NUMBER	SET OF FACTORS	
1	{1}	exactly one factor
2	{1, 2}	**exactly two factors**
3	{1, 3}	**exactly two factors**
4	{1, 2, 4}	three factors
5	{1, 5}	**exactly two factors**
6	{1, 2, 3, 6}	four factors
7	{1, 7}	**exactly two factors**
8	{1, 2, 4, 8}	four factors
9	{1, 3, 9}	three factors
10	{1, 2, 5, 10}	four factors

Note that 1 has exactly one factor. Note that the numbers 2, 3, 5, and 7 have **exactly two factors,** namely 1 and the number itself. The numbers 4, 6, 8, 9, and 10 have more than two factors.

Numbers that have exactly two factors are very important in mathematics. They are called **prime numbers**.

A natural number p is prime if and only if p has exactly two different factors, namely p and 1.

Prime

As examples, 2, 3, 5, and 7 are prime numbers.

Note that 1 is not a prime, since it has only one factor.

A natural number n is composite if and only if n has more than two different factors.

Composite

As examples, 4, 6, 12, 15, and 21 are composites.

Note from the definitions that the special number 1 is neither prime nor composite.

SAMPLE PROBLEM 1
The set of prime numbers less than 20.

SOLUTION
{2, 3, 5, 7, 11, 13, 17, 19}

SAMPLE PROBLEM 2
The set of composite numbers less than 20.

SOLUTION
{4, 6, 8, 9, 10, 12, 14, 15, 16, 18}

SAMPLE PROBLEM 3
The set of odd composite factors of 90.

SOLUTION
1. List the factors of 90. {1, 2, 3, 5, 6, 9, 10, 15, 18, 30, 45, 90}
2. Select the odd factors. {1, 3, 5, 9, 15, 45}
3. Reject 1 and all prime factors. {9, 15, 45}

The set of odd composite factors of 90 is {9, 15, 45}.

EXERCISES 1.3B

List the members of each set described.

1. Prime numbers greater than 20 and less than 30. **1.** ________
2. Composite numbers greater than 20 and less than 30. **2.** ________
3. Even prime numbers. **3.** ________
4. Odd prime numbers. **4.** ________
5. Prime factors of 35. **5.** ________
6. Composite factors of 35. **6.** ________
7. Composite factors of 66. **7.** ________
8. Prime factors of 66. **8.** ________
9. Numbers neither prime nor composite. **9.** ________
10. Numbers both prime and composite. **10.** ________

EXERCISES 1.3 S

1–20. List the members of each set described.

1. The set of primes greater than 30 and less than 50. 1. ________
2. The set of composites greater than 35 and less than 45. 2. ________
3. The set of even multiples of 11. 3. ________
4. The set of odd multiples of 11. 4. ________
5. The set of prime factors of 154. 5. ________
6. The set of composite factors of 154. 6. ________
7. The odd factors of 70. 7. ________
8. The even factors of 70. 8. ________
9. The composite factors of 61. 9. ________
10. The set of even prime numbers. 10. ________
11. The factors of 54. 11. ________
12. The prime factors of 54. 12. ________
13. The composite factors of 54. 13. ________
14. The even factors of 54. 14. ________
15. The odd factors of 54. 15. ________
16. The multiples of 1. 16. ________
17. The odd multiples of 8. 17. ________
18. The odd factors of 130. 18. ________
19. The prime factors of 130. 19. ________
20. The composite factors of 130. 20. ________

NAME ________ DATE ________ COURSE ________

21–30. Write each number as a product of two natural numbers in as many ways as possible.

SAMPLE PROBLEM
54

SOLUTION
$54 = 1 \cdot 54 = 2 \cdot 27 = 3 \cdot 18 = 6 \cdot 9$

21. 35

22. 28

23. 12

24. 36

25. 45

26. 60

27. 90

28. 75

29. 42

30. 30

21. ____________________

22. ____________________

23. ____________________

24. ____________________

25. ____________________

26. ____________________

27. ____________________

28. ____________________

29. ____________________

30. ____________________

GROUPING SYMBOLS, COMBINED OPERATIONS 1.4

A. EVALUATION: ONE SET OF GROUPING SYMBOLS

Note that the expressions listed below are evaluated in two different ways: first, by doing the operations as we read them from left to right and, second, by doing the operations in the opposite order.

Expression	*Doing the Operations*	
	Left to right	*Opposite order*
$12 \div 3 \times 2$	$4 \times 2 =$ **8**	$12 \div 6 =$ **2**
$24 \div 6 + 2$	$4 + 2 =$ **6**	$24 \div 8 =$ **3**
$15 - 10 \div 5$	$5 \div 5 =$ **1**	$15 - 2 =$ **13**
$12 - 4 \times 3$	$8 \times 3 =$ **24**	$12 - 12 =$ **0**

Note that for each problem, we get a different answer if we do the operations in a different order.

We would like each expression we write to represent exactly one number.

Therefore, we use grouping symbols and order rules to indicate the order in which the operations are to be done.

ORDER RULES

1. The operation within the set of grouping symbols is performed first.
2. Unless grouping symbols indicate otherwise, the operations are done in the following order:
   first, squaring and/or cubing, as read from left to right.
   second, multiplication and/or division, as read from left to right.
   third, addition and/or subtraction, as read from left to right.
3. The operations within grouping symbols are done in the order stated in rule 2.

GROUPING SYMBOLS

The grouping symbols used most often are

Parentheses	( )
Brackets	[ ]
Braces	{ }
Bar (also called vinculum)	——
as used in division:	$\dfrac{x+3}{2}$

Usually, the parentheses are used when only one set of grouping symbols is needed. This is because the parentheses are the easiest to write. However, in this section, all four types of grouping symbols are used in order to provide practice with them.

SAMPLE PROBLEM 1

Evaluate $20 - 2(3 + 4)$.

SOLUTION

Do the operation inside the parentheses first.

Multiply before subtracting, following the order rules.

$$\begin{aligned} 20 - 2(\mathbf{3 + 4}) &= 20 - 2(\mathbf{7}) \\ &= 20 - \mathbf{14} \\ &= 6 \end{aligned}$$

SAMPLE PROBLEM 2

Evaluate $15 - 8 - 3$.

SOLUTION

Do subtraction as read from left to right.

$$\mathbf{15 - 8} - 3 = \mathbf{7} - 3 = 4$$

SAMPLE PROBLEM 3

Evaluate $15 - [8 - 3]$.

SOLUTION

Do the operation inside the brackets first.

$$15 - [\mathbf{8 - 3}] = 15 - \mathbf{5} = 10$$

SAMPLE PROBLEM 4

Evaluate $\frac{4^2 - 6}{5}$.

SOLUTION

The division bar is a grouping symbol, and therefore, the operations in the numerator must be done before the division by 5.

Do squaring before subtraction, by order rules.

$$\begin{aligned} \frac{\mathbf{4^2} - 6}{5} &= \frac{\mathbf{16} - 6}{5} \\ &= \frac{10}{5} \\ &= 2 \end{aligned}$$

SAMPLE PROBLEM 5

Evaluate $6 + 3\{7 - 5\}$.

SOLUTION

The subtraction inside braces is done first, then the multiplication by 3, and **last** the addition of 6.

$$6 + 3\{2\} = 6 + 6 = 12$$

Note that braces and brackets are also used to indicate multiplication; $5[3] = 5\{3\} = 5(3) = 15$.

EXERCISES 1.4A

Evaluate.

1. $10 - 4 - 2$
2. $18 - 6 + 5$
3. $10 - (4 - 2)$
4. $18 - (6 + 5)$
5. $9 - 2(3 + 1)$
6. $3 + 4(5 - 2)$
7. $2(5^2)$
8. $(2 \cdot 5)^2$
9. $\dfrac{30}{2 + 3}$
10. $\dfrac{30}{5 - 2}$
11. $\dfrac{16 - 2(3)}{2}$
12. $\dfrac{16 + 2^3}{8}$
13. $4(8 - 5)^2$
14. $4 + (8 - 5)^2$
15. $(24 - 4)^2$
16. $24 - 4^2$
17. $15 + 3[7 - 2]$
18. $4[3^2 + 3^3]$
19. $5\{5 + 5^2\}$
20. $3\{3^2 - 7\}$
21. $8 - [10 - 2]$
22. $8\{10 - 6 - 4\}$
23. $6^2 + \dfrac{6 + 4}{6 - 4}$
24. $\dfrac{9^2 - 11}{9 - 2}$

1. __________
2. __________
3. __________
4. __________
5. __________
6. __________
7. __________
8. __________
9. __________
10. __________
11. __________
12. __________
13. __________
14. __________
15. __________
16. __________
17. __________
18. __________
19. __________
20. __________
21. __________
22. __________
23. __________
24. __________

B. TRANSLATING COMBINED OPERATIONS

When a second operation is performed on a sum or difference, there must be a way to indicate that the sum or difference is considered as a single number. Usually, parentheses (   ) are used, or the division bar is used when the operation of division is involved.

SAMPLE PROBLEMS 1–11

Express each of the following in symbols.

WORDS	SYMBOLS
1. Twice the sum of x and 5.	1. $2(x + 5)$
2. One third the difference of x decreased by 5.	2. $\frac{x - 5}{3}$ or $\frac{1}{3}(x - 5)$
3. 5 less than the product of 2 and x.	3. $2x - 5$
4. 3 times the quotient of x divided by 5.	4. $3\frac{x}{5}$ or $\frac{3x}{5}$
5. The sum of x and the difference when 2 is subtracted from y.	5. $x + (y - 2)$
6. The difference when the sum of y and 2 is subtracted from x.	6. $x - (y + 2)$
7. The square of the sum of x and 5.	7. $(x + 5)^2$
8. The cube of the difference of x less than y.	8. $(y - x)^3$
9. The cube of the sum of x and 2.	9. $(x + 2)^3$
10. The square of the difference when 2 is subtracted from y.	10. $(y - 2)^2$
11. The quotient obtained when the sum of a and 4 is divided by the product of 4 and a.	11. $\frac{a + 4}{4a}$

EXERCISES 1.4B

Express each of the following in symbols.

1. Five is added to a number x and then the sum is multiplied by 3. — 1. ______
2. A number x is multiplied by 3 and then 5 is added to the product. — 2. ______
3. A number x is subtracted from 8 and then the difference is divided by 4. — 3. ______
4. A number x is divided by 4 and then 8 is subtracted from the quotient. — 4. ______
5. A number x is multiplied by 5 and then the product is squared. — 5. ______
6. A number x is squared and then the result is multiplied by 5. — 6. ______
7. The number 8 is subtracted from y and then the difference is subtracted from x. — 7. ______
8. A number y is subtracted from a number x and then 8 is subtracted from the difference. — 8. ______
9. Three is added to a number x. Then the sum is multiplied by 2. Then this product is subtracted from a number y. — 9. ______
10. Two is subtracted from a number y. Then this difference is multiplied by the sum of 3 and a number x. — 10. ______
11. Nine times the sum of x and 3. — 11. ______
12. One half the difference when 3 is subtracted from x. — 12. ______
13. One fifth the sum of y and 2. — 13. ______
14. Four times the difference of y decreased by 6. — 14. ______
15. Three times the quotient of y divided by 8. — 15. ______
16. The sum of 4 and the product of 5 and x. — 16. ______
17. Five times the sum of x and 4. — 17. ______
18. The difference when 1 is subtracted from the product of 6 and y. — 18. ______
19. Six times the difference when 1 is subtracted from y. — 19. ______
20. The difference when the sum of t and 6 is subtracted from s. — 20. ______
21. The sum of x and the difference when y is subtracted from 8. — 21. ______
22. The sum of x and the quotient of 3 divided by x. — 22. ______
23. The difference when the product of 2 and r is subtracted from 6. — 23. ______
24. The sum of 5 and the product of 9 and z. — 24. ______

25. The product of x and the difference when 3 is subtracted from y. 25. ________

26. The quotient when the sum of 4 and z is divided by 9. 26. ________

27. The product of t multiplied by the sum of t and 8. 27. ________

28. The square of the product of z and 7. 28. ________

29. The cube of the sum of x and y and z. 29. ________

30. The sum of the square of n and the square of 3. 30. ________

31. The quotient when 5 is divided by the sum of n and 3. 31. ________

32. The sum of q and the product of 7 and q. 32. ________

33. One ninth the product of 4 and t. 33. ________

34. One ninth the cube of t. 34. ________

35. The quotient when the difference of 6 less than n is divided by 2. 35. ________

C. EVALUATION: SEVERAL SETS OF GROUPING SYMBOLS

When two or more sets of grouping symbols occur in a problem, the operations are performed as follows:

First, do the operation or operations within the innermost set of grouping symbols.

Then, do the indicated operation or operations within the remaining innermost set of grouping symbols.

Within each set of grouping symbols, follow the order of operations as stated in Section 1.4A.

SAMPLE PROBLEM 1

Evaluate $8 - [30 - (2 + 3)^2]$.

SOLUTION

First, do inside the innermost grouping symbols (the parens).

$8 - [30 - \mathbf{(2 + 3)}^2] = 8 - [30 - \mathbf{(5)}^2]$

Inside brackets, square before subtracting.

$= 8 - [30 - \mathbf{25}]$

$= 8 - 5$

$= 3$ (answer)

SAMPLE PROBLEM 2

Evaluate $2\{19 - 3[5^2 - 2(7 + 3)]\}$.

SOLUTION

First, do inside the innermost grouping symbols.

$2\{19 - 3[5^2 - 2\mathbf{(7 + 3)}]\} = 2\{19 - 3[5^2 - 2\mathbf{(10)}]\}$

Square and multiply, before subtracting.

$= 2\{19 - 3[\mathbf{25 - 20}]\}$

$= 2\{19 - 3[\mathbf{5}]\}$

Multiply before subtracting.

$= 2\{19 - \mathbf{15}\}$

Subtract inside braces before the final multiplication.

$= 2\{\mathbf{4}\}$

$= 8$ (answer)

SAMPLE PROBLEM 3

Evaluate $\dfrac{(8 + 7)(8 - 5)}{2 \cdot 7 - 5}$.

SOLUTION

The division is done last. The operations above the bar and below the bar must be done first.

Do inside parens first.
Multiply before subtracting.

$$\frac{(8 + 7)(8 - 5)}{2 \cdot 7 - 5} = \frac{\mathbf{(15)(3)}}{\mathbf{14} - 5}$$

$$= \frac{45}{9} = 5 \text{ (answer)}$$

When evaluating a numerical expression, some of the steps may be done mentally and may be omitted in the written solution, according to one's experience and the complexity of the problem. However, omissions should be done mentally, they should be reasonable in order to avoid mistakes, and clear solutions should be written, so that they can be understood by another reader.

When two or more sets of grouping symbols are needed, any type can be used, and the sets can be written in any order desired. Usually parentheses are used first because they are the easiest to write, then brackets, and finally braces, which are considered the most difficult to write.

NOTE TO STUDENT:
The secret word in algebra is PRACTICE. The more problems you do, the better you understand.

EXERCISES 1.4C

Evaluate.

1. $50 - [6^2 - (3 + 1)^2]$

2. $\dfrac{7(7 + 3)}{5(9 - 2)}$

3. $6[(2^2 - 3) + 5]$

4. $8[(4 + 5)(3 + 2)]$

5. $\dfrac{(5 + 1)(5 + 2)}{2^2 - 1}$

6. $2[16 - 5(40 - 38)]$

7. $[6 + (3 - 1)][5 + (3 - 1)]$

8. $\dfrac{(12 - 4)(5 - 3)}{12 - 4(5 - 3)}$

9. $(5^2 - 4^2)(9 - [7 - 2])$

10. $\dfrac{1}{4}[53 - (2^3 - 3)]$

11. $7[(5 + 6)(5 - 2)]$

12. $7\{[5 + 6] - [5 - 2]\}$

13. $10 - \{8 - [6 - (4 - 2)]\}$

14. $2\{15 - [5 - 2(5 - 3)]\}$

15. $\{12 + 3[(5 - 3)(5 - 2)]\}^2$

16. $2 + 2\{2 + 2[2 + 2(2)]\}$

1. __________
2. __________
3. __________
4. __________
5. __________
6. __________
7. __________
8. __________
9. __________
10. __________
11. __________
12. __________
13. __________
14. __________
15. __________
16. __________

1–20. Evaluate.

1. $10(5 - 2)$

2. $(10)(5) - 2$

3. $10(5 \cdot 2)$

4. $(10 \cdot 5)(10 \cdot 2)$

5. $(10 - 2)(5 - 2)$

6. $10 - 2(5 - 2)$

7. $15 - [10 - 5]$

8. $[15 - 10] - 5$

9. $18 - 3(2 + 3)$

10. $18 - 3(2) + 3$

11. $(18 - 3)(2 + 3)$

12. $\dfrac{125 - 27}{5 - 3}$

13. $\dfrac{125}{5} - \dfrac{27}{3}$

14. $\dfrac{2(8 + 7)}{2 + 3}$

15. $\dfrac{10^3 - 5^3}{10^2 + 5(10 + 5)}$

16. $19 - (10 - 3 \cdot 2)$

17. $19 - 10 - 3 \cdot 2$

18. $50 - 10(7 - 4)$

19. $5(8 - 5 - 1)$

20. $\dfrac{(9 - 2)4}{9 - 2(4)}$

1. ______
2. ______
3. ______
4. ______
5. ______
6. ______
7. ______
8. ______
9. ______
10. ______
11. ______
12. ______
13. ______
14. ______
15. ______
16. ______
17. ______
18. ______
19. ______
20. ______

NAME ______ DATE ______ COURSE ______

21–45. Express in symbols.

21. The difference when the product of x and 3 is subtracted from the cube of x. **21.** ______

22. The sum when the product of 5 and z is added to 7. **22.** ______

23. One third the product of 8 and y. **23.** ______

24. The product of 9 times the difference when s is subtracted from 9. **24.** ______

25. The sum of x and the difference of x decreased by 6. **25.** ______

26. The difference when the sum of x and 6 is subtracted from y. **26.** ______

27. The difference when x less than 6 is subtracted from 12. **27.** ______

28. Twice the sum of t and 3. **28.** ______

29. The difference when the sum of 4 and n is subtracted from n. **29.** ______

30. The product of 2 and the product of x and y. **30.** ______

31. The quotient when the product of x and y is divided by 2. **31.** ______

32. The product of 5 and the square of y. **32.** ______

33. The product of 6 and the cube of x. **33.** ______

34. One less than the square of x. **34.** ______

35. The sum of 1 and the cube of y. **35.** ______

36. The difference when the square of y is subtracted from the square of x. **36.** ______

37. The square of the difference when y is subtracted from x. **37.** ______

38. The sum of the square of x and the square of 5. **38.** ______

39. The sum of the cube of y and the cube of 2. **39.** ______

40. The difference when y decreased by z is subtracted from x. **40.** ______

41. The sum when y decreased by z is added to x. **41.** ______

42. The product of x and the difference obtained by subtracting 3 from x. **42.** ______

43. The sum of x and the product of 3 times the difference when 3 is subtracted from x. **43.** ______

44. One half the product of 5 and the sum of y and 3. **44.** ______

45. The quotient when the sum of x and y is divided by twice the difference obtained by subtracting y from x. **45.** ______

46–60. Evaluate.

46. $\dfrac{(5^2 - 5)(5^2 + 5)}{5^3 - 5^2}$

47. $8\left\{\dfrac{3(4^2 + 6)}{2^2 + 7}\right\}$

48. $3^2(7 - 4)^2$

49. $(7 + 4)^2 - (7 + 3)^2$

50. $5(7 - 3)^2$

51. $[5(7 - 3)]^2$

52. $5(6 - 3) - 6(5 - 3)$

53. $4^2 - [6 - (20 - 18)]$

54. $\dfrac{7^2 - 4^2}{7 - 4}$

55. $\dfrac{7^2 - 4^2}{7 + 4}$

46. ______________________

47. ______________________

48. ______________________

49. ______________________

50. ______________________

51. ______________________

52. ______________________

53. ______________________

54. ______________________

55. ______________________

56. $10 - \{8 - [6 - (4 - 2)]\}$

57. $10 - \{8 + [6 - (4 + 2)]\}$

58. $2(5 - 2)^2$

59. $[2(5 - 2)]^2$

60. $2\{8 - 2[6 - 2(5 - 3)]\}$

56. ______________________

57. ______________________

58. ______________________

59. ______________________

60. ______________________

EQUALITY, SUBSTITUTION 1.5

A. EQUALITY, SUBSTITUTION (ONE VARIABLE)

An equation is a statement having the form $A = B$, which is read "A equals B." The statement $A = B$ means that A and B are names of the same number.

An equation does not have to be a true statement. As examples,

$$6 + 4 = 10 \text{ is a true statement}$$

$$6 + 4 = 15 \text{ is a false statement}$$

Instead of saying $6 + 4 = 15$ is false, we say $6 + 4$ does not equal 15. In symbols, we write $6 + 4 \neq 15$.

The symbol $\neq$ means "does not equal."

Some statements are true just because of the form they have. Two cases involving the equality relation are so important that they have special names. These are the **symmetric** and **reflexive axioms** of equality. (An axiom is a statement that is accepted without proof.)

REFLEXIVE AXIOM	SYMMETRIC AXIOM
$A = A$	If $A = B$, then $B = A$
EXAMPLES OF THE REFLEXIVE AXIOM:	EXAMPLES OF THE SYMMETRIC AXIOM:
$3 = 3$	If $6 + 4 = 10$, then $10 = 6 + 4$
$x = x$	If $10 = x + 3$, then $x + 3 = 10$
$2x + 1 = 2x + 1$	

The **reflexive axiom** means that we call an equation true whenever **both sides of the equation are identical.**

The **symmetric axiom** means that **we can exchange the sides of an equation** without changing the truth or falsity of the equation.

The **substitution axiom** is another important axiom about equality. Informally, this axiom states that **any quantity may be substituted for its equal.**

> THE SUBSTITUTION AXIOM
>
> If $A = B$, then A may replace B or B may replace A in an algebraic expression without changing the number that is being named or in an algebraic statement without changing the truth or falsity of the statement.

SAMPLE PROBLEM 1

Evaluate $2x - \frac{x+5}{3}$ for $x = 7$.

SOLUTION

1. Rewrite the expression, removing the letter and holding its place with open parentheses.

$$2x - \frac{x+5}{3}$$

$$2(\ ) - \frac{(\ )+5}{3}$$

2. Insert the given value of the letter within the parentheses.

$$2(7) - \frac{(7)+5}{3}$$

3. Do the indicated operations and write the answer as a single number.

$$= 14 - \frac{12}{3}$$
$$= 14 - 4$$
$$= 10 \text{ (answer)}$$

SAMPLE PROBLEM 2

Evaluate $\frac{2y^2 + 9y - 5}{2y - 1}$ for $y = 4$.

NOTE TO STUDENT:
Be sure you write the parentheses. This will help you avoid errors.

SOLUTION

1. Rewrite expression.

$$\frac{2y^2 + 9y - 5}{2y - 1}$$

2. Remove y, holding its place.

$$\frac{2(\ )^2 + 9(\ ) - 5}{2(\ ) - 1}$$

3. Insert 4 within the parentheses.

$$\frac{2(4)^2 + 9(4) - 5}{2(4) - 1}$$

4. Do the operations.

$$= \frac{2(16) + 36 - 5}{8 - 1}$$
$$= \frac{32 + 36 - 5}{7}$$
$$= \frac{68 - 5}{7} = \frac{63}{7} = 9 \text{ (answer)}$$

Note in Sample Problem 2 that the exponent 2 of $2y^2$ acts only on y. If $y = 4$, then $2y^2 = 2(y^2) = 2yy = 2(4^2) = 2(16) = 32$.

The squaring of 4 must be done before the multiplication by 2. Compare $2y^2$ with $(2y)^2$.

$$(2y)^2 = (2y)(2y) = 2 \cdot 2yy$$

For $y = 4$, then $(2y)^2 = (2 \cdot 4)^2 = 8^2 = 64$

NOTE TO STUDENT:
$2x^2$ is not the same as $(2x)^2$. Watch this!

EXERCISES 1.5A

Evaluate.

1. $5x^2 - 2x - 2$ for $x = 3$

2. $6y + \dfrac{2y^2}{y - 4}$ for $y = 5$

3. $(x^2 + 4)(2x - 3)$ for $x = 4$

4. $\dfrac{3y^2 - 2y - 8}{3y + 4}$ for $y = 3$

5. $4(5x) - 3(5 - x)$ for $x = 4$

6. $5z - \left[z - \dfrac{z}{3}\right]$ for $z = 9$

7. $7\{4 + 12q\}$ for $q = 2$

8. $q\{4 + 12q\}$ for $q = 3$

9. $\dfrac{6r - 16}{r}$ for $r = 4$

10. $t^2 - t + 3$ for $t = 7$

11. $2 + \dfrac{5x}{2}$ for $x = 6$

12. $3s + \dfrac{s^2 - 5}{2}$ for $s = 9$

13. $n(n - 3)$ for $n = 5$

14. $z^2 - 2[z - 3]$ for $z = 8$

15. $\dfrac{65 - y}{y + 7}$ for $y = 2$

16. $3n^2 + \dfrac{n}{n - 7}$ for $n = 14$

17. $5n^2\left(\dfrac{5n}{7 - n}\right)$ for $n = 2$

18. $4t + \{16 - t^3\}$ for $t = 2$

19. $4t - \{16 - t^3\}$ for $t = 2$

20. $7y + \dfrac{y^2 - 3}{y - 3}$ for $y = 5$

1. ____________
2. ____________
3. ____________
4. ____________
5. ____________
6. ____________
7. ____________
8. ____________
9. ____________
10. ____________
11. ____________
12. ____________
13. ____________
14. ____________
15. ____________
16. ____________
17. ____________
18. ____________
19. ____________
20. ____________

B. SUBSTITUTION, SEVERAL VARIABLES

SAMPLE PROBLEM 1

Evaluate $\dfrac{x^2 - 2xy + y^2}{x - y}$ for $x = 7$ and $y = 2$.

SOLUTION

1. Replace one variable, say x, by open parentheses. $\dfrac{(\;)^2 - 2(\;)y + y^2}{(\;) - y}$
2. Insert the value for x. $\dfrac{(7)^2 - 2(7)y + y^2}{(7) - y}$
3. Replace the other variable, y, by open parentheses. $\dfrac{(7)^2 - 2(7)(\;) + (\;)^2}{(7) - (\;)}$
4. Insert the value for y. $\dfrac{(7)^2 - 2(7)(2) + (2)^2}{(7) - (2)}$
5. Evaluate. $= \dfrac{49 - 28 + 4}{5}$

$$= \frac{21 + 4}{5} = \frac{25}{5} = 5$$

SAMPLE PROBLEM 2

Evaluate $\dfrac{3x^2 - 5y^2 - 1}{xy}$ for $x = 3$ and $y = 2$.

SOLUTION

1. Replace one variable, say y, by open parentheses. $\dfrac{3x^2 - 5(\;)^2 - 1}{x(\;)}$
2. Insert the value for y. $\dfrac{3x^2 - 5(2)^2 - 1}{x(2)}$
3. Replace the other variable, x, by open parentheses. $\dfrac{3(\;)^2 - 5(2)^2 - 1}{(\;)(2)}$
4. Insert the value for x. $\dfrac{3(3)^2 - 5(2)^2 - 1}{(3)(2)}$
5. Evaluate. $= \dfrac{3(9) - 5(4) - 1}{(3)(2)}$

$$= \frac{27 - 20 - 1}{6}$$

$$= \frac{6}{6} = 1$$

With practice, both substitutions may be made at the same time. Also, some steps may be done mentally and may be omitted from the written solution. These omissions should be reasonable, and they should be clearly understood by both the writer and any reader of the written work.

EXERCISES 1.5B

Evaluate each of the following.

1. $x - [2y - x]$ for $x = 4$ and $y = 3$ 1. ________

2. $a^2 + 4ab + b^2$ for $a = 9$ and $b = 5$ 2. ________

3. $2m(n - m)$ for $m = 2$ and $n = 7$ 3. ________

4. $5xy^2 - 3x^2y$ for $x = 3$ and $y = 5$ 4. ________

5. $3(a - b)^2$ for $a = 10$ and $b = 7$ 5. ________

6. $\dfrac{m^2 - n^2}{m - n}$ for $m = 6$ and $n = 5$ 6. ________

7. $30 - [s^2 - \{3t - s\}]$ for $s = 3$ and $t = 2$ 7. ________

8. $\dfrac{10ab}{a + b}$ for $a = 7$ and $b = 3$ 8. ________

9. $\dfrac{a^2b + ab^2}{a + b}$ for $a = 7$ and $b = 3$ 9. ______

10. $y^2 + \dfrac{x^3 + y^3}{x + y}$ for $x = 5$ and $y = 4$ 10. ______

11. $\dfrac{x^2 - 3xy - 10y^2}{x - 5y}$ for $x = 20$ and $y = 3$ 11. ______

12. $\dfrac{2x^2 - 3y^2 + 24}{2xy}$ for $x = 6$ and $y = 4$ 12. ______

13. $(a^2 + b^2)(a^2 - b^2)$ for $a = 4$ and $b = 3$ 13. ______

14. $(3a^2 - 10b)(a + 5b)$ for $a = 10$ and $b = 8$ 14. ______

15. $(x + y)(x^2 - xy + y^2)$ for $x = 3$ and $y = 2$ 15. ______

16. $\dfrac{A^3 - B^3}{A - B}$ for $A = 10$ and $B = 5$ 16. ______

1–26. Evaluate.

1. $\frac{x+3}{2} - \frac{x^2+x}{5x}$ for $x = 9$

2. $[3m - 2] + m^2$ for $m = 6$

3. $7(r^2 - 4) - r(r^2 - 4)$ for $r = 3$

4. $\frac{8(x+7)}{2x}$ for $x = 4$

5. $30 - \frac{y^2 - y}{6}$ for $y = 7$

6. $5y - 2(y + 3)$ for $y = 4$

7. $12 - 3(x - 2)$ for $x = 6$

8. $4x^2 + x - 7$ for $x = 3$

9. $3y^2 - 11y$ for $y = 5$

10. $(x + 3)(x + 5)$ for $x = 4$

11. $x + 3(x + 5)$ for $x = 4$

12. $(3y - 5)(y - 3)$ for $y = 5$

13. $3y - 5(y - 3)$ for $y = 5$

14. $[2(x - 7)]^2 - 2(x - 7)^2$ for $x = 10$

1. ____________________
2. ____________________
3. ____________________
4. ____________________
5. ____________________
6. ____________________
7. ____________________
8. ____________________
9. ____________________
10. ____________________
11. ____________________
12. ____________________
13. ____________________
14. ____________________

NAME ____________________ DATE ____________________ COURSE ____________________

15. $3\{y^2 - z^2\}$ for $y = 8$ and $z = 7$

16. $a^2b - b^2a - 3ab$ for $a = 6$ and $b = 2$

15. ______

16. ______

17. $r^3s + r^2s - rs$ for $r = 3$ and $s = 5$

18. $(x + y)(x^2 - xy + y^2)$ for $x = 6$ and $y = 4$

17. ______

18. ______

19. $(x - y)(x^2 + xy + y^2)$ for $x = 5$ and $y = 3$

20. $(x + 3y)(x - 3y)$ for $x = 8$ and $y = 2$

19. ______

20. ______

21. $(x + 3y)(x + 3y)$ for $x = 4$ and $y = 2$

22. $(x^2 + 2)(y^2 - 3)$ for $x = 3$ and $y = 4$

21. ______

22. ______

23. $\dfrac{xy - 4x + 5y - 20}{(x + 5)(y - 4)}$ for $x = 2$ and $y = 7$

24. $\dfrac{2x^2 + 10y^2 - 9xy}{x - 2y}$ for $x = 10$ and $y = 3$

23. ______

24. ______

25. $a^2 + b^2 + c^2 - abc$ for $a = 2$, $b = 3$, and $c = 4$

26. $(x + y - 2z)^2$ for $x = 1$, $y = 4$, and $z = 2$

25. ______

26. ______

1.6 FORMULAS

There are many problems in mathematics, science, business, and other areas that can be solved by using formulas. A **formula** is an equation whose letters have special meanings. Substitution is used to find the value of one letter when the values of the other letters are given.

SAMPLE PROBLEM 1

Find 30% of 50.

Use the formula $A = \frac{pN}{100}$.

Find A for $p = 30$ and $N = 50$.

SOLUTION

$$A = \frac{30(50)}{100} = 15$$

30% of 50 is 15.

SAMPLE PROBLEM 2

What percent of 48 is 12?

Use the formula $p = \frac{100B}{A}$.

Find p for $A = 48$ and $B = 12$.

SOLUTION

$$p = \frac{100(12)}{48} = 25$$

25% of 48 is 12.

SAMPLE PROBLEM 3

Find the perimeter of a rectangle whose sides are 5 inches wide and 8 inches long.
Use the formula $P = 2(a + b)$.
Find P for $a = 5$ and $b = 8$.

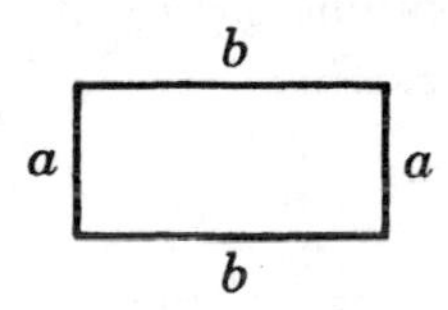

SOLUTION

$P = 2(5 + 8) = 26$ inches

SAMPLE PROBLEM 4

Find the amount in an account after 4 years when \$6000 is invested at the simple interest rate of 8% per year.

Use the formula $A = P + \frac{Pnr}{100}$ and find A for $P = 6000$, $n = 4$, and $r = 8$.

SOLUTION

$$A = 6000 + \frac{6000(4)(8)}{100} = 6000 + 1920 = 7920 \text{ dollars}$$

NOTE TO STUDENT:
Do not panic! These problems are very easy. All you need to know is how to read algebra and how to substitute.

SAMPLE PROBLEM 5

Find the Celsius temperature if the Fahrenheit temperature is 77°.

Use $C = \dfrac{5(F - 32)}{9}$ and find C for $F = 77$.

SOLUTION

$$C = \frac{5(77 - 32)}{9} = \frac{5(45)}{9} = 25$$

The Celsius temperature is 25°.

SAMPLE PROBLEM 6

The leading baseball team of a certain league has won 26 games and lost 12 games. Another team of the league has won 23 games and lost 11 games. How many games behind the leader is the other team?

Use the formula $G = \dfrac{W - L - (w - \ell)}{2}$.

Find G for $W = 26$, $L = 12$, $w = 23$, and $\ell = 11$.

SOLUTION

$$G = \frac{26 - 12 - (23 - 11)}{2} = \frac{14 - 12}{2} = \frac{2}{2} = 1$$

The other team is one game behind the leader.

1–4. Solve by using the formula $A = \frac{pN}{100}$.

1. What is 20% of 45?
Find A for $p = 20$ and $N = 45$.

1. ____________

2. What is 18% of $500?
Find A for $p = 18$ and $N = 500$.

2. ____________

3. Find the sales tax on a purchase of $200 if the tax rate is 6%.
Find A for $p = 6$ and $N = 200$.

3. ____________

4. One rule in budget making states that the maximum monthly rent should be 25% of the monthly salary. What is the maximum monthly rent for a monthly salary of $1200?
Find A for $p = 25$ and $N = 1200$.

4. ____________

5–8. Solve by using the formula $p = \frac{100\,B}{A}$.

5. What percent of 75 is 30?
Find p for $A = 75$ and $B = 30$.

5. ____________

6. What percent of 36 is 108?
Find p for $A = 36$ and $B = 108$.

6. ____________

7. Pure gold is rated 24K (carats).
What percent of gold is in a piece of jewelry rated 12K?
Find p for $A = 24$ and $B = 12$.

7. ____________

8. A person whose weight was 150 pounds lost 18 pounds on a diet. What percent of the weight was the loss?
Find p for $A = 150$ and $B = 18$.

8. ____________

NAME ____________ DATE ____________ COURSE ____________

9–10. Solve by using the formula $P = 2(a + b)$.

9. Find the perimeter of a rectangle whose sides are 15 meters and 12 meters long.
Find P for $a = 15$ and $b = 12$.

9. ______________

10. Find the amount of fencing needed to enclose a rectangular field whose sides measure 100 feet and 150 feet.
Find P for $a = 100$ and $b = 150$.

10. ______________

11–12. Solve by using the formula $C = \frac{5(F - 32)}{9}$.

11. The temperature of the water in a hot water tank is 140°F. Find the corresponding Celsius temperature.
Find C for $F = 140$.

11. ______________

12. On a certain day, the outdoor temperature was 50°F. What was the temperature in degrees Celsius?
Find C for $F = 50$.

12. ______________

13–16. Sports. Solve by using the formula $G = \frac{W - L - (w - \ell)}{2}$ and the table below.

Basketball team	*Games won*	*Games lost*
Golds	30	9
Reds	29	8
Whites	27	8
Blues	25	12
Greens	22	15

13. How many games behind the leader are the Reds?
Find G for $W = 30$, $L = 9$, $w = 29$, and $\ell = 8$.

13. ______________

14. How many games behind the leader are the Whites?
Find G for $W = 30$, $L = 9$, $w = 27$, and $\ell = 8$.

14. ______________

15. How many games behind the leader are the Blues?
Find G for $W = 30$, $L = 9$, $w = 25$, and $\ell = 12$.

15. ______________

16. How many games behind the leader are the Greens?
Find G for $W = 30$, $L = 9$, $w = 22$, and $\ell = 15$.

16. ______________

17. **Mathematics.** Find the sum of all the natural numbers from 1 to 50.
Use $s = \dfrac{n(n+1)}{2}$ and find s for $n = 50$.

17. ______________

18. **Averages.** Find the average of the three grades 86, 91, and 99.
Use $A = \dfrac{a+b+c}{3}$ and find A for $a = 86$, $b = 91$, and $c = 99$.

18. ______________

19. **Salaries.** The hourly salary of an employee increased from \$4 an hour to \$6 an hour. What percent of the old salary was the increase?
Use $p = \dfrac{100(B-A)}{A}$ and find p for $A = 4$ and $B = 6$.

19. ______________

20. **Sports.** A football team won 21 games and lost 4 games. What percent of the games played did it win?
Use $p = \dfrac{100W}{W+L}$ and find p for $W = 21$ and $L = 4$.

20. ______________

21. **Geometry.** The formula for the perimeter P of a square is $P = 4s$.
The formula for the area A of a square is $A = s^2$.

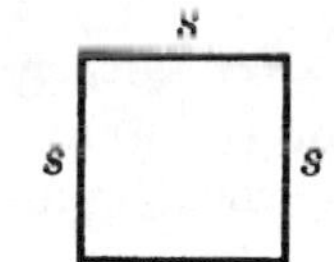

a. Find P for $s = 7$.

b. Find A for $s = 7$.

21. a. ______________
b. ______________

22. **Geometry.** The area A of a trapezoid is given by the formula

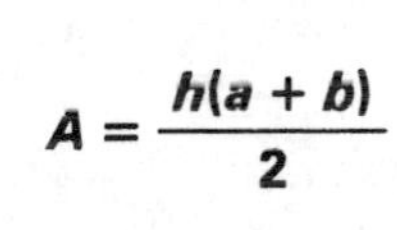

$$A = \frac{h(a+b)}{2}$$

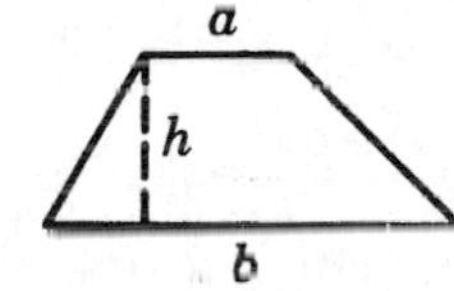

where h is the height, a is the length of one base, and b is the length of the other base.
Find A for $h = 7$, $a = 9$, and $b = 15$.

22. ______________

23. **Business.** The **Rule of 72** is used in **business** to find the number of years, n, that it takes for a sum of money to double in value when invested at a yearly rate of $p\%$.
The Rule of 72 is as follows.

$$n = \frac{72}{p} \quad \text{and} \quad p = \frac{72}{n}$$

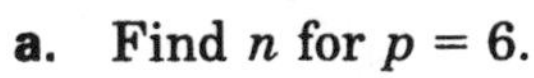

a. Find n for $p = 6$.

b. Find n for $p = 9$.

c. Find p for $n = 4$.

d. Find p for $n = 3$.

23. a. ______________
b. ______________
c. ______________
d. ______________

24. **Investments.** Find the amount of money, A, in an account at the end of 1 yr if $12,000 was invested at the yearly rate of 9%.

Use $A = P + \frac{rP}{100}$ and find A for $P = 12{,}000$ and $r = 9$.

24. ____________

25. **Investments.** For a person who pays t% income tax, an investment in tax-free bonds earning s% is related to a taxable investment earning r% by the formula

$$r = \frac{100\,s}{100 - t}$$

a. Find r for $t = 30$ and $s = 7$.

b. Find r for $t = 50$ and $s = 7$.

25. a. ____________

b. ____________

26. **Coins deposited in a bank.** The total amount A in cents for P pennies, N nickels, D dimes, and Q quarters is given by the formula

$$A = P + 5N + 10D + 25Q$$

Find A for $P = 75$, $N = 35$, $D = 18$, and $Q = 12$.

26. ____________

27. **Cost of first-class mail.** The cost C in cents to send a letter weighing n ounces by first-class mail is given by the formula

$$C = 20 + 17(n - 1)$$

Find C for $n = 5$.

27. ____________

28. The **ideal weight** W in pounds **for a man** whose height is H inches is given by the formula

$$W = 105 + 6(H - 60)$$

Find the ideal weight for a man 5 ft 9 in tall.
Find W for $H = 5(12) + 9 = 69$.

28. ____________

29. The **ideal weight** W in pounds **for a woman** whose height is H inches is given by the formula

$$W = 100 + 5(H - 60)$$

Find the ideal weight for a woman 5 ft 4 in tall.
Find W for $H = 64$.

29. ____________

30. **Concrete** is sold by the cubic yard. The number n of cubic yards of concrete needed to cover an area of A square feet to a depth of d inches is given by

$$n = \frac{dA}{12(27)}$$

Find n for $d = 3$ in and $A = 540$ sq ft.

30. ______________

31. **Low income energy assistance.** To qualify for a certain federal tax credit, the maximum income I in dollars of a family of n persons is given by

$$I = 4738 + 1525(n - 1)$$

a. Find I for $n = 1$.

b. Find I for $n = 3$.

31. a. ______________

b. ______________

32. **Loans.** The total amount A in dollars that is paid for an object having a cash purchase of P dollars is given by

$$A = D + 12yM$$

where M is the monthly loan payment in dollars, D is the down payment in dollars, and y is the number of years of the loan. Find A for a car whose cash price is \$12,500. Use $D = 5{,}000$, $y = 5$, and $M = 250$.

32. ______________

33. Using the formula in Exercise 32, find A for a house whose cash price is \$80,000. Use $D = 23{,}000$, $y = 30$, and $M = 460$.

33. ______________

34. **Cars, stopping distance.** The distance D in feet for a car to come to a stop safely when traveling r m.p.h. is given by

$$D = \frac{r^2 + 22r}{20}$$

a. Find D for $r = 30$.

b. Find D for $r = 50$.

34. a. ______________

b. ______________

35. **Salaries.** A person works n hours a week at a base rate of B dollars per hour. Overtime is paid at a rate of $1\frac{1}{2}$ for those hours over 40. The total weekly salary S is given by

$$S = 40B + \frac{3(n - 40)B}{2}$$

Find S for $B = 8$ and $n = 49$.

35. ______________

36. **Sports.** The score S of a football team making T touchdowns, C conversions, and F field goals is given by

$$S = 6T + C + 3F$$

Find the score S for $T = 3$, $C = 2$, and $F = 2$.

36. ____________

37. **Training heart rate (pulse).** The maximum pulse T after exercising for a person A years old with a resting pulse R is given by

$$T = R + \frac{4(220 - A - R)}{5}$$

Find T for $A = 30$ and $R = 60$.

37. ____________

38. In reference to Exercise 37, the minimum pulse T after exercising is given by

$$T = R + \frac{3(220 - A - R)}{5}$$

Find T for $A = 30$ and $R = 60$.

38. ____________

39. **Temperature conversion.** The temperature C in degrees Celsius is related to the temperature F in degrees Fahrenheit by the formula

$$F = \frac{9C}{5} + 32$$

a. Find F for $C = 20$.

b. Find F for $C = 35$.

c. Find F for $C = 100$.

39. **a.** ____________

b. ____________

c. ____________

40. **Electronics.** The total resistance R in ohms for two resistors having resistances of A ohms and B ohms, respectively, is given by the formula below when the resistors are connected in parallel.

$$R = \frac{AB}{A + B}$$

Find R for $A = 10$ and $B = 15$.

40. ____________

SAMPLE TEST FOR CHAPTER 1

1–15. Express each of the following in symbols. (See Section 1.1.)

1. The product of 8 and x. **1.** ______

2. The sum of n and 6. **2.** ______

3. The difference when 4 is subtracted from y. **3.** ______

4. The quotient when x is divided by 3. **4.** ______

5. The square of t. **5.** ______

6. The product of d and c. **6.** ______

7. The sum of d and c. **7.** ______

8. c subtracted from d. **8.** ______

9. c divided by d. **9.** ______

10. One third of b. **10.** ______

11. The product of 1 and x. **11.** ______

12. The sum of x and 0. **12.** ______

13. x divided by 1. **13.** ______

14. 0 divided by y. **14.** ______

15. y decreased by 0. **15.** ______

NAME ______ DATE ______ COURSE ______

16–18. List the members of the set described. (See Section 1.2.)

16. The set of digits less than 4. **16.** ______________

17. The set of whole numbers w for which $\frac{w}{6}$ is a whole number. **17.** ______________

18. The set of natural numbers n for which $\frac{20}{n}$ is a natural number. **18.** ______________

19–22. List the elements in each described subset of the set of natural numbers. (See Section 1.2.)

19. The factors of 42. **19.** ______________

20. The multiples of 1. **20.** ______________

21. The multiples of 9. **21.** ______________

22. The factors of 6 greater than 7. **22.** ______________

23–26. List the elements in each described subset of the set of natural numbers. (See Section 1.3.)

23. The set of odd numbers between 100 and 110. **23.** ______________

24. The set of even factors of 48. **24.** ______________

25. The primes greater than 40 but less than 50. **25.** ______________

26. The even composite factors of 20. **26.** ______________

27–31. Express each of the following in symbols. (See Section 1.4.)

27. 7 times the sum of n and 8. **27.** ____________

28. 5 less than the product of 3 and x. **28.** ____________

29. The square of the difference when 8 is subtracted from y. **29.** ____________

30. The difference when the cube of the sum of x and 2 is subtracted from y. **30.** ____________

31. The quotient when the product of x and 2 is divided by the sum of x and 2. **31.** ____________

32–35. Evaluate. (See Section 1.4.)

32. $30 - \{3^2 - 2^3\}$

33. $10 - \{9 - (5 - 2)\}$

32. ____________

33. ____________

34. $(7 + [6 - 4])(7 - [6 - 4])$

35. $\dfrac{(4 + 1)^2 - (3 + 1)^2}{(4 - 1) - (3 - 1)}$

34. ____________

35. ____________

36–40. Evaluate. (See Section 1.5.)

36. $\dfrac{x + 2}{x - 2}$ for $x = 4$ **36.** ____________

37. $(y - 6)(y + 2)$ for $y = 8$ **37.** ____________

38. $n^2 - 5(n - 2)$ for $n = 6$ **38.** ____________

39. $\dfrac{x^2 - (y - 3)^2}{x + y - 3}$ for $x = 4$ and $y = 5$ **39.** ____________

40. $(x - y)(x^2 + xy + y^2)$ for $x = 7$ and $y = 4$ **40.** ____________

41–46. Solve by using formulas. (See Section 1.6.)

41. The sales price of a chair decreased from \$50 to \$30. What percent of the original price was the decrease?
Use $p = \dfrac{100(A - B)}{A}$ and find p for $A = 50$ and $B = 30$.

41. ______________________

42. A steel ball is dropped from the top of a building. It hits the ground 9 seconds later. How many feet high is the building?
Use $s = 16t^2$ and find s for $t = 9$.

42. ______________________

43. A hockey team has won 21 games and lost 14 games. What percent of the games played has it won?
Use $P = \dfrac{100W}{W + L}$ and find P for $W = 21$ and $L = 14$.

43. ______________________

44. How much masking tape is needed to go around a rectangular ceiling of a room that is 15 feet long and 9 feet wide?
Use $P = 2(a + b)$ and find P for $a = 15$ and $b = 9$.

44. ______________________

45. In medicine, the dosage C of medicine for a child is often determined by Young's Rule:

$$C = \frac{yA}{y + 12}$$

where A is the adult dosage and y is the age of the child in years.
Find C for $y = 3$ and $A = 250$.

45. ______________________

46. What is the weight W in pounds of an astronaut who is 12,000 miles from the surface of the earth?
Use $W = \dfrac{16(144)}{(d + 4)^2}$ and find W for $d = 12$.

46. ______________________

Signed Numbers

ALGEBRA IN ACTION: RECREATION

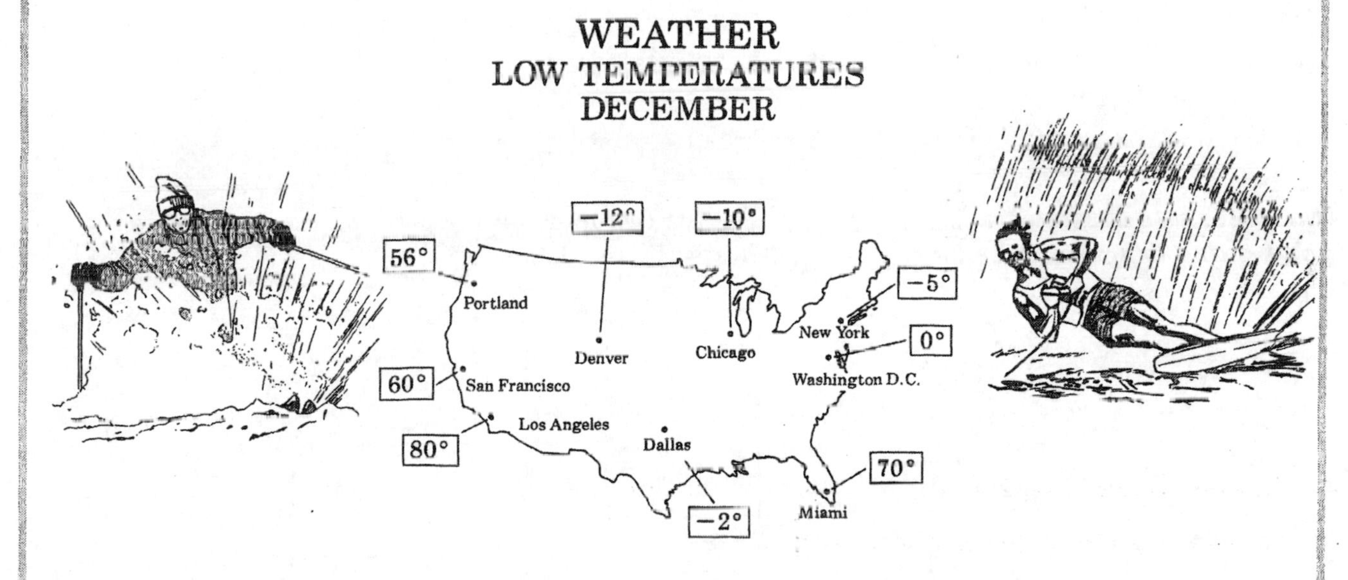

TIDE TABLE (Low P.M.)

Water City

1	Wed	12:31	1.6
2	Thurs	1:24	0.8
3	Fri	2:15	0.0
4	Sat	3:03	-0.6
5	Sun	3:52	-1.0
6	Mon	4:42	-1.2
7	Tues	5:34	-1.1

Signed numbers give us useful information for our recreational life.

A knowledge of tides and temperatures tell us the best places and times for fishing, skiing, or swimming.

Chapter 2 Objectives

Objectives	Typical Problems

State the meaning of a negative number or zero for a given meaning of the corresponding positive number.

1–2. State the meaning if +6 means a gain of 6 pounds.
1. −6 means ____.
2. 0 means ____.

Name the coordinate of a point on a horizontal number line.

3–5. Name the coordinate of the point on the number line indicated by the dot.

3.

4.
0 1

5.
−5 5

Name the coordinate of a point on a vertical number line.

6–8. Name the coordinate of the point on the number line indicated by the dot.

6.
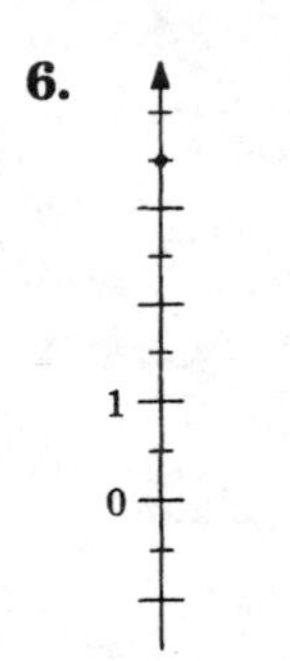

7.
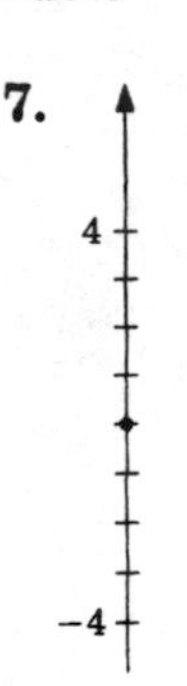

8.
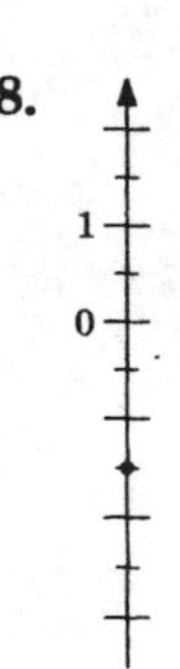

Draw the graph of a number on a horizontal number line.

9. Graph each number on a horizontal number line.
a. 2 b. −2

Draw the graph of a number on a vertical number line.

10. Graph each number on a vertical number line.
a. $1\frac{3}{4}$ b. −3

Simplify expressions involving absolute value symbols.

11–12. Simplify.
11. $|-5| + |-3|$
12. $-(|-9| - |-5|)$

Insert one of the relation symbols, <, >, or =, between two numbers to form a true statement.

13–15. Insert <, >, or = between each pair of numbers to form a true statement.
13. 5, 8
14. −5, −8
15. $|-5|$, 5

Write an order statement in symbols.

16–17. Write in symbols.
16. The sum of x and y is positive.
17. The product of 3 and x is less than 12.

Add signed numbers.

18–22. Do each addition.

18. $(-6) + (-9)$
19. $2 + (-8)$
20. $-3 + 7$
21. $-8 + 0$
22. $8 + (-8)$

Subtract signed numbers.

23–26. Do each subtraction.

23. $24 - 50$
24. $-12 - 8$
25. $-9 - (-2)$
26. $7 - (-10)$

Multiply signed numbers.

27–30. Do each multiplication.

27. $8(-6)$
28. $-8(-6)$
29. $(-4)^2$
30. $(-3)(+4)$

Divide signed numbers.

31–36. Do each operation, if possible.

31. $\dfrac{72}{-9}$ **34.** $\dfrac{8-8}{8}$
32. $\dfrac{-32}{4}$ **35.** $\dfrac{2-6}{6-2}$
33. $\dfrac{-65}{-5}$ **36.** $\dfrac{2-6}{6-6}$

Evaluate expressions involving signed numbers.

37–38. Evaluate.

37. $20 - 3(-7 + 4)$
38. $-2(-5)^2$

Evaluate expressions by substituting integral values.

39–40. Evaluate.

39. $5x^2 + 3x$ for $x = -4$
40. $\dfrac{AB}{A + B}$ for $A = -10$, $B = 14$

ANSWERS to Typical Problems

1. A loss of 6 pounds **2.** No gain or loss **3.** 5 **4.** −5 **5.** 0 **6.** $3\frac{1}{2}$ **7.** 0 **8.** $-1\frac{1}{2}$

9.

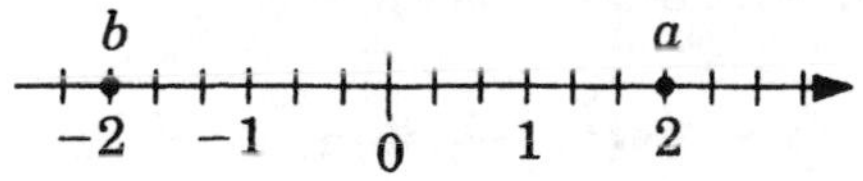

10.

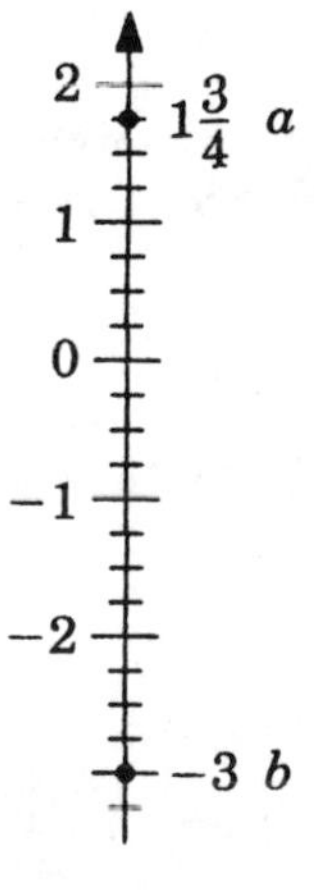

11. 8 **12.** −4 **13.** $5 < 8$ **14.** $-5 > -8$ **15.** $|-5| = 5$ **16.** $x + y > 0$
17. $3x < 12$ **18.** −15 **19.** −6 **20.** 4 **21.** −8 **22.** 0 **23.** −26
24. −20 **25.** −7 **26.** 17 **27.** −48 **28.** 48 **29.** 16 **30.** −12
31. −8 **32.** −8 **33.** 13 **34.** 0 **35.** −1 **36.** Undefined **37.** 29
38. −50 **39.** 68 **40.** −35

INTEGERS, RATIONAL NUMBERS

A. SIGNED NUMBERS, INTEGERS, INTERPRETATIONS

The counting numbers were the first numbers to be invented by man in order to answer the question, "How many?"

Later, fractions were developed to provide greater accuracy in measurement.

As civilization progressed, a need arose to measure quantities that were opposite in nature. Signed numbers were created to meet this need.

$-x$ is the opposite of x

Signed numbers are numbers that are tagged with a + or − sign. The numbers x and $-x$ are called **opposites,** or **additive inverses,** of each other. For example, $+5$ and -5 are opposites of each other. The number $+5$ is called a **positive number,** and the number -5 is called a **negative number.** It is agreed that $+5$ and 5 are different names for the same number. In general, if p is a positive number, then $+p = p$. Similarly, $+n = n$ and $+5 = 5$.

$+x = x$

The number **zero, 0,** is neither positive nor negative. Zero is located midway between the positive numbers and the negative numbers on a scale of measurement.

The difference between two natural numbers is not always a natural number. For example, although $7 - 2$ is the natural number 5, $2 - 7$ is not a natural number, and $7 - 7$ is not a natural number. Actually, as we shall see later, $2 - 7 = -5$ and $7 - 7 = 0$. The numbers -5 and 0 are called **integers.**

By including 0 and the negatives of the natural numbers in the universal set of numbers, the answer to every subtraction problem will be a number in the universal set.

The set of integers, J, is the set consisting of the natural numbers, the number 0 (zero), and the negatives of the natural numbers. In symbols,

$$J = \{\ldots, -5, -4, -3, -2, -1, 0, 1, 2, 3, 4, 5, \ldots\}$$

The natural numbers are also called the **positive integers.** The negatives of the natural numbers are called the **negative integers.** The integer zero, 0, is neither positive nor negative.

For Sample Problems 1–3, state the meaning of −25 if +25 has the value given.

SAMPLE PROBLEM 1
A gain of $25.

SOLUTION
−25 means a loss of $25.

SAMPLE PROBLEM 2
A temperature of 25° above 0° Celsius.

SOLUTION
−25 means a temperature of 25° below 0° Celsius.

SAMPLE PROBLEM 3
A velocity of 25 miles per hour of a car moving due east.

SOLUTION
−25 means a velocity of 25 miles per hour of a car moving due west.

SAMPLE PROBLEM 4
State the meaning of 0 if +30 means the distance of a car 30 miles due north of a city and if −30 means the distance of a car 30 miles due south of the city.

SOLUTION
0 means the car is neither north nor south of the city but is in the city.

EXERCISES 2.1A

State the meaning of each of the following numbers for the given interpretation.

1. −50 if +50 means a bank deposit of $50. 1. ______
2. −8 if +8 means a gain of 8 yards in a football game. 2. ______
3. −200 if +200 means an altitude 200 feet above sea level. 3. ______
4. −45 if +45 means a velocity of 45 m.p.h. in the north direction. 4. ______
5. 0 if +4 means a gain in weight of 4 pounds and −4 means a loss in weight of 4 pounds. 5. ______
6. $-\frac{3}{4}$ if $+\frac{3}{4}$ means an increase of 75¢ in the price of a stock. 6. ______
7. −1250 if +1250 means A.D. 1250. 7. ______

8. -20 if $+20$ means a force of 20 pounds downward. 8. ______

9. 0 if $+60$ means a rotation of 60° counterclockwise and -60 means a rotation of 60° clockwise. 9. ______

10. -8 if $+8$ means a current of 8 amps for a charging battery. 10. ______

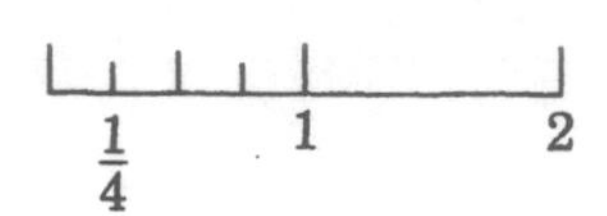

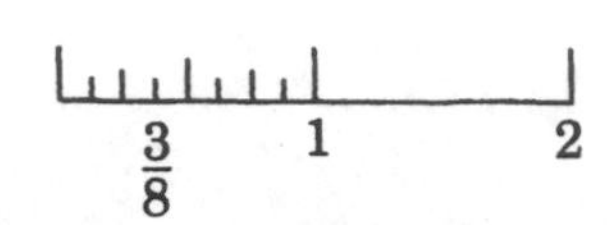

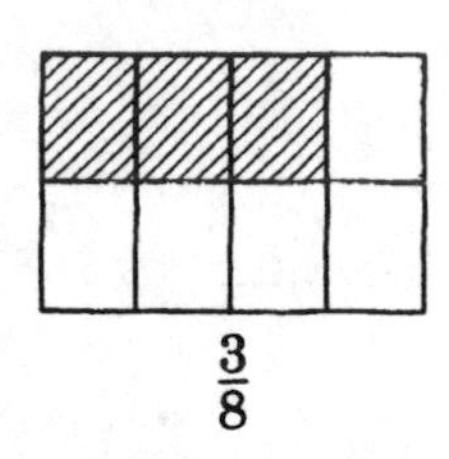

FIGURE 2.1

B. RATIONAL NUMBERS, NUMBER LINES

When one natural number is divided by another natural number, the quotient is not always a natural number. In arithmetic, these quotients are called **fractions.** Each fraction is thought of as the **number of subunits of a basic unit.** Many visual models of fractions are possible, such as those shown in Figure 2.1.

In algebra, we keep this meaning of a fraction. The fractions of arithmetic are called the **positive fractions.** Their negatives, called the **negative fractions,** are also included in the number system of algebra.

The set that includes all the integers and the positive and negative fractions is called the set of **rational numbers.**

> **The set of rational numbers, Q, may also be described as the set of numbers that can be written as the quotient of two integers, $\frac{n}{d}$, where n and d are integers and $d \neq 0$.**

Examples of rational numbers are 0, 5, -5, $\frac{2}{3}$, $-\frac{2}{3}$, $\frac{-15}{7}$, $\frac{5}{-2}$.

Note that $0 = \frac{0}{1}$ and $5 = \frac{5}{1}$ and $-5 = \frac{-5}{1}$. In general, $\frac{n}{1} = n$. In other words, n and $\frac{n}{1}$ name the same number.

The integers are those rational numbers $\frac{n}{d}$ where n is an integer and $d = 1$.

Number lines

We get a useful picture by using the names of numbers as the names of points on a straight line.

A point is selected as the starting point. This point is called the **origin** and is given the name 0 (zero).

A unit of measurement and a direction, called the positive direction, are selected. With this unit of length, points are marked off in succession in the positive direction. For a horizontal line, this direction is usually to the right, as shown in Figure 2.2.

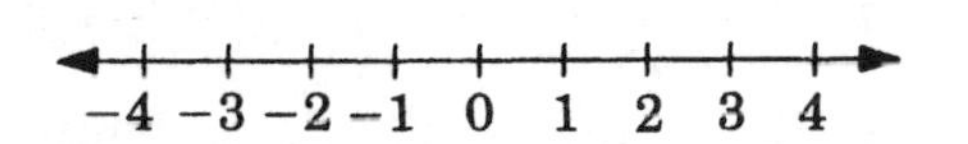

FIGURE 2.2
A horizontal number line

The arrow at the right indicates that the line continues indefinitely in this direction.

The negative integers are assigned to the points in the opposite direction in a similar manner. This direction is to the left in Figure 2.2.

The arrow at the left indicates that the line continues indefinitely in this direction.

A **number line** is a line whose points are named by using numbers.

The **coordinate** of a point on a number line is the number that names the point.

The **graph** of a number is the point on the number line that is named by the number.

The **origin** is the point whose coordinate is 0 (zero).

A number line does not have to be horizontal. A vertical arrangement is often used, as shown in Figure 2.3. In this arrangement the positive direction is usually upward.

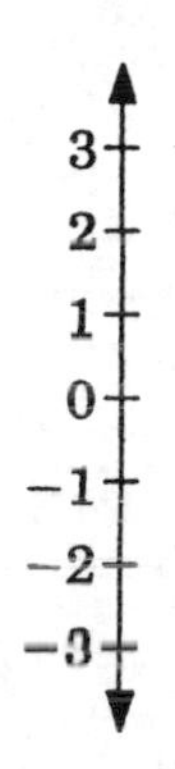

FIGURE 2.3
A vertical number line

All the rational numbers can be identified as coordinates on a number line by considering subdivisions of the basic unit and by using the concept of opposites (Figure 2.4).

The notation $A(3)$ is used to designate the point whose geometric name is A and whose algebraic name is 3. Similarly, $B(-4)$ means the point whose geometric name is B and whose algebraic name is -4.

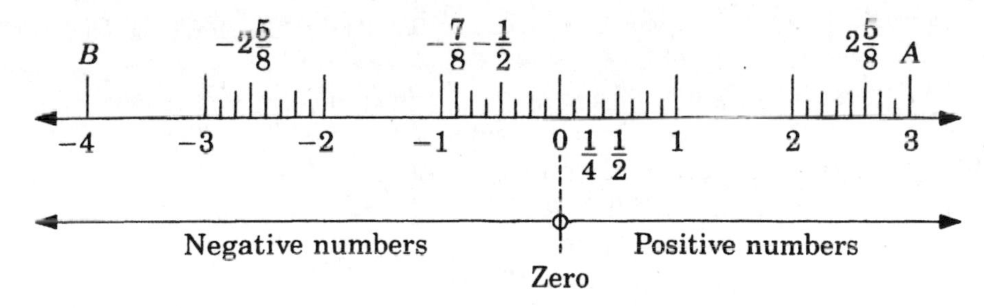

FIGURE 2.4
Rational numbers on a number line

EXERCISES 2.1B

SAMPLE PROBLEM 1

State the coordinates of points A, B, C, and D on the number line below.

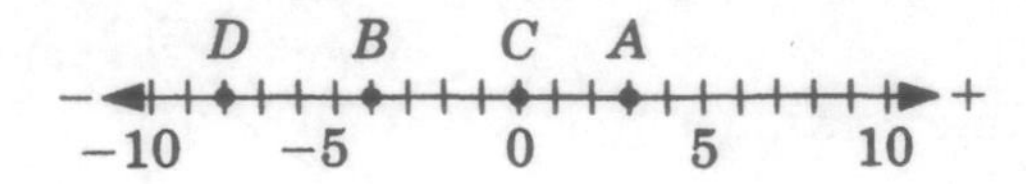

SOLUTION

$A(3)$, $B(-4)$, $C(0)$, $D(-8)$

1–5. State the coordinates of the points A, B, C, and D on each number line.

1. A B C D
−10 −5 0 5 10

1. $A(\ \ )$, $B(\ \ )$, $C(\ \ )$, $D(\ \ )$

2. B A D C
−10 −5 5 10

2. $A(\ \ )$, $B(\ \ )$, $C(\ \ )$, $D(\ \ )$

3. B A D C
−10 −5 10

3. $A(\ \ )$, $B(\ \ )$, $C(\ \ )$, $D(\ \ )$

4. D C A B
−2 −1 0 1 2

4. $A(\ \ )$, $B(\ \ )$, $C(\ \ )$, $D(\ \ )$

5. D C A B
-1 $-\frac{1}{2}$ $\frac{1}{2}$ 1

5. $A(\ \ )$, $B(\ \ )$, $C(\ \ )$, $D(\ \ )$

6–15. Graph each of the following points on a horizontal number line whose positive direction is to the right. Graph five points on one number line. Place a solid dot at the correct position on the number line, and write the letter name of the point above the dot.

6. $A(6)$ **7.** $B(-6)$ **8.** $C(13)$ **9.** $D(8)$ **10.** $E(-8)$

−5 0 5

11. $A(4)$ **12.** $B(-4)$ **13.** $C(-6)$ **14.** $D(0)$ **15.** $E(6)$

−5 5

16–17. State the coordinates of points A, B, C, and D on the given number line.

SAMPLE PROBLEM 2

SOLUTION

$A(+7)$
$B(+2)$
$C(-2)$
$D(-7)$

16.

17.

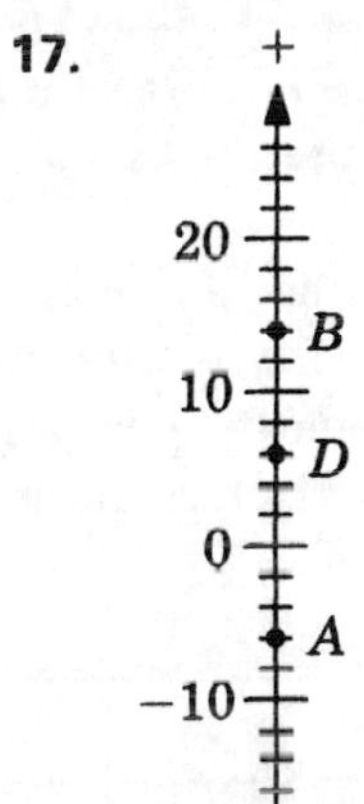

16. $A(\quad)$
$B(\quad)$
$C(\quad)$
$D(\quad)$

17. $A(\quad)$
$B(\quad)$
$C(\quad)$
$D(\quad)$

18–32. Graph each of the following points on a vertical number line whose positive direction is upward. Graph five points on one number line. Place a solid dot at the correct position on the number line, and write the letter name of the point to the right of the dot.

18. $A(-3)$

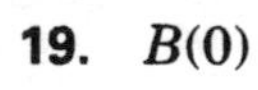

19. $B(0)$

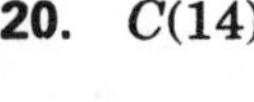

20. $C(14)$

21. $D(3)$

22. $E(-7)$

23. $A(150)$

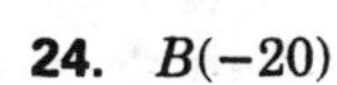

24. $B(-20)$

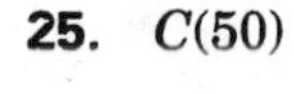

25. $C(50)$

26. $D(-70)$

27. $E(0)$

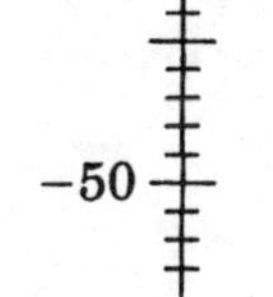

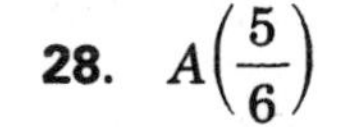

28. $A\left(\frac{5}{6}\right)$

29. $B\left(\frac{-5}{6}\right)$

30. $C\left(1\frac{1}{6}\right)$

31. $D\left(-1\frac{1}{6}\right)$

32. $E\left(\frac{-1}{3}\right)$

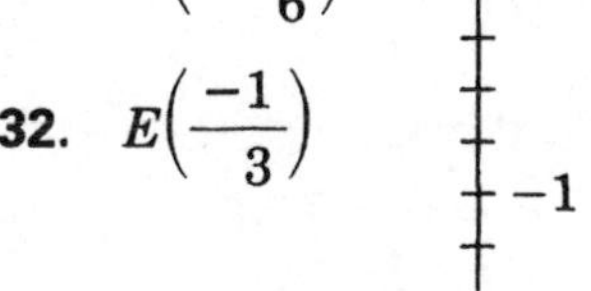

C. ABSOLUTE VALUE

Every nonzero rational number has the following two parts:

1. A quality indicated by a + sign or a − sign (the + sign may be either written or understood).
2. A quantity, a positive amount.

The **absolute value** of a nonzero number is the positive quantity of the number.

In symbols, we write the absolute value of a number as $|x|$. As examples, $|5| = 5$ and $|-5| = 5$. Also $|0| = 0$.

NOTE TO STUDENT:
$|x|$ means "strip the sign from x." Do not confuse $|x|$ with (x) or $[x]$ or $\{x\}$. The symbols | | give different instructions for what is to be done than do the grouping symbols you met earlier.

If p is a positive number, then

$$|p| = p \quad \text{and} \quad |-p| = p \quad \text{and} \quad |0| = 0$$

On a number line, we can think of the absolute value of a number as its distance from the origin.

The points 6 and −6, the opposite of 6, are the **same distance** from the origin. Distance is always a positive number or zero. However, 6 and −6 lie on **opposite sides** of the origin. See Figure 2.5 and recall that $+6 = 6$.

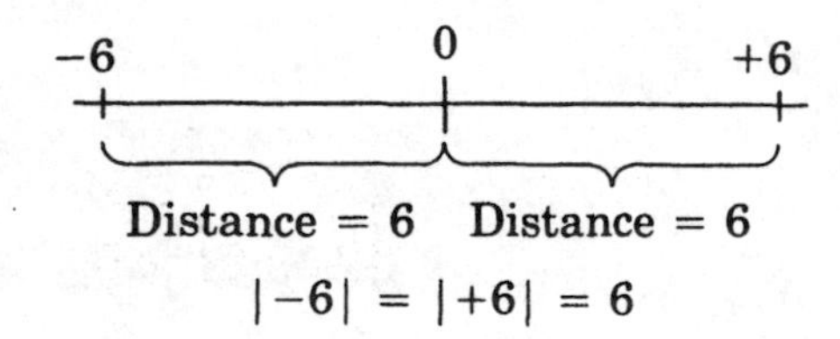

$|-6| = |+6| = 6$

FIGURE 2.5

On a number line, the + or − **quality** of a number is shown by a **direction.** The **quantity** is interpreted as the **distance** from the origin.

In the table below, p is a positive number, and the geometric meaning is for a horizontal number line with its positive direction to the right.

Number	*Quality*	*Quantity*	*Geometric meaning*
6	+ (positive)	6	A distance of 6 units to the right of 0
−6	− (negative)	6	A distance of 6 units to the left of 0
p	+ (positive)	p	A distance of p units to the right of 0
$-p$	− (negative)	p	A distance of p units to the left of 0

The distance of a point from the origin is one application of absolute value. Absolute value is also useful for explaining how to add, subtract, multiply, and divide integers.

SAMPLE PROBLEM 1	SAMPLE PROBLEM 2
Simplify $\|-7\|$.	Simplify $\|-3\| + \|-6\|$.
SOLUTION $\|-7\| = 7$	**SOLUTION** $\|-3\| + \|-6\| = 3 + 6 = 9$

SAMPLE PROBLEM 1
Simplify $|-7|$.

SOLUTION
$|-7| = 7$

SAMPLE PROBLEM 2
Simplify $|-3| + |-6|$.

SOLUTION
$|-3| + |-6| = 3 + 6 = 9$

SAMPLE PROBLEM 3
Simplify $|-8| - |+3|$.

SOLUTION
$|-8| - |+3| = 8 - 3 = 5$

SAMPLE PROBLEM 4
Simplify $3|-5|$.

SOLUTION
$3|-5| = 3(5) = 15$

SAMPLE PROBLEM 5
Simplify $\frac{|-12|}{3}$.

SOLUTION
$\frac{|-12|}{3} = \frac{12}{3} = 4$

SAMPLE PROBLEM 6
Simplify $\frac{|+5|}{|-5|}$.

SOLUTION
$\frac{|+5|}{|-5|} = \frac{5}{5} = 1$

EXERCISES 2.1C

1–22. Simplify each of the following. (Rewrite each as a single integer.)

1. $|6|$ **2.** $|-7|$

3. $|0|$ **4.** $|-4|$

5. $|-8|$ **6.** $|+9|$

7. $|-13| + |5|$ **8.** $|-1| + |-2|$

9. $|5| + |-2|$ **10.** $|10| - |-7|$

11. $|-10| - |-3|$ **12.** $|-9| - |+5|$

1. ______
2. ______
3. ______
4. ______
5. ______
6. ______
7. ______
8. ______
9. ______
10. ______
11. ______
12. ______

13. $2|-6|$

14. $7|-2|$

15. $|-5| \cdot |-4|$

16. $3|7| - 7|-3|$

17. $-|-5|$

18. $3|-4| - 2|-6|$

19. $\frac{|-9|}{3}$

20. $\frac{|48|}{|-6|}$

21. $\frac{10}{|-10|}$

22. $\frac{|-8|}{|-2|}$

13. ______
14. ______
15. ______
16. ______
17. ______
18. ______
19. ______
20. ______
21. ______
22. ______

23–26. Write each of the following in symbols.

23. The absolute value of $+20$. **23.** ______

24. The absolute value of 0. **24.** ______

25. The absolute value of -20. **25.** ______

26. The absolute value of x. **26.** ______

27–32. Answer each question.

27. Does $|x| = x$ if $x = +8$? **27.** ______

28. Does $|x| = x$ if $x = 0$? **28.** ______

29. Does $|x| = x$ if $x = -8$? **29.** ______

30. Does $|x| = -x$ if $x = -8$? **30.** ______

31. Does $|x| = x$ if x is positive or zero? **31.** ______

32. Does $|x| = -x$ if x is negative?
(Remember to read $-x$ as the opposite of x.) **32.** ______

D. ORDER RELATIONS

Given two sets of objects, there is a natural way to compare the number of objects in one set with those in the other. Either the two sets have the same number of objects, or the first set has fewer objects than the second set, or the first set has more objects than the second set.

If a is the number of objects in the first set and if b is the number of objects in the second set, then there are three possibilities:

1. a is equal to b. In symbols, $a = b$.
2. a is less than b. In symbols, $a < b$.
3. a is greater than b. In symbols, $a > b$.

These three types of comparisons are called **relations**. They include the equal relation $=$ and the order relations $<$ and $>$. As examples

$2 < 7$ (read "two is less than seven")

$8 > 5$ (read "eight is greater than five")

NOTE TO STUDENT:
For the relation symbols $<$ and $>$ the smaller part of the symbol is closest to the smaller number and the larger part of the symbol is closest to the larger number. The symbol "opens up" toward the larger number.

A number line shows a natural way to order the set of rational numbers. A number to the left of another number on a horizontal number line is less than the number to its right. Similarly, the number on the right is greater than the number on its left.

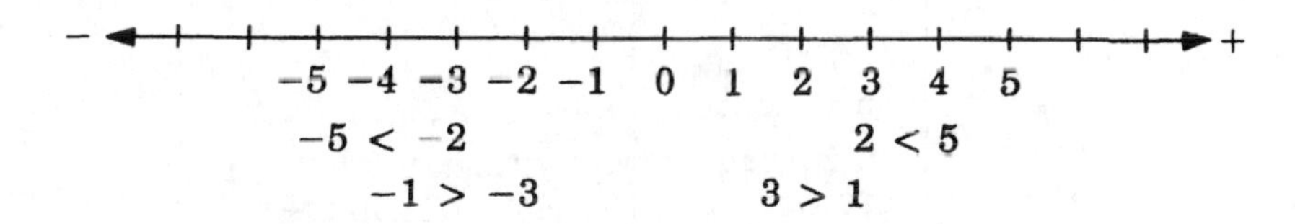

These ideas are summarized in the following table.

Relation in symbols	*Algebraic meaning*	*Geometric meaning (for a horizontal number line with positive direction to the right)*
$a < b$	a is less than b	point a is left of point b
$a > b$	a is greater than b	point a is right of point b
$a = b$	a equals b	point a is the same as point b

EXERCISES 2.1D

1–10. Insert an order symbol or an equal sign between each given pair of numbers, so that the resulting statement is true. State the algebraic meaning of the relation, and state the geometric meaning.

SAMPLE PROBLEM 1

+6 ____ +2

SOLUTION

+6 > +2

+6 is greater than +2.

Point +6 is to the right of point +2.

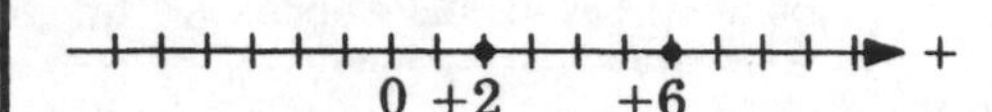

SAMPLE PROBLEM 2

−6 ____ −2

SOLUTION

−6 < −2

−6 is less than −2.

Point −6 is to the left of point −2.

SAMPLE PROBLEM 3

−5 ____ 9

SOLUTION

−5 < 9

−5 is less than 9.

Point −5 is to the left of point 9.

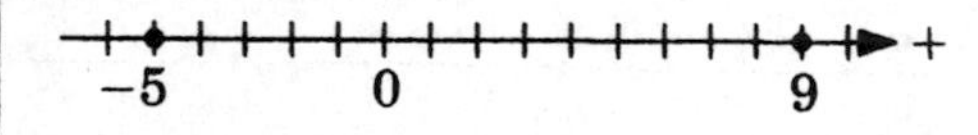

SAMPLE PROBLEM 4

0 ____ −4

SOLUTION

0 > −4

0 is greater than −4.

Point 0 is to the right of point −4.

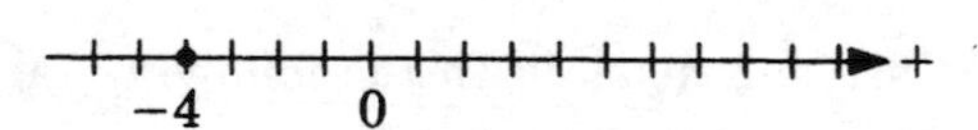

1. +4 ____ +7

2. +8 ____ +3

3. −4 ____ −7

4. −8 ____ −3

5. −2 ____ +5

6. +2 ____ −5

7. 0 ____ −4

8. |−6| ____ 6

9. 6 ____ −3

10. −6 ____ 0

1. __________

2. __________

3. __________

4. __________

5. __________

6. __________

7. __________

8. __________

9. __________

10. __________

11–20. Translate each verbal statement into an algebraic (symbolic) statement.

SAMPLE PROBLEM 5 The product of x and y is positive. **SOLUTION** $xy > 0$	**SAMPLE PROBLEM 6** 8 more than the sum of r and s is negative. **SOLUTION** $r + s + 8 < 0$
SAMPLE PROBLEM 7 The product of 3 and x is greater than 12. **SOLUTION** $3x > 12$	**SAMPLE PROBLEM 8** Twice the sum of n and 4 is less than 15. **SOLUTION** $2(n + 4) < 15$

11. The sum of x and 7 is greater than 12. **11.** ____________

12. The product of y and 5 is less than 30. **12.** ____________

13. The sum of n and t is negative. **13.** ____________

14. 4 more than x is greater than 10. **14.** ____________

15. 5 less than y is less than 8. **15.** ____________

16. The absolute value of x is positive. **16.** ____________

17. Twice the sum of 6 and t is positive. **17.** ____________

18. The product of b and a is negative. **18.** ____________

19. y is less than the sum of $3x$ and 9. **19.** ____________

20. The sum of $3x$ and 9 is greater than y. **20.** ____________

EXERCISES 2.1S

1–10. State the meaning of each of the following for the given interpretation.

1. −30 if +30 means 30° longitude west of Greenwich. **1.** ____________

2. −20 if +20 means a temperature of 20° above zero. **2.** ____________

3. 0 if +10 means 10° north latitude and −10 means 10° south latitude. **3.** ____________

4. −5 if +5 means 5 yards gained. **4.** ____________

5. 0 if +3 and −3 mean an increase and a decrease, respectively. **5.** ____________

6. −100 if +100 means a profit of $100. **6.** ____________

7. −15 if +15 means an acceleration of 15 m.p.h. **7.** ____________

8. −24 if +24 means the distance of an image 24 cm in front of a mirror. **8.** ____________

9. −32 if +32 means 32 points won in a card game where points may be won or lost. **9.** ____________

10. −17 if +17 means a score of 17 points above the average score on a certain psychology test. **10.** ____________

11–16. State the coordinates of the points *A*, *B*, *C*, and *D* on each number line.

11. 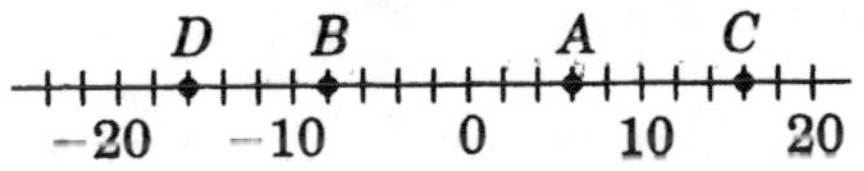

11. *A*( ), *B*( ), *C*( ), *D*( )

12. 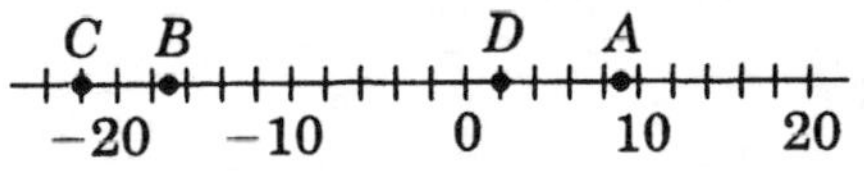

12. *A*( ), *B*( ), *C*( ), *D*( )

13. 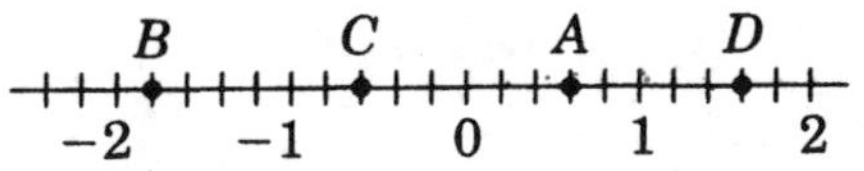

13. *A*( ), *B*( ), *C*( ), *D*( )

NAME ____________ DATE ____________ COURSE ____________

14.

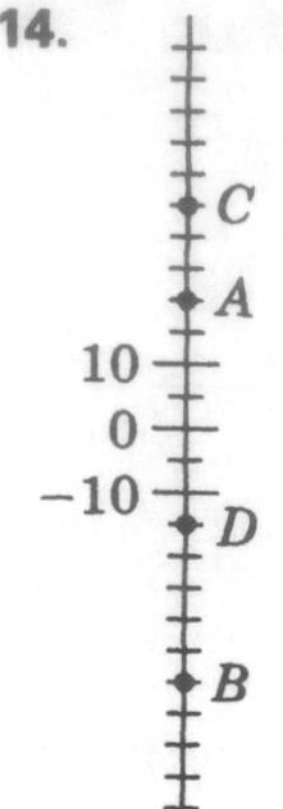

15.

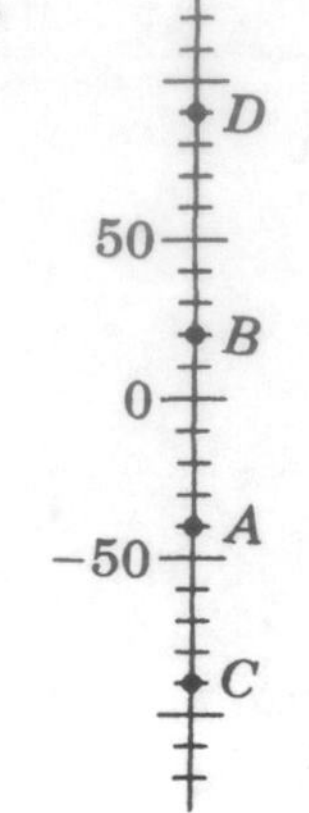

16.

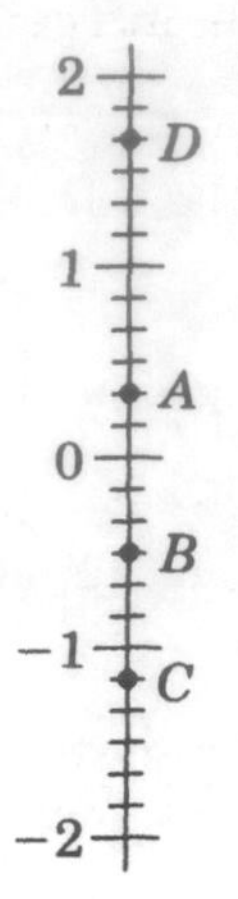

14. $A(\ )$ $B(\ )$ $C(\ )$ $D(\ )$

15. $A(\ )$ $B(\ )$ $C(\ )$ $D(\ )$

16. $A(\ )$ $B(\ )$ $C(\ )$ $D(\ )$

17–36. Graph each point on the number line indicated.

17. $A(-25)$ **18.** $B(25)$ **19.** $C(-30)$ **20.** $D(15)$ **21.** $E(-10)$

22. $A(250)$ **23.** $B(-150)$ **24.** $C(-25)$ **25.** $D(100)$ **26.** $E(50)$

−50 0 50

27. $A(-10)$

28. $B(-50)$

29. $C(20)$

30. $D(30)$

31. $E(-40)$

32. $A(0)$

33. $B(8)$

34. $C(-4)$

35. $D(-12)$

36. $E(4)$

37–50. Simplify. (Rewrite as a single integer.)

37. $|-2| + |-7|$

38. $|+81| - |+61|$

39. $|10| + |-7|$

40. $|-10| - |3|$

41. $|-10| + |-3|$

42. $|12| - |-5|$

43. $4|-4|$

44. $|-5|4$

45. $|-2| \cdot |+30|$

46. $|5| \cdot |-8|$

47. $4|-5| - 5|-3|$

48. $20 - 2|-3| \cdot |-2|$

49. $\dfrac{|-35|}{|5|}$

50. $\dfrac{|-8|}{8}$

37. ______

38. ______

39. ______

40. ______

41. ______

42. ______

43. ______

44. ______

45. ______

46. ______

47. ______

48. ______

49. ______

50. ______

51–60. Insert an order symbol or an equal sign between each given pair of numbers, so that the resulting statement is true. State the algebraic meaning of the relation, and state the geometric meaning.

51. 2 ___ −3

52. 4 ___ 0

53. 0 ___ 5

54. 7 ___ 5

55. −12 ___ −4

56. |−3| ___ +3

57. −12 ___ 0

58. |−7| ___ −7

59. −8 ___ 4

60. −4 ___ −9

51. ______________

52. ______________

53. ______________

54. ______________

55. ______________

56. ______________

57. ______________

58. ______________

59. ______________

60. ______________

61–66. Translate each verbal statement into an algebraic statement.

61. The product of t and 6 is less than 24.

62. The sum of x and y is greater than the product of x and y.

63. The product of 4 and y is negative.

64. The absolute value of the product of a and b is positive.

65. 7 less than x is greater than y.

66. 8 more than x is less than y.

61. ______________

62. ______________

63. ______________

64. ______________

65. ______________

66. ______________

ADDITION OF SIGNED NUMBERS

A. ADDITION ON A NUMBER LINE

What meaning should we have for the sum of two integers? We could define the sum any way we wanted to. However, a definition would not be very useful unless it agreed with the world around us. Also, we want to keep the number properties we already have. It would be very difficult and confusing to change the rules in the middle of the game.

To determine what rules we want, we begin by examining one of the earliest uses, namely, business transactions. We shall use a **positive number for a gain** and a **negative number for a loss.**

A gain of $5 followed by a gain of $2 results in a gain of $7. $(+5) + (+2) = +7$

A **loss** of $5 followed by a **loss** of $2 results in a **loss** of $7. $(-5) + (-2) = -7$

A gain of $5 followed by a **loss** of $2 results in a gain of $3. $(+5) + (-2) = +3$

A **loss** of $5 followed by a gain of $2 results in a **loss** of $3. $(-5) + (+2) = -3$

A **loss** of $5 followed by no gain or loss results in a **loss** of $5. $(-5) + 0 = -5$

A gain of $5 followed by a **loss** of $5 results in no gain or loss. $(+5) + (-5) = 0$

A number line may also be used as an aid in understanding. In the following, let us think of a **positive number as a gain** in yardage on a football play and a **negative number as a loss** in yardage. We start at the origin and **move to the right for a positive number** and **move to the left for a negative number.**

Gain 3 yd and gain 7 yd
Result: gain 10 yd

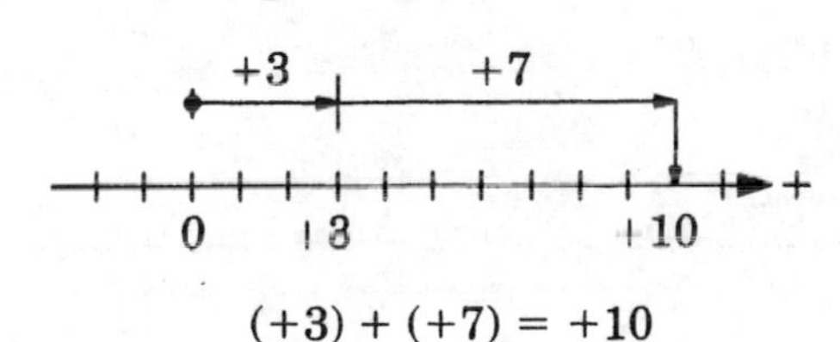

$(+3) + (+7) = +10$

Lose 3 yd and lose 7 yd
Result: lose 10 yd

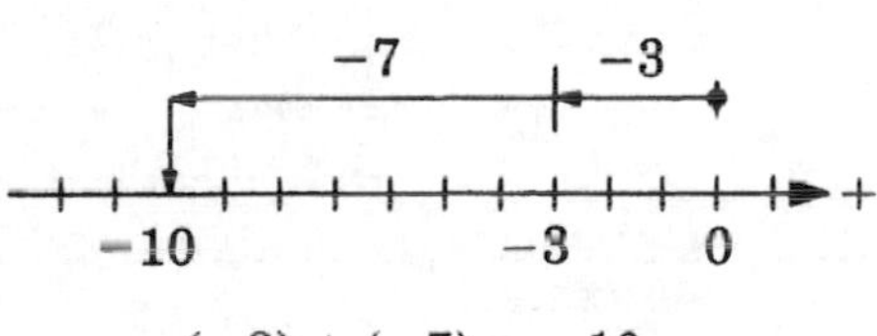

$(-3) + (-7) = -10$

Gain 3 yd and lose 7 yd
Result: lose 4 yd

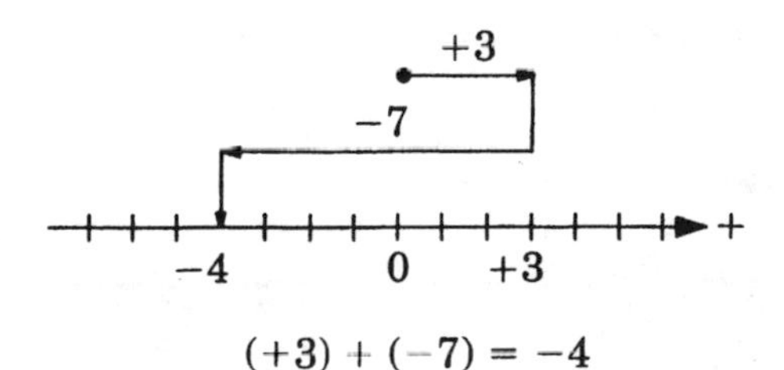

$(+3) + (-7) = -4$

Lose 3 yd and gain 7 yd
Result: gain 4 yd

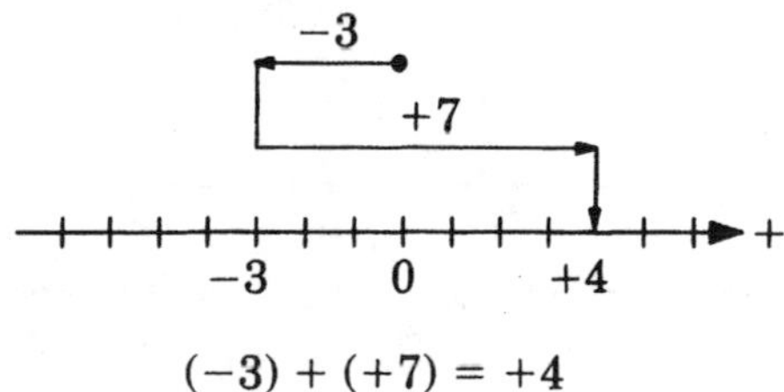

$(-3) + (+7) = +4$

No gain or loss and a loss of 7 yd
Result: lose 7 yd

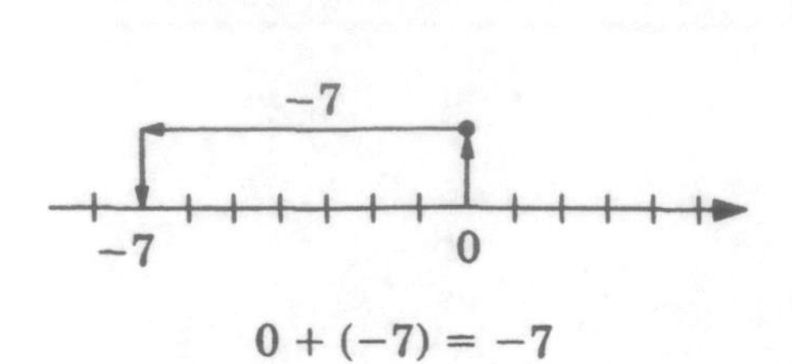

$0 + (-7) = -7$

Lose 3 yd and gain 3 yd
Result: no gain or loss

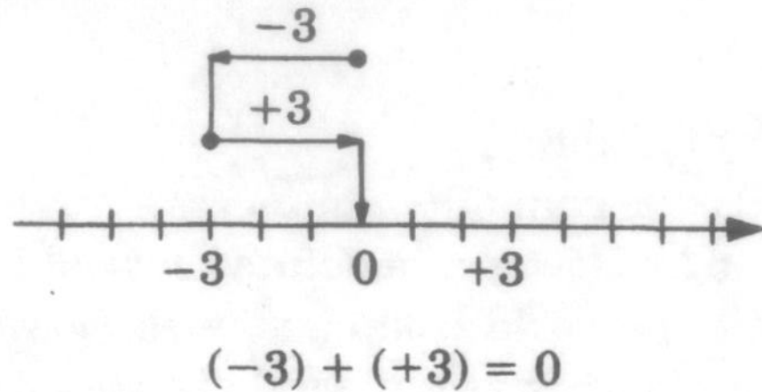

$(-3) + (+3) = 0$

EXERCISES 2.2A

Express the statement in symbols, and do the addition, using a positive number for a gain and a negative number for a loss. (Think of the meaning. Use a number line application if you find this helpful.)

SAMPLE PROBLEM 1

A gain of $8 followed by a loss of $3.

SOLUTION

$(+8) + (-3) = +5$

OPTIONAL AID

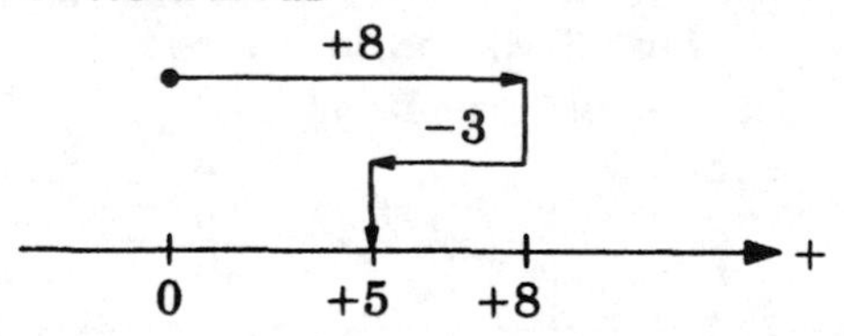

SAMPLE PROBLEM 2

A gain of $4 followed by a loss of $10.

SOLUTION

$(+4) + (-10) = -6$

OPTIONAL AID

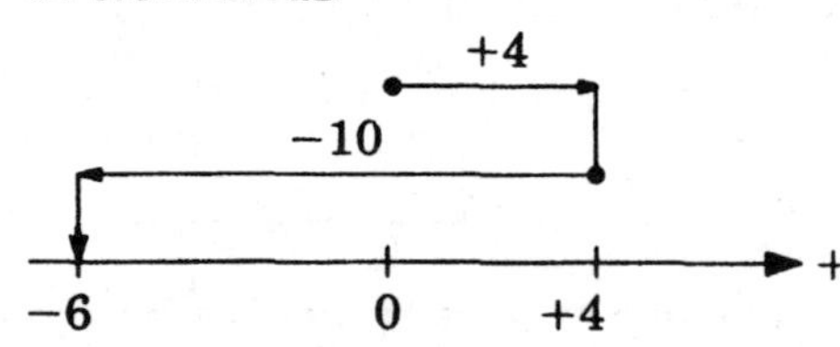

SAMPLE PROBLEM 3

A loss of 5 inches followed by a loss of 7 inches.

SOLUTION

$(-5) + (-7) = -12$

OPTIONAL AID

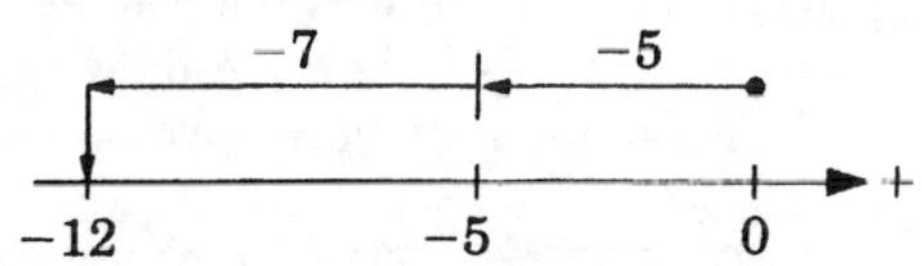

SAMPLE PROBLEM 4

A loss of 4 yards on a football play followed by a gain of 4 yards.

SOLUTION

$-4 + 4 = 0$

OPTIONAL AID

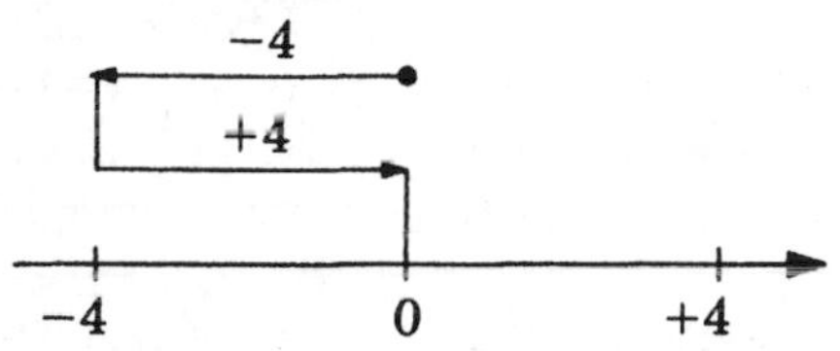

1. A gain of $6 followed by a gain of $3. 1. ______
2. A gain of $8 followed by a loss of $6. 2. ______
3. A loss of $4 followed by a loss of $5. 3. ______
4. A loss of $7 followed by a gain of $10. 4. ______
5. A loss of $7 followed by a gain of $3. 5. ______
6. A gain of $5 followed by a loss of $9. 6. ______
7. A gain of $8 followed by a loss of $8. 7. ______
8. A loss of $6 followed by neither a gain nor a loss. 8. ______
9. A gain of 3 yards on a football play followed by a gain of 7 yards on the next play. 9. ______
10. A loss of 8 yards on a football play followed by a loss of 7 yards on the next play. 10. ______
11. A gain of 6 yards on a football play followed by a loss of 15 yards on the next play. 11. ______
12. A gain of 14 yards on a football play followed by a loss of 10 yards on the next play. 12. ______
13. A loss of 4 yards on a football play followed by neither a gain nor a loss on the next play. 13. ______
14. No gain or loss in yardage on one football play followed by no gain or loss in yardage on the next play. 14. ______
15. A loss of 5 yards followed by a gain of 5 yards 15. ______

B. ADDITION OF SIGNED NUMBERS

We are now ready to state the rules for the addition of signed numbers. With these rules, the number properties of the set of natural numbers are also valid for the set of integers and the set of rational numbers. In particular, the sum of two integers is an integer. Also, the sum of two rational numbers is a rational number.

First let us summarize some results from the previous section.

Same signs	*Different signs*	*Sums involving zero*
$(+5) + (+2) = +7$	$(+5) + (-2) = +3$	$-5 + 0 = -5$
$(+3) + (+7) = +10$	$(+3) + (-7) = -4$	$0 + (-7) = -7$
$(-5) + (-2) = -7$	$(-5) + (+2) = -3$	$(+5) + (-5) = 0$
$(-3) + (-7) = -10$	$(-3) + (+7) = +4$	$(-3) + (+3) = 0$

Now, let us generalize and thereby form our addition rules.

SUMS INVOLVING ZERO

1. The sum of 0 and any number is the number.
2. The sum of any number and its opposite is 0.

$x + 0 = x$ and $0 + x = x$

$x + (-x) = 0$ and $-x + x = 0$

SUMS NOT INVOLVING ZERO

To add two signed numbers

1. If the signs are the same, add their absolute values and put their common sign before this sum.
2. If the signs are different, subtract their absolute values and put the sign of the number having the larger absolute value before this difference.

$(+x) + (+y) = +(x + y)$
$(-x) + (-y) = -(x + y)$

$(+x) + (-y) = +(x - y)$ if $x > y$
$(+x) + (-y) = -(y - x)$ if $y > x$

SAMPLE PROBLEMS 1–11	SOLUTIONS	
Add:		Signs are the same:
1. $(+7) + (+2)$	$= +(7 + 2) = +9 = 9$	1. Add absolute values.
2. $(-5) + (-3)$	$= -(5 + 3) = -8$	2. Use common sign.
3. $(+7) + (-3)$	$= +(7 - 3) = +4 = 4$	Signs are different:
4. $(+3) + (-7)$	$= -(7 - 3) = -4$	1. Subtract absolute values.
5. $(-7) + (+2)$	$= -(7 - 2) = -5$	2. Use sign of larger absolute.
6. $(-12) + (+15)$	$= +(15 - 12) = 3$	value.
7. $(+7) + (-7)$	$= 0$	The sum of any number and its
8. $(-50) + (+50)$	$= 0$	opposite is 0.
9. $(-6) + 0$	$= -6$	The sum of 0 and any number is
10. $0 + (-9)$	$= -9$	that number.
11. $0 + 0$	$= 0$	

The rules for addition are valid for any interpretation or application. If, at first, you have difficulty in understanding or using the rules, use an application from the previous section. Choose the interpretation that makes the most sense to you. Then do enough practice problems so that you are able to add quickly with very little effort.

EXERCISES 2.2B

Add.

1. $(+3) + (+5)$
2. $(+7) + (+4)$
3. $(-2) + (-8)$
4. $(-43) + (-51)$
5. $(+11) + (+13)$
6. $(-3) + (-4)$
7. $(-6) + (-8)$
8. $(-11) + (-15)$
9. $(+20) + (+50)$
10. $(-25) + (-75)$
11. $(+4) + (-3)$
12. $(+8) + (-11)$
13. $(-12) + (+5)$
14. $(-17) + (+12)$
15. $(+15) + (-8)$
16. $(+10) + (-15)$
17. $(-11) + (+6)$
18. $(-7) + (+10)$
19. $(+1) + (-1)$
20. $(-6) + (+6)$
21. $0 + (-5)$
22. $(-4) + 0$

1. ________ **2.** ________
3. ________ **4.** ________
5. ________ **6.** ________
7. ________ **8.** ________
9. ________ **10.** ________
11. ________ **12.** ________
13. ________ **14.** ________
15. ________ **16.** ________
17. ________ **18.** ________
19. ________ **20.** ________
21. ________ **22.** ________

We use certain conventions in writing sums of signed numbers. The fewer symbols we use, the easier it is to read an expression and understand what it says.

CONVENTIONS	EXAMPLES
1. Write $+n$ as n.	$(+7) + (+2) = 7 + 2$ $(+7) + (-2) = 7 + (-2)$
2. Use parentheses only to enclose a negative number following an addition symbol.	$(-8) + (+3) = -8 + 3$ $(-8) + (-3) = -8 + (-3)$

23. $-9 + (-10)$ **24.** $-45 + (-55)$

25. $100 + (-25)$ **26.** $-25 + 100$

27. $-17 + 7$ **28.** $7 + (-16)$

29. $-8 + 0$ **30.** $0 + (-8)$

31. $-8 + 8$ **32.** $8 + (-8)$

33. $0 + \left(\frac{-3}{4}\right)$ **34.** $-8\frac{2}{5} + 0$

35. $-6\frac{2}{3} + 6\frac{2}{3}$ **36.** $\frac{7}{8} + \left(-\frac{7}{8}\right)$

37. $8 + (-15)$ **38.** $-18 + 30$

39. $25 + (-16)$ **40.** $-25 + (-16)$

41. $-20 + 17$ **42.** $25 + (-35)$

43. $-9 + (-16)$ **44.** $-12 + 7$

45. $39 + (-40)$ **46.** $-10 + (-40)$

47. $-12 + 8$ **48.** $-60 + 90$

49. $50 + (-30)$ **50.** $30 + (-50)$

23. ________ 24. ________
25. ________ 26. ________
27. ________ 28. ________
29. ________ 30. ________
31. ________ 32. ________
33. ________ 34. ________
35. ________ 36. ________
37. ________ 38. ________
39. ________ 40. ________
41. ________ 42. ________
43. ________ 44. ________
45. ________ 46. ________
47. ________ 48. ________
49. ________ 50. ________

EXERCISES 2.2S

Express in symbols and do the addition.

1–6. Use a positive number for a deposit in a bank account and a negative number for a withdrawal.

1. A deposit of \$100, then a withdrawal of \$20. **1.** ______

2. A deposit of \$35, then a deposit of \$40. **2.** ______

3. A withdrawal of \$25, then a deposit of \$15. **3.** ______

4. A withdrawal of \$12, then a withdrawal of \$30. **4.** ______

5. A deposit of \$250, then a withdrawal of \$400. **5.** ______

6. A withdrawal of \$46, then a deposit of \$70. **6.** ______

7–12. Use a positive number for an increase and a negative number for a decrease.

7. An increase in weight of 3 pounds followed by an increase of 6 pounds. **7.** ______

8. A decrease in weight of 4 pounds followed by a decrease of 1 pound. **8.** ______

9. An increase in weight of 2 pounds followed by a decrease of 5 pounds. **9.** ______

10. An increase in temperature of 10° followed by a decrease of 3°. **10.** ______

11. An increase in temperature of 8° followed by a decrease of 12°. **11.** ______

12. A decrease in temperature of 23° followed by an increase of 6°. **12.** ______

13–40. Add.

13. $(-11) + 10$

14. $(-11) + (-10)$

15. $(-4) + 4$

16. $(+13) + 0$

17. $8 + (-7)$

18. $8 + (-9)$

13. ______

14. ______

15. ______

16. ______

17. ______

18. ______

NAME ______ DATE ______ COURSE ______

19. $(-7) + (-6)$

20. $(+7) + (-6)$

21. $(-7) + (+6)$

22. $(-13) + (+13)$

23. $(-13) + 14$

24. $(-5) + 4$

25. $(+8) + (-8)$

26. $(+7) + (+5)$

27. $(+25) + (+75)$

28. $(-30) + (-70)$

29. $(-3) + (+2)$

30. $(-2) + (-3)$

31. $10 + (-10)$

32. $-8 + 8$

33. $-50 + 35$

34. $75 + (-25)$

35. $(+32) + 0$

36. $0 + (-32)$

37. $-67 + (-33)$

38. $-120 + 20$

39. $49 + (-79)$

40. $98 + (-18)$

19. ____________________

20. ____________________

21. ____________________

22. ____________________

23. ____________________

24. ____________________

25. ____________________

26. ____________________

27. ____________________

28. ____________________

29. ____________________

30. ____________________

31. ____________________

32. ____________________

33. ____________________

34. ____________________

35. ____________________

36. ____________________

37. ____________________

38. ____________________

39. ____________________

40. ____________________

SUBTRACTION OF SIGNED NUMBERS 2.3

A. SUBTRACTION ON A NUMBER LINE

In arithmetic, subtraction is defined by means of addition. For example, $7 - 2$ is the number that is added to the subtrahend 2 to get 7. In symbols, $2 + (7 - 2) = 7$. We know that $2 + 5 = 7$ and thus $7 - 2 = 5$.

> ***m* − *s* is the number that is added to the subtrahend *s* to obtain the minuend *m*.**
>
> $$s + (m - s) = m$$

Sample Problems 1–5 show the use of the number line for subtraction problems.

SAMPLE PROBLEM 1

$7 - 2$

SOLUTION

What do we add to 2 to get 7? We start at 2.
Now how do we move on the number line to reach 7?

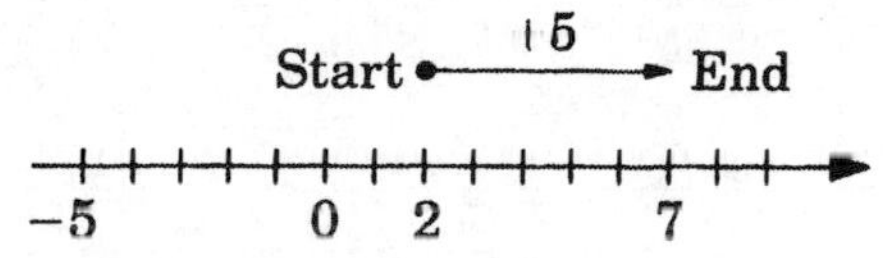

We see from the number line that we must move in the positive direction 5 units.

Therefore, $7 - 2 = 5$. minuend − subtrahend = difference

CHECK

$2 + 5 = 7$ subtrahend + difference = minuend

SAMPLE PROBLEM 2

$1 - 5$

SOLUTION

What do we add to 5 to get 1? We start at 5.
How do we move to reach 1?

End ← −4 — Start

−5 0 1 5

We move in the negative direction 4 units.
Therefore, $1 - 5 = -4$.

CHECK

$5 + (-4) = 1$

SAMPLE PROBLEM 3

$-2 - 4$

SOLUTION

What do we add to 4 to get -2? We start at 4. How do we move to reach -2?

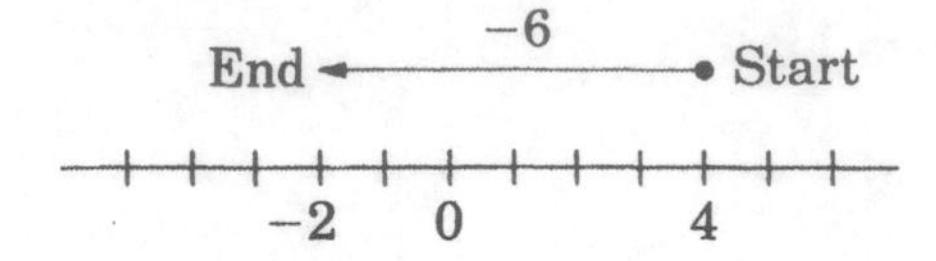

Move 6 units in the negative direction. Therefore, $-2 - 4 = -6$.

CHECK

$4 + (-6) = -2$

SAMPLE PROBLEM 4

$2 - (-5)$

SOLUTION

What do we add to -5 to get 2? We start at -5. How do we move to reach 2?

+7
Start End
−5 0 2

Move 7 units in the positive direction. Therefore, $2 - (-5) = 7$.

CHECK

$-5 + 7 = 2$

SAMPLE PROBLEM 5

$-8 - (-3)$

SOLUTION

What do we add to -3 to get -8? We start at -3. How do we move to reach -8?

−5
End Start
−8 −3 0 3

We move in the negative direction 5 units. Therefore, $-8 - (-3) = -5$.

CHECK

$-3 + (-5) = -8$

EXERCISES 2.3A

1–10. Subtract by using a number line.

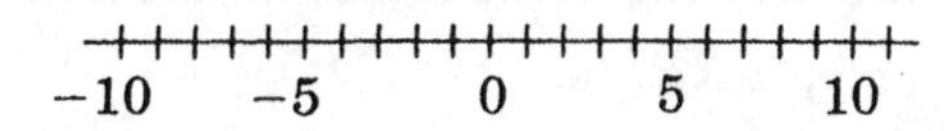

1. $8 - 2$ **2.** $6 - 4$ **1.** ________ **2.** ________

3. $2 - 5$ **4.** $3 - 7$ **3.** ________ **4.** ________

5. $-3 - 4$ **6.** $-1 - (-6)$ **5.** ________ **6.** ________

7. $5 - (-3)$ **8.** $-7 - (-5)$ **7.** ________ **8.** ________

9. $0 - 9$ **10.** $0 - (-9)$ **9.** ________ **10.** ________

11–20. Find the distance between the given points on a number line. The distance between $A(a)$ and $B(b)$ is given by $|a - b|$.

SAMPLE PROBLEM 6

$A(7)$ and $B(-2)$

SOLUTION

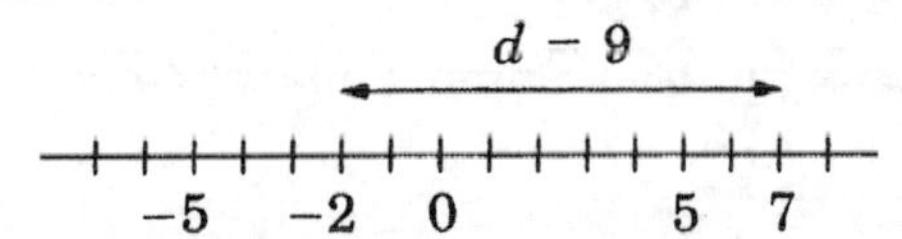

The distance between $A(7)$ and $B(-2)$ is $|7 - (-2)| = |7 + 2| = 9$.

11. $A(8)$ and $B(1)$ **12.** $A(6)$ and $B(2)$ **11.** ________ **12.** ________

13. $A(9)$ and $B(0)$ **14.** $A(0)$ and $B(-8)$ **13.** ________ **14.** ________

15. $A(2)$ and $B(-4)$ **16.** $A(5)$ and $B(-7)$ **15.** ________ **16.** ________

17. $A(-1)$ and $B(-5)$ **18.** $A(-3)$ and $B(-9)$ **17.** ________ **18.** ________

19. $A(4)$ and $B(-4)$ **20.** $A(0)$ and $B(-6)$ **19.** ________ **20.** ________

21–30. Find the directed distance from A to B.
Directed distance is used in mathematics, science, engineering, and other areas of application.
The directed distance on a number line from $A(a)$ to $B(b)$ is $b - a$.

SAMPLE PROBLEM 7	SAMPLE PROBLEM 8
From $A(7)$ to $B(-2)$	From $A(-6)$ to $B(-1)$
SOLUTION	SOLUTION
$a = 7$ and $b = -2$	$a = -6$ and $b = -1$
$b - a = -2 - 7 = -9$	$b - a = -1 - (-6) = 5$
The directed distance is -9, 9 units in the negative direction.	The directed distance is $+5$, 5 units in the positive direction.

21. From $A(3)$ to $B(7)$

22. From $A(8)$ to $B(2)$

23. From $A(4)$ to $B(-3)$

24. From $A(-1)$ to $B(9)$

25. From $A(0)$ to $B(5)$

26. From $A(0)$ to $B(-5)$

27. From $A(-2)$ to $B(-5)$

28. From $A(-6)$ to $B(-4)$

29. From $A(9)$ to $B(0)$

30. From $A(-8)$ to $B(0)$

21. ________ **22.** ________
23. ________ **24.** ________
25. ________ **26.** ________
27. ________ **28.** ________
29. ________ **30.** ________

B. ADDITIVE INVERSES (OPPOSITES)

The **additive inverse,** or **opposite,** of a number plays a key role in subtraction.

$x + (-x) = 0$
$-x + x = 0$

The sum of a number and its opposite is zero.

When graphed on a number line, a number and its opposite are the same distance from the origin, but they lie on opposite sides of the origin.

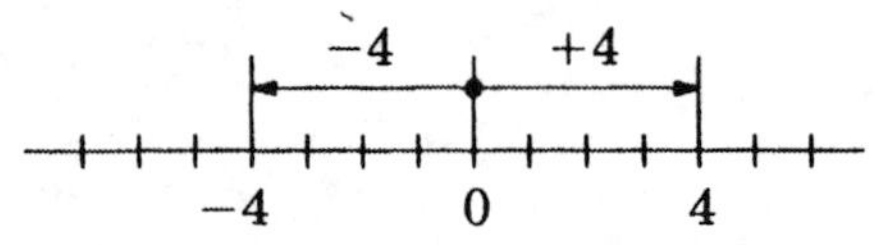

If $x = 4$, the opposite of x is -4 and $-x = -4$.
If $x = -4$, the opposite of x is 4 and $-x = 4$.

SAMPLE PROBLEMS 1–5

For each value of x, find its opposite, $-x$.

1.	2.	3.	4.	5.
$x = 3$	$x = 5$	$x = -5$	$x = -3$	$x = 0$
$-x = -3$	$-x = -5$	$-x = 5$	$-x = 3$	$-x = 0$
		$-(-5) = 5$	$-(-3) = 3$	$-0 = 0$

Note that $-x$ can be a positive number or a negative number or zero.

It is important to read and think of $-x$ as "the opposite of x" or as "the additive inverse of x." Do **not** say "negative x" for $-x$. Note that sometimes $-x$ is positive.

Now consider $-(-4)$. We read this as "the opposite of -4." Thus $-(-4) = 4$. This is true, in general, for any integer x and for any rational number x.

The opposite of (the opposite of x) is x. $-(-x) = x$

EXERCISES 2.3B

1–5. State the opposite of x for each given value of x.

1. $x = 6$ **2.** $x = 1$ **3.** $x = 0$ **4.** $x = -1$ **5.** $x = -6$

6–10. Find $-x$ for each given value of x.

6. $x = 10$, $-x =$ **7.** $x = -10$, $-x =$ **8.** $x = 0$, $-x =$ **9.** $x = -2$, $-x =$ **10.** $x = 2$, $-x =$

11–14. Find $-(-x)$ for each given value of x.

11. $x = 7$, $-(-x) =$ **12.** $x = 9$, $-(-x) =$

13. $x = -9$, $-(-x) =$ **14.** $x = -8$, $-(-x) =$

15–18. Write as a single integer.

15. $-(-6) =$ **16.** $-(-2) =$

17. $-(-1) =$ **18.** $-(-(-3)) =$

19–25. State the opposite of the subtrahend for each given subtraction. (Recall for $m - s$, the number on the right, s, is the subtrahend.)

19. $12 - 8$ **20.** $3 - 9$

21. $-5 - 7$ **22.** $4 - (-6)$

23. $-2 - (-5)$ **24.** $-8 - (-7)$

25. $0 - (-1)$

C. SUBTRACTION

There is a fast and easy way to subtract by using the concept of opposites.

Instead of subtracting the subtrahend, we add the opposite of the subtrahend.

$$9 - \mathbf{4} = 9 + (\mathbf{-4}) = 5$$
$$8 - \mathbf{5} = 8 + (\mathbf{-5}) = 3$$
$$10 - \mathbf{3} = 10 + (\mathbf{-3}) = 7$$
$$a - \boldsymbol{b} = a + (\boldsymbol{-b})$$

(subtrahend) (opposite of subtrahend)

Now let us do some new types of subtractions by adding the opposite of the subtrahend. Then let us check the results by adding the difference obtained to the subtrahend.

Positive subtrahends

$2 - \mathbf{7} = 2 + (\mathbf{-7}) = -5$	Check. $7 + (-5) = 2$
$5 - \mathbf{8} = 5 + (\mathbf{-8}) = -3$	$8 + (-3) = 5$
$-3 - \mathbf{4} = -3 + (\mathbf{-4}) = -7$	$4 + (-7) = -3$
$-6 - \mathbf{2} = -6 + (\mathbf{-2}) = -8$	$2 + (-8) = -6$

Negative subtrahends

$4 - (\mathbf{-5}) = 4 + (\mathbf{+5}) = 9$	Check. $(-5) + 9 = 4$
$-4 - (\mathbf{-5}) = -4 + (\mathbf{+5}) = 1$	$(-5) + 1 = -4$
$-8 - (\mathbf{-3}) = -8 + (\mathbf{+3}) = -5$	$(-3) + (-5) = -8$

Note, in all of these special cases, that subtracting by adding the opposite (additive inverse) of the subtrahend gives the same answer as does ordinary subtraction in arithmetic. This is true in general. Since this property gives us an easy way to find differences, we use it as the definition of subtraction.

DEFINITION OF SUBTRACTION

$$m - s = m + (-s)$$

for all integers m and s and for all rationals m and s.

When we write a subtraction problem, we change the subtraction symbol, −, to the addition symbol, +, **and** we change the sign of the subtrahend.

We write	$m - s$	but we think	$m + (-s)$
	from m subtract s		to m add the opposite of s

Sample Problems 1–6. Subtract, by adding the opposite of the subtrahend.

SAMPLE PROBLEM 1

$7 - 2$

SOLUTION

Think: "Subtract 2" means "add the opposite of 2."

$7 - 2 = 7 + (-2)$ — Change subtraction to addition and

$= 5$ — change the sign of the subtrahend, 2.

SAMPLE PROBLEM 2

$2 - 7$

SOLUTION

Think: "Subtract 7" means "add the opposite of 7."

$2 - 7 = 2 + (-7)$ — Change subtraction to addition and

$= -5$ — change the sign of the subtrahend, 7.

SAMPLE PROBLEM 3

$-3 - 6$

SOLUTION

$-3 - 6 = -3 + (-6)$ — Change subtraction to addition and

$= -9$ — change the sign of 6, the subtrahend.

SAMPLE PROBLEM 4

$4 - (-3)$

SOLUTION

Think: "Subtract -3" means "add the opposite of -3."

$4 - (-3) = 4 + (-(-3))$ — Change subtraction to addition and

$= 4 + 3$ — change the sign of -3, the subtrahend.

$= 7$

SAMPLE PROBLEM 5

$-5 - (-8)$

SOLUTION

Think: "Subtract -8" means "add the opposite of -8."

$-5 - (-8) = -5 + (+8)$ — Change subtraction to addition and

$= -5 + 8$ — change the sign of -8, the subtrahend.

$= 3$

SAMPLE PROBLEM 6

$-6 - (-2)$

SOLUTION

$-6 - (-2) = -6 + (+2)$ — Change subtraction to addition and

$= -6 + 2$ — change the sign of -2, the subtrahend.

$= -4$

EXERCISES 2.3C

1–30. Subtract (add the opposite of the subtrahend).

1. $5 - 3$ **2.** $11 - (-10)$

3. $4 - 6$ **4.** $12 - 10$

5. $6 - (-4)$ **6.** $5 - 9$

7. $-4 - 3$ **8.** $-2 - 8$

9. $3 - (-13)$ **10.** $-3 - (-13)$

11. $-5 - (-7)$ **12.** $-8 - (-2)$

13. $-1 - 1$ **14.** $7 - (-7)$

15. $-6 - 6$ **16.** $4 - 4$

17. $-5 - (-5)$ **18.** $0 - 7$

19. $0 - (-6)$ **20.** $-3 - (-3)$

21. $17 - 29$ **22.** $-12 - (-12)$

23. $15 - (-15)$ **24.** $0 - (-30)$

25. $-12 - 18$ **26.** $-35 - 0$

27. $46 - 29$ **28.** $29 - 46$

29. $40 - (-15)$ **30.** $-64 - 28$

1. ________ 2. ________
3. ________ 4. ________
5. ________ 6. ________
7. ________ 8. ________
9. ________ 10. ________
11. ________ 12. ________
13. ________ 14. ________
•15. ________ 16. ________
17. ________ 18. ________
19. ________ 20. ________
21. ________ 22. ________
23. ________ 24. ________
25. ________ 26. ________
27. ________ 28. ________
29. ________ 30. ________

31–40. Find the difference between the high temperature and the low temperature recorded for each of the cities listed below.

	City	*High*	*Low*
31.	Los Angeles	63	55
32.	Anchorage	−9	−16
33.	Minneapolis	15	−5
34.	Bismarck	15	0
35.	New York City	10	−5
36.	Washington, D.C.	12	0
37.	Chicago	0	−10
38.	Dallas	4	−2
39.	Denver	−4	−12
40.	Fairbanks	−11	−20

31. ________ 32. ________
33. ________ 34. ________
35. ________ 36. ________
37. ________ 38. ________
39. ________ 40. ________

41–50. Use the tide table shown.

41–46. Find the difference in feet between the P.M. low tide and the A.M. low tide for each date. Subtract the number of feet for the A.M. tide from the number of feet for the P.M. tide. Explain any negative answer.

TIDE TABLE
JULY 1984

	LOW TIDE				HIGH TIDE			
	AM	Ft.	PM	Ft.	AM	Ft.	PM	Ft.
1 Su	7:53	-1.5	7:47	3.0	1:00	6.4	3:29	4.9
2 M	8:38	-1.3	8:50	2.8	1:49	6.1	4:11	5.0
3 Tu	9:25	-0.9	10:00	2.5	2:45	5.6	4:53	5.2
4 W	10:17	-0.4	11:17	2.0	3:47	5.1	5:36	5.4
5 Th	11:08	0.3	....	....	4:55	4.5	6.23	5.7
6 Fr	12:33	1.4	12:03	1.0	6.27	4.0	7:08	6.0
7 Sa	1:42	0.7	1:00	1.6	8:00	3.8	7.55	6.2
8 Su	2:41	0.1	1:58	2.2	9:32	3.9	8:41	6.4

41. July 1 **42.** July 2 **41.** ________ **42.** ________

43. July 3 **44.** July 4 **43.** ________ **44.** ________

45. July 6 **46.** July 7 **45.** ________ **46.** ________

47–50. Find the difference in feet between the P.M. high tide and the A.M. high tide for each date. (Subtract A.M. from P.M. as in Problems 41–46.)

47. July 1 **48.** July 3 **47.** ________ **48.** ________

49. July 4 **50.** July 5 **49.** ________ **50.** ________

51–76. The table at the right shows the number of hours to be added to the London time of day to find the standard time (in 24-hour time) at the places listed.

Change in hours	*City*
+10	Sydney
+9	Tokyo
+8	Hong Kong
+7	Bangkok
+6	Calcutta
+5	Bombay
+4	Volgograd
+3	Moscow
+2	Athens
+1	Paris, Rome
0	London
−1	Iceland
−2	Azores
−3	Rio de Janeiro
−4	Santiago
−5	New York
−6	Chicago
−7	Denver
−8	San Francisco
	Los Angeles
−9	Dawson (Yukon)
−10	Honolulu
	Anchorage

51–58. When it is noon, 12 P.M., in London, find the local standard time for each city.
(To 12 add the given number for the city.)

51. Athens **52.** Hong Kong **51.**______ **52.**______

53. Rio de Janeiro **54.** New York **53.**______ **54.**______

55. San Francisco **56.** Paris **55.**______ **56.**______

57. Tokyo **58.** Honolulu **57.**______ **58.**______

59–66. When it is noon, 12 P.M., standard time, in the given city, find the local time in London.
(From 12 subtract the given number for the city.)

59. Rome **60.** Santiago **59.**______ **60.**______

61. Moscow **62.** Chicago **61.**______ **62.**______

63. Los Angeles **64.** Sydney **63.**______ **64.**______

65. Bangkok **66.** Denver **65.**______ **66.**______

67–76. Find the difference in local times when one travels from the first city given to the second city. Using the table, subtract the table number for the first city from the table number for the second city. State whether the travel is eastward (positive direction) or westward (negative direction).

67. Tokyo to Moscow. **68.** New York to Paris. **67.** ________ **68.** ________

69. Hong Kong to London. **70.** Chicago to Rome. **69.** ________ **70.** ________

71. Rio de Janeiro to Los Angeles. **71.** ________

72. San Francisco to London. **72.** ________

73. New York to Honolulu. **74.** Athens to Denver. **73.** ________ **74.** ________

75. Sydney to Calcutta. **76.** Los Angeles to Santiago. **75.** ________ **76.** ________

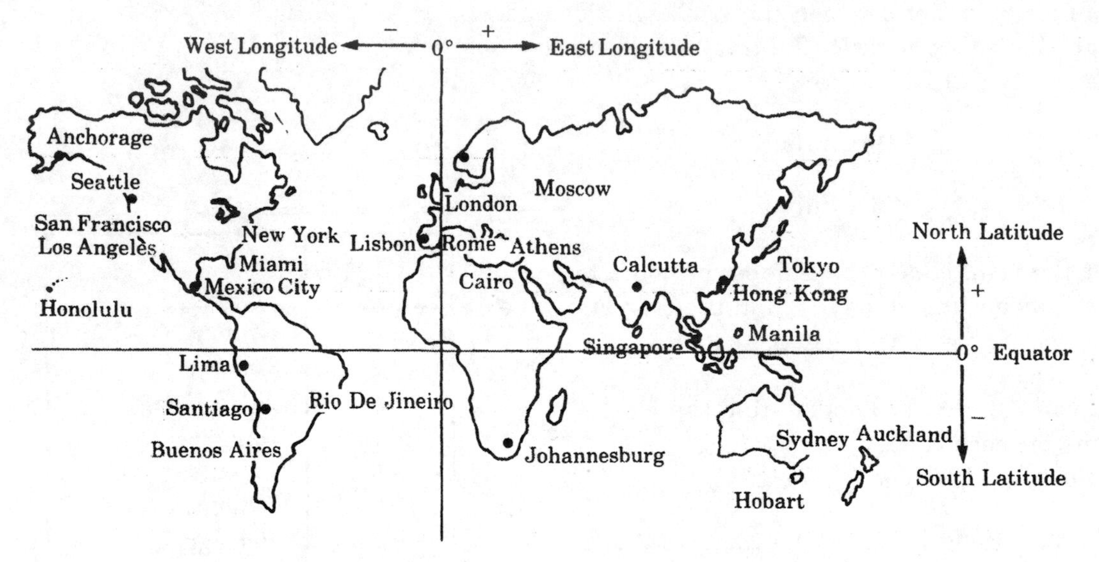

77–86. Find the difference in longitudes when traveling from the first city to the second. State whether the travel is eastward or westward. A positive number means degrees east longitude, and a negative number means degrees west longitude.
(Subtract the first longitude given from the second.)

77. From Hong Kong, +114, to Manila, +121. **77.** ________

78. From Athens, +24, to San Francisco, −122. **78.** ________

79. From Lisbon, −10, to Milan, +9. **79.** ________

80. From Boston, −71, to Denver, −105. **80.** ________

81. From Sydney, +151, to Singapore, +104. **81.** ________

82. From Los Angeles, −118, to Miami, −80. **82.** ________

83. From Amsterdam, +5, to Dublin, −6. 83. ____________

84. From Dallas, −97, to Seattle, −122. 84. ____________

85. From Honolulu, −158, to Reno, −120. 85. ____________

86. From Denver, −105, to Mexico City, −99. 86. ____________

LATITUDE AND LONGITUDE

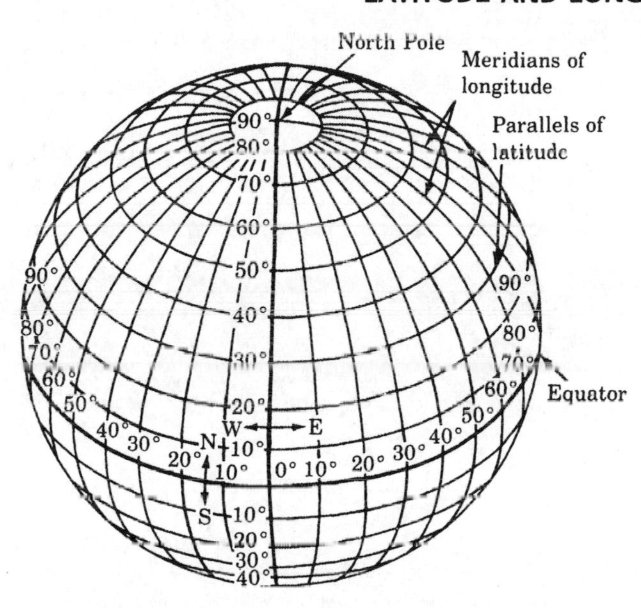

The figure at the left shows the parallels of latitude and the meridians of longitude. The parallels of latitude are circles parallel to the Equator. The north latitudes go from 0° at the Equator to 90° at the North Pole. The south latitudes go from 0° at the Equator to 90° at the South Pole.

The meridians of longitude are great circles passing through both the North Pole and the South Pole. The meridian through Greenwich, England, has been selected as 0°. East longitudes go eastward from Greenwich from 0° to 180°. West longitudes go westward from Greenwich from 0° to 180°. 180° east longitude = 180° west longitude.

87–96. Find the difference in latitudes when traveling from the first city to the second. State whether the travel is northward or southward. A positive number means degrees north latitude. A negative number means degrees south latitude.

87. From Miami, +26, to New York, +41. 87. ____________

88. From Anchorage, +61, to San Francisco, +38. 88. ____________

89. From Denver, +40, to Rio de Janeiro, −23. 89. ____________

90. From Rio de Janeiro, −23, to Panama City, +9. 90. ____________

91. From Lima, −12, to Santiago, −33. 91. ____________

92. From Hobart (Tasmania), −43, to Sydney, −34. 92. ____________

93. From Cairo, +30, to Johannesburg, −26. 93. ____________

94. From London, +52, to Athens, +38. 94. ____________

95. From Paris, +49, to Oslo, +60. 95. ____________

96. From Tokyo, +36, to Auckland, −37. 96. ____________

EXERCISES 2.3 S

1–40. Subtract.

1. $-4 - 6$	**2.** $7 - 0$	**1.** ______	**2.** ______
3. $-5 - (-5)$	**4.** $-5 - 5$	**3.** ______	**4.** ______
5. $5 - (-5)$	**6.** $0 - 4$	**5.** ______	**6.** ______
7. $-3 - 0$	**8.** $0 - (-2)$	**7.** ______	**8.** ______
9. $9 - 9$	**10.** $-9 - 0$	**9.** ______	**10.** ______
11. $5 - 0$	**12.** $-4 - (-9)$	**11.** ______	**12.** ______
13. $-8 - 8$	**14.** $-5 - (-6)$	**13.** ______	**14.** ______
15. $6 - 4$	**16.** $6 - (-4)$	**15.** ______	**16.** ______
17. $6 - 0$	**18.** $0 - 6$	**17.** ______	**18.** ______
19. $4 - (-6)$	**20.** $4 - 6$	**19.** ______	**20.** ______
21. $40 - 15$	**22.** $-40 - 15$	**21.** ______	**22.** ______
23. $40 - (-15)$	**24.** $-40 - (-15)$	**23.** ______	**24.** ______
25. $27 - 39$	**26.** $-52 - 88$	**25.** ______	**26.** ______
27. $63 - (-72)$	**28.** $-34 - (-90)$	**27.** ______	**28.** ______
29. $0 - 25$	**30.** $0 - (-78)$	**29.** ______	**30.** ______
31. $-35 - 65$	**32.** $35 - 65$	**31.** ______	**32.** ______
33. $-40 - (-40)$	**34.** $40 - (-40)$	**33.** ______	**34.** ______
35. $9 - 16$	**36.** $9 - (-16)$	**35.** ______	**36.** ______
37. $-50 - (-20)$	**38.** $-50 - 20$	**37.** ______	**38.** ______
39. $-1 - 1$	**40.** $0 - 1$	**39.** ______	**40.** ______

NAME ______________ DATE ______________ COURSE ______________

41–44. On a certain day, a newspaper listed the closing prices in dollars for the day and the net changes in dollars for the day of different stocks, as shown below. For each stock, subtract the "net" from the "close" to find the closing price for the previous day. Note, $\frac{1}{4}$ means 25¢ and $\frac{1}{8}$ means $12\frac{1}{2}$¢.

	Stock	*Close*	*Net*		
41.	Texaco	$29\frac{3}{4}$	$+2\frac{3}{4}$	**41.**	____________
42.	General Motors	48	+1	**42.**	____________
43.	Coca-Cola	$9\frac{1}{8}$	$-4\frac{5}{8}$	**43.**	____________
44.	Pan Am	4	$-\frac{1}{4}$	**44.**	____________

45–50. The elevations are given below for each geographic location. A positive number means an elevation above sea level. A negative number means an elevation below sea level.

For each pair, subtract the elevation of the first location from that of the second. State the difference in elevations and whether one travels upwards (+) or downwards (−) when going from the first to the second.

45. Kathmandu, Nepal; +4000 ft
Mount Everest; +29,028 ft

45. ____________

46. Mount Elbert, Rockies, Colorado; +14,420 ft
Denver, Colorado; +5280 ft

46. ____________

47. Brawley, California; −119 ft
El Centro, California; −45 ft

47. ____________

48. Brawley, California; −119 ft
Death Valley, California; −282 ft

48. ____________

49. Mount Kilimanjaro, Africa; +19,321 ft
Qattara Depression, Egypt; −436 ft

49. ____________

50. Dead Sea (surface level); −1292 ft
Jerusalem; +2370 ft

50. ____________

MULTIPLICATION AND DIVISION OF SIGNED NUMBERS 2.4

A. MULTIPLICATION OF SIGNED NUMBERS

What rules should we have for the product of two signed numbers?

In arithmetic 3×4 means the sum of three fours, that is,

$$3 \times 4 = 4 + 4 + 4$$

We want to keep this meaning. Therefore,

$$(+3)(+4) = (+4) + (+4) + (+4) = +12$$

and

$$(+3)(-4) = (-4) + (-4) + (-4) = -12$$

In Chapter 1 we noted that the order in which we write two factors does not change the value of the product. We want to keep this property.

$$(-3)(+4) = 4(-3) = (-3) + (-3) + (-3) + (-3) = -12$$

If we think of $(-3)(+4)$ as the subtraction of three fours,

$$\begin{aligned}(-3)(+4) &= -4 - 4 - 4 \\ &= (-4) + (-4) + (-4) = -12\end{aligned}$$

Using this idea we can think of $(-3)(-4)$ as the subtraction of three negative fours.

$$\begin{aligned}(-3)(-4) &= -(-4) - (-4) - (-4) \\ &= +(+4) + (+4) + (+4) \\ (-3)(-4) &= +12\end{aligned}$$

For these special cases, we note the following:

The product of two numbers having the same sign is positive.

The product of two numbers having different signs is negative.

We will accept these properties for all integers and for all rational numbers.

For the special numbers 0 and 1, we keep the properties that $\boldsymbol{x \cdot 0 = 0}$ **and** $\boldsymbol{x \cdot 1 = x}$. There is also a special property for -1.

Note that $(-1)5 = -5$ and $(-1)7 = -7$. In general, $\boldsymbol{-1(x) = -x}$.

The multiplication properties are summarized below, where x and y are any two integers or any two rational numbers.

$x \cdot 0 = 0$ and $0 \cdot x = 0$
$x \cdot 1 = x$ and $1 \cdot x = x$
$-1(x) = -x$

MULTIPLICATION PROPERTIES INVOLVING 0 AND 1

The product of any number and 0 is 0.
The product of any number x and 1 is x.
The product of any number x and -1 is $-x$.

$(-x)(-y) = +xy = xy$
$x(-y) = -xy$
$-x(y) = -xy$

MULTIPLICATION PROPERTIES FOR NONZERO NUMBERS

The product of two numbers having the same sign is positive (+).
The product of two numbers having different signs is negative (−).

EXERCISES 2.4A

1–40. Multiply.

SAMPLE PROBLEMS 1–10	SOLUTIONS
1. $(+5)(+3) = +15 = 15$	The product of two integers having the same sign is positive.
2. $(-5)(-3) = +15 = 15$	
3. $(+5)(-3) = -15$	The product of two integers having different signs is negative.
4. $(-5)(+3) = -15$	
5. $(-3)(0) = 0$	If 0 is a factor, the product is 0.
6. $(0)(-3) = 0$	
7. $(-1)5 = -5$	The product of -1 and x is $-x$.
8. $6(-1) = -6$	
9. $(+5)^2 = (+5)(+5) = +25 = 25$	
10. $(-5)^2 = (-5)(-5) = +25 = 25$	

1. $(+5)(-2)$ **2.** $(-5)(-2)$ **1.** ________ **2.** ________

3. $(+2)(-5)$ **4.** $(+5)(+2)$ **3.** ________ **4.** ________

5. $(-3)(-9)$ **6.** $(+8)(-5)$ **5.** ________ **6.** ________

7. $(+4)(+1)$ **8.** $(-4)(+1)$ **7.** ________ **8.** ________

9. $(-1)(-1)$	**10.** $(-4)(-9)$	**9.** ________	**10.** ________
11. $(-7)(+8)$	**12.** $(-4)(25)$	**11.** ________	**12.** ________
13. $(-3)(-3)$	**14.** $(-25)(-4)$	**13.** ________	**14.** ________
15. $(+8)(+125)$	**16.** $(+125)(-8)$	**15.** ________	**16.** ________
17. $(16)(-4)$	**18.** $(+3)(-3)$	**17.** ________	**18.** ________
19. $(-10)(100)$	**20.** $(100)(-100)$	**19.** ________	**20.** ________
21. $5(-6)$	**22.** $4(-25)$	**21.** ________	**22.** ________
23. $-6(7)$	**24.** $-2(9)$	**23.** ________	**24.** ________
25. $(-9)(-9)$	**26.** $(60)(60)$	**25.** ________	**26.** ________
27. $(-10)^2$	**28.** $(+20)^2$	**27.** ________	**28.** ________
29. $(-12)(0)$	**30.** $(0)(-8)$	**29.** ________	**30.** ________
31. $(-1)(6)$	**32.** $(-1)(10)$	**31.** ________	**32.** ________
33. $-1(9)$	**34.** $-1(12)$	**33.** ________	**34.** ________
35. $8(-1)$	**36.** $15(-1)$	**35.** ________	**36.** ________
37. $-1(-7)$	**38.** $-1(0)$	**37.** ________	**38.** ________
39. $0(-20)$	**40.** $0 \cdot 0$	**39.** ________	**40.** ________

B. DIVISION OF INTEGERS

In arithmetic $\frac{12}{3} = 4$ because $3 \cdot 4 = 12$ and 4 is the only number we can multiply 3 by to get 12. We keep this idea in defining the quotient of any two integers n and d.

DEFINITION OF DIVISION

$$\frac{n}{d} = q \text{ if and only if } n = dq \text{ for exactly one integer } q$$

As examples,

$$\frac{+20}{+4} = +5, \quad \text{since} \quad (+4)(+5) = +20$$

$$\frac{-20}{-4} = +5, \quad \text{since} \quad (-4)(+5) = -20$$

$$\frac{+20}{-4} = -5, \quad \text{since} \quad (-4)(-5) = +20$$

$$\frac{-20}{+4} = -5, \quad \text{since} \quad (+4)(-5) = -20$$

$$\frac{0}{-4} = 0, \quad \text{since} \quad (-4)(0) = 0$$

Suppose we let $\frac{0}{0} = 1$, since $0 \cdot 1 = 0$. We could also let $\frac{0}{0} = 2$, since $0 \cdot 2 = 0$. Then it would follow that $1 = 2$. This is a contradiction. To avoid this, we leave $\frac{0}{0}$ undefined. Each symbol representing a number must represent exactly one number. Otherwise, we have contradictions.

However, $\frac{4}{0}$ is undefined, since there is no integer q for which $0 \cdot q = 4$ because $0 \cdot q = 0$ for all integers q.

$\frac{0}{0}$ is undefined for another reason. Since $0 \cdot q = 0$ for all integers q, one and only one integer cannot be selected as the quotient.

The indicated quotient of two nonzero integers is not always an integer. For example, $\frac{2}{7}$ and $\frac{-3}{6}$ are not integers.

$\frac{n}{0}$ is undefined

$\frac{0}{d} = 0$ for $d \neq 0$

$\frac{n}{1} = n$

$\frac{n}{n} = 1$

DIVISION PROPERTIES INVOLVING 0 AND 1

Division by zero is undefined.
Zero divided by a nonzero integer is zero.
Any integer divided by one is that integer.
Any nonzero integer divided by itself is one.

$\frac{+n}{+d} = \frac{n}{d}$ and $\frac{-n}{-d} = \frac{n}{d}$

$\frac{-n}{d} = -\frac{n}{d}$ and $\frac{n}{-d} = -\frac{n}{d}$

DIVISION PROPERTIES FOR NONZERO INTEGERS

The quotient of two numbers having the same sign is positive.
The quotient of two numbers having different signs is negative.

SAMPLE PROBLEMS 1–8

1. $\dfrac{+12}{+3} = +4$
2. $\dfrac{-12}{-3} = +4$
3. $\dfrac{+12}{-3} = -4$
4. $\dfrac{-12}{+3} = -4$
5. $\dfrac{0}{-3} = 0$
6. $\dfrac{3}{0}$ and $\dfrac{-3}{0}$ are undefined
7. $\dfrac{-12}{1} = -12$
8. $\dfrac{-12}{-12} = 1$

EXERCISES 2.4B

1–30. Divide. If an indicated quotient is undefined, write "undefined."

1. $\dfrac{+15}{+5}$
2. $\dfrac{+15}{+3}$
3. $\dfrac{+15}{-3}$
4. $\dfrac{-15}{-5}$
5. $\dfrac{-42}{+21}$
6. $\dfrac{+30}{-3}$
7. $\dfrac{-45}{-9}$
8. $\dfrac{-21}{+7}$
9. $\dfrac{+8}{-2}$
10. $\dfrac{-8}{+4}$
11. $\dfrac{42}{-7}$
12. $\dfrac{-1000}{10}$
13. $\dfrac{-100}{-10}$
14. $\dfrac{-7}{-7}$
15. $\dfrac{7}{-7}$
16. $\dfrac{-21}{21}$
17. $\dfrac{-5}{-5}$
18. $\dfrac{0}{+3}$
19. $\dfrac{0}{-5}$
20. $\dfrac{-200}{8}$
21. $\dfrac{25}{0}$
22. $\dfrac{-18}{0}$

1. ________ 2. ________
3. ________ 4. ________
5. ________ 6. ________
7. ________ 8. ________
9. ________ 10. ________
11. ________ 12. ________
13. ________ 14. ________
15. ________ 16. ________
17. ________ 18. ________
19. ________ 20. ________
21. ________ 22. ________

23. $\frac{69}{69}$

24. $\frac{0}{0}$

25. $\frac{0}{75}$

26. $\frac{4000}{-200}$

27. $\frac{-60}{15}$

28. $\frac{-100,000}{-1000}$

29. $\frac{-50}{-1}$

30. $\frac{-30}{1}$

23. ________ 24. ________

25. ________ 26. ________

27. ________ 28. ________

29. ________ 30. ________

C. RECIPROCALS, RATIONAL NUMBERS (OPTIONAL)

While the quotient of two integers is not always an integer, we shall see in this section that the quotient of two nonzero rational numbers is always a rational number.

In arithmetic, the product of two fractions is a fraction whose numerator is the product of the numerators of the factors and whose denominator is the product of the denominators of the factors. For example,

$$\frac{2}{3} \cdot \frac{5}{7} = \frac{2 \cdot 5}{3 \cdot 7} = \frac{10}{21}$$

We use this arithmetic meaning for our definition of the product of two nonzero rational numbers. In the definition, $a \neq 0$, $b \neq 0$, $c \neq 0$, and $d \neq 0$, and a, b, c, and d are integers.

DEFINITION: PRODUCT OF RATIONAL NUMBERS

$$\frac{a}{b} \cdot \frac{c}{d} = \frac{ac}{bd}$$

$n = \frac{n}{1}$

$\frac{n}{n} = 1$

Now note that $3\left(\frac{1}{3}\right) = \frac{3}{1} \cdot \frac{1}{3} = \frac{3}{3} = 1$. The numbers 3 and $\frac{1}{3}$ are called **reciprocals** because their product is 1; $\frac{1}{3}$ is the reciprocal of 3, and 3 is the reciprocal of $\frac{1}{3}$.

Similarly, $\frac{-1}{2}$ is the reciprocal of -2 because

$$-2\left(\frac{-1}{2}\right) = \frac{-2}{1} \cdot \frac{-1}{2} = \frac{2}{2} = 1$$

$\frac{3}{2}$ is the reciprocal of $\frac{2}{3}$ because $\frac{3}{2} \cdot \frac{2}{3} = \frac{6}{6} = 1$.

$\frac{-5}{7}$ is the reciprocal of $\frac{-7}{5}$ because $\left(\frac{-5}{7}\right)\left(\frac{-7}{5}\right) = \frac{35}{35} = 1$.

In general, every nonzero rational number has a reciprocal.

Every nonzero rational number x has exactly one reciprocal $\frac{1}{x}$ (also called the multiplication inverse of x) such that the product of a number and its reciprocal is 1.

RECIPROCAL
$x\left(\frac{1}{x}\right) = \frac{1}{x}(x) = 1$
for $x \neq 0$

If $x = 5$, then $\frac{1}{x} = \frac{1}{5}$ since $5\left(\frac{1}{5}\right) = 1$.

For $x = \frac{1}{5}$ the reciprocal of x is 5. In symbols, we have $\frac{1}{x} = \frac{1}{\frac{1}{5}}$. In general, **the reciprocal of the reciprocal of a number is that number.**

$\frac{1}{\frac{1}{x}} = x$ for $x \neq 0$

Now note that $\frac{2}{3} = \frac{2}{1} \cdot \frac{1}{3} = 2\left(\frac{1}{3}\right)$.

In words, the **division** of 2 by 3 **is the same as the multiplication** of 2 by $\frac{1}{3}$, the reciprocal of 3. This idea can be used for an alternate division definition, for rational numbers r and s.

ALTERNATE DEFINITION OF DIVISION

$$\frac{r}{s} = r\left(\frac{1}{s}\right) \text{ for } s \neq 0$$

Division and multiplication are called **inverse operations** because they are related by the definition of division.

Now, by definition, the product of two rational numbers is always a rational number. The quotient is defined as the product of two rational numbers. Therefore, **the quotient of any two nonzero rational numbers is always a rational number.**

EXERCISES 2.4C

1–8. Find the reciprocal of each number.

SAMPLE PROBLEMS 1–5	SOLUTIONS
1. 6	$\frac{1}{6}$ is the reciprocal of 6 because $6\left(\frac{1}{6}\right) = 1$.
2. $\frac{1}{8}$	8 is the reciprocal of $\frac{1}{8}$ because $\frac{1}{8}(8) = 1$.
3. -5	$\frac{-1}{5}$ is the reciprocal of -5 because $(-5)\left(\frac{-1}{5}\right) = 1$.
4. $\frac{5}{8}$	$\frac{8}{5}$ is the reciprocal of $\frac{5}{8}$ because $\left(\frac{5}{8}\right)\left(\frac{8}{5}\right) = 1$.
5. $\frac{-4}{3}$	$\frac{-3}{4}$ is the reciprocal of $\frac{-4}{3}$ because $\left(\frac{-4}{3}\right)\left(\frac{-3}{4}\right) = 1$.

1. 4 **2.** $\frac{1}{5}$ **1.** ________ **2.** ________

3. -3 **4.** $\frac{-1}{7}$ **3.** ________ **4.** ________

5. $\frac{3}{5}$ **6.** $\frac{-4}{9}$ **5.** ________ **6.** ________

7. $-\frac{11}{2}$ **8.** -1 **7.** ________ **8.** ________

9–16. Find $\frac{1}{x}$, the reciprocal of x, for each.

SAMPLE PROBLEMS 6–9	SOLUTIONS
6. $x = 9$	$\frac{1}{x} = \frac{1}{9}$ and $9\left(\frac{1}{9}\right) = 1$
7. $x = -10$	$\frac{1}{x} = \frac{-1}{10}$ and $(-10)\left(\frac{-1}{10}\right) = 1$
8. $x = \frac{1}{3}$	$\frac{1}{x} = 3$ and $3\left(\frac{1}{3}\right) = 1$
9. $x = \frac{-7}{9}$	$\frac{1}{x} = \frac{-9}{7}$

9. $x = 6$

10. $x = -6$

9. ________ **10.** ________

11. $x = \frac{1}{6}$

12. $x = \frac{-1}{6}$

11. ________ **12.** ________

13. $x = \frac{2}{3}$

14. $x = \frac{-2}{3}$

13. ________ **14.** ________

15. $x = 1$

16. $x = -1$

15. ________ **16.** ________

17–24. Find each product by using the alternate definition of division.

SAMPLE PROBLEMS 10–13	SOLUTIONS
10. $12\left(\frac{1}{3}\right)$	$12\left(\frac{1}{3}\right) = \frac{12}{3} = 4$
11. $-35\left(\frac{1}{7}\right)$	$-35\left(\frac{1}{7}\right) = \frac{-35}{7} = -5$
12. $60\left(\frac{-1}{10}\right)$	$60\left(\frac{-1}{10}\right) = \frac{60}{-10} = -6$
13. $-48\left(\frac{-1}{6}\right)$	$-48\left(\frac{-1}{6}\right) = \frac{-48}{-6} = 8$

Multiplying by the reciprocal of a number is the same as dividing by the number.

$$r\left(\frac{1}{s}\right) = \frac{r}{s}$$

17. $6\left(\frac{1}{2}\right)$

18. $-40\left(\frac{1}{5}\right)$

17. ________ **18.** ________

19. $56\left(\frac{-1}{8}\right)$

20. $-15\left(\frac{-1}{3}\right)$

19. ________ **20.** ________

21. $-8\left(\frac{-1}{8}\right)$

22. $\left(\frac{-1}{4}\right)(-4)$

21. ________ **22.** ________

23. $-1\left(\frac{4}{5}\right)$

24. $-1\left(\frac{-2}{3}\right)$

23. ________ **24.** ________

25–32. Find each product.

SAMPLE PROBLEMS 14–16

14. $\frac{2}{3} \cdot \frac{5}{8}$

15. $\frac{-4}{5}\left(\frac{5}{7}\right)$

16. $\left(\frac{-3}{4}\right)\left(\frac{-2}{9}\right)$

SOLUTIONS

$\frac{2 \cdot 5}{3 \cdot 8} = \frac{10}{24} = \frac{5 \cdot \mathbf{2}}{12 \cdot \mathbf{2}} = \frac{5}{12} \cdot \frac{\mathbf{2}}{\mathbf{2}} = \frac{5}{12} \cdot \mathbf{1} = \frac{5}{12}$

$\frac{-4(5)}{5(7)} = \frac{-4 \cdot 5}{7 \cdot 5} = \frac{-4}{7}$

$\frac{(-3)(-2)}{(4)(9)} = \frac{6}{36} = \frac{1(6)}{6(6)} = \frac{1}{6}$

25. $\frac{3}{5} \cdot \frac{7}{6}$

26. $\frac{-8}{9}\left(\frac{9}{4}\right)$

27. $\frac{3}{8}\left(\frac{-5}{3}\right)$

28. $\left(\frac{-5}{10}\right)\left(\frac{-7}{10}\right)$

29. $\left(\frac{-2}{25}\right)\left(\frac{-5}{6}\right)$

30. $18\left(\frac{-5}{12}\right)$

31. $-10\left(\frac{3}{4}\right)$

32. $\left(\frac{-7}{8}\right)(-20)$

25. ________ **26.** ________

27. ________ **28.** ________

29. ________ **30.** ________

31. ________ **32.** ________

33–40. Find each quotient.

SAMPLE PROBLEMS 17–19

17. $\frac{15}{8} \div \frac{3}{4}$

18. $-30 \div \frac{5}{6}$

19. $\frac{-3}{8} \div \frac{-3}{2}$

SOLUTIONS

$\frac{15}{8} \cdot \frac{4}{3} = \frac{60}{24} = \frac{12(5)}{12(2)} = \frac{5}{2}$

$\frac{-30}{1} \cdot \frac{6}{5} = \frac{-5(6)(6)}{5} = -36$

$\frac{-3}{8} \cdot \frac{-2}{3} = \frac{6}{24} = \frac{1}{4}$

Multiply by the reciprocal of the divisor.
(Invert the divisor and multiply.)

33. $\frac{1}{2} \div \frac{2}{3}$

34. $\frac{-25}{8} \div \frac{5}{2}$

35. $\frac{3}{5} \div \left(\frac{-6}{5}\right)$

36. $\left(\frac{-3}{4}\right) \div \left(\frac{-3}{8}\right)$

37. $-20 \div \frac{5}{2}$

38. $\frac{12}{5} \div (-10)$

39. $-1 \div \left(\frac{-7}{8}\right)$

40. $\left(\frac{-7}{15}\right) \div \left(\frac{-7}{15}\right)$

33. ________ **34.** ________

35. ________ **36.** ________

37. ________ **38.** ________

39. ________ **40.** ________

1–40. Multiply.

1. $6(0)$

2. $(-5)(0)$

3. $-2(-54)$

4. $125(-8)$

5. $(-25)(-4)$

6. $(0)(5)$

7. $(0)(-6)$

8. $(-7)(0)$

9. $(0)(0)$

10. $(+1)(-15)$

11. $(-25)(1)$

12. $(-1)(-10)$

13. $(-46)(100)$

14. $(-58)(-1000)$

15. $(+6)^2$

16. $(+8)^2$

17. $(-7)^2$

18. $(-9)^2$

19. $(-30)(+20)$

20. $(+50)(-60)$

21. $6(-25)$

22. $-7(40)$

23. $-9(-800)$

24. $-16(-9)$

25. $-2(-3)^2$

26. $(-2 \cdot 3)^2$

27. $(0)(-5)^2$

28. $(3 \cdot 0)(-7)$

29. $(-4)(-4)^2$

30. $(-6 \cdot 6)(-6)$

31. $-1(8)$

32. $(-1)(7)$

33. $(15)(-1)$

34. $20(-1)$

35. $(-1)(-1)$

36. $-1(29)$

37. $(-1)(-25)$

38. $-1(0)$

39. $5(-4)^2$

40. $[5(-4)]^2$

1.	______	**2.**	______
3.	______	**4.**	______
5.	______	**6.**	______
7.	______	**8.**	______
9.	______	**10.**	______
11.	______	**12.**	______
13.	______	**14.**	______
15.	______	**16.**	______
17.	______	**18.**	______
19.	______	**20.**	______
21.	______	**22.**	______
23.	______	**24.**	______
25.	______	**26.**	______
27.	______	**28.**	______
29.	______	**30.**	______
31.	______	**32.**	______
33.	______	**34.**	______
35.	______	**36.**	______
37.	______	**38.**	______
39.	______	**40.**	______

NAME ______________ DATE ______________ COURSE ______________

41–66. Divide. If an indicated quotient is undefined, write "undefined."

41. $\frac{352}{-11}$

42. $\frac{65}{-5}$

43. $\frac{800}{-10}$

44. $\frac{-68}{17}$

45. $\frac{0}{9}$

46. $\frac{-84}{-6}$

47. $\frac{-63}{-9}$

48. $\frac{100{,}000}{-100}$

49. $\frac{-10{,}000}{-10}$

50. $\frac{-1{,}000{,}000}{-10{,}000}$

51. $\frac{-72}{+8}$

52. $\frac{-25}{25}$

53. $\frac{0}{-4}$

54. $\frac{-6}{-6}$

55. $\frac{19}{-1}$

56. $\frac{100}{-25}$

57. $\frac{-27}{1}$

58. $\frac{-1000}{-8}$

59. $\frac{-16}{-1}$

60. $\frac{-10{,}000}{16}$

61. $\frac{-60}{0}$

62. $\frac{0}{0}$

63. $\frac{-7}{-7}$

64. $\frac{-7}{1}$

65. $\frac{-1}{0}$

66. $\frac{0}{-1}$

41. ______________

42. ______________

43. ______________

44. ______________

45. ______________

46. ______________

47. ______________

48. ______________

49. ______________

50. ______________

51. ______________

52. ______________

53. ______________

54. ______________

55. ______________

56. ______________

57. ______________

58. ______________

59. ______________

60. ______________

61. ______________

62. ______________

63. ______________

64. ______________

65. ______________

66. ______________

67–70. A survey taken by the American Council for Better Broadcasts asked people to rate certain television shows by using the following rating system:

+2 for excellent, +1 for good, −1 for fair, −2 for poor.

Using this system, the table below shows some possible results from a total of 100 persons.
Find the rating for each show as follows:
a. Find nr for each value of r.
b. Add the four products found in **a.**
c. Divide the sum of the four products by 100.
The answer to **c** is the rating for the show.

	Number of people, n			
Rating r	Show A	Show B	Show C	Show D
+2	45	8	16	3
+1	35	10	30	8
−1	15	38	46	14
−2	5	44	8	75

67. Show A

68. Show B

69. Show C

70. Show D

67. ____________

68. ____________

69. ____________

70. ____________

71–76. In chemistry, a subscript at the lower right of a symbol for an atom indicates the number of atoms. For example, sulfuric acid is expressed in symbols as H_2SO_4, meaning two atoms of hydrogen (H), one atom of sulfur (S), and four atoms of oxygen (O). The valence of an ion, a positive or negative integer indicating how the ion combines chemically, is the sum of the valences of the atoms composing the ion.

SAMPLE PROBLEM

Sulfate ion, SO_4; valence of S = +6, valence of O = −2.

SOLUTION

$$\begin{aligned}\text{Valence of } SO_4 &= (\text{valence of S}) + 4(\text{valence of O})\\ &= +6 + 4(-2) = 6 - 8 = -2.\end{aligned}$$

71. Nitrate ion, NO_3; valence of N = +5, valence of O = −2. **71.** ______________

72. Phosphate ion, PO_4; valence of P = +5, valence of O = −2. **72.** ______________

73. Bichromate ion, Cr_2O_7; valence of Cr = +5, valence of O = −2. **73.** ______________

74. Ammonium ion, NH_4; valence of N = −3, valence of H = +1. **74.** ______________

75. Permanganate ion, MnO_4; valence of Mn = +7, valence of O = −2. **75.** ______________

76. Carbonate ion, CO_3; valence of C = +4, valence of O = −2. **76.** ______________

77–80. Moments are used to find the center of gravity of an object. Designers of automobiles, boats, airplanes, furniture, and other objects use moments to obtain stability.

A moment M is a measure of the tendency of an object to rotate owing to a force F applied at a directed distance d from a fixed position.

$$M = Fd$$

For a horizontal wire, rod, or beam,

direction to the right is positive and to the left, negative.

a force upwards is positive, and a force downwards is negative.

a counterclockwise (CCW) moment is positive.

a clockwise (CW) moment is negative.

EXAMPLES

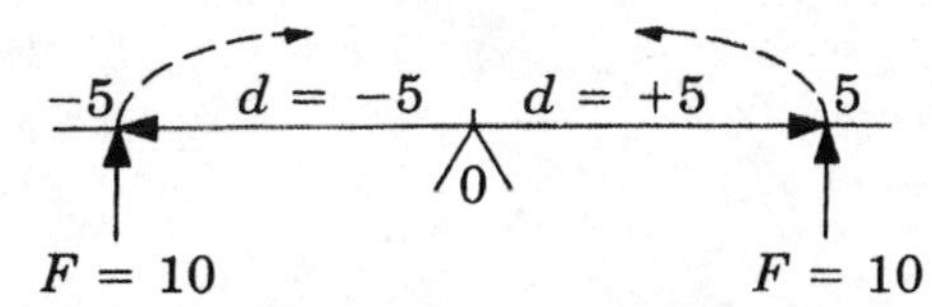

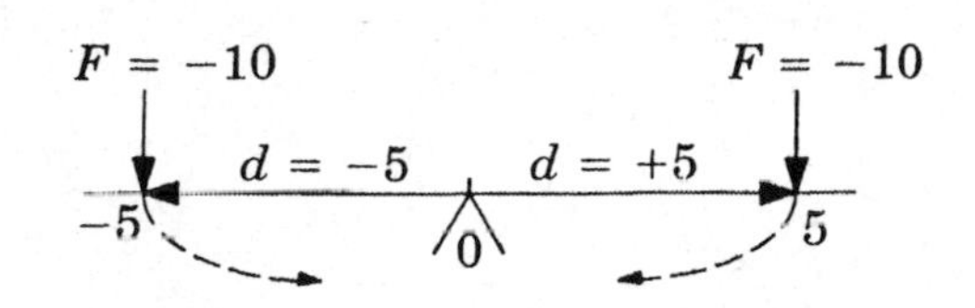

$M = (10)(-5)$ $= -50$ clockwise rotation (CW)	$M = (10)(5)$ $= 50$ counterclockwise rotation (CCW)	$M = (-10)(-5)$ $= 50$ counterclockwise rotation (CCW)	$M = (-10)(5)$ $= -50$ clockwise rotation (CW)

Find the moments and state if the rotation is clockwise (CW) or counterclockwise (CCW).

77. $F = +40$ lb
$d = +6$ ft

78. $F = +80$ grams
$s = -50$ cm

79. $F = -1000$ grams
$d = +200$ cm

80. $F = -90$ lb
$d = -4$ ft

77. ____________

78. ____________

79. ____________

80. ____________

81–84. If two electrically charged bodies, having charges of x and y, respectively, are in a certain medium at a certain distance apart, then a measure f of the force of attraction or repulsion between the bodies is given by $f = xy$.
If f is positive, the charges repel.
If f is negative, the charges attract.
Find f for each of the following and state whether the charges repel or attract.

81. $x = +500$ and $y = +300$

82. $x = +400$ and $y = -100$

83. $x = -200$ and $y = +600$

84. $x = -300$ and $y = -200$

81. ____________

82. ____________

83. ____________

84. ____________

COMBINED OPERATIONS, EVALUATIONS 2.5

A. COMBINED OPERATIONS

The order rules are restated here as a convenience in doing combined operations on signed numbers.

ORDER RULES

1. The operations within the innermost set of grouping symbols are done first, following the procedure in Step 2. (Recall that the bar used in division is a grouping symbol.)
2. Unless grouping symbols indicate otherwise, the operations are done in the following order:
   First, squaring and/or cubing, as read from left to right.
   Second, multiplication and/or division, as read from left to right.
   Third, addition and/or subtraction, as read from left to right.

Note that $-5 = (-1)(5)$ and $-8 = (-1)(8)$. In general,

$$-x = (-1)x$$

It is often useful in calculations to replace $-x$ by $(-1)x$.

Recall the meaning of $-x$ when x is a negative number. For example, if $x = -6$, then $-x = -(-6) = (-1)(-6) = 6$.

In general, $-(-x)$ is the opposite (additive inverse) of $-x$.

Since x is the opposite of $-x$,

$$-(-x) = x$$

The opposite of (the opposite of x) is x

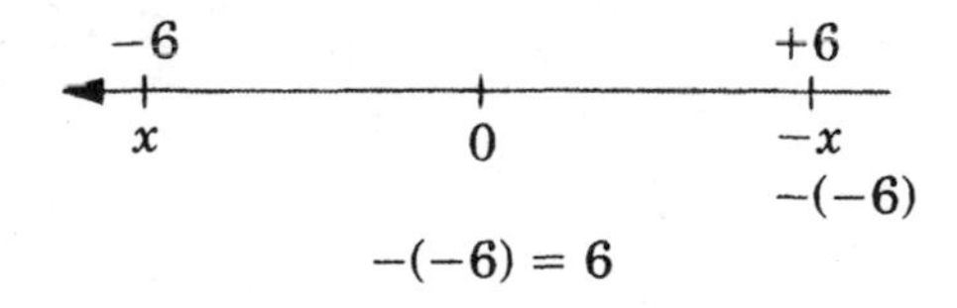

$-(-6) = 6$

SAMPLE PROBLEM 1

Evaluate $20 - 3(-7 + 4)$.

SOLUTION

Do inside parentheses first. $20 - 3(-7 + 4) = 20 - 3(-3)$

Multiply before subtracting. $= 20 - (-9) = 20 + 9 = 29$

SAMPLE PROBLEM 2

Evaluate $\frac{20 - 50}{-7 + 1}$.

SOLUTION

Since the bar is a grouping symbol, the division is done last. $\frac{20 - 50}{-7 + 1} = \frac{-30}{-6} = 5$

SAMPLE PROBLEM 3

Evaluate $-(-2)^3 - (-4)^2$.

SOLUTION

Square and cube first. $-(-2)^3 - (-4)^2 = -(-8) - (16)$

Use $-(-x) = x$. $= 8 - 16$

$= -8$

SAMPLE PROBLEM 4

Evaluate $4 - 2(3 - 6)(4 - 9)$.

SOLUTION

Do inside parentheses first. $4 - 2(-3)(-5)$

Multiply before subtracting. $= 4 - (-6)(-5) = 4 - (30) = -26$

(Multiply from left to right.)

EXERCISES 2.5A

Evaluate.

1. $5 - (-6 + 6)$ 2. $2(-6) + (-4)(-3)$ 1. ________ 2. ________

3. $5(4 - 4)$ 4. $-6(-8 + 8)$ 3. ________ 4. ________

5. $\dfrac{7 - 7}{3}$ 6. $-2(3 - 2) - (-3)(2 - 3)$ 5. ________ 6. ________

7. $\dfrac{3 - 4}{4 - 3}$ 8. $\dfrac{6 - 2}{2 - 6}$ 7. ________ 8. ________

9. $(0 - 2)(5 - 3)$ 10. $0 - 3(2 - 4)$ 9. ________ 10. ________

11. $(-3)(-3)(-3)$ 12. $(-5)(-5)(-5)(-5)$ 11. ________ 12. ________

13. $-5(7 - 11)$ 14. $7 - [3 + 3(-3)]$ 13. ________ 14. ________

15. $8 - [2(9) - 4(4)]$ 16. $\left(\dfrac{7 - 3}{3 - 5}\right)\left(\dfrac{-7 + 7}{4 - 6}\right)$ 15. ________ 16. ________

17. $\left(\dfrac{-4 - 6}{5 - 3}\right)\left(\dfrac{11 - 4}{5 + 2}\right)$ 18. $3\left(\dfrac{21 - 12}{-8 + 5}\right)$ 17. ________ 18. ________

19. $-5\left(\dfrac{5 - 3}{3 - 5}\right)$ 20. $9 - (5 - 7)$ 19. ________ 20. ________

21. $9 - 5 - 7$ 22. $(-5)(-6)(-4)$ 21. ________ 22. ________

23. $-(-3)^2 - 4^2$ 24. $(-10)^3 + (-10)^2$ 23. ________ 24. ________

25. $2(-5)^2 - 3(-5)$ 26. $(-6)^2 - 4(2)(-3)$ 25. ________ 26. ________

27. $\dfrac{100(-40)(-24)}{-40 - 24}$ 28. $\dfrac{-25 - 45 - 20}{3}$ 27. ________ 28. ________

29. $(-8)^2 - 6^2 - (-8)(-6)$ 30. $6(-4) - 3(-10)^2$ 29. ________ 30. ________

B. EVALUATION

It is sometimes necessary to substitute one or more negative values in an expression or a formula. The following examples illustrate such evaluations.

SAMPLE PROBLEM 1

Evaluate $b^2 - 4ac$ for $b = -2$, $a = 3$, $c = -5$.

SOLUTION

Remove a letter. Hold its place with open parentheses. Insert the value. $\quad b^2 - 4ac = (-2)^2 - 4(3)(-5)$

Square, then multiply. $\quad = 4 - 4(-15)$

Subtract last. $\quad = 4 - (-60) = 4 + 60 = 64$

SAMPLE PROBLEM 2

Evaluate $C = \dfrac{5(F - 32)}{9}$ when C is degrees Celsius and $F = -22°$ Fahrenheit.

SOLUTION

Remove letter and hold its place. $\quad C = \dfrac{5(F - 32)}{9} = \dfrac{5((\quad) - 32)}{9}$

Insert value. $\quad = \dfrac{5(-22 - 32)}{9}$

Do inside parentheses first. $\quad = \dfrac{5(-54)}{9}$

Division can be done before multiplication. $\quad = 5(-6) = -30°$ Celsius

SAMPLE PROBLEM 3

Evaluate $P = \dfrac{100(p + q)}{pq}$ for $p = 10$, $q = -20$.

SOLUTION

$$P = \frac{100(p + q)}{pq}$$

$$= \frac{100(10 + (-20))}{(10)(-20)} = \frac{100(-10)}{-200} = \frac{-1000}{-200} = 5$$

EXERCISES 2.5B

1–22. Evaluate.

1. $5(3x - 4)$ for $x = -2$

2. $20 - (5x - 6)$ for $x = -6$

3. $6x - 5(x + 2)$ for $x = -5$

4. $25 - (10 + 3x)$ for $x = -4$

5. $2y - (y - 3)$ for $y = -7$

6. $10 - (8 - 5y)$ for $y = -3$

7. $32 - (3t + 4)$ for $t = -8$

8. $5t - 3(t + 8)$ for $t = -5$

9. $5x^2 + 3x$ for $x = -4$

10. $4x - 2x^2$ for $x = -5$

11. $(A + B)^2$ for $A = -2$ and $B = -4$

12. $(A - B)^2$ for $A = -10$ and $B = -2$

13. $A^2 + B^2$ for $A = -2$ and $B = -4$

14. $A^2 - B^2$ for $A = -10$ and $B = -2$

15. $(A + B)(A - B)$ for $A = -8$ and $B = 3$

16. $(A + 2B)(A - 3B)$ for $A = -6$ and $B = 4$

17. $2x^2 - xy - 6y^2$ for $x = -5$ and $y = -2$

18. $4x^2 + 2xy - y^2$ for $x = -3$ and $y = 4$

19. $\dfrac{xy}{x + y}$ for $x = 2$ and $y = -3$

20. $\dfrac{x + y}{x - y}$ for $x = -2$ and $y = -3$

1. ____________
2. ____________
3. ____________
4. ____________
5. ____________
6. ____________
7. ____________
8. ____________
9. ____________
10. ____________
11. ____________
12. ____________
13. ____________
14. ____________
15. ____________
16. ____________
17. ____________
18. ____________
19. ____________
20. ____________

21. xyz for $x = -2$, $y = -3$, $z = -5$

22. $2xy - 3xz$ for $x = 4$, $y = -5$, $z = -2$

21. ________

22. ________

23–24. Temperature conversion. $C = \dfrac{5(F - 32)}{9}$

23. For $F = -4$

24. For $F = -22$

23. ________

24. ________

25–26. Temperature conversion. $F = \dfrac{9C + 160}{5}$

25. For $C = 10$

26. For $C = -20$

25. ________

26. ________

27–28. Average of three numbers. $A = \dfrac{a + b + c}{3}$

27. For $a = -25$, $b = -40$, $c = -43$

28. For $a = 40$, $b = -50$, $c = -20$

27. ________

28. ________

29–30. Part of quadratic formula. $D = b^2 - 4ac$

29. For $a = 4$, $b = 4$, $c = -2$

30. For $a = 5$, $b = -4$, $c = 2$

29. ________

30. ________

31–32. Lenses. $P = \dfrac{100(p + q)}{pq}$

31. For $p = 25$, $q = -4$

32. For $p = 5$, $q = -10$

31. ________

32. ________

33–34. Mirrors, photography. $f = \dfrac{pq}{p + q}$

33. For $p = 45$, $q = -180$

34. For $p = 40$, $q = -30$

33. ________

34. ________

EXERCISES 2.5 S

Evaluate.

1. $12 - 2(-6 - 4)$
2. $(-6 + 4)^2$
3. $(-6 + 4)^3$
4. $\dfrac{3(6 - 10)}{-6}$
5. $\dfrac{-2(5 - 8)}{0 - 3}$
6. $(9 - 8)(8 - 7)(7 - 6)$
7. $15 - 6(5 - 2)$
8. $-8 - 2(-4 + 7)$
9. $\dfrac{9 - 16}{3 - 4}$
10. $\dfrac{25 - 4}{5 - 8}$
11. $(-2)(-3)(-4)$
12. $\dfrac{(-3)(-4)(-5)}{-10}$
13. $2(-10)^2$
14. $-2(10)^2$
15. $(-4)^3$
16. $-(-4)^3$
17. $(-5)^2 - (-3)^2$
18. $(-10)^3 + (-5)^3$
19. $(-1)^2 - 4(-2)(-1)$
20. $(5 - 9)^2 - (3 - 5)^2$
21. $(9 - 10)(6 - 8)(4 - 6)$
22. $2 - 7 - (0 - 4)$

1. ______
2. ______
3. ______
4. ______
5. ______
6. ______
7. ______
8. ______
9. ______
10. ______
11. ______
12. ______
13. ______
14. ______
15. ______
16. ______
17. ______
18. ______
19. ______
20. ______
21. ______
22. ______

NAME ______ DATE ______ COURSE ______

23. $3(-4)^2 + 5(-4)$

24. $10 - 2[3 - 6(1 - 3)]$

25. $(-9)^2 - 4(3)(-2)$

26. $5(-3)^3 - (-2)^3$

27. $\left(\dfrac{1-9}{2-4}\right)\left(\dfrac{7-10}{10-7}\right)$

28. $\dfrac{-5-6-7}{-2-3-4}$

29. $\dfrac{-5-(6-7)}{-2-(3-4)}$

30. $\dfrac{-8-(-3-5)}{-8-3-5}$

31. **Average of four numbers.**
$\dfrac{x+y+z+w}{4}$ for $x = 9$, $y = -14$, $z = -18$, $w = 15$

32. **Trigonometry.**
$a^2 + b^2 - ab$ for $a = 6$, $b = -3$

33. **Analytic geometry.**
$\dfrac{3x + 4y - 24}{5}$ for $x = -6$, $y = 3$

34. **Temperature conversion.**
$F = \dfrac{9C}{5} + 32$
Find F for $C = -5°$.

35. **Corrective lenses.**
$P = \dfrac{100(p+q)}{pq}$
Find P for $p = 100$ and $q = -25$.

36. **Acoustics.**
$F = \dfrac{fV}{v + V}$
Find F for $f = 250$, $v = -1320$, and $V = 1100$.

37. **Radio and television.**
$r = \dfrac{R(E - G)}{E}$
Find r for $R = 2$, $E = -12$, and $G = -9$.

38. **Mirrors and photography.**
$f = \dfrac{pq}{p + q}$
Find f for $p = 6$ and $q = -3$.

23. ____________
24. ____________
25. ____________
26. ____________
27. ____________
28. ____________
29. ____________
30. ____________
31. ____________
32. ____________
33. ____________
34. ____________
35. ____________
36. ____________
37. ____________
38. ____________

SAMPLE TEST FOR CHAPTER 

1–2. Rewrite each of the following as a single integer. (See Section 2.1.)

1. $|-8| + |-6|$ **2.** $|-20| - |+9|$

1. ________ **2.** ________

3–4. Translate into algebraic symbols. (See Section 2.1.)

3. Six less than y is greater than 5. **4.** Nine more than y is negative.

3. ________ **4.** ________

5–8. Insert the symbol < or the symbol > between each given pair of integers, so that the resulting statement is true. State the algebraic meaning of the resulting statement. (See Section 2.1.)

5. $+4$ —— $+8$ **6.** $+4$ —— -8

5. ________ **6.** ________

7. -4 —— -8 **8.** -8 —— -4

7. ________ **8.** ________

9–14. Add. (See Section 2.2.)

9. $(-17) + (-40)$ **10.** $(-28) + (+15)$

9. ________ **10.** ________

11. $(+28) + (-15)$ **12.** $0 + (-12)$

11. ________ **12.** ________

13. $(-150) + 125$ **14.** $(-78) + (+78)$

13. ________ **14.** ________

15–20. Subtract. (See Section 2.3.)

15. $(+80) - (+100)$ **16.** $(-17) - (+8)$

15. ________ **16.** ________

17. $(-18) - (-6)$ **18.** $(-17) - (-17)$

17. ________ **18.** ________

19. $40 - 85$ **20.** $56 - 56$

19. ________ **20.** ________

21–26. Multiply. (See Section 2.4.)

21. $(-8)(+125)$ **22.** $(-25)(-6)$

21. ________ **22.** ________

23. $(+23)(-27)$ **24.** $(-35)(-1)$

23. ________ **24.** ________

25. $(0)(-34)$ **26.** $(+1)(-17)$

25. ________ **26.** ________

NAME ________ DATE ________ COURSE ________

27–34. Divide. (See Section 2.4.)

27. $\frac{-72}{-12}$

28. $\frac{-36}{+4}$

29. $\frac{+39}{-3}$

30. $\frac{0}{-39}$

31. $\frac{-23}{-23}$

32. $\frac{-8}{1}$

33. $\frac{9}{-1}$

34. $\frac{-1}{0}$

27. ________ **28.** ________

29. ________ **30.** ________

31. ________ **32.** ________

33. ________ **34.** ________

35–42. Evaluate. (See Section 2.5.)

35. $(-4)(-38)(-25)$

36. $\frac{5^2 - (-3)^2}{2 - 6}$

35. ____________________

36. ____________________

37. $(-6)^2 - 4(2)(-1)$

38. $(10 - 2)(10 - 16)$

37. ____________________

38. ____________________

39. $5x - 2(x + 4)$ for $x = -6$

39. ____________________

40. $2x^2 + 6x$ for $x = -3$

40. ____________________

41. $\frac{x - y}{x + y}$ for $x = 5$ and $y = -7$

41. ____________________

42. $\frac{ab}{a + b}$ for $a = 6$ and $b = -10$

42. ____________________

SAMPLE TEST FOR CHAPTERS 

1–10. Express each of the following in symbols.

1. The product of 5 and y.	**1.**	______
2. The sum of x and 7.	**2.**	______
3. The quotient when x is divided by 4.	**3.**	______
4. The difference when 8 is subtracted from y.	**4.**	______
5. The square of t.	**5.**	______
6. The cube of r.	**6.**	______
7. 6 times the sum of x and 3.	**7.**	______
8. 4 less than the square of n.	**8.**	______
9. One third the sum of y and 9.	**9.**	______
10. The cube of the product of 2 and x.	**10.**	______

11–18. List the members of the set described.

11. The set of natural numbers less than 6.	**11.**	______
12. The set of even numbers less than 10.	**12.**	______
13. The multiples of 15.	**13.**	______
14. The odd multiples of 15.	**14.**	______
15. The factors of 99.	**15.**	______
16. The prime factors of 99.	**16.**	______
17. The even factors of 99.	**17.**	______
18. The composite factors of 99.	**18.**	______

NAME ______ DATE ______ COURSE ______

19–32. Evaluate.

19. $8 - 2(1 - 7)$

20. $(8 - 2)(1 - 7)$

19. ______________

20. ______________

21. $(2 - 5)(2 - 6)(2 - 7)$

22. $2 - 5[2 - 6(2 - 7)]$

21. ______________

22. ______________

23. $-5(6 - 9)^2$

24. $(-6)^2 - 4(-6)$

23. ______________

24. ______________

25. $\dfrac{2^3 + 3^2}{2^3 - 3^2}$

26. $\dfrac{10 - 2(3 - 8)}{5}$

25. ______________

26. ______________

27. $3x - 2(x - 4)$ for $x = -4$

28. $2y^2 - 5y - 4$ for $y = -2$

27. ______________

28. ______________

29. $ab - 2a + 3b$ for $a = -3, b = -2$

29. ______________

30. $\dfrac{-b + d}{2a}$ for $a = 5, b = -2, d = 8$

30. ______________

31. $\dfrac{x^2 + 5x}{x^2 - 25}$ for $x = 4$

31. ______________

32. $\dfrac{pq}{p + q}$ for $p = -15, q = 10$

32. ______________

Linear Equations in One Variable

ALGEBRA IN ACTION: THE HOME

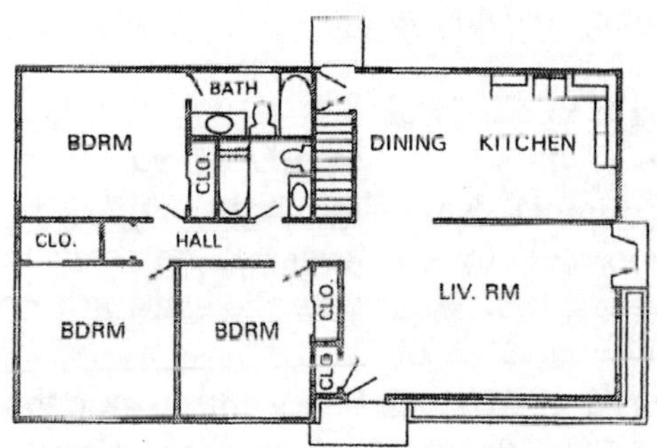

Formulas and equations can be used to determine the amount of paint, carpet, concrete, or ground cover needed to do a job in or around the home.

The rectangle is a common shape found both inside and outside the home.

Rectangles: $P = 2W + 2L$
$A = WL$

Painting: $C = \frac{pA}{G}$

Carpeting: $C = \frac{pA}{9}$

Concrete: $C = \frac{Ad}{324}$

Ground cover: $C = \frac{pA}{n}$

Chapter 3 Objectives

Objectives / **Typical Problems**

Rewrite the terms of a sum or the factors of a product in conventional form.

1–2. Rewrite in conventional form.

1. $(b - 7) + (a + 3)$
2. $(-7y)(-2xy)$

Use the distributive axiom to remove parentheses and collect like terms.

3–4. Remove parentheses and collect like terms.

3. $8(x - 2) - 3(x - 4)$
4. $2x(x + 3) - 5(x + 3)$

Solve simple linear equations in one variable.

5–10. Solve and check.

5. $x + 7 = 3$
6. $5 - t = 8$
7. $3y - 8 = 7$
8. $\dfrac{x + 5}{5} = -6$
9. $x - 6 = -(6 - x)$
10. $x - 6 = x + 6$

Solve a general linear equation in one variable.

11–12. Solve and check.

11. $7(y - 2) - 3(y - 4) = 18$
12. $x - 2(x + 2) = 16 + 2(x - 4)$

Solve a stated formula problem.

13. At the end of t seconds, the height H in feet of a projectile that is fired upwards with a velocity of v feet per second is given by

$$H = vt - 16t^2$$

Find v for $t = 5$ sec and $H = 400$ ft.

Solve a literal equation.

14. Solve $2x - 5y = 10$ for y.
15. Solve $S = a + (n - 1)d$ for n.

Solve a stated number problem.

16. 3 times the difference when 5 is subtracted from a number is 9 more than the number. Find the number.

Solve a stated percentage problem.

17. The 15% commission made by a salesperson on the sale of furniture amounted to \$180. What was the sales price of the furniture?
18. Paint originally priced at \$20 a gallon went on sale for \$12 a gallon. What percent of the original price was the decrease of \$8?

Solve a stated age problem.

19. Sara's aunt is 5 times as old as Sara. In 3 years Sara's aunt will be 4 times as old as Sara. Find the present ages of Sara and her aunt.

Solve a stated geometric problem.

20. The length of a rectangle is 6 ft less than 3 times its width. The perimeter of the rectangle is 68 ft. Find the width and length of the rectangle.

ANSWERS to Typical Problems

1. $a + b - 4$ **2.** $14xy^2$ **3.** $5x - 4$ **4.** $2x^2 + x - 15$ **5.** -4 **6.** -3 **7.** 5 **8.** -35 **9.** Q, or all x **10.** $\varnothing$ **11.** 5 **12.** -4 **13.** 160 fps **14.** $y = \dfrac{2x - 10}{5}$ **15.** $n = \dfrac{S - a + d}{d}$ **16.** $3(x - 5) = x + 9$; 12 **17.** \$1200 **18.** 40% **19.** $5x + 3 = 4(x + 3)$; 9, 45 **20.** $2W + 2(3W - 6) = 68$; 10 ft, 24 ft

An equation may be a true statement, a false statement, or an open statement. As examples,

$5 + 4 = 9$ is a true statement

$3 + 4 = 9$ is a false statement

$x + 4 = 9$ is an open statement

An **open statement** is one that contains one or more variables. It is neither true nor false, but it becomes either true or false when each variable is replaced by a numerical value.

For $x = 5$, $x + 4 = 9$ becomes $5 + 4 = 9$, a true statement.

For $x = 3$, $x + 4 = 9$ becomes $3 + 4 = 9$, a false statement.

A **solution** of an equation in one variable is a value for which the equation becomes true when the variable is replaced by this value. As an example, 5 is a solution of $x + 4 = 9$.

Equations such as $x = 4$, $x + 4 = 9$, $3x - 7 = 8$, $3y - 2(y - 3) = 0$, and $2(t - 4) = 9$ are examples of a **linear equation in one variable.** Finding the solutions of a linear equation in one variable is the major concern of this chapter.

The solution process involves the use of certain axioms and theorems. These will be discussed first. In mathematics, there are three major types of general statements: definitions, axioms, and theorems.

A **definition** is a precise explanation of the meaning of a term. An **axiom** is a statement that is assumed to be true without proof. A **theorem** is a statement that is proved.

EQUALITY, COMMUTATIVE, ASSOCIATIVE AXIOMS

A. EQUALITY, COMMUTATIVE AXIOMS

In Chapter 1, we saw that the **reflexive axiom** stated that any equation having the form $A = A$ is accepted as a true statement. In solving an equation, we try to reach the equation $x = a$. For example, for $x = 5$, we know that the solution is 5. By substituting 5 for x, we get $5 = 5$, a true statement.

We also met the **symmetric axiom** in Chapter 1. **If $A = B$, then $B = A$.** In other words, the **sides of an equation can be exchanged.** In solving an equation, it is often useful to exchange sides. For example,

$15 = 2x - 3$ can be rewritten as $2x - 3 = 15$

In Chapter 1, we also noted that the order in which we write two terms of a sum or two factors of a product does not change the result.

$a + b = b + a$

The commutative axiom for addition states that the order of the terms of a sum may be changed without changing the value of the sum.

For example, $3 + 5 = 5 + 3$ and $4 + x = x + 4$.

$ab = ba$

The commutative axiom for multiplication states that the order of the factors of a product may be changed without changing the value of the product.

For example, $3 \cdot 5 = 5 \cdot 3$ and $x5 = 5x$.
Also, $(4 + n)7 = 7(4 + n) = 7(n + 4)$.

Subtraction and division are *not* commutative operations. For example, $7 - 5 \neq 5 - 7$ and $\frac{6}{2} \neq \frac{2}{6}$. (The symbol $\neq$ is read "does not equal.")

We noted in Chapter 2 that the sum or product of any two integers is an integer. Also, the sum or product of any two rational numbers is a rational number. This means, in particular, that an expression such as

$$x + 3 \qquad \text{or} \qquad 5x \qquad \text{or} \qquad 4x - 7$$

can be treated as a single number.

The algebraic conventions stated in Chapter 1 are restated below for convenience.

ALGEBRAIC CONVENTIONS

1. Letters are written in alphabetical order, when possible.
2. In a product, a single letter or a single numeral is written to the left of an expression enclosed by grouping symbols.
3. In a sum, a term involving letters is written to the left of a constant or numerical term.
4. In a product, a numerical factor is written to the left of another factor.

EXERCISES 3.1A

1–18. Rewrite each of the following by using the symmetric axiom. (Exchange sides.)

SAMPLE PROBLEM 1	SAMPLE PROBLEM 2
$8 = 3x - 7$	$x - 7 = 5x + 3$
SOLUTION	SOLUTION
$3x - 7 = 8$	$5x + 3 = x - 7$

1. $5 = x$

2. $-3 = x$

3. $9 = 2x - 7$

4. $16 = 3x + 1$

5. $10 = 6 - x$

6. $0 = \frac{x}{4} + 5$

7. $7 - x = x + 5$

8. $x + 2 = 4x - 13$

9. $10 = x + y$

10. $6 = 2x - 3y$

11. $4 - x = -x + 4$

12. $-1(x) = -x$

13. $a(b + c) = ab + ac$

14. $ab - ac = a(b - c)$

15. $-1(x - y) = y - x$

16. $1 = \frac{x}{x}$

17. $b - a = -1(a - b)$

18. $x = \frac{x}{1}$

1. ______________
2. ______________
3. ______________
4. ______________
5. ______________
6. ______________
7. ______________
8. ______________
9. ______________
10. ______________
11. ______________
12. ______________
13. ______________
14. ______________
15. ______________
16. ______________
17. ______________
18. ______________

19–30. Use the commutative axiom(s) to rewrite each of the following in conventional form.

SAMPLE PROBLEM 3

$(3 + x)2$

SOLUTION

First, we treat $3 + x$ as a single number.

$(3 + x)2 = 2(3 + x)$	commutative axiom, multiplication
$2(3 + x) = 2(x + 3)$	commutative axiom, addition $(3 + x = x + 3)$

The conventional form is $2(x + 3)$.

SAMPLE PROBLEM 4

$4 + yx$

SOLUTION

First, we treat yx as a single number.

$4 + yx = yx + 4$	commutative axiom, addition
$= xy + 4$	commutative axiom, multiplication $(yx = xy)$

The conventional form is $xy + 4$.

19. dc

20. $d + c$

21. $x7$

22. $5 + x$

23. $(y + 5)2$

24. $2 + yx$

25. $3 + (y + x)$

26. $(6 + x)3$

27. $(5x + 4) + 3x$

28. $(yx)8$

29. $(c + b)a$

30. $(2y - 9) + 4y$

19. ______________

20. ______________

21. ______________

22. ______________

23. ______________

24. ______________

25. ______________

26. ______________

27. ______________

28. ______________

29. ______________

30. ______________

B. ASSOCIATIVE AXIOMS, REARRANGEMENT

The associative axiom for addition states that the terms of a sum may be regrouped without changing the value of the sum.

$a + (b + c) = (a + b) + c$

For example,

$$65 + (35 + 87) = (65 + 35) + 87$$
$$= 100 + 87 = 187$$

The associative axiom for multiplication states that the factors of a product may be regrouped without changing the value of the product.

$a(bc) = (ab)c$

For example,

$$4(25 \cdot 56) = (4 \cdot 25)56$$
$$= (100)56 = 5600$$

Subtraction and division are *not* associative operations. For example, $15 - (10 - 2) \neq (15 - 10) - 2$, since $7 \neq 3$.

Also, $24 \div (6 \div 2) \neq (24 \div 6) \div 2$, since $8 \neq 2$.

However, by treating subtraction as an addition of an opposite, we can apply the associative axiom when subtraction is involved. Recall that $a - b = a + (-b)$.

$$(15 - 10) - 2 = [15 + (-10)] + (-2) = 15 + [-10 + (-2)]$$
$$(15 - 10) - 2 = 15 + (-10 - 2)$$

When the commutative and associative axioms for addition are combined, they imply that the **terms of a sum can be rearranged** in any order without changing the value of the sum.

When the commutative and associative axioms for multiplication are combined, they imply that the **factors of a product can be rearranged** in any order without changing the value of the product.

EXERCISES 3.1B

1–10. Regroup each of the following by using an associative axiom. Then compute the value of both the original and the final expression.

SAMPLE PROBLEM 1	SAMPLE PROBLEM 2
$3 + (7 + 9)$	$4(25 \cdot 9)$
SOLUTION	SOLUTION
$3 + (7 + 9) = (3 + 7) + 9$ $3 + 16 = 10 + 9$ $19 = 19$	$4(25 \cdot 9) = (4 \cdot 25)9$ $4(225) = (100)9$ $900 = 900$
SAMPLE PROBLEM 3	SAMPLE PROBLEM 4
$(87 + 85) + 15$	$(17 \cdot 8)125$
SOLUTION	SOLUTION
$(87 + 85) + 15 = 87 + (85 + 15)$ $172 + 15 = 87 + 100$ $187 = 187$	$(17 \cdot 8)125 = 17(8 \cdot 125)$ $(136)125 = 17(1000)$ $17{,}000 = 17{,}000$

1. $4 + (6 + 7)$ **2.** $(3 + 2) + 6$

3. $(3 + 9) + 5$ **4.** $25 + (75 + 87)$

5. $3(7 \cdot 4)$ **6.** $2(50 \cdot 8)$

7. $(21 \cdot 4)5$ **8.** $(57 \cdot 25)4$

9. $(45 + 55) + 87$ **10.** $(96 \cdot 8)125$

1. ________ **2.** ________
3. ________ **4.** ________
5. ________ **6.** ________
7. ________ **8.** ________
9. ________ **10.** ________

11–20. Rearrange the terms of a sum and the factors of a product to comply with the conventions stated in this section.

SAMPLE PROBLEM 5	SAMPLE PROBLEM 6	SAMPLE PROBLEM 7
$b + 5 + a$	$x3y$	$(5 + x)4$
SOLUTION	SOLUTION	SOLUTION
$a + b + 5$	$3xy$	$4(x + 5)$
SAMPLE PROBLEM 8	SAMPLE PROBLEM 9	
$(8b)(2ca)$	$7 + (yx + 2z)$	
SOLUTION	SOLUTION	
$16abc$	$xy + 2z + 7$	

11. $2 + y + x$

12. $c4d$

13. $sr5$

14. $(6 + x)7$

15. $(y + 2 + x)5$

16. $y + 3 + x$

17. $(2y)(3x)$

18. $(3q)(5p)$

19. $(y + x)2$

20. $5 + yx + z$

11. ________ 12. ________
13. ________ 14. ________
15. ________ 16. ________
17. ________ 18. ________
19. ________ 20. ________

21–40. Rearrange the terms so that letter terms are written to the left of numerical terms and letters are in alphabetical order.

SAMPLE PROBLEM 10	SAMPLE PROBLEM 11
$5x + 4 - 3x$	$5x - 3y - 2x + 7y$
SOLUTION	**SOLUTION**
$5x + 4 + \mathbf{(-3x)}$	$5x + \mathbf{(-3y)} + \mathbf{(-2x)} + 7y$
$5x + (-3x) + 4$	$5x + (-2x) + (-3y) + 7y$
$5x - 3x + 4$	$5x - 2x - 3y + 7y$

Note that subtraction was replaced by addition of the opposite, $a - b = a + (-b)$. Note also that the final result is written in terms of subtraction by using the same definition of subtraction.

21. $7x + 2 + 5x$

22. $4y + 6 + 8y$

23. $6 + 8x - 10y$

24. $8y + 5 - 2y$

25. $2x + y + 4x + 3y$

26. $9x - 2y - 5x + 6y$

27. $3x + 7 - x - 10$

28. $6t - 9 + 8 - 2t$

29. $c - a + b$

30. $c + a - b$

31. $c - a - b$

32. $4 - x$

21. ________________
22. ________________
23. ________________
24. ________________
25. ________________
26. ________________
27. ________________
28. ________________
29. ________________
30. ________________
31. ________________
32. ________________

33. $6 - y$

34. $-5 + x$

35. $y - 2 + x + 7$

36. $b - 3 + a - 4$

37. $-y + x$

38. $2t - 5 + 4t$

39. $8t - r - 6t + 3r$

40. $x - y - x + y$

33. ____________________

34. ____________________

35. ____________________

36. ____________________

37. ____________________

38. ____________________

39. ____________________

40. ____________________

EXERCISES 3.1S

1–10. Rewrite each of the following by using the symmetric axiom. (Exchange sides.)

1. $15 = 2x + 1$

2. $0 = 3y - 12$

3. $y = 6y - 25$

4. $2x = 3x + 7$

5. $ab + ac = a(b + c)$

6. $a(b - c) = ab - ac$

7. $y - x = -1(x - y)$

8. $5x - 2x = (5 - 2)x$

9. $8(x + y) = 8x + 8y$

10. $-1(5 - t) = t - 5$

1. __________
2. __________
3. __________
4. __________
5. __________
6. __________
7. __________
8. __________
9. __________
10. __________

11–20. Use the commutative axiom(s) to rewrite each of the following in compliance with the stated algebraic conventions.

11. $4 + dc$

12. $(yx)8$

13. $4 + (d + c)$

14. $(c - d)4$

15. $5(b + a)$

16. $(y + x)3$

17. $y(3x)$

18. $ts + 7$

19. $t + sr$

20. $b(4 + a)$

11. ______ 12. ______

13. ______ 14. ______

15. ______ 16. ______

17. ______ 18. ______

19. ______ 20. ______

21–30. Apply an associative axiom to each of the following, and then compute the value of both the original and the final expressions.

21. $15 + (10 + 5)$

22. $8 + (7 + 4)$

21. __________

22. __________

NAME __________ DATE __________ COURSE __________

23. $(68 + 35) + 65$

24. $(59 + 49) + 51$

25. $3(4 \cdot 5)$

26. $4(25 \cdot 9)$

27. $(47 \cdot 2)50$

28. $(92 \cdot 4)25$

29. $8(125 \cdot 25)$

30. $43 + (57 + 39)$

23. ______
24. ______
25. ______
26. ______
27. ______
28. ______
29. ______
30. ______

31–40. Rearrange the terms of a sum and the factors of a product to comply with the conventions stated in this section.

31. $yx2$

32. $d + (b + c)$

33. $y + 5 + x$

34. zxy

35. $n(1 + m)$

36. $n7k$

37. $m5n$

38. $(5y)(3x)$

39. $(4 + x + y)3$

40. $(b + a)6$

31. ______
32. ______
33. ______
34. ______
35. ______
36. ______
37. ______
38. ______
39. ______
40. ______

41–48. Rearrange the terms so that letter terms are written to the left of numerical terms and letters are in alphabetical order.

41. $6x + 9 + 2x$

42. $7y + 4 - 3y$

43. $10 - 5y + 2x$

44. $y - 1 - x$

45. $4x - 6 + 5x - 2$

46. $8x + 5 - 3x - 8$

47. $5x + 3y - 2x - 6y$

48. $6a + 6b - 5a - 7b$

41. ______
42. ______
43. ______
44. ______
45. ______
46. ______
47. ______
48. ______

DISTRIBUTIVE PROPERTIES 3.2

The distributive axiom and related distributive properties play an important role in the solution of many linear equations. They are used for removing parentheses and collecting like terms, the topics of this section.

A. COLLECTING LIKE TERMS

Like terms are terms whose letter factors are identical. Like terms can be collected by using the distributive axiom.

Like terms have the same letters. Also, the exponents on a letter must be the same.

$3x$ and $-7x$ are **like terms** and

$$3x - 7x = 3x + (-7)x = [3 + (-7)]x = -4x$$

$3x^2$ and $5x^2$ are **like terms** and

$$3x^2 + 5x^2 = (3 + 5)x^2 = 8x^2$$

$2xy$ and $7xy$ are **like terms** and

$$2xy - 7xy = 2xy + (-7)xy = [2 + (-7)]xy = -5xy$$

$2xy$ and $7yx$ are **like terms** and

$$2xy + 7yx = 2xy + 7xy = (2 + 7)xy = 9xy$$

The terms $3x^2$ and $5x$ are **unlike terms**, and $3x^2 + 5x$ cannot be combined by adding 3 and 5. Similarly, $3x$ and 5 are unlike terms, and the sum $3x + 5$ must be left as a sum of two terms. As another example, $5x - 6y$ must be left as a difference of two unlike terms.

The process of collecting like terms can be extended to three or more terms.

$$4y + 3y + y = 4y + 3y + 1y = (4 + 3 + 1)y = 8y$$
$$6x - 2x - 5x = (6 - 2 - 5)x = -1x = -x$$

The distributive properties that follow are useful for collecting like terms.

DISTRIBUTIVE PROPERTIES FOR COLLECTING LIKE TERMS

$$ba + ca = (b + c)a$$
$$ba - ca = (b - c)a$$
$$ad + bd + cd = (a + b + c)d$$

EXERCISES 3.2A

1–10. Rewrite as a single term.

SAMPLE PROBLEM 1
$3x + 4x$

SOLUTION
$(3 + 4)x = 7x$

SAMPLE PROBLEM 2
$5x - x$

SOLUTION
$5x - 1x = (5 - 1)x = 4x$

SAMPLE PROBLEM 3
$7x + 5x + x$

SOLUTION
$(7 + 5 + 1)x = 13x$

SAMPLE PROBLEM 4
$9x - 10x + 2x$

SOLUTION
$(9 - 10 + 2)x = (1)x = x$

1. $5x + 2x$

2. $6y - 3y$

3. $-3x + 5x$

4. $6y - 5y$

5. $2x - 2x$

6. $-7z + z$

7. $y + y$

8. $3x + 2x + 5x$

9. $7y - 4y - 2y$

10. $7z - 5z + 6z$

1. ________ 2. ________

3. ________ 4. ________

5. ________ 6. ________

7. ________ 8. ________

9. ________ 10. ________

11–24. Simplify each of the following by collecting like terms.

SAMPLE PROBLEM 5
Simplify $5x + 2 - 3x - 5$.

SOLUTION
$(5x - 3x) + (2 - 5) = 2x - 3$

SAMPLE PROBLEM 6
Simplify $2x^2 - 15x + 6x - 15$.

SOLUTION

$$\begin{aligned} 2x^2 + (-15x + 6x) - 15 &= 2x^2 + (-9x) - 15 \\ &= 2x^2 - 9x - 15 \end{aligned}$$

SAMPLE PROBLEM 7
Simplify $x^2 + 3xy - 2yx + y^2$.

SOLUTION

$$\begin{aligned} x^2 + (3xy - 2yx) + y^2 &= x^2 + (3xy - 2xy) + y^2 \\ &= x^2 + (3 - 2)xy + y^2 \\ &= x^2 + 1xy + y^2 \\ &= x^2 + xy + y^2 \end{aligned}$$

11. $4x + 6 - 2x - 1$

12. $6x - 4 - 5x + 2$

13. $4y - 3y - 6$

14. $7x + 5 - 8x$

15. $10 - x + 5x$

16. $x^2 - 3x + 2x - 6$

17. $3y^2 - y - 6y + 2$

18. $2x + 3y + 3x$

19. $2s + 2t - s - t$

20. $3x - y + 2y - 3x$

21. $5xy + 4yx$

22. $-yx - 4xy$

23. $x^2 + 4xy - 5yx + y^2$

24. $3xy - 4xz + 5zx + 6yx$

11. ________ 12. ________

13. ________ 14. ________

15. ________ 16. ________

17. ________ 18. ________

19. ________ 20. ________

21. ________ 22. ________

23. ________ 24. ________

B. REMOVING PARENTHESES

The distributive properties $a(b + c) = ab + ac$ and $a(b - c) = ab - ac$ state that the products $a(b + c)$ and $a(b - c)$ can be rewritten as expressions that do not contain parentheses. This process, called removing parentheses, is useful in solving certain linear equations.

Note that $-5 = (-1)5$ and $-8 = (-1)8$. In general,

$$-c = (-1)c$$

Recall from the definition of subtraction,

$$b - c = b + (-c)$$

Although we write $b - c$ as a subtraction, we think of it as the addition of the opposite of c.

$$x - 3 \quad \text{means} \quad x + (-3)$$

$$2y - 4 \quad \text{means} \quad 2y + (-4)$$

Examples 1–4 show how the distributive axiom is used to remove parentheses.

EXAMPLE 1

Definition of subtraction	$4(x - 3) = 4(x + [-3])$
Distributing	$= 4(x) + 4(-3)$
Simplifying products	$= 4x + (-12)$
Rewriting as a subtraction	$= 4x - 12$

EXAMPLE 2

Definition of subtraction	$-6(2y - 4) = -6(2y + [-4])$
Distributing	$= -6(2y) + (-6)(-4)$
Simplifying products	$= -12y + 24$

EXAMPLE 3

Using $-c = (-1)c$	$-(2x + 5) = (-1)(2x + 5)$
Distributing	$= (-1)(2x) + (-1)(5)$
Simplifying products	$= -2x + (-5)$
Rewriting as a subtraction	$= -2x - 5$

EXAMPLE 4

Using $-c = (-1)c$	$-(y - 2) = (-1)(y - 2)$
Definition of subtraction	$= (-1)(y + [-2])$
Distributing	$= (-1)y + (-1)(-2)$
Simplifying products	$= -y + 2$

With practice, most of these steps can be done mentally.

In summary, the following distributive properties are valid:

DISTRIBUTIVE PROPERTIES FOR REMOVING PARENTHESES

$$a(b + c) = ab + ac$$
$$a(b - c) = ab - ac$$
$$-(b + c) = -b - c$$
$$-(b - c) = -b + c$$
$$a(b + c + d) = ab + ac + ad$$

EXERCISES 3.2B

Remove parentheses.

SAMPLE PROBLEMS 1–7	SOLUTIONS
1. $\mathbf{5}(x + 6)$	$= \mathbf{5}x + \mathbf{5}(6) = 5x + 30$
2. $\mathbf{3}(x - 2)$	$= \mathbf{3}x - \mathbf{3}(2) = 3x - 6$
3. $\mathbf{-2}(x + 4)$	$= \mathbf{-2}x + (\mathbf{-2})(4) = -2x - 8$
4. $\mathbf{-4}(2y - 7)$	$= \mathbf{-4}(2y) + (\mathbf{-4})(-7) = -8y + 28$
5. $-(x + 5)$	$= (\mathbf{-1})(x + 5) = \mathbf{-x - 5}$
6. $-(2y - 3)$	$= (\mathbf{-1})(2y - 3) = \mathbf{-2y + 3}$
7. $2(x + y - 5)$	$= 2x + 2y - 10$

1. $2(x + 4)$ **2.** $5(x - 3)$ **1.** ________ **2.** ________

3. $-3(x + 6)$ **4.** $-4(x - 1)$ **3.** ________ **4.** ________

5. $6(2x + 3)$ **6.** $7(3x + 4)$ **5.** ________ **6.** ________

7. $4(3y - 5)$ **8.** $y(y - 4)$ **7.** ________ **8.** ________

9. $-(6x + 7)$ **10.** $-9(2r + 1)$ **9.** ________ **10.** ________

11. $-8(t - 2)$ **12.** $-(4y - 5)$ **11.** ________ **12.** ________

13. $-(6y - 9)$ **14.** $-(5t + 2)$ **13.** ________ **14.** ________

15. $5(x + y - 3)$ **16.** $4(2x - 3y + 2)$ **15.** ________ **16.** ________

17. $3(x + 5y + 7)$ **18.** $10(5x - 6y - 8)$ **17.** ________ **18.** ________

19. $-3(x + 4y - 1)$ **20.** $-7(2x - 5y + 6)$ **19.** ________ **20.** ________

21. $-(x + y + 1)$ **22.** $-(x - y + 1)$ **21.** ________ **22.** ________

23. $-(2x - y - 2)$ **24.** $-(4x + 9y - 8)$ **23.** ________ **24.** ________

25. $2x(3x^2 - 2x - 1)$ **26.** $-x(x^2 + 3x + 4)$ **25.** ________ **26.** ________

27. $2x(x - 4)$ **28.** $3y(y + 5)$ **27.** ________ **28.** ________

29. $-4t(t + 7)$ **30.** $-5n(1 - n)$ **29.** ________ **30.** ________

C. REMOVING PARENTHESES AND COLLECTING LIKE TERMS

In solving some linear equations, it is necessary both to remove parentheses and to collect like terms. This can be done by using one or more of the distributive properties that are listed in the summary found in Section 3.2B.

EXERCISES 3.2C

Remove parentheses and collect like terms.

SAMPLE PROBLEM 1

$4x + 2(x - 5)$

SOLUTION

$4x + 2x - 10 = 6x - 10$

SAMPLE PROBLEM 2

$8 - (y + 6 - 2x)$

SOLUTION

$8 - y - 6 + 2x = 2x - y + 2$

SAMPLE PROBLEM 3

$3(x - 4) - 2(x - 7)$

SOLUTION

$$3x - 12 - 2x + 14 = (3x - 2x) + (-12 + 14)$$
$$= x + 2$$

1. $10 + 3(x - 2)$

2. $3x - (5x - 4)$

3. $7 + 3(y - 4)$

4. $2 - (6t - 5)$

5. $8x - 9(x - 4)$

6. $3(2y - 4) - 4y$

7. $6z - (9z - 5)$

8. $5 - (7 - 8n)$

9. $9 + (x - 4)$

10. $2(3 - x) - 4(x - 3)$

11. $3y - 2(x - 2y)$

12. $2y^2 - 3(y - 2) - y^2$

13. $5(a - b) + 2(2a + 3b)$

14. $4(3x - 2y) - 3(4x - 3y)$

15. $7(r + s) - (r - s)$

16. $-(y - x)$

17. $x^2 + x - (x^2 - x)$

18. $-3(r + s - 2t)$

19. $5x - (7y + 5x - 4)$

20. $\dfrac{5 - 4x - 5}{-4}$

1. ________ 2. ________
3. ________ 4. ________
5. ________ 6. ________
7. ________ 8. ________
9. ________ 10. ________
11. ________ 12. ________
13. ________ 14. ________
15. ________ 16. ________
17. ________ 18. ________
19. ________ 20. ________

EXERCISES 3.2S

1–10. Rewrite as a single term.

1. $5x + x$ **2.** $7y - y$

3. $4y - 7y$ **4.** $-4x + 3x$

5. $-5x + 5x$ **6.** $4n + 3n$

7. $x + x$ **8.** $15x - 2x - 12x$

9. $2t - 4t + 2t$ **10.** $12k - 5k + 3k$

1. ________ 2. ________
3. ________ 4. ________
5. ________ 6. ________
7. ________ 8. ________
9. ________ 10. ________

11–30. Simplify.

11. $5x + 7 + 3x - 9$ **12.** $3y - y + 5$

13. $6x - 2 - 6x$ **14.** $10 + 2x - 12$

15. $6 - 2t - 3t$ **16.** $y^2 + 3y - 2y - 6$

17. $5k^2 + 5k - 8k - 8$ **18.** $3x + 5y - 5x$

19. $5s + 3t - 3s - 3t$ **20.** $6n + 3m + 3n$

21. $7xy - 6yx$ **22.** $-yx - 4xy$

23. $r^2 - 7rs - sr - s^2$ **24.** $2ab + ac - 2ca - ba$

25. $6xyz - 4yzx - 5zxy$ **26.** $x^2 + x + x$

27. $y + y + xy$ **28.** $t + t - 6$

29. $x + 6 + x - 6$ **30.** $8 - y - 3 - y$

11. ________ 12. ________
13. ________ 14. ________
15. ________ 16. ________
17. ________ 18. ________
19. ________ 20. ________
21. ________ 22. ________
23. ________ 24. ________
25. ________ 26. ________
27. ________ 28. ________
29. ________ 30. ________

31–50. Remove parentheses.

31. $3(y + 6)$ **32.** $-4(x + 7)$

33. $7(x - 9)$ **34.** $-6(x - 8)$

35. $2x(x - 5)$ **36.** $-3y(4 - y)$

37. $-(5x + 8)$ **38.** $-(6y - 9)$

39. $5(2x - 3y + 6)$ **40.** $-8(4x + 5y - 2)$

31. ________ 32. ________
33. ________ 34. ________
35. ________ 36. ________
37. ________ 38. ________
39. ________ 40. ________

NAME ________ DATE ________ COURSE ________

41. $-(r - s - t)$

42. $-(r - 2s + 1)$

43. $2x(x^2 - x - 1)$

44. $-4y(2 + y - y^2)$

45. $-a(a + b + c)$

46. $ab(a^2 - ab - b^2)$

47. $xy(x^2 - 5xy - 6y^2)$

48. $-2xy(x + 7y - 1)$

49. $4x(x^2 - 2x - 1)$

50. $-2t(3t^2 + 4t - 5)$

41. ______ 42. ______

43. ______ 44. ______

45. ______ 46. ______

47. ______ 48. ______

49. ______ 50. ______

51–70. Remove parentheses and collect like terms.

51. $6x - 2(x + 5)$

52. $12 + 4(y - 5)$

53. $8 - 5(x + 2)$

54. $9t - 7(t + 1)$

55. $7x - 6(x - 1)$

56. $5y - (3y + 4)$

57. $6 - (4n + 9)$

58. $4x - (3 + 3x)$

59. $4(x - 3) - 3(x - 2)$

60. $2x + 3(x + 2y)$

61. $x^2 - 2(x + 3) + 5x$

62. $3(x + y) - 2(x - y)$

63. $8(r - 2s) - 3(2r + s)$

64. $5(x + 2y) - 2(2x + 5y)$

65. $c + d - (d - c)$

66. $-(9 - x)$

67. $-2(x - y + z)$

68. $7 - (y - 4x - 5)$

69. $-(2x + 3 - 3x)$

70. $-(9 - 8x + 7y)$

51. ______ 52. ______

53. ______ 54. ______

55. ______ 56. ______

57. ______ 58. ______

59. ______ 60. ______

61. ______ 62. ______

63. ______ 64. ______

65. ______ 66. ______

67. ______ 68. ______

69. ______ 70. ______

SOLUTION, EQUIVALENCE AXIOMS, EQUATIONS 3.3

A. ONE-STEP EQUATIONS

A **solution of an equation in one variable** is a constant such that the equation becomes a true statement when the variable is replaced by the constant.

A solution of an equation is also called a **root** of the equation. An equation may have no roots, exactly one root, or more than one root.

The **solution set** of an equation in one variable is the set of all solutions of the equation.

Equivalent equations are equations that have the same solution set.

The solution set of an equation having the form $x = a$ is $\{a\}$.

For each pair of equations in Examples 1–2, determine whether the equations are equivalent or not. The set stated with each problem contains all the solutions to both equations.

EXAMPLE 1

$2x - 6 = 2$ and $x = 4$; $\{2, 4, 6, 8, 10\}$

Solution

The only solution of $x = 4$ is 4.
For $x = 4$ and $2x - 6 = 2$,

$$2(4) - 6 = 2$$
$$8 - 6 = 2$$
$$2 = 2, \text{ true; 4 is a solution.}$$

For $x = 2$, 6, 8, or 10, it may be seen that $2x - 6 = 2$ is false.
Thus, the equations are equivalent, and $\{4\}$ is the solution set of each.

EXAMPLE 2

$x^2 = 9$ and $x = 3$; $\{3, -3\}$

Solution

The solution set of $x = 3$ is $\{3\}$.
For $x = 3$ and $x^2 = 9$, $3^2 = 9$, $9 = 9$, true; 3 is a solution.
For $x = -3$ and $x^2 = 9$, $(-3)^2 = 9$, $9 = 9$, true; -3 is a solution.
The solution set of $x^2 = 9$ is $\{3, -3\}$.
Since the set $\{3\}$ is not the same as $\{3, -3\}$, the equations are not equivalent.

The **basic idea** in solving a linear equation is to change the given equation into an equivalent equation having the form $x = a$.

Method. Use an equivalence axiom that involves the operation which is inverse to the one stated in the problem.

Inverse operations

1. Addition and subtraction are inverse operations of each other.
2. Multiplication and division are inverse operations of each other.

Equivalence axioms. (These axioms state which operations may be performed on an equation without changing its solution set.) Let $A = B$ be an equation in one (or more) variable(s). Let C be any rational number.

1. **Addition axiom.** $A + C = B + C$ is equivalent to $A = B$. (The same number may be added to each side.)
2. **Subtraction axiom.** $A - C = B - C$ is equivalent to $A = B$. (The same number may be subtracted from each side.)
3. **Multiplication axiom.** $AC = BC$ where $C \neq 0$ is equivalent to $A = B$. (Each side may be multiplied by a number different from zero.)
4. **Division axiom.** $\frac{A}{C} = \frac{B}{C}$ where $C \neq 0$ is equivalent to $A = B$. (Each side may be divided by a number different from zero.)

EXERCISES 3.3A

Solve each equation, state which equivalence axiom was used, and check the answer by substituting into the original equation.

SAMPLE PROBLEMS 1–5	SOLUTIONS	CHECK
1. $x - 4 = 5$	$(x - 4) + 4 = 5 + 4$ $x = 9$ Addition axiom	$9 - 4 = 5$ $5 = 5$, true

Note that 4 must be ***added*** to $x - 4$ to ***undo*** the operation of ***subtracting*** 4 from x. Adding 4 to each side produces the equivalent equation $x = 9$ where 9 can be recognized as the solution.

2. $x + 3 = 8$	$(x + 3) - 3 = 8 - 3$ $x = 5$ Subtraction axiom	$5 + 3 = 8$ $8 = 8$, true

Note that 3 must be ***subtracted*** from $x + 3$ to ***undo*** the operation of ***adding*** 3 to x. Subtracting 3 from each side produces the equivalent equation $x = 5$ where 5 can be recognized as the solution.

3. $\frac{x}{4} = 3$	$\frac{x}{4}(4) = 3(4)$ $x = 12$ Multiplication axiom	$\frac{12}{4} = 3$ $3 = 3$, true

Note that $\frac{x}{4}$ must be ***multiplied*** by 4 to ***undo*** the operation of ***dividing*** x by 4. Multiplying each side by 4 produces the equivalent equation $x = 12$ where 12 can be recognized as the solution.

4. $-3x = 21$	$\frac{-3x}{-3} = \frac{21}{-3}$ $x = -7$ Division axiom	$-3(-7) = 21$ $21 = 21$, true

Note that $-3x$ must be ***divided*** by -3 to ***undo*** the operation of ***multiplying*** x by -3. Dividing each side by -3 produces the equivalent equation $x = -7$ where -7 can be recognized as the solution.

5. $-x = 6$	$\frac{(-1)x}{-1} = \frac{6}{-1}$ $x = -6$ Division axiom	$-(-6) = 6$ $6 = 6$, true

Note that $-x = (-1)x$. Since $(-1)(-x) = (-1)(-1)x = x$, the multiplication theorem can also be used for this special case.

$$(-1)(-1)x = (-1)6 \quad \text{and} \quad x = -6$$

1. $x - 3 = 9$ **2.** $x + 4 = 8$

3. $8x = 24$ **4.** $\frac{x}{3} = 12$

5. $x + 7 = 2$ **6.** $x - 1 = -3$

1. ____________
2. ____________
3. ____________
4. ____________
5. ____________
6. ____________

7. $-x = 16$

8. $\frac{x}{5} = 10$

9. $x + 1 = 1$

10. $-x = -4$

11. $-3x = 18$

12. $\frac{x}{-2} = 6$

13. $y + 5 = -8$

14. $y - 6 = -8$

15. $y - 3 = -3$

16. $-5t = -20$

17. $\frac{-t}{4} = -8$

18. $5u = 0$

19. $-x = 50$

20. $-10x = 4000$

7. ____________
8. ____________
9. ____________
10. ____________
11. ____________
12. ____________
13. ____________
14. ____________
15. ____________
16. ____________
17. ____________
18. ____________
19. ____________
20. ____________

B. MULTIPLE-STEP EQUATIONS

BASIC SOLUTION METHOD

1. **Analyze the equation to determine the last operation that was done on the number named by the letter.**
2. **Do the inverse operation on each side.**
3. **Repeat Steps 1 and 2 until you have the form $x = a$.**
4. **Check the result by substituting into the original equation the value that has been found and by doing the operations in the exact order in which they are indicated.**

SAMPLE PROBLEM 1

Solve and check $3x - 7 = 11$.

ANALYSIS

First, x was multiplied by 3. Last, 7 was subtracted from the product.

We want to undo the operations that were done on x. Our goal is to obtain an equivalent equation having the form $x = a$ where a is the solution.

First, we **add 7** to each side to undo the **last** operation of **subtracting 7.**

Then, we divide each side by 3 to undo the operation of multiplying x by 3.

SOLUTION

$$3x - 7 = 11$$

Add 7 to each side. $\quad 7 = 7$

Divide each side by 3. $\quad \frac{3x}{3} = \frac{18}{3}$

Simplify each side. $\quad x = 6$

CHECK

Write the original equation. $\quad 3x - 7 = 11$

Replace x by the value found. $\quad 3(6) - 7 = 11$

Do the operations in specified order. $\quad 18 - 7 = 11$

True, the result checks. $\quad 11 = 11$

The solution is 6.

Note to the Student

Here is **another idea** you can use to help you with the analysis. Replace x by any number, say 2 or 5.
Evaluate the expression involving x.
Note which operation is done first, second, . . . , and last.
To solve the equation, **first** you must **undo** the **last operation.**

Here is **an analogy** that might help.
Suppose you are wrapping a gift. You put a small box inside a medium-sized box. Then you put these two boxes inside a large box.

To **unwrap** the gift, you open the boxes in the **reverse order.** First, you open the large box, then the medium-sized one, and last the small box.

SAMPLE PROBLEM 2

Solve and check $\dfrac{2x + 40}{5} = 12.$

ANALYSIS

We read that a number x was multiplied by 2, and then 40 was added. Last, the sum was divided by 5.
To undo the division by 5, we multiply each side by 5.

SOLUTION

Multiply each side by 5. $$5\left(\frac{2x + 40}{5}\right) = 5(12)$$

Simplify. $$2x + 40 = 60$$

Subtract 40 from each side. $$\begin{array}{rcr} 2x + 40 & = & 60 \\ -40 & & -40 \\ \hline 2x & = & 20 \end{array}$$

Divide each side by 2. $$\frac{2x}{2} = \frac{20}{2}$$

Simplify. $$x = 10$$

CHECK

Write the original equation. $$\frac{2x + 40}{5} = 12$$

Replace x by the value found. $$\frac{2(10) + 40}{5} = 12$$

Do the operations following the Order Rules. $$\frac{20 + 40}{5} = 2$$

$$\frac{60}{5} = 12$$

True, the result checks. $$12 = 12$$

The solution is 10.

Note: If the final statement of the check is false, the problem must be reworked. An error was made in the solution process.

Note to the Student

Need some help?

How would you evaluate $\dfrac{2x + 40}{5}$ for a given value of x?

Mentally, replace x by any value, say 5.
List the operations in the order they would be done.
Then list the inverse operations and reverse the order.

	Operations	Inverse Operation	Reversed Order	
	Multiply by 2.	Divide by 2.	→ Multiply by 5. ←	FIRST TO UNDO
	Add 40.	Subtract 40.	Subtract 40.	
LAST DONE →	Divide by 5. →	Multiply by 5. ↗	Divide by 2.	

SAMPLE PROBLEM 3

Solve and check $38 = 18 - 5x$.

SOLUTION

Exchange sides (if $A = B$, then $B = A$). $\quad 18 - 5x = 38$

Use $a - b = a + (-b)$. $\quad 18 + (-5)x = 38$

Subtract 18 from each side to undo the last operation, an addition of 18. $\quad \underline{-18 \qquad -18}$

$(-5)x = 20$

Divide each side by -5. $\quad \frac{(-5)x}{-5} = \frac{20}{-5}$

Simplify. $\quad x = -4$

CHECK

Write the original equation. $\quad 38 = 18 - 5x$

Replace x. $\quad 38 = 18 - 5(-4)$

Do the operations using the Order Rules. $\quad 38 = 18 - (-20)$

True, the result checks. $\quad 38 = 18 + 20 = 38$

The solution is -4.

EXERCISES 3.3B

Solve and check.

1. $3x - 5 = 7$
2. $5x + 42 = 7$
3. $\frac{x}{5} - 6 = 3$
4. $\frac{x}{3} + 7 = 5$
5. $\frac{x + 15}{4} = 6$
6. $\frac{x - 7}{2} = 3$

1. ____________________

2. ____________________

3. ____________________

4. ____________________

5. ____________________

6. ____________________

7. $42 = 6x - 30$

8. $6 = 2x + 14$

9. $\frac{5y}{2} + 8 = 3$

10. $\frac{3y + 7}{2} = 8$

11. $\frac{5y - 4}{8} = 2$

12. $\frac{2y}{3} - 3 = 9$

13. $32 - 4t = 12$

14. $30 = 12 - 6t$

15. $5 - \frac{u}{2} = 9$

16. $\frac{5 - u}{2} = 9$

17. $\frac{3(x - 2) + 9}{4} = 6$

18. $3\left(\frac{x + 4}{2} - 1\right) = 15$

19. $5\left(\frac{x}{2} + 7\right) - 4 = 6$

20. $\frac{4x - 5}{3} + 2 = 7$

7. ______

8. ______

9. ______

10. ______

11. ______

12. ______

13. ______

14. ______

15. ______

16. ______

17. ______

18. ______

19. ______

20. ______

EXERCISES 3.3 S

Solve and check.

1. $x + 5 = 12$
2. $y - 7 = 8$
3. $\frac{x}{5} = 15$
4. $5x = 30$
5. $x - 4 = -9$
6. $-6y = 42$
7. $t + 8 = 4$
8. $\frac{x}{-4} = 20$
9. $\frac{u}{-2} = -12$
10. $x + 1 = 0$
11. $y + 6 = 6$
12. $-3x = -24$
13. $-5t = 0$
14. $-x = -30$
15. $-10x = 20$
16. $\frac{u}{-10} = 20$
17. $-10v = -20$
18. $r + 5 = -5$
19. $s - 5 = -5$
20. $-x = 6$

1. ______
2. ______
3. ______
4. ______
5. ______
6. ______
7. ______
8. ______
9. ______
10. ______
11. ______
12. ______
13. ______
14. ______
15. ______
16. ______
17. ______
18. ______
19. ______
20. ______

NAME ______ DATE ______ COURSE ______

21. $3x + 7 = 25$

22. $2x + 15 = 35$

23. $5x - 17 = 23$

24. $13 = 4x - 15$

25. $7 + \frac{x}{5} = 4$

26. $\frac{x + 4}{3} = 3$

27. $\frac{3x - 7}{5} = 4$

28. $2y + 9 = 9$

29. $\frac{3y}{5} + 26 = 5$

30. $\frac{20 - t}{3} = 7$

31. $9 - \frac{2t}{3} = 5$

32. $8 = \frac{x}{3} + 5 + 6$

33. $\frac{2x - 7}{5} + 1 = 4$

34. $\frac{17 - 3x}{4} = 8$

35. $2 = \frac{14 - 4x}{7}$

36. $2 = 14 - \frac{4x}{7}$

21. ______________________

22. ______________________

23. ______________________

24. ______________________

25. ______________________

26. ______________________

27. ______________________

28. ______________________

29. ______________________

30. ______________________

31. ______________________

32. ______________________

33. ______________________

34. ______________________

35. ______________________

36. ______________________

GENERAL LINEAR EQUATIONS 3.4

A. EQUATIONS EQUIVALENT TO $x = a$

GENERAL PROCEDURE FOR SOLVING A LINEAR EQUATION

1. Remove parentheses.
2. Collect the like terms on the left side of the equation. Collect the like terms on the right side of the equation.
3. Add or subtract the same terms to each side to obtain the form $ax = b$, if possible.
4. For the form $ax = b$, where $a \neq 0$, divide each side by a.
5. Check by substituting into the original equation and doing the operations on each side according to the Order Rules.

SAMPLE PROBLEM 1

Solve and check $4(x - 2) = 19 - (2x - 3)$.

SOLUTION

Remove parentheses. $4x - 8 = 19 - 2x + 3$

Collect like terms. $4x - 8 = 22 - 2x$

Add $2x$ to each side. $6x - 8 = 22$

Add 8 to each side. $6x = 30$

Divide each side by 6. $\frac{6x}{6} = \frac{30}{6}$

$x = 5$

CHECK

$4(x - 2) = 19 - (2x - 3)$

$4(5 - 2) = 19 - (2 \cdot 5 - 3)$

$4(3) = 19 - (10 - 3)$

$12 = 19 - 7$

$12 = 12$

The solution is 5.

SAMPLE PROBLEM 2

Solve and check $x + 7 - 2(x + 2) = 4(x - 8)$.

SOLUTION

	$x + 7 - 2(x + 2) = 4(x - 8)$
Remove parentheses.	$x + 7 - 2x - 4 = 4x - 32$
Collect like terms.	$-x + 3 = 4x - 32$
Exchange sides.*	$4x - 32 = -x + 3$
Add 32 to each side.	$4x = -x + 35$
Add x to each side.	$5x = 35$
Divide each side by 5.	$x = 7$

CHECK

$x + 7 - 2(x + 2) = 4(x - 8)$

$7 + 7 - 2(7 + 2) = 4(7 - 8)$

$14 - 2(9) = 4(-1)$

$14 - 18 = -4$

$-4 = -4$, true; value checks.

The solution is 7.

*This step is optional. It is useful to exchange sides when the coefficient of the variable on the right side is positive and greater than the coefficient of the variable on the left side. This avoids working with a minus sign, a common source of error.

EXERCISES 3.4A

Solve and check by completing the format.

1. Given: $4(2x - 5) = 5x - (3x + 2)$ **1.** ______________

1. Remove parentheses.
2. Collect like terms.
3. Add or subtract terms on each side to obtain $ax = b$.
4. Solve $ax = b$.
5. Check.

2. Given: $5 + 4x = 5(x - 2) + 8$ **2.** ______________

1. Remove parentheses.
2. Collect like terms.
3. Add or subtract terms on each side to obtain $ax = b$.
4. Solve $ax = b$.
5. Check.

3. $20 - (5x - 6) = 6 - 10(x + 1)$

4. $3x - 5 = 3 - x$

5. $4x - 7 = 7x + 2$

6. $2(y - 3) = 3(13 - y)$

7. $10 - (y + 4) = 2y - 15$

8. $7x - 5(x + 2) = 6$

9. $18x - 4(2x - 8) = 12$

10. $5x - 3 = 2(2x + 5)$

11. $5 - 2z = 24 - 3(z + 2)$

12. $2(x + 4) = 5x - 2(x - 3)$

3. ________

4. ________

5. ________

6. ________

7. ________

8. ________

9. ________

10. ________

11. ________

12. ________

B. EQUATIONS NOT EQUIVALENT TO $x = a$

Not all linear equations can be reduced to the form $x = a$. It is also possible for the solution set to be the empty set or the universal set. Examples of these cases are shown in Sample Problems 1 and 2.

SAMPLE PROBLEM 1

Solve and check $3(x + 5) = 25 - (10 - 3x)$.

SOLUTION

	$3(x + 5) = 25 - (10 - 3x)$
Remove parentheses.	$3x + 15 = 25 - 10 + 3x$
Collect like terms.	$3x + 15 = 15 + 3x$

Because $3x + 15 = 15 + 3x$ for all values of x, since it is an example of the commutative axiom for addition, the original equation is also true for all x. The solution set is Q, the set of rational numbers.

If the commutative axiom were not apparent, and if the equivalence axioms were applied, then

	$3x + 15 = 15 + 3x$
Subtract 15 from each side.	$3x = 3x$
Subtract $3x$ from each side.	$0 = 0$

Certainly at this point, it can be noted that $0 = 0$ is true and that it is true for all x.

CHECK

Select any value for x, say $x = 2$.

$3(x + 5) = 25 - (10 - 3x)$
$3(2 + 5) = 25 - (10 - 3 \cdot 2)$
$3(7) = 25 - 4$
$21 = 21$, and 2 checks.

SAMPLE PROBLEM 2

Solve and check $6x - 5(x - 2) = x + 26$.

SOLUTION

	$6x - 5(x - 2) = x + 26$
Remove parentheses.	$6x - 5x + 10 = x + 26$
Collect like terms.	$x + 10 = x + 26$
Subtract 10 from each side.	$x = x + 16$
Subtract x from each side.	$0 = 16$

Since $0 = 16$ is false, this means that the original equation is also false for all values of x. The solution set is $\varnothing$, the empty set.

CHECK

Since there is no number that can be substituted to obtain a true statement, another method must be used. One way is to rework the problem, very slowly and carefully, to see if the same result is obtained a second time.

> **If the original equation is equivalent to $0 = 0$, then the solution set is Q, the set of rational numbers. We may write Q, or all x.**

> **If the original equation is equivalent to a false statement, such as $0 = c$, where $c \neq 0$, then the solution set is $\varnothing$, the empty set. We write $\varnothing$.**

EXERCISES 3.4B

Solve and check.

1. $x + 2 = 5 + x$

2. $2x - (x - 3) = x + 3$

3. $2(3x + 6) = 3(2x + 4)$

4. $x = 8 - (5 - x)$

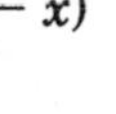

5. $6x + 2(x + 7) = 8x + 7$

6. $8x - 4(2x - 3) = 10$

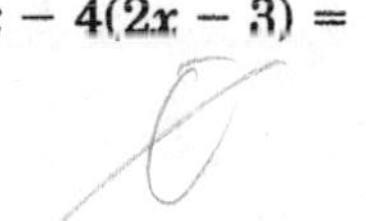

7. $2(3x - 1) - 3(2x - 1) = 1$

8. $9 - 2(x + 4) = 1 - 2x$

9. $-(5 - 3x) = 3x - 5$

10. $5(8x + 7) - 8(5x - 2) = 45$

1. ________________

2. ________________

3. ________________

4. ________________

5. ________________

6. ________________

7. ________________

8. ________________

9. ________________

10. ________________

11. $x - 3 = x$

12. $x - 3 = 3 - x$

13. $x - 3 = -(3 - x)$

14. $x - 3 = -3 - x$

15. $2(x + 4) = x + 6$

16. $2(x + 4) = x + 8$

17. $2(x + 4) = 2x + 4$

18. $2(x + 4) = 2x + 8$

19. $2(3x - 5) = 5(3x - 2)$

20. $2(3x - 5) = 3(2x - 5)$

21. $2(3x - 5) = 5x - 7$

22. $2(3x - 5) = 6x - 10$

11. ______________________

12. ______________________

13. ______________________

14. ______________________

15. ______________________

16. ______________________

17. ______________________

18. ______________________

19. ______________________

20. ______________________

21. ______________________

22. ______________________

EXERCISES 3.4 S

Solve and check.

1. $6(x + 9) = 5x - 2(x - 6)$
2. $2(x - 5) = 10 - (8 - 5x)$
3. $2x + 19 = 5x + 4$
4. $5x + 4 = 3x - 8$
5. $5(2y + 2) - 2(5y - 1)$
6. $7 - (y - 5) - 3 - 2y$
7. $4x - 3(x - 5) = 25$
8. $5(x - 4) - 6(x + 1) = 4$
9. $8x + 7 = 5(2x - 3)$
10. $7 - 3z = 12z - 8(z - 7)$

1. __________
2. __________
3. __________
4. __________
5. __________
6. __________
7. __________
8. __________
9. __________
10. __________

NAME __________ DATE __________ COURSE __________

11. $8x - (2x + 2) = 2(3x - 1)$ **11.** ______

12. $3(5t + 6) - 2(10t - 3) = 24$ **12.** ______

13. $32 - (7x + 3) = 8(3 - x)$ **13.** ______

14. $3(n + 2) - 4(n - 4) = 5(n - 10)$ **14.** ______

15. $7(n - 20) - 5(n - 22) = 2n$ **15.** ______

16. $5x - 3(x - 2) = 2(x + 3)$ **16.** ______

FORMULA PROBLEMS, LITERAL EQUATIONS 3.5

A. FORMULA PROBLEMS

There are many problems that can be solved by substituting one or more known values into a formula and then solving the equation for the unknown value.

SAMPLE PROBLEM 1

A total, T, of 890¢ was deposited in a bank.
It consisted of nickels, dimes, and quarters.
There were 24 nickels and 32 dimes.
How many quarters were there?

SOLUTION

Use the formula $5N + 10D + 25Q = T$
where N is the number of nickels, D is the number of dimes, and Q is the number of quarters.
From the problem, $T = 890$, $N = 24$, and $D = 32$.

Substitute. $5(24) + 10(32) + 25Q = 890$

Solve. $440 + 25Q = 890$

$25Q = 450$

$Q = 18$

There were 18 quarters.

SAMPLE PROBLEM 2

A student received the grades of 88, 96, and 85 on the first three tests. What grade should the student get on the fourth test so that the average, A, of the four tests will be 90?

Use $\dfrac{a + b + c + d}{4} = A$.

SOLUTION

$a = 88$, $b = 96$, $c = 85$, and $A = 90$.

Substitute. $\dfrac{88 + 96 + 85 + d}{4} = 90$

Solve. $269 + d = 360$

$d = 91$

The student should get a grade of 91.

SAMPLE PROBLEM 3

The perimeter of a rectangle is 68 inches.
The width is 15 inches. What is the length?

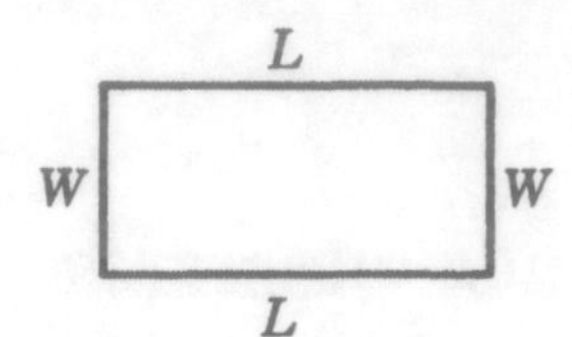

SOLUTION

Use the formula $P = 2W + 2L$, with $P = 68$ and $W = 15$.

Substitute. $2(15) + 2L = 68$

Solve. $2L = 38$

$L = 19$

The length is 19 inches.

EXERCISES 3.5A

1. The length, L, of a rectangle is 25 cm.
   The perimeter, P, is 86 cm.
   Find the width, W, of the rectangle.
   Use $2W + 2L = P$.

   1. ______________

2. A rectangular picture frame is to be made from a piece of wood molding 32 inches long.
   The width, W, is to be 6 inches.
   Find the length.
   Use $2W + 2L = P$, with $P = 32$.

   2. ______________

3. A total, T, of 600 cents was deposited in a bank in the form of nickels, dimes, and quarters. There were 15 dimes, D, and 14 quarters, Q. How many nickels, N, were there?
   Use $5N + 10D + 25Q = T$.

   3. ______________

4. A student received the grades of 80 and 92 on the first two tests. What must the grade be on the third test so that the average, A, of the three tests is 85?
   Use $\dfrac{a + b + c}{3} = A$.

   4. ______________

5. A student received the grades of 74, 78, and 83 on the first three tests. What should the grade be on the fourth test so that the average, A, of the four tests will be 80?
   Use $\dfrac{a + b + c + d}{4} = A$.

   5. ______________

6. A trip of 165 miles, d, is to be made traveling at an average rate, r, of 55 m.p.h. How many hours, t, will the trip take?
   Use $d = rt$.

   6. ______________

7. A piece of gold jewelry is made of 75% (P%) gold. What is the carat rating, K, of this jewelry?
Use $P = \dfrac{25K}{6}$.

7. ______________

8. The cost, C, for 6 yards, n, of material was $24. What was the price, p, per yard?
Use $C = np$.

8. ______________

9. The size, R, of a ring is related to the number of inches, L, in the circumference of the finger by the formula

$$R = 8L - 10$$

What is the circumference of a finger for the ring size 6?

9. ______________

10. The cost, C, to mail a package first class was 207 cents. How many ounces, n, did the package weigh?
Use $C = 20 + 17(n - 1)$.

10. ______________

11. A 12-foot-long piece of low fencing is to be used to enclose a flower garden having the shape of a square.
   a. How long, s, is the length of each side?
   Use $P = 4s$, with $P = 12$.

   b. Find the area, A, of the garden. Use $A = s^2$.

11. a. ______________
    b. ______________

12. The cost, C, of covering an area of A sq ft with grass seed is given by

$$C = \frac{pA}{n}$$

where p is the price per bag of lawn seed and n is the area in square feet that can be covered by one bag. Find A for $C = \$40$, $p = \$2$, and $n = 9$ sq ft.

12. ______________

B. LITERAL EQUATIONS

A **literal equation** is an equation that contains two or more letters. For some applications, it is useful to solve such an equation for one of the letters. To do this, we treat the letter that is to be solved for as the variable, and we treat all other letters as constants. The methods used for solving linear literal equations are the same as those used for solving linear equations in one variable.

SAMPLE PROBLEM 1

Solve $7x + 4y = 28$ for y.

SOLUTION

$$7x + 4y = 28$$
$$-7x \qquad\qquad -7x$$

Subtract $7x$ from each side. $\quad 4y = 28 - 7x$

Divide each side by 4. $\quad \dfrac{4y}{4} = \dfrac{28 - 7x}{4}$

$$y = \frac{28 - 7x}{4}$$

SAMPLE PROBLEM 2

Solve $C = \dfrac{5(F - 32)}{9}$ for F.

SOLUTION

A. Mental analysis

1. Determine what was done to the number F.
   First, 32 was subtracted.
   Second, the difference was multiplied by 5 and divided by 9.
2. Perform the inverse operations in the reverse order.

B. Written solution

$$C = \frac{5}{9}(F - 32)$$

Multiply each side by 9. $\quad 9C = 5(F - 32)$

Divide each side by 5. $\quad \dfrac{9C}{5} = F - 32$

Add 32 to each side. $\quad \dfrac{9C}{5} + 32 = F$

Exchange sides. $\quad F = \dfrac{9C}{5} + 32$

Alternate solution

Remove parens. $\quad 9C = 5F - 160$

Exchange sides. $\quad 5F - 160 = 9C$

Add 160. $\quad 5F = 9C + 160$

Divide by 5. $\quad F = \dfrac{9C + 160}{5}$

SAMPLE PROBLEM 3

Solve $C = \dfrac{k - 5n}{k}$ for n.

SOLUTION

Multiply each side by k.	$Ck = k - 5n$
Add $5n$ to each side.	$Ck + 5n = k$
Subtract Ck from each side.	$5n = k - Ck$
Divide each side by 5.	$n = \dfrac{k - Ck}{5}$

EXERCISES 3.5B

Solve each equation for the variable indicated.

1. $y = x + 4$ for x
2. $y = 3x$ for x
3. $x = y - 5$ for y
4. $x = \dfrac{y}{2}$ for y
5. $x + y = 6$ for y
6. $x - y = 6$ for x
7. $x - y = 8$ for y
8. $2x + y = 8$ for x
9. $S = C + P$ for P
10. $rn = 72$ for n
11. $N = B - C$ for B
12. $A = \dfrac{H}{T}$ for H
13. $d = rt$ for t
14. $A = bh$ for b
15. $P = a + b + c$ for c
16. $I = Prt$ for r

1. ____________
2. ____________
3. ____________
4. ____________
5. ____________
6. ____________
7. ____________
8. ____________
9. ____________
10. ____________
11. ____________
12. ____________
13. ____________
14. ____________
15. ____________
16. ____________

17. $S = \dfrac{A + L}{2}$ for L

18. $P = 2W + 2L$ for L

19. $p = \dfrac{100B}{A}$ for B

20. $A = \dfrac{bh}{2}$ for h

21. $2x + y = 10$ for y

22. $x - 3y = 6$ for x

23. $2x + y = 10$ for x

24. $2x + 3y = 12$ for y

25. $x - y = 1$ for y

26. $y - x = 1$ for x

27. $4x - 5y = 20$ for x

28. $4x - 5y = 20$ for y

29. $3x + 4y + 12 = 0$ for y

30. $ax + by + c = 0$ for y

31. $A = \dfrac{a + b + c}{3}$ for c

32. $V = abc$ for c

33. $D = \dfrac{C - S}{n}$ for C

34. $R = 8L - 10$ for L

35. $C = 20 + 17(n - 1)$ for n

36. $L = a + (n - 1)d$ for n

37. $T = m(g - a)$ for a

38. $D = \dfrac{P - C}{3}$ for C

39. $F = 6(N - 14) + 470$ for N

40. $I = \dfrac{p(n + 1)L}{200}$ for p

17. ______________________

18. ______________________

19. ______________________

20. ______________________

21. ______________________

22. ______________________

23. ______________________

24. ______________________

25. ______________________

26. ______________________

27. ______________________

28. ______________________

29. ______________________

30. ______________________

31. ______________________

32. ______________________

33. ______________________

34. ______________________

35. ______________________

36. ______________________

37. ______________________

38. ______________________

39. ______________________

40. ______________________

1. A total amount, A, of \$17,800 is to be collected for the sale of a car. A is the sum of a down payment, D, of \$5800 and 60 monthly payments of M dollars each. Find M, the monthly payment.
Use $D + 60M = A$.

1. ____________

2. As a guideline to buying on credit, the maximum safe debt load, D, should be one third the difference between the annual take-home pay, P, and the annual cost, C, of food, housing, and clothing.

$$D = \frac{P - C}{3}$$

Find P for $D = \$5000$ and $C = \$9800$.

2. ____________

3. A worker receives a weekly paycheck of P dollars for n hours of work. The worker is paid a basic rate of b dollars per hour and "time and one half" for the hours over 40 that are worked.

$$P = 40b + \frac{3(n - 40)b}{2}$$

How many hours, n, were worked if the paycheck, P, was \$588 and the basic rate, b, was \$12 per hour?

3. ____________

4. A baseball team won 13 games, W, with a winning percentage, $P\%$, of 65%. How many games did it lose?
Use $\frac{100W}{W + L} = P$.

4. ____________

5. For the score, S, of a football team, a touchdown, T, scores 6 points, a conversion, C, 1 point, and a field goal, F, 3 points.

$$6T + C + 3F = S$$

A team had a score of 32 points. It made 3 touchdowns and 2 conversions. How many field goals did it make?

5. ____________

6. In preparing a $P\%$ solution from n grams of a pure drug,

$$n = \frac{PV}{100}$$

where V is the volume in milliliters to be prepared. Find the volume, V, for a 4% boric acid solution made from 20 grams of boric acid crystals.

6. ____________

NAME ____________ DATE ____________ COURSE ____________

7. The cost, C, to carpet a rectangular area with width, W ft, and length, L ft, for carpeting costing p dollars per square yard is given by

$$C = \frac{pWL}{9}$$

a. Find C for $W = 10$ ft, $L = 15$ ft, and $p = \$12$.
b. Find p for $C = \$192$, $W = 9$ ft, and $L = 12$ ft.

7. a. ______________

b. ______________

8. The cost, C, to paint an area of A sq ft with paint costing p dollars a gallon when 1 gal of paint covers G sq ft is given by

$$C = \frac{pA}{G}$$

a. Find C for $A = 1800$ sq ft, $p = \$9$, and $G = 450$ sq ft.
b. Find A for $C = \$24$, $p = \$8$, and $G = 400$ sq ft.

8. a. ______________

b. ______________

9. The number, C, of cubic yards of concrete needed to cover an area of A square feet to a depth of d inches is given by

$$C = \frac{Ad}{324}$$

a. Find C for $A = 405$ sq ft and $d = 4$ in.
b. Find d for $A = 216$ sq ft and $C = 4$ cu yd.

9. a. ______________

b. ______________

10. When the fuel tank of a car is full, an odometer reading of m miles is taken. After the car has been driven a certain distance, the fuel tank is filled with G gallons of fuel. Then a second odometer reading of M miles is taken. The miles per gallon, mpg, for the car is given by

$$mpg = \frac{M - m}{G}$$

a. Find mpg for $M = 25{,}972$ mi, $m = 25{,}740$ mi, $G = 8$ gal.
b. Find M for $m = 6489$ mi, $G = 12$ gal, and $mpg = 18$.

10. a. ______________

b. ______________

11. To repay a small loan of L dollars in n months, the borrower must pay $\frac{L}{n}$ dollars a month plus $p\%$ of the unpaid balance. The total interest paid is given by

$$I = \frac{p(n + 1)L}{200}$$

a. Find I for $p\% = 2\%$, $L = \$450$, and $n = 9$ months
b. Find p for $I = \$117$, $n = 12$ mo, and $L = \$600$.

11. a. ______________

b. ______________

12. The formula given in this problem is used as a guideline in deciding if one should invest in tax-free bonds. For a person in the $t\%$ income tax bracket, a tax-free bond having a yield rate of $s\%$ will produce the same net income as a taxable investment having a yield rate of $r\%$ if

$$r = \frac{100s}{100 - t}$$

a. Find r for $t = 30$ and $s = 7$.
b. Find s for $t = 50$ and $r = 10$.
c. Find t for $r = 10$ and $s = 6$.

12. **a.** ____________
b. ____________
c. ____________

13–30. Solve each equation for the variable indicated.

13. $3x + y = 7$ for y

14. $2x - y = 6$ for y

13. ____________

14. ____________

15. $2x + 3y + 12 = 0$ for x

16. $5x - 2y - 10 = 0$ for x

15. ____________

16. ____________

17. $mx - y + b = 0$ for y

18. (Ohm's law)
$E = IR$ for I

17. ____________

18. ____________

19. (Perimeter of rectangle)
$P = 2W + 2L$ for W

20. (Area of trapezoid)
$A = \frac{1}{2}h(a + b)$ for b

19. ____________

20. ____________

21. (Arithmetic progression)
$L = a + (n - 1)d$ for d

22. (Ideal gas law)
$PV = RT$ for R

21. ____________

22. ____________

23. (Business profit)
$P = S - C$ for C

24. (Expansion of gases)
$L = a(1 + ct)$ for c

25. (Rockets, missiles)
$L = \dfrac{cdAv^2}{2}$ for A

26. (Psychology—intelligence quotient)
$Q = \dfrac{100M}{C}$ for M

27. (Chemistry—ionization)
$K = \dfrac{a^2c}{1 - a}$ for c

28. (Business—installments)
$M = \dfrac{C(12 + nr)}{12n}$ for r

29. (Simple interest)
$A = P + Prt$ for t

30. (Photography)
$(a + b)f = ab$ for f

23. ______________

24. ______________

25. ______________

26. ______________

27. ______________

28. ______________

29. ______________

30. ______________

NUMBER AND PERCENTAGE PROBLEMS 3.6

A. NUMBER PROBLEMS

In many cases, an applied problem is stated in words that can be translated into symbols forming an equation. The solution of the equation then provides an answer to the problem. Number relations are basic to many problems in many areas, such as business, economics, science, and engineering.

Some number relations occur quite often in applications. Knowing how to solve a general number problem provides preparation for solving problems dealing with specific subject matter.

The table below lists common expressions that occur in number problems. How to write these expressions in symbols is also shown.

Basic translations

x added to y	$x + y$	x subtracted from y	$y - x$
the sum of x and y	$x + y$	the difference between x and y when x is subtracted from y	$y - x$
x plus y	$x + y$	x minus y	$x - y$
x more than y	$x + y$	x less than y	$y - x$
x increased by y	$x + y$	x decreased by y	$x - y$
the product of x and y	xy	the quotient of x divided by y	$\frac{x}{y}$
x times y	xy		
x multiplied by y	xy	x divided by y	$\frac{x}{y}$
twice x, x doubled	$2x$	one half of x	$\frac{x}{2}$ or $\frac{1}{2}(x)$
thrice x, x tripled	$3x$	two thirds of x	$\frac{2x}{3}$ or $\frac{2}{3}(x)$
is, are, is equal to, is the same as, becomes, was, will be: =			

One way of solving stated number problems is described by the following method.

Method of short statements

1. **Choose a letter, such as x or n, as the name of the number whose value is to be found.**
2. **Rewrite the problem as shortly and simply as possible.**
3. **Translate the words into symbols. Use the list of basic translations provided, if necessary.**
4. **Solve the resulting linear equation.**

SAMPLE PROBLEM 1

One third the sum of a certain number increased by 4 is 12 less than twice this number. Find the number.

SOLUTION

1. Let x = the number.
2. Rewrite problem as a short statement. $\frac{1}{3}$(sum of x increased by 4) = 12 less than $2x$
3. Translate into symbols. $\frac{1}{3}(x + 4) = 2x - 12$
4. Solve.

$$\begin{aligned} x + 4 &= 3(2x - 12) \\ 3(2x - 12) &= x + 4 \\ 6x - 36 &= x + 4 \\ 5x &= 40 \\ x &= 8 \end{aligned}$$

SAMPLE PROBLEM 2

7 more than the product of a number and 5 is twice the difference of the number decreased by 4. Find the number.

SOLUTION

1. Let n = the number.
2. Rewrite the problem as a short statement. 7 more than $5n$ = 2(n decreased by 4)
3. Translate into symbols. $7 + 5n = 2(n - 4)$
4. Solve.

$$\begin{aligned} 7 + 5n &= 2n - 8 \\ 3n &= -15 \\ n &= -5 \end{aligned}$$

SAMPLE PROBLEM 3

When the sum of a number and 6 is subtracted from 15, the result is 3. Find the number.

SOLUTION

1. Let $x =$ the number.
2. $(x + 6)$ subtracted from $15 = 3$
3. $15 - (x + 6) = 3$
4. $15 - x - 6 = 3$

$$9 - x = 3$$
$$-x = -6$$
$$x = 6$$

EXERCISES 3.6A

1. 5 more than a certain number is equal to 5 times the difference obtained when 3 is subtracted from the number. Find the number. 1. ______

2. 1 more than 4 times a certain number is equal to 3 less than twice the number. Find the number. 2. ______

3. 4 times the difference of a number decreased by 4 is the same as 5 times the difference of the number decreased by 6. Find the number. 3. ______

4. When 10 times a number is decreased by twice the difference when 1 is subtracted from the number, the result is equal to 58. Find the number. 4. ______

5. 5 increased by the product of 4 and a certain number is the same as 17 decreased by twice the number. Find the number. 5. ______

$\frac{1}{100} = 0.01$

$$\begin{array}{r} 0.01 \\ 100\overline{)1.00} \\ 1\,00 \\ \hline \end{array}$$

B. PERCENTAGE PROBLEMS

Percent means **parts per hundred.**

$$35\% \quad \text{means} \quad \frac{35}{100}$$

$$p\% \quad \text{means} \quad \frac{p}{100}$$

Percentages are also written in decimal form. As an example, $35\% = \frac{35}{100} = 35(0.01) = 0.35$. Also, $4\% = \frac{4}{100} = 4(0.01) = 0.04$.

Basic percentage problems have the form

$$p\% \text{ of } A \text{ is } B$$

In symbols,

$$\frac{p}{100}(A) = B \quad \text{or} \quad \frac{pA}{100} = B$$

Note that the word **"of"** means **"multiply."**

EXERCISES 3.6B

1–12. Find the percentage.

SAMPLE PROBLEM 1

Find 35% of 500.

SOLUTION

$$\frac{35}{100}(500) = \frac{35(500)}{100} = 175$$

ALTERNATE SOLUTION

$0.35(500) = 175.00 = 175$

35% of 500 is 175.

SAMPLE PROBLEM 2

Find 6% of $25.

SOLUTION

$$\frac{6}{100}(25) = \frac{6(25)}{100} = \frac{150}{100} = 1.50$$

ALTERNATE SOLUTION

$(6)(0.01)(25) = 0.06(25) = 1.50$

6% of $25 is $1.50.

SAMPLE PROBLEM 3

Find the 5% sales tax on the $1200 purchase of a refrigerator.

SOLUTION

1. Rewrite as a short statement. — Find 5% of 1200.
2. Rewrite in symbols, and do the operations. — $\frac{5(1200)}{100} = 60$

The 5% sales tax on $1200 is $60.

SAMPLE PROBLEM 4
Find the 20% down payment on a house costing $90,000.

SOLUTION

1. Rewrite as a short statement. — Find 20% of 90,000.
2. Rewrite in symbols, and do the operations. — $\frac{20(90{,}000)}{100} = 18{,}000$

The down payment is $18,000.

1. Find 25% of 600.

2. Find 5% of $30.

3. Find 40% of 85.

4. Find 12% of $4000.

5. Find the 6% sales tax on the $24 purchase of a pair of jeans.

6. Find the 8% sales tax on the $500 purchase of a color TV set.

7. Find the 10% tip for a restaurant bill of $46.

8. Find the 15% tip for a taxi bill of $18.

9. Find the 20% down payment on the purchase of a car costing $8000.

10. Find the 5% down payment on the purchase of a home costing $85,000.

11. Find the 11% interest on $2500.

12. Find the 7.5% interest on $6000.

1. ____________
2. ____________
3. ____________
4. ____________
5. ____________
6. ____________
7. ____________
8. ____________
9. ____________
10. ____________
11. ____________
12. ____________

13–24. Find the number.

SAMPLE PROBLEM 5

6% of what number is 15?

SOLUTION

1. Rewrite in symbols. $6\% \cdot N = 15$
2. Change the percentage to a common fraction or a decimal. $\frac{6}{100}(N) = 15$ $\quad$ $0.06(N) = 15$
3. Solve the equation. $6N = 1500$, $N = 250$ $\quad$ $N = \frac{15}{0.06} \cdot \frac{100}{100} = \frac{1500}{6} = 250$

6% of 250 is 15.

SAMPLE PROBLEM 6

A 5% increase raised the monthly salary of an employee by $60. What was the old monthly salary?

SOLUTION

1. Rewrite in short form. 5% of what number is 60?
2. Write in symbols. $\frac{5N}{100} = 60$
3. Solve the equation. $5N = 6000$ and $N = 1200$

The old salary was $1200.

13. 15% of what number is 24?

14. 60% of what number is 480?

15. 40% of some number is 10. Find the number.

16. 3% of some number is 36. Find the number.

17. An 8% increase raised the weekly salary of an employee by $24. What was the old weekly salary?

18. A 10% decrease reduced the employees of a company by 700. How many employees were there before the decrease?

19. A student received a grade of 84% by answering 21 questions correctly. How many questions were on the test?

20. A baseball team won 60% of the games it played. The team won 21 games. How many games did it play?

13. ____________

14. ____________

15. ____________

16. ____________

17. ____________

18. ____________

19. ____________

20. ____________

21. A sewing machine is advertised at 30% off for a savings of $240. What was the original price of the machine?

22. How much money must be invested at 9% to earn $1800 interest?

21. ______________

22. ______________

23. A family budgets $350 a month for food. This is 20% of the monthly take-home pay. Find the monthly take-home pay.

24. The 10% rebate on the purchase of a car amounted to $715. What was the original cost of the car?

23. ______________

24. ______________

25–36. Find the percentage.

SAMPLE PROBLEM 7

What percent of 16 is 12?

SOLUTION

1. Rewrite in symbols. $\frac{p}{100}(16) = 12$
2. Solve the equation. $\frac{16p}{100} = 12$ and $p = \frac{1200}{16} = 75$

75% of 16 is 12.

SAMPLE PROBLEM 8

Over a 5-year period, the average payment on a home rose from $320 to $592. What percent of the old price is the increase of $272?

SOLUTION

1. Rewrite as a short statement. What percent of 320 is 272?
2. Write in symbols. $\frac{p(320)}{100} = 272$
3. Solve the equation. $p = \frac{27{,}200}{320} = 85$

The increase over the old price is 85%.

25. What percent of 12 is 3?

26. What percent of 1200 is 96?

25. ______________

26. ______________

27. An hourly salary increased from $10 to $12. What percent of the old salary was the $2 increase?

28. A tax of $14 was added to a purchase of $350. What percent of the purchase was the tax?

27. ______________

28. ______________

29. The price of a dress was reduced from $80 to $48. What percent of the original price was the $32 reduction?

30. Pure gold is rated as 24K (carats). What percent gold is in a piece of jewelry rated 18K?

31. A football team played 40 games and won 22 games. What percent of the games played did it win?

32. On a test of 80 questions, a student answered 72 questions correctly. What percent of the 80 questions did the student answer correctly?

33. The IRS allows a tax credit of $300 for the first $2000 spent by a taxpayer on residential energy conservation. What percent of the $2000 is this credit?

34. In a survey of 9000 persons, 3420 stated they were Republicans. What percent of the 9000 persons were Republicans?

35. A bank requires a $2500 down payment for a loan on the purchase of a new car costing $12,500. What percent of the cost of the car is the down payment?

36. The social security tax on a salary of $2000 was $133. What percent of the salary was this tax?

29. ____________

30. ____________

31. ____________

32. ____________

33. ____________

34. ____________

35. ____________

36. ____________

EXERCISES 3.6 S

1. When the product of 3 and the result of subtracting a certain number from 7 is increased by 12, the result is equal to the product of 8 and the number. Find the number.

1. ____________

2. One half the sum of 7 and a certain number is equal to 1 more than the given number. Find the number.

2. ____________

3. 5 more than the product of 3 times a given number is the same as 3 less than the product of 5 times the given number. Find the number.

3. ____________

4. 4 is the sum of a certain number and twice the difference obtained when 5 is subtracted from 3 times the number. Find the number.

4. ____________

5. 6 times the difference of a number decreased by 3 equals 5 times the difference of the number minus 5. Find the number.

5. ____________

6. 7 less than a certain number is 3 more than twice the number. Find the number.

6. ____________

NAME ____________ DATE ____________ COURSE ____________

7. Find 80% of 45.

8. Find 3% of $700.

7. ____________

8. ____________

9. 15% of what number is 90?

10. 125% of what number is 40?

9. ____________

10. ____________

11. What percent of 80 is 120?

12. What percent of 50 is 7?

11. ____________

12. ____________

13. The sale price of a lawnmower marked 40% off is 60% of the regular price of $300. Find the sale price.

14. A 30% decrease reduced the number of government employees in a certain area by 75. How many employees were there before the reduction?

13. ____________

14. ____________

15. The 10¢ pay phone call was increased to 20¢. What percent of the old cost was the increase of 10¢?

16. The Hertz Corporation found that for every 20 cars sold in 1983, 14 were used cars. What percent of the cars sold were used cars?

15. ____________

16. ____________

17. The sales tax in Omaha, Nebraska, is 4.5%. Find the sales tax on a $48 tire bought in Omaha.

18. The AAA suggests that this year a family of four should budget 3% more for a vacation trip by automobile than last year. What would be the increase for a trip that cost $800 last year?

17. ____________

18. ____________

19. The 15% commission made by a salesperson on the sale of furniture amounted to $180. What was the sales price of the furniture?

20. Paint originally priced at $20 a gallon went on sale for $12 a gallon. What percent of the original price was the decrease of $8?

19. ____________

20. ____________

AGE PROBLEMS (OPTIONAL), GEOMETRIC PROBLEMS 3.7

A. COMPONENT EXPRESSIONS

Age problems have little practical value, but they are useful in developing techniques that can be used in solving applied problems. As puzzles, these problems have fascinated people for hundreds of years. Number problems that are dated as early as 1650 B.C. are found in the Rhind papyrus, an Egyptian manuscript, and age problems occur in the *Greek Anthology*, a Greek work dated at about A.D. 500. Even today, age and number problems appear on the puzzle pages of various newspapers.

The following table summarizes common phrases used in age problems.

Age now	Age k years ago	Age k years from now	k years younger	k years older
a	$a-k$	$a+k$	$a-k$	$a+k$

EXERCISES 3.7A

1–14. Express each of the following in symbols.

1. The age of a person 5 years ago if his present age is 30. **1.** ______

2. The age of a person 5 years ago if his present age is x. **2.** ______

3. The age of a person x years ago if he is now 20. **3.** ______

4. The age of a person 12 years from now if his present age is 25. **4.** ______

5. The age of a person 12 years from now if his present age is x. **5.** ______

6. The age of a person x years from now if his present age is 24. **6.** ______

7. The age of a person 7 years younger than a man 40 years old. **7.** ______

8. The age of a person x years younger than a man 40 years old. **8.** ______

9. The age of a person 7 years younger than a man x years old. 9. ____________

10. The age of a person 10 years older than a child 8 years old. 10. ____________

11. The age of a person 10 years older than a child y years old. 11. ____________

12. The age of a person x years older than a child 8 years old. 12. ____________

13. The age of a person 6 years younger than a man $x + 3$ years old. 13. ____________

14. The age of a person 6 years older than a child whose age is $2x$. 14. ____________

15–18. Complete each table.

15.

Person	*Age now*	*Age 3 yr ago*	*Age 7 yr from now*
Girl	x		
Sister	$x - 5$		
Mother	$4x$		

16.

Building	*Age now*	*Age 10 yr ago*	*Age 50 yr from now*
Museum	x		
Church	$x + 20$		
Library	$x - 30$		

17.

Person	*Age now*	*Age 6 yr ago*	*Age 9 yr from now*
Boy	14		
Brother, x yr younger			
Sister, y yr older			

18.

Object	*Age now*	*Age 8 yr ago*	*Age 4 yr from now*
Table	x		
Chair, twice as old as table			
Desk, 5 yr older than table			

B. AGE PROBLEMS

Method

1. Let x = the present age of a person or object. (The equation is usually simplest when x is the present age or the youngest age.)
2. Make a table of the ages for the times mentioned in the problem.

3–4. Form an equation by translating a sentence from the problem, using expressions from the table.

5. Solve the equation.

SAMPLE PROBLEM 1

10 years from now John will be twice as old as he was 7 years ago. How old is John now?

Age now	*Age 10 yr from now*	*Age 7 yr ago*
x	$x + 10$	$x - 7$

SOLUTION

1. Let x = John's age now.
2. Make a table as follows:
3. Rewrite problem as a short statement.

   (John's age 10 yr from now) will be twice (John's age 7 yr ago).
4. Translate into symbols, using the table in Step 2.

   $$(x + 10) = 2(x - 7)$$
5. Solve the equation.

   $$2(x - 7) = x + 10$$
   $$2x - 14 = x + 10$$
   $$x = 24$$

John is 24 years old.

SAMPLE PROBLEM 2

A boy's sister is 3 years younger than he is. The boy's father is 26 years older than the boy. Four years ago, the father's age was 3 times the sum of the ages of his two children then. Find the present age of the father and his two children.

SOLUTION

Person	*Age now*	*Age 4 years ago*
Boy	x	$x - 4$
Sister	$x - 3$	$(x - 3) - 4 = x - 7$
Father	$x + 26$	$(x + 26) - 4 = x + 22$

(Father's age 4 years ago) was 3 times (sum of ages of children 4 years ago)

$$x + 22 = 3(x - 4 + x - 7)$$
$$x + 22 = 3(2x - 11)$$
$$6x - 33 = x + 22$$
$$5x = 55$$
$$x = 11,\ x - 3 = 8,\ x + 26 = 37$$

The father is 37. The boy is 11, and his sister is 8.

EXERCISES 3.7B

1. 20 years from now, a man will be 3 times as old as he was 8 years ago. Find his present age.

1. ______________

2. Nine years ago, Mary was one half as old as she is now. How old is Mary now?

2. ______________

3. Bill's brother is 2 years younger than Bill. Bill's sister is 5 years older than Bill. Five years ago, the sum of the ages of Bill and his brother was equal to their sister's age then. How old is Bill now?

3. ______________

4. When Mr. and Mrs. Smith were married, Mr. Smith was 2 years older than Mrs. Smith. On their 25th wedding anniversary, the sum of their ages was 100. How old were Mr. and Mrs. Smith when they were married?

4. ______________

5. A man is now 3 times as old as his daughter. Fourteen years from now the man will be twice the age of his daughter then. How old is the man now?

5. ______________

C. GEOMETRIC PROBLEMS

BASIC TECHNIQUE

1. **Make a well-labeled sketch of the figure(s) in the problem.**
2. **Let a letter such as x equal the measurement of one unknown quantity.**
3. **Express other unknown measurements in terms of x.**
4. **Find an equation by using a geometric formula that applies.**
5. **Solve the equation.**
6. **Check in the original problem.**

Geometric figures and formulas

Triangle

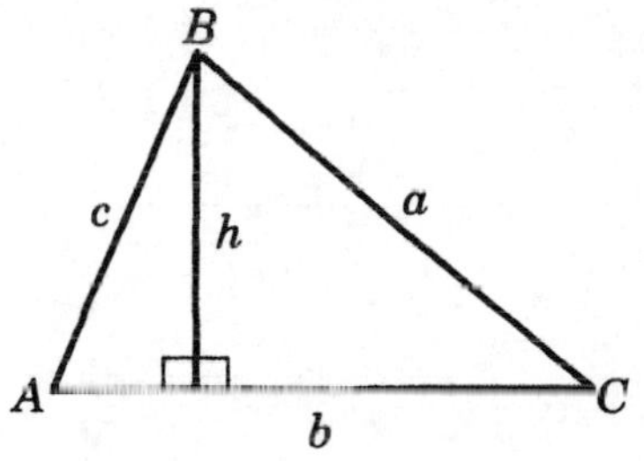

Rectangle

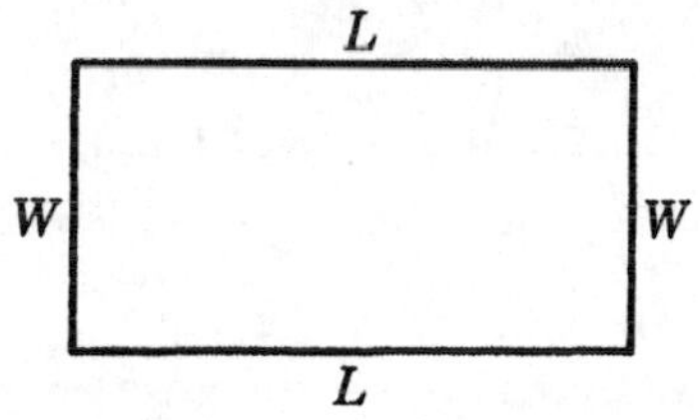

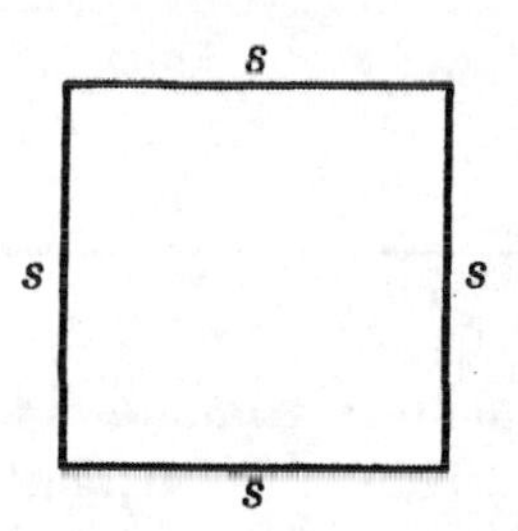

Square

	Triangle	Rectangle	Square
Angles.	$A + B + C = 180°$		
Perimeter.	$P = a + b + c$	$P = 2W + 2L$	$P = 4s$
Area.	$A = \dfrac{bh}{2}$	$A = WL$	$A = s^2$

EXERCISES 3.7C

1–6. Solve each perimeter problem.

SAMPLE PROBLEM 1

The length of a rectangle is 8 cm longer than its width. The perimeter of the rectangle is 112 cm. Find the width and length of the rectangle.

SOLUTION

1. Make sketch.
2. Rewrite each statement in symbols. $L = W + 8$, $P = 112$
3. Substitute in formula. $2W + 2L = P$
4. Solve equation.

$$2W + 2(W + 8) = 112$$
$$4W + 16 = 112$$
$$4W = 96$$
$$W = 24 \text{ cm, width}$$
$$\text{and } L = W + 8 = 32 \text{ cm, length}$$

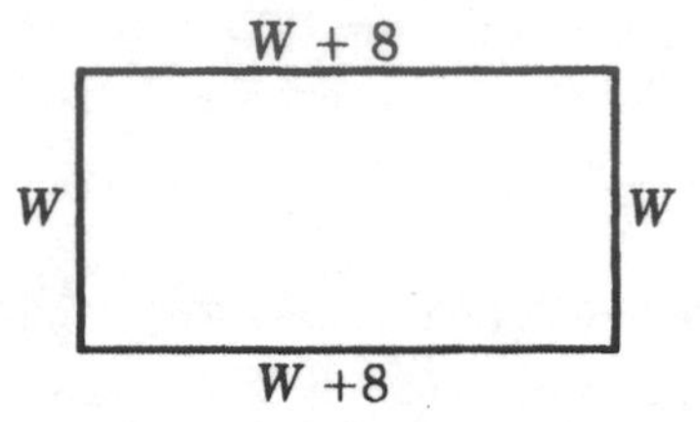

SAMPLE PROBLEM 2

The longest side of a triangle is twice the length of the shortest side. The third side of the triangle is 3 ft longer than the shortest side. The perimeter of the triangle is 23 ft. Find the length of each side of the triangle.

SOLUTION

Let a = length of shortest side.

1. Make sketch.
2. Rewrite in symbols. $b = 2a$ and $c = a + 3$, $P = 23$
3. Substitute in formula. $a + b + c = P$
4. Solve equation.

$$a + 2a + (a + 3) = 23$$
$$4a + 3 = 23$$
$$4a = 20$$
$$a = 5 \text{ ft}$$
$$b = 2a = 10 \text{ ft}$$
$$c = a + 3 = 8 \text{ ft}$$

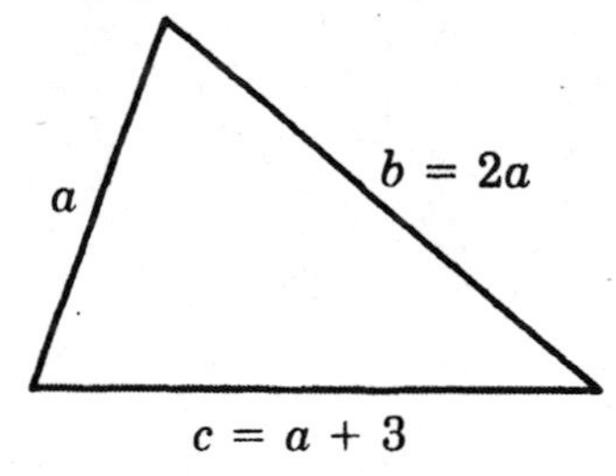

SAMPLE PROBLEM 3

A piece of wire 48 in long is to be bent into the shape of a rectangle whose length is 3 times its width. Find the dimensions of the rectangle.

SOLUTION

Let W = the width in inches.

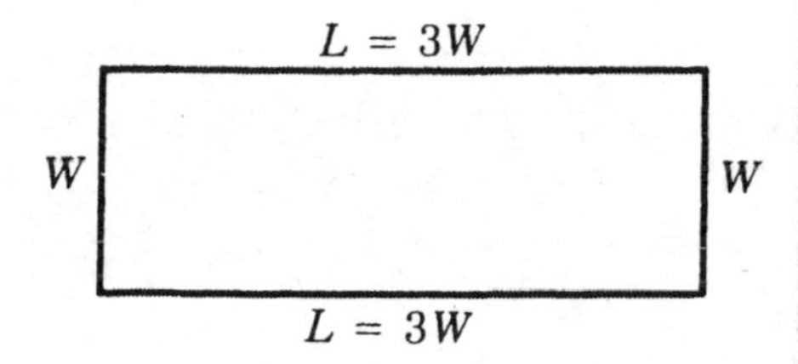

1. Make sketch.
2. Rewrite problem in symbols. $P = 48$ and $L = 3W$
3. Substitute in formula.

$$2W + 2L = 48$$
$$2W + 2(3W) = 48$$

4. Solve equation.

$$8W = 48$$
$$W = 6 \text{ in, width}$$
$$3W = 18 \text{ in, length}$$

The width and length are the dimensions of the rectangle.

1. The length of a rectangle is 9 in longer than its width. The perimeter of the rectangle is 50 in. Find the width and length.

1. 

$L = ?$

$W = x$

2. The length of a rectangle is 4 times its width. The perimeter of the rectangle is 150 meters. Find the width and length.

2. 

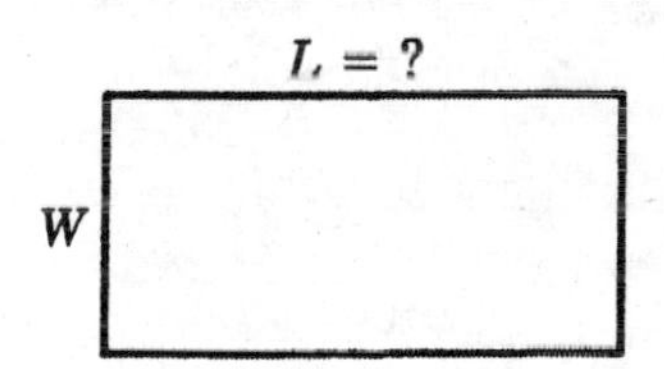

3. One side of a triangle is 10 ft longer than the shortest side. The third side of the triangle is 3 times as long as the shortest side. The perimeter of the triangle is 30 ft. Find the length of each side of the triangle.

3. 

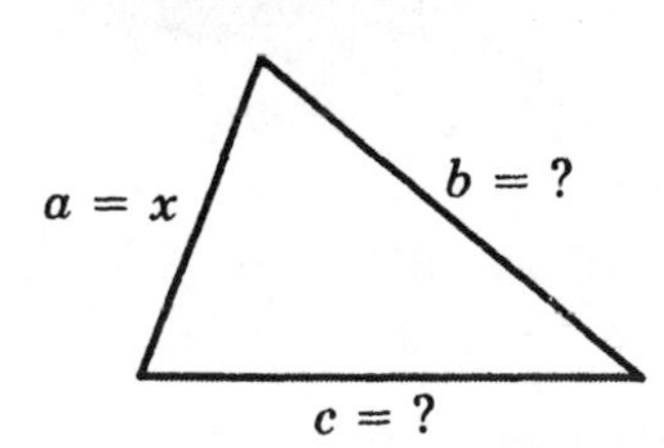

4. The perimeter of a triangle is 35 cm. Two of the sides are equal in length. The length of the third side is 5 cm less than twice the length of one of the equal sides. Find the length of each side of the triangle.

4. 

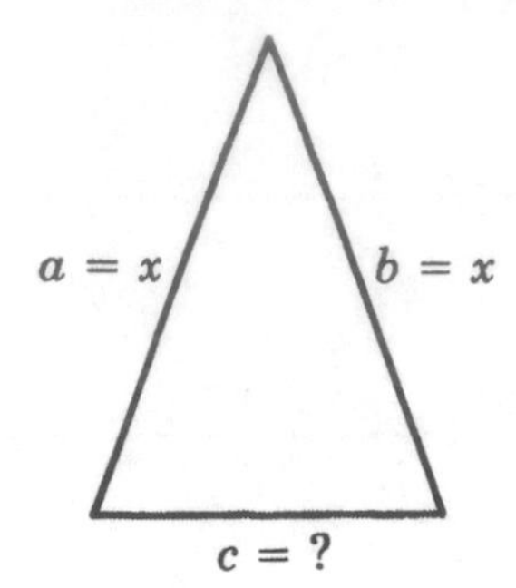

5. A farmer uses 180 ft of fencing to enclose a rectangular area that is 10 ft longer than it is wide. Find the length and width of the rectangular area.

5. 

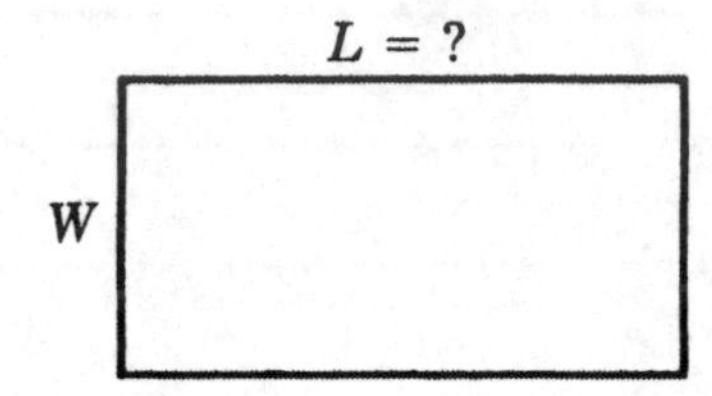

6. A piece of wire 36 cm long is to be bent into the shape of a triangle. The longest side is to be 6 cm longer than the shortest side, and the third side is to be 3 cm longer than the shortest side. Find the length of each side of the triangle.

6. 

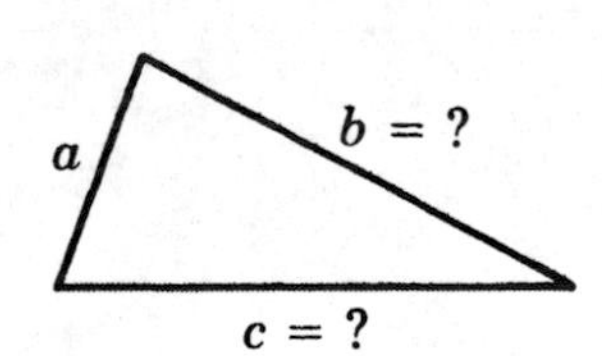

7–10. Solve each area problem.

SAMPLE PROBLEM 4

The cross section of an irrigation ditch has the shape of an isosceles triangle (two equal sides). Its base is 12 in, and its depth is 7 in. By how much should the depth be increased so that the resulting area of the cross section will be 78 sq in?

SOLUTION

Let x = amount to increase depth.
Then the base, $b = 12$, and the height, $h = x + 7$.
The area, $A = 78$.

Use $\frac{bh}{2} = A$ and substitute. $\quad \frac{12(x + 7)}{2} = 78$

Solve. $\quad 6x + 42 = 78$

$6x = 36$

$x = 6$

The depth should be increased by 6 in.

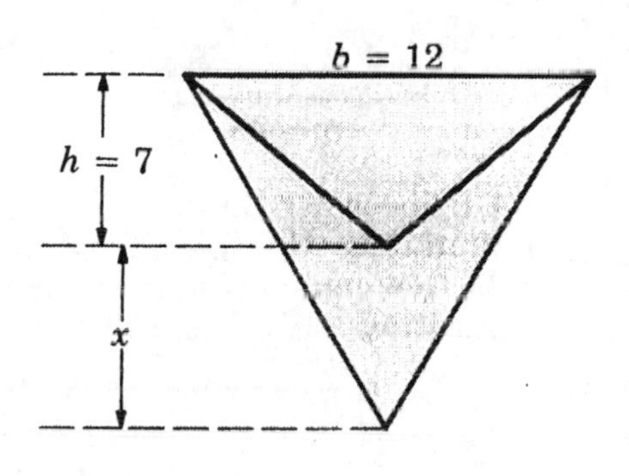

SAMPLE PROBLEM 5

A rectangular piece of glass has a width (W) of 20 cm and a length (L) of 28 cm. How much should its length be decreased so that the resulting area (A) is 480 sq cm?

SOLUTION

Let x = amount to decrease length.
Then $W = 20$ and $L = 28 - x$ and $A = 480$.

Use $WL = A$ and substitute. $\quad 20(28 - x) = 480$

Solve. $\quad 560 - 20x = 480$

$-20x = -80$

$x = 4$

The length should be decreased by 4 cm.

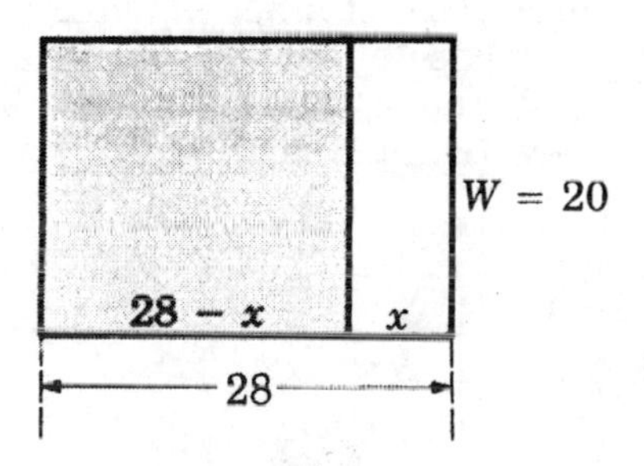

7. A window was designed to have the shape of a rectangle with a length of 75 in and a height of 10 in. To provide for more light, the window must be made higher so that the new area will be 900 sq ft. How much higher should the window be?

7. ________________

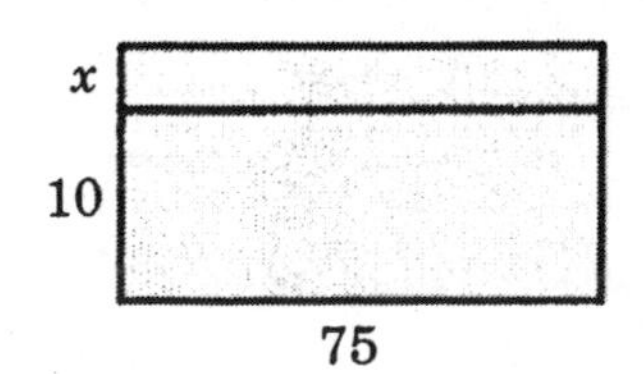

8. A triangular corner lot has a base of 28 meters and a height of 30 meters. By how much should its height be decreased so that the resulting area of the lot is 364 square meters?

8. 

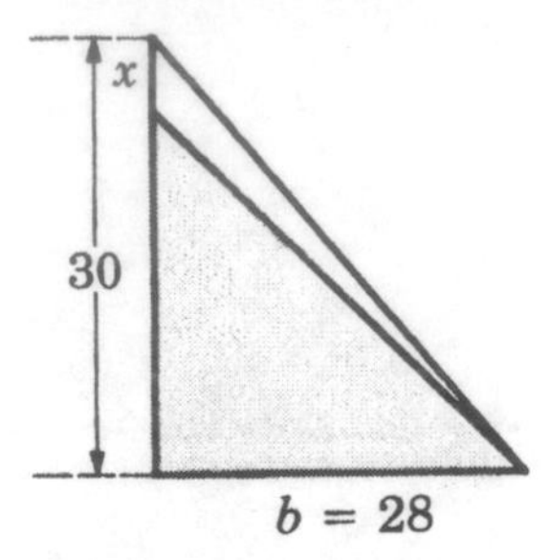

9. A homeowner has a rectangular strip of land 18 in wide and 90 in long. He has enough plants to cover an area of 1350 sq in. How much should the length be decreased so that the resulting strip has an area of 1350 sq in?

9. ____________

W = 18

x

90

10. To determine how much dirt fill to order to decrease the slope of a piece of land, a triangular cross section is used as shown. The base of the existing triangle is 3 ft, and the height is 6 ft. How much should the base be increased so that the area of the new triangle is 15 sq ft?

10. 

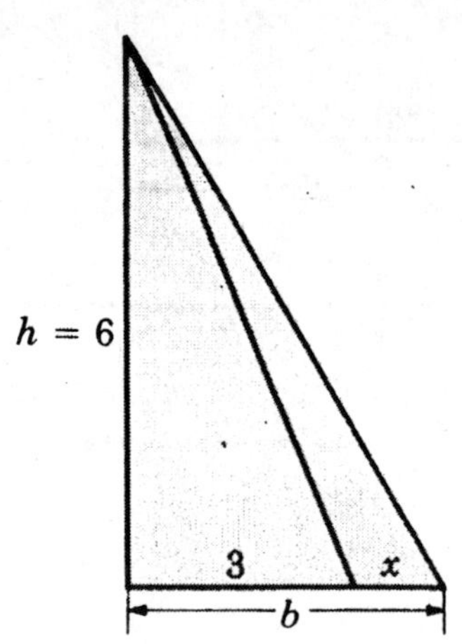

EXERCISES 3.7S

1. Fourteen years ago, Jack was one third as old as he will be 8 years from now. Find Jack's present age.

1. ________________

2. In 20 years, a building will be twice as old as it is now. In how many years will the building be 3 times as old as it is now?

2. ________________

3. When David was born, his father had a case of Amontillado wine laid down for him. This wine will be ready to drink 14 years from now, when David's age equals 3 times his present age. How old is David now?

3. ________________

4. Anne has a sister 3 years younger and a sister twice her age. If the sum of the ages of three girls is 29, how old is Anne?

4. ________________

5. Vanessa has a book that is 54 years older than she is. In 5 years from now, the book will be 4 times as old as Vanessa. How old is Vanessa?

5. ________________

NAME ____________ DATE ____________ COURSE ____________

6. Find the length of a rectangle whose length is twice its width and whose perimeter is 54 meters.

6. ____________

7. The length of a rectangle is 12 ft longer than its width. The perimeter of the rectangle is 60 ft. Find the width and length of the rectangle.

7. ____________

8. Thirty inches of metal casing are to be used to make a rectangular frame for a picture. The length of the frame is to be 3 inches less than twice the width. Find the width and length of the frame.

8. ____________

9. Ninety feet of wood molding are used around the perimeter of a rectangular room. If the room is twice as long as it is wide, find the dimensions of the room.

9. ____________

10. One side of a triangle is 8 inches less than 3 times the shortest side. The third side of the triangle is 1 inch more than twice the shortest side. The perimeter of the triangle is 35 inches. Find the length of each side of the triangle.

10. ____________

11. On a blueprint, each of the two equal sides of a triangle is twice as long as the third side. If 160 cm of metal stripping is specified to go around this triangle, find the length of each side.

11. ______________

12. A construction company used 360 lineal ft of chain link fencing to enclose a plot of land in the shape of a right triangle. The hypotenuse (the longest side) was 30 ft less than twice the length of the shortest side. The third side of the triangle was 30 ft longer than the shortest side. Find the lengths of the sides of the plot of land.

12. ______________

13. A piece of plastic has the shape of a triangle whose sides have lengths 8 dm, 9 dm, and 10 dm, respectively. Each side is to be increased by the same amount so that the perimeter of the resulting triangle is twice that of the original triangle. Find how much each side should be increased.

13. ______________

14. A rectangle has a 6-ft base and a 4-ft altitude. How much should the altitude be increased so that the new area is 42 sq ft?

14. ______________

15. A triangle has an 8-meter base and a 6-meter altitude. How much should the base be decreased so that the new area is 12 square meters?

15. ______________

16. A rectangular piece of wood is 100 cm long and 50 cm wide. A craftsperson has enough tiles on hand to cover an area of 4000 sq cm. By how much should the length of this wood be decreased so that the resulting area is 4000 sq cm?

16. 

17. A triangular window has a base of 4 ft and a height of 3 ft. By how much should the height be increased so that the new area is 10 sq ft?

17. 

18. A 36-in by 48-in rectangular picture is to have a frame of uniform width. The perimeter of the framed picture determines the length of the material needed for the frame. Find the width of the frame if the perimeter of the framed picture is 176 in.

18. ______

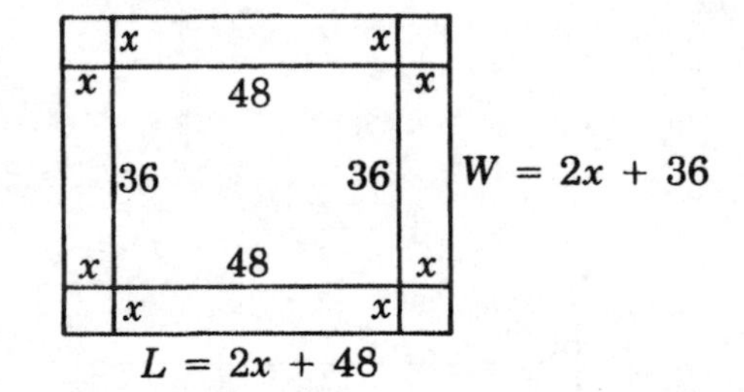

19. A path of uniform width is to be made around a 5-meter by 8-meter rectangular garden. The perimeter of the resulting rectangular area is to be 42 meters. Find the width of the path.

19. ______

20. The altitude of a triangle is 8 in shorter than its base. If the altitude is increased by 6 in, the area is increased by 60 sq in. Find the base and altitude of the original triangle.

20. ______

SAMPLE TEST FOR CHAPTER 3

1–4. Use the closure, commutative, and associative axioms to rewrite each of the following in conventional form. (See Section 3.1.)

1. $6 + sr$ **2.** $(3 + x)9$

3. $4 + y + x$ **4.** $c9x$

1. ________ 2. ________
3. ________ 4. ________

5–8. Collect like terms. (See Section 3.2.)

5. $10 - 2x - 3x - 12$ **6.** $2x^2 + 2x - 5x - 5$

7. $x^2 - 4yx + 2xy - 8y^2$ **8.** $6x^2y + 4xy^2 - 5yx^2$

5. ________ 6. ________
7. ________ 8. ________

9–12. Remove parentheses and collect like terms. (See Section 3.2.)

9. $30 - 4(x + 5)$ **10.** $5(x - 2) - (x - 6)$

11. $3x(2x - 1) - 4(2x - 1)$ **12.** $6x(x - 2y) + y(x - 2y)$

9. ________ 10. ________
11. ________ 12. ________

13–18. Solve and check. (See Section 3.3.)

13. $x - 35 = 65$ **14.** $x + 17 = 13$

15. $-6x = 48$ **16.** $\frac{-x}{5} = 45$

17. $25 - 4x = 5$ **18.** $\frac{6x + 18}{5} = 12$

13. ________ 14. ________
15. ________ 16. ________
17. ________ 18. ________

19–22. Solve and check. (See Section 3.4.)

19. $7x - 3(x + 5) = x + 3$ **20.** $4(x - 2) - 6(x - 4) = 26$

21. $2(6 - x) = 16 - 2(x + 2)$ **22.** $2x - (x - 1) = 5 - (5 - x)$

19. ________
20. ________
21. ________
22. ________

NAME ________ DATE ________ COURSE ________

23–25. Solve for the indicated letter. (See Section 3.5.)

23. $s = \dfrac{a + b + c}{2}$ for c

24. $9x - 2y = 18$ for y

23. ______

24. ______

25. Find the lower base, b, of a trapezoid if its area, A, = 40 sq ft, its height, h, = 5 ft, and its upper base, a, = 7 ft.

25. ______

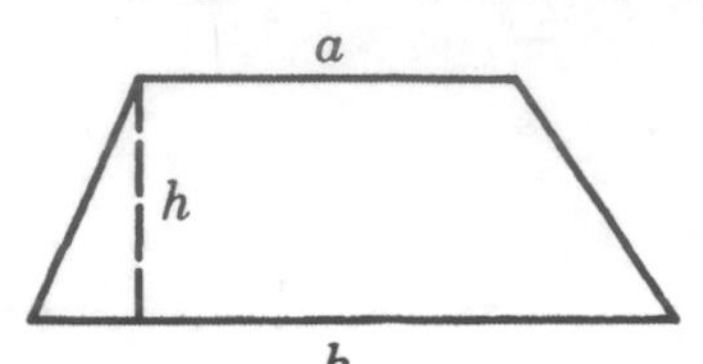

Use $A = \dfrac{h(a + b)}{2}$.

26. 6 times the sum of a number and 5 is 10 less than the number. Find the number. (See Section 3.6.)

26. ______

27–28. Solve each percentage problem. (See Section 3.6.)

27. What percent of the \$6000 price of a car is the \$1800 down payment?

28. A 5% raise increased the weekly salary of a worker by \$16. What was the weekly salary before the raise?

27. ______

28. ______

29. In 3 years Jay and Kay will celebrate their golden wedding anniversary (50th). Kay would not reveal her age. However, Jay said that he is 4 years older than Kay, and on their golden wedding day, the sum of their ages will be 3 times the sum of their ages on the day that they were married. How old are Jay and Kay? (See Section 3.7.)

29. ______

30. To carpet a rectangular room whose width is 4 ft less than its length, 45 ft of tackless stripping are needed. The stripping is placed at all edges of the floor except along the 3-ft-wide doorway. Find the dimensions of the room. (See Section 3.7.)

30. ______

Operations on Polynomials

ALGEBRA IN ACTION: WORK AND CAREERS

Algebra is used in many occupations, such as art, chemistry, architecture, medicine, and construction work.

Architects use algebra to find various measurements, such as the dimensions of a circular arch.

Circular Arch: $s^2 = 8rx - 4x^2$

Chapter 4 Objectives

Objectives	Typical Problems
Find the product of monomials.	1. Multiply. $(-3x^3y)(4x^2y^2)$
Add and subtract polynomials.	2. Add. $(5x^2 - 4x + 7) + (x^2 + 5x - 9)$ 3. Subtract. $(6x - 10y + 12) - (x - 10y - 2)$
Multiply polynomials.	4. Multiply. $-3x^2(2x^2 - 4x + 5)$ 5. Multiply. $(x - 4)(x^2 + 2x - 2)$ 6. Multiply. $(3x - 2)(2x + 5)$
Find the greatest common monomial factor of a polynomial.	7. Factor. $16x^3 - 40x^2 + 8x$
Factor a difference of squares.	8. Factor. $9x^2 - 16y^2$
Factor a simple trinomial.	9. Factor. $x^2 - 2x - 48$
Factor a perfect square trinomial.	10. Factor. $16x^2 - 56xy + 49y^2$
Factor a general trinomial.	11. Factor. $5x^2 + 7x - 6$
Completely factor special polynomials over the integers.	12–15. Factor, if possible. 12. $x^4 - 24x^2 - 25$ 13. $12x^3 - 14x^2 - 10x$ 14. $4x^4 - 64$ 15. $81x^4 - 72x^2 + 16$
Solve special polynomial equations by factoring.	16–18. Solve. 16. $x^2 = 36$ 17. $4x^2 - 28x = 0$ 18. $x^2 + x = 30$

ANSWERS to Typical Problems

1. $-12x^5y^3$ **2.** $6x^2 + x - 2$ **3.** $5x + 14$ **4.** $-6x^4 + 12x^3 - 15x^2$ **5.** $x^3 - 2x^2 - 10x + 8$ **6.** $6x^2 + 11x - 10$ **7.** $8x(2x^2 - 5x + 1)$ **8.** $(3x - 4y)(3x + 4y)$ **9.** $(x + 6)(x - 8)$ **10.** $(4x - 7y)^2$ **11.** $(5x - 3)(x + 2)$ **12.** $(x - 5)(x + 5)(x^2 + 1)$ **13.** $2x(2x + 1)(3x - 5)$ **14.** $4(x - 2)(x + 2)(x^2 + 4)$ **15.** $(3x - 2)^2(3x + 2)^2$ **16.** $x = \pm 6$ **17.** $x = 0$ or $x = 7$ **18.** $x = 5$ or $x = -6$

POLYNOMIALS

A. PRODUCT OF MONOMIALS

A **monomial** is a single numeral, a single letter, or an indicated product of constants and variables.

Examples of monomials are 5, x, $\frac{1}{2}x^3$, $-2xy^2$, and x^2y^3.

Expressions such as $\frac{2}{5x}$ and $3\sqrt{x}$ are **not** monomials. Note that each of these is **not** a **product** of numeral or letter factors.

Monomials are multiplied by rearranging the factors so that the numerical factors are written first (at the left) and the literal factors are written from left to right in alphabetical order, using the exponential form so that no letter is repeated.

The definition of x^n, where n is a natural number, is used in the multiplication of monomials.

DEFINITION OF x^n

If n is a natural number, x^n is defined as

$$x^1 = x \quad \text{and} \quad x^n = x \cdot x \ldots x \text{ (}n\text{ factors)}$$

As examples,

$$x^4 = xxxx \quad \text{and} \quad y^5 = yyyyy$$

SAMPLE PROBLEM 1

Multiply $(5x^3)(-2x^2)$.

SOLUTION

Rearrange factors. $\quad (5x^3)(-2x^2) = (5)(-2)(xxx)(xx)$

Simplify. $\quad = -10x^5$

SAMPLE PROBLEM 2
Multiply $(3xy^2)(8x^3y^4)$.

SOLUTION

Rearrange factors. $(3xy^2)(8x^3y^4) = (3)(8)(x)(x^3)(y^2)(y^4)$

Simplify. $= 24(xxxx)(yy)(yyyy)$

$= 24x^4y^6$

EXERCISES 4.1A

Multiply.

1. $(4x)(5x^3)$
2. $(-6y^2)(7y^4)$
3. $(-5y^3)(-8y^2)$
4. $(-25x^4)(-4x^3)$
5. $(9x^5)(-4x)$
6. $(7x^2)(7x^2)$
7. $(8x^3)(8x^3)$
8. $(-6y^3)(-6y^3)$
9. $(-9y^2)(-9y^2)$
10. $(x^2y)(xy^2)$
11. $(2xy)(-3x^2y)(4xy^2)$
12. $(2x)(2x)(2x)$
13. $(-3x^2)(-3x^2)(-3x^2)$
14. $-(5xy)^2$
15. $(-5xy)^2$
16. $(-4x^2)^3$
17. $-4(x^2)^3$
18. $3x(5x^2)^2$
19. $10x^2(2x)^3$
20. $-7x(-2x^2)^3$

1. ______
2. ______
3. ______
4. ______
5. ______
6. ______
7. ______
8. ______
9. ______
10. ______
11. ______
12. ______
13. ______
14. ______
15. ______
16. ______
17. ______
18. ______
19. ______
20. ______

B. POLYNOMIALS

A **polynomial** is a monomial or an algebraic sum of monomials.

As examples,

$5x^2 - 2x + 3$ is a polynomial in the variable x.

$2x + 3y - 6$ is a polynomial in two variables: x and y.

$x^2 + 2y^2 - z^2$ is a polynomial in three variables: x, y, and z.

A polynomial is written in **descending powers** of a variable when the highest power of the variable is written first, at the left, the second highest power is written next, and so on, with the lowest power written last, at the right.

A polynomial is written in **ascending powers** of a variable when it is written in the reverse of the order for descending powers.

As examples,

$5x^3 + 7x^2 - 6x + 3$ is written in descending powers of x.

$8 - 9y - y^5$ is written in ascending powers of y.

$4x^2y + 9xy^3 + y^2$ is written in descending powers of x.

EXERCISES 4.1B

Rewrite each of the following in descending powers of x.

SAMPLE PROBLEM 1	SAMPLE PROBLEM 2	SAMPLE PROBLEM 3
$7 - 2x^2 - 5x + 4x^3$	$6 + 5x - x^2$	$x^2y^4 + x^4y^3 - 9y^6 - 4xy^5$
SOLUTION	**SOLUTION**	**SOLUTION**
$4x^3 - 2x^2 - 5x + 7$	$-x^2 + 5x + 6$	$x^4y^3 + x^2y^4 - 4xy^5 - 9y^6$

1. $8 - 5x + 3x^2$
2. $x + x^3 - 1 - x^2$
3. $14x^2 - 49 - x^4$
4. $4x + 6x^2 + 7x^3 + 5x^4$
5. $x^2 - y^2 + 2x - 4y - 3$
6. $xy^2 - 8x^3y + y^4$
7. $9y^2 - 4x^2 + 8x - 6y - 10$
8. $1 - x - y - x^3 - y^3$
9. $5 - x^2 + 36y^2 + 6x - 72y$
10. $2x^3y - 7xy^3 + 4x^2y^2$

1. ______
2. ______
3. ______
4. ______
5. ______
6. ______
7. ______
8. ______
9. ______
10. ______

C. SUM OF POLYNOMIALS

A sum of two or more polynomials can often be simplified by first writing each polynomial in descending (or ascending) powers of one variable and then rearranging the terms so that like terms can be combined.

EXERCISES 4.1C

Simplify.

SAMPLE PROBLEM 1

Simplify $(3x^2 - 4 - 6x) + (8x - 7 + 2x^2)$.

SOLUTION

Rearrange terms.	$(3x^2 + 2x^2) + (-6x + 8x) + (-4 - 7)$
Collect like terms.	$5x^2 + 2x - 11$

SAMPLE PROBLEM 2

Simplify $(6x - 8y + 10) + (8y - 4x - 12)$.

SOLUTION

Rearrange terms.	$(6x - 4x) + (-8y + 8y) + (10 - 12)$
Collect like terms.	$2x + 0 + (-2) = 2x - 2$

1. $(5x^2 - 7x + 9) + (3x^2 - 2x - 8)$ — 1. ____________
2. $(4x - 5y + 8) + (3x + 2y - 5)$ — 2. ____________
3. $(8 - x^3 - x) + (x^3 + 1 - 4x)$ — 3. ____________
4. $(x + y - 6) + (y - x - 6)$ — 4. ____________
5. $(4x - 12y - 8) + (6x + 12y + 6)$ — 5. ____________
6. $(x^3 - x^2 + x) + (x^2 - x + 1)$ — 6. ____________
7. $(5x - 4y + 3) + (5y - 5x - 10)$ — 7. ____________
8. $(y + 3z - 4x) + (5z - 6y - 4x)$ — 8. ____________
9. $(y^4 + y - 3y^2 - 3) + (2y^2 + 7 - y)$ — 9. ____________
10. $(x^3 - 2x^2y + xy^2) + (2xy^2 - y^3 - x^2y)$ — 10. ____________

D. DIFFERENCE OF POLYNOMIALS

Informally, subtracting is done by changing the sign of the subtrahend and adding. It is important to remember, however, that when the subtrahend is a polynomial,

each term of the subtrahend must have its sign changed.

EXERCISES 4.1D

Simplify.

SAMPLE PROBLEM 1

Simplify $(8x^2 + 5x - 10) - (2x^2 + 7x - 5)$.

SOLUTION

Use $a - b = a + (-1)b$.	$(8x^2 + 5x - 10) + \mathbf{(-1)(2x^2 + 7x - 5)}$
Change sign of each term of b.	$(8x^2 + 5x - 10) + \mathbf{(-2x^2 - 7x + 5)}$
Rearrange terms.	$(8x^2 - 2x^2) + (5x - 7x) + (-10 + 5)$
Collect like terms.	$6x^2 - 2x - 5$

SAMPLE PROBLEM 2

Simplify $(5x - 2y - 7) - (3y + 5x - 8)$.

SOLUTION

Use $a - b = a + (-1)b$.	$(5x - 2y - 7) + \mathbf{(-1)(3y + 5x - 8)}$
Change sign of each term of b.	$(5x - 2y - 7) + \mathbf{(-3y - 5x + 8)}$
Rearrange terms.	$(5x - 5x) + (-2y - 3y) + (-7 + 8)$
Collect like terms.	$0 - 5y + 1$
	$-5y + 1$

1. $(9x^2 + 6x + 10) - (4x^2 + 2x + 7)$ 1. ______
2. $(8x^2 - 5x + 4) - (7x^2 - 2x + 8)$ 2. ______
3. $(3x - 6y - 1) - (4x + 3y - 6)$ 3. ______
4. $(2y - 5x + 7) - (5x - 2y - 8)$ 4. ______
5. $(x^4 - 12x^2 + 36) - (16x^2 - 8x + 1)$ 5. ______
6. $25 - (x^2 + 14x + 49)$ 6. ______
7. $(x + y - z) - (x - y + z)$ 7. ______
8. $(2x^2y - 3x^2y^2 - 6xy^2) - (4xy^2 - 3x^2y - x^2y^2)$ 8. ______
9. $(5x^2 - 6y^2 + 7) - (5x^2 + 6y^2 - 7)$ 9. ______
10. $(x^2 - xy + x) - (yx - y^2 + y)$ 10. ______

EXERCISES 4.1S

1–12. Multiply.

1. $(3x^2)(7x^4)$ **2.** $(8y^5)(-2y)$

3. $(-6t^3)(-5t^5)$ **4.** $(-4xy^2)(4x^2y)$

5. $(9x^2)(9x^2)$ **6.** $(10x^2)^2$

7. $(-5y^3)(-5y^3)$ **8.** $(-12y^3)^2$

9. $(5rs)(-2r^2s)(-3rs^2)$ **10.** $(-4t^2)(-4t^2)(-4t^2)$

11. $(-6t^2)^3$ **12.** $-2xy(-2xy)^2$

1. ______
2. ______
3. ______
4. ______
5. ______
6. ______
7. ______
8. ______
9. ______
10. ______
11. ______
12. ______

13–20. Write in descending powers of x.

13. $5 - 2x + 3x^2$ **14.** $1 + x^3 - 6x$

15. $10x^2 - 25 - x^4$ **16.** $4x - 2x^3 + x^5 - 10$

17. $x^4 - x - 4x^3 + 4$ **18.** $y^2 + 6y + x^2 + 4xy$

19. $5x - 2 - x^3$ **20.** $x^2y^2 + y^4 + x^4$

13. ______
14. ______
15. ______
16. ______
17. ______
18. ______
19. ______
20. ______

NAME ______ DATE ______ COURSE ______

21–40. Simplify.

21. $(4x^2 + 5x - 6) + (2x^2 - 4x - 3)$ **21.** ____________

22. $(1 - 7y - 8y^2) + (5y + 3y^2 - 1)$ **22.** ____________

23. $(x^3 + 2x^2 + x) + (x^2 + 2x + 1)$ **23.** ____________

24. $(2x - 3y - 5) + (3x + 3y - 10)$ **24.** ____________

25. $(4y - 8 - 2y^2) + (y^3 - 4y + 2y^2)$ **25.** ____________

26. $(y - x + 4) + (x + y - 7)$ **26.** ____________

27. $(2x - y + 3z - 5) + (2y - 3x - 3z - 4)$ **27.** ____________

28. $(x^2 + xy + y^2) + (y^2 - yx - 2x^2)$ **28.** ____________

29. $(x^3 - 3x^2y + 9xy^2) + (27y^3 - 9xy^2 + 3x^2y)$ **29.** ____________

30. $(x^4 + 2x^3 + 2x^2) + (-2x^3 - 4x^2 - 4x) + (2x^2 + 4x + 4)$ **30.** ____________

31. $(4x + 2y - 5z - 10) - (x + 5y - 5z + 15)$ **31.** ____________

32. $36x^2 - (y^2 - 10xy + 25x^2)$ **32.** ____________

33. $(x^3 + 7x^2 + 49x) - (7x^2 + 49x - 343)$ **33.** ____________

34. $(5x^2 - 3xy - 4y^2) - (4x^2 - 2xy + 3y^2)$ **34.** ____________

35. $(x^2 + 2xy - y^2) - (y^2 - 2yx + x^2)$ **35.** ____________

36. $(x^2y^2 - 18xy + 81) - (36y^2 - 18yx + x^2)$ **36.** ____________

37. $(1 - xy - xz) - (yz - tx)$ **37.** ____________

38. $(5 - x - y) - (x + y - 5)$ **38.** ____________

39. $(x^3 - 6x^2 + 9x) - (3x^2 - 18x + 27)$ **39.** ____________

40. $1 - (8 - x - y + z)$ **40.** ____________

PRODUCTS OF POLYNOMIALS 4.2

A. PRODUCT OF POLYNOMIAL AND MONOMIAL

The distributive axiom is used to multiply a polynomial by a monomial.

Each term of the polynomial is multiplied by the monomial.

SAMPLE PROBLEM 1

Multiply $5x^2(2x^3 - 3x^2 - 4x + 1)$.

SOLUTION

Distribute monomial to each term of polynomial.

$$\mathbf{5x^2}(2x^3 - 3x^2 - 4x + 1) = (\mathbf{5x^2})(2x^3) + (\mathbf{5x^2})(-3x^2) + (\mathbf{5x^2})(-4x) + (\mathbf{5x^2})(1)$$

Simplify each product.

$$= 10x^5 - 15x^4 - 20x^3 + 5x^2$$

SAMPLE PROBLEM 2

Multiply $4by(7y^2 - 8y + 1)$.

SOLUTION

Distribute monomial to each term of polynomial.

$$\mathbf{-4by}(7y^2 - 8y + 1) = (\mathbf{-4by})(7y^2) + (\mathbf{-4by})(-8y) + (\mathbf{-4by})(1)$$

Simplify each product.

$$= -28by^3 + 32by^2 - 4by$$

EXERCISES 4.2A

Multiply.

1. $2x(7x^2 - 3x - 4)$
2. $-4x(5x - 2y + 7)$
3. $5x^2(x^2 - 2xy + y^2)$
4. $6y^2(2y^2 - 3xy - x^2)$
5. $6xy(x^2 - 4xy + 3y^2)$
6. $7a(a^2 - 5ab + 1)$

1. ______________
2. ______________
3. ______________
4. ______________
5. ______________
6. ______________

7. $-9b(b^3 - b^2 - 6)$

8. $8x^2y^2(4x - 5y - 6)$

9. $-15a^2b^2(4a^2 - 4a + 1)$

10. $100x^2(0.05x^2 + 1.50x - 4.25)$

7. ____________________

8. ____________________

9. ____________________

10. ____________________

B. PRODUCT OF BINOMIALS, FOIL METHOD

A **binomial** is a polynomial having two terms. As examples,

$$3x + 4,\ 2x - 5y,\ x^2 - 9, \text{ and } x^2 - y^3 \text{ are binomials}$$

There is a special method, called the **FOIL method**, that is very useful for finding the product of two binomials.

Multiplying, by using the distributive axiom,

$$(A + B)(C + D) = A(C + D) + B(C + D)$$
$$= AC + AD + BC + BD$$

Note the following about the four terms of the product polynomial:

1. The **first** term, AC, is the product of the **first** terms of the factors.
2. The **last** term, BD, is the product of the **last** terms of the factors.
3. The **two middle terms** form the **sum** of the **inner product** and the **outer product** as shown below.

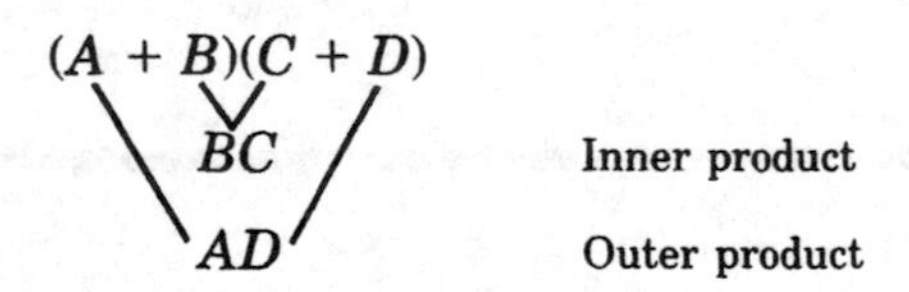

The word FOIL is a useful device for remembering the terms of this special product: F means first terms product, O means outer product, I means inner product, and L means last terms product.

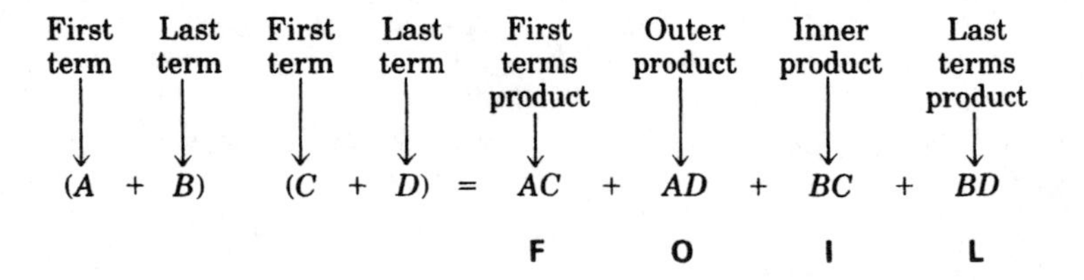

EXERCISES 4.2B

Find each product by using the FOIL method.

SAMPLE PROBLEM 1	SAMPLE PROBLEM 2
$(4x + 5)(2x - 3)$	$(3x - 8y)(2x - 5y)$
SOLUTION	**SOLUTION**
F O I L	F O I L
$4(2)x^2 + 4(-3)x + 5(2)x + 5(-3)$	$(3x)(2x) + (3x)(-5y) + (-8y)(2x) + (-8y)(-5y)$
$8x^2 - 12x + 10x - 15$	$6x^2 - 15xy - 16xy + 40y^2$
$8x^2 - 2x - 15$	$6x^2 - 31xy + 40y^2$
SAMPLE PROBLEM 3	**SAMPLE PROBLEM 4**
$(r - 2t)(s + 4t)$	$(y^3 - 8)(y - 5)$
SOLUTION	**SOLUTION**
F O I L	F O I L
$(r)(s) + (r)(4t) + (-2t)(s) + (-2t)(4t)$	$y^3(y) + y^3(-5) + (-8)y + (-8)(-5)$
$rs + 4rt - 2st - 8t^2$	$y^4 - 5y^3 - 8y + 40$

F O I L

1. $(2x + 5)(3x + 4)$ **1.** ________

2. $(4x - 3)(5x - 1)$ **2.** ________

3. $(6y + 7)(2y - 5)$ **3.** ________

4. $(6y - 7)(2y + 5)$ **4.** ________

5. $(6y - 7)(2y - 5)$ **5.** ________

6. $(6y + 7)(2y + 5)$ **6.** ________

7. $(5x - 2y)(3x + 4y)$ 7. ________

8. $(5x + 4y)(3x - 2y)$ 8. ________

9. $(5x + 2y)(3x - 4y)$ 9. ________

10. $(5x - 4y)(3x + 2y)$ 10. ________

11. $(4x^2 - 1)(x^2 + 1)$ 11. ________

12. $(9y^2 - 4)(4y^2 - 1)$ 12. ________

13. $(x + 3)(y + 2)$ 13. ________

14. $(x - 4)(y - 5)$ 14. ________

15. $(x + 7)(y - 1)$ 15. ________

16. $(x - 6)(y + 4)$ 16. ________

17. $(x^2 - 1)(x - 1)$ 17. ________

18. $(y^2 + 5)(y - 6)$ 18. ________

19. $(x^3 + 1)(x - 5)$ 19. ________

20. $(y^3 - 2)(y^2 + 4)$ 20. ________

C. PRODUCT OF TWO POLYNOMIALS, VERTICAL FORMAT

The product of two polynomials can be written as a single polynomial by applying the distributive axiom.

Each term of one polynomial must be multiplied by each term of the other polynomial.

EXERCISES 4.2C

Multiply, using a vertical format.

SAMPLE PROBLEM 1

Multiply $(2x + y)(3x^2 - 2xy - 4y^2)$.

SOLUTION

Vertical format (a vertical format is often used if one polynomial has three or more terms):

Note that like terms are put in the same column.

$$\begin{array}{lrrrrl}
 & 3x^2 & - 2xy & - 4y^2 & & \\
 & 2x & + y & & & \\
\hline
\text{Distribute } 2x. & 6x^3 & - 4x^2y & - 8xy^2 & & 2x(3x^2 - 2xy - 4y^2) \\
\text{Distribute } y. & & 3x^2y & - 2xy^2 & - 4y^3 & y(3x^2 - 2xy - 4y^2) \\
\hline
\text{Collect like terms.} & 6x^3 & - x^2y & - 10xy^2 & - 4y^3 &
\end{array}$$

SAMPLE PROBLEM 2

Multiply $(x + y - 5)(x - y + 5)$.

SOLUTION

Vertical format:

$$\begin{array}{lll} & x + y - 5 & \\ & x - y + 5 & \\ \hline \textbf{Distribute } x. & x^2 + xy - 5x & x(x + y - 5) \\ \textbf{Distribute } -y. & \quad - xy \qquad - y^2 + 5y & -y(x + y - 5) \\ \textbf{Distribute 5.} & \qquad + 5x \qquad + 5y - 25 & 5(x + y - 5) \\ \hline \textbf{Collect like terms.} & x^2 + 0 + 0 - y^2 + 10y - 25 & \end{array}$$

The product is $x^2 - y^2 + 10y - 25$.

1. $(x + 3)(x^2 + 4x + 5)$

$$\begin{array}{r} x^2 + 4x + 5 \\ x + 3 \\ \hline \end{array}$$

2. $(x - y)(x^2 - xy - 2y^2)$

$$\begin{array}{r} x^2 - xy - 2y^2 \\ x - y \\ \hline \end{array}$$

1. ______________

2. ______________

3. $(x - 6)(x^2 + 6x + 36)$

$$\begin{array}{r} x^2 + 6x + 36 \\ x - 6 \\ \hline \end{array}$$

4. $(y + 2)(y^2 + 4y + 4)$

$$\begin{array}{r} y^2 + 4y + 4 \\ y + 2 \\ \hline \end{array}$$

3. ______________

4. ______________

5. $(x - 1)(x^3 + x^2 + x + 1)$

6. $(x + y + z)(x + y + z)$

5. ______________

6. ______________

7. $(x^2 + 3x + 2)(x^2 - 3x + 2)$

8. $(y^2 - 4y - 5)(y^2 + 4y - 5)$

7. ______________

8. ______________

9. $(y^2 - 1)(y^2 + y - 2)$

10. $(2x + y - 3z)^2$

9. ______________

10. ______________

1–10. Multiply.

1. $-12rs(4r^2 - 4rs + 1)$
2. $-15st(t^2 - 5t + 4)$
3. $-r(r^3 - r^2 + r - 1)$
4. $-c^2d(c^2 + c - d^2 - d)$
5. $4x^3(2x^3 + x^2 - 2x - 1)$
6. $8y^3(1 - y - y^2 - y^3)$
7. $8y^2(1.25y^2 - 0.75y + 0.25)$
8. $-xy(1 - x + y - xy)$
9. $-7st(s + t - s^2 - t^2)$
10. $5ab^2c^3(a + 2b - 2c + 5)$

1. ______
2. ______
3. ______
4. ______
5. ______
6. ______
7. ______
8. ______
9. ______
10. ______

11–26. Multiply, using the FOIL method.

11. $(3x - 4)(x + 2)$
12. $(3x + 4)(x - 2)$
13. $(7x - 2y)(7x + 2y)$
14. $(7x - 2y)(7x - 2y)$
15. $(5x + 2)(5x - 2)$
16. $(x^2 + y^2)(x - y)$
17. $(y + 2x)(x - 2y)$
18. $(4y - 5x)(5x - 4y)$
19. $(x + 2y)(2y + x)$
20. $(7x - 6y)(7x - 6y)$

11. ______
12. ______
13. ______
14. ______
15. ______
16. ______
17. ______
18. ______
19. ______
20. ______

NAME ______ DATE ______ COURSE ______

21. $(a^2 - b^2)(a^2 + b^2)$

22. $(x^2 + 4)(x^2 - 4)$

23. $(x^3 + 1)(x^3 - 1)$

24. $(8x^3 - 27)(8x^3 + 27)$

25. $(4x - 7y)(x + 6)$

26. $(x - 8)(x^2 - 9)$

21. ______

22. ______

23. ______

24. ______

25. ______

26. ______

27–36. Multiply, using a vertical format.

27. $(y - 2)(2y^2 - 3y + 1)$

$$\begin{array}{r} 2y^2 - 3y + 1 \\ y - 2 \\ \hline \end{array}$$

28. $(x + 4y)(y^2 - 5xy - x^2)$

$$\begin{array}{l} y^2 - 5xy - x^2 \\ 4y + x \\ \hline \end{array}$$

27. ______

28. ______

29. $(x + 5)(x^2 - 5x + 25)$

$$\begin{array}{r} x^2 - 5x + 25 \\ x + 5 \\ \hline \end{array}$$

30. $(y - 4)(y^2 - 8y + 16)$

$$\begin{array}{r} y^2 - 8y + 16 \\ y - 4 \\ \hline \end{array}$$

29. ______

30. ______

31. $(r + 2s - 5)(r + 2s + 5)$

$$\begin{array}{r} r + 2s + 5 \\ r + 2s - 5 \\ \hline \end{array}$$

32. $(x^2 - 2x + 2)(x^2 + 2x + 2)$

$$\begin{array}{r} x^2 + 2x + 2 \\ x^2 - 2x + 2 \\ \hline \end{array}$$

31. ______

32. ______

33. $(y - 5)(4y^3 + 2y - 1)$

34. $(x^2 + 2)(x^2 - 3y - 1)$

33. ______

34. ______

35. $(x - y - z)(x - y - z)$

36. $(3a - 5b + c)^2$

35. ______

36. ______

MONOMIAL FACTORS 4.3

FACTORING

The sum, difference, or product of two polynomials is always a polynomial. However, this is not the case for the quotient of two polynomials. Note that the set of integers also has this property.

$$\frac{2x(x^2 + 5)}{2x} = x^2 + 5 \quad \text{and} \quad x^2 + 5 \text{ is a polynomial}$$

$$\frac{x}{x + 5} \text{ is not a polynomial}$$

The quotient is a polynomial when the divisor (denominator) is a factor of the dividend (numerator).

$2x(x^2 + 5)$ is called the factored form of the product $2x^3 + 10x$.

To factor a polynomial means to write the polynomial as a product of factors. We may think of this process as a form of division.

COEFFICIENTS

To discuss factoring, we need to talk about the **coefficients** of a polynomial. The numerical factors of the terms of a polynomial are called the coefficients of the polynomial.

For example, for the polynomial $5x^2 - 6x + 4$,

5 is the coefficient of x^2,

-6 is the coefficient of x, and

4 is the coefficient of 1

The coefficient of the highest power of the variable is also called the **leading coefficient.**

The term not involving the variable is called the **constant term.** For factoring, we will require the coefficients of all polynomials to be integers. When we do this, we say we are **factoring over the integers.**

MONOMIAL FACTORS

The first step in factoring a polynomial is to find the common monomial factors, if there are any. The distributive axiom is used for this type of factoring.

Examples of the **distributive axiom:**

$$AB + AC = A(B + C)$$

$$AB + AC + AD = A(B + C + D)$$

After a common monomial factor has been found, it is important to check that the **greatest common monomial factor, GCMF,** has been found; that is, the nonmonomial factor of the product should not have a common monomial factor.

For example,

$$2x(3x^2 - 15x + 21) \text{ is a factored form for } 6x^3 - 30x^2 + 42x$$

Although $2x$ is a common factor, $2x$ is not the greatest common monomial factor. Note that 3 is also a common factor.

$$\begin{aligned} 2x(3x^2 - 15x + 21) &= 2x(3)(x^2 - 5x + 7) \\ &= 6x(x^2 - 5x + 7) \end{aligned}$$

Now the terms of $x^2 - 5x + 7$ do not have a common factor, and $6x$ is the GCMF of $6x^3 - 30x^2 + 42x$.

EXERCISES 4.3

Completely factor.
(Write as a product of the greatest common monomial factor and a polynomial.)

SAMPLE PROBLEM 1
$5x^4 + 20x^2$

SOLUTION

1. Find the GCMF.
   a. Factor each term into prime factors. — $5x^4 = 5xxxx$, $20x^2 = 2(2)(5)xx$
   b. Select the factors common to each term. — $5xx$
2. Rewrite given polynomial as $AB + AC$ where A = GCMF. — $\mathbf{5x^2}(x^2) + \mathbf{5x^2}(4)$
3. Use $AB + AC = A(B + C)$. — $\mathbf{5x^2}(x^2 + 4)$

$5x^2(x^2 + 4)$ is the completely factored form of $5x^4 + 20x^2$.

SAMPLE PROBLEM 2
$3x^2y^3 - 18x^2y^2 + 3xy^2$

SOLUTION

1. The common numerical factor is 3.
2. The common power of x is x.
3. The common power of y is y^2.
4. The greatest common monomial factor is $3xy^2$, the product of 3, x, and y^2.

$$\begin{aligned} 3x^2y^3 - 18x^2y^2 + 3xy^2 &= \mathbf{3xy^2}(xy) - \mathbf{3xy^2}(6x) + \mathbf{3xy^2}(1) \\ &= \mathbf{3xy^2}(xy - 6x + 1) \end{aligned}$$

1. $12x^3 + 6x^2$

2. $18y^4 - 12y^3$

3. $10xy + 15x^2y^2$

4. $14xy^2 - 21x^2y$

5. $16x^3 + 8x$

6. $18x^4 + 9x^2$

7. $16x^3y^2 - 40x^2y^3$

8. $54x^3y^3 + 9x^2y^5$

9. $15at^4 + 75at^2$

10. $24c^2s^2 - 42c^3s$

11. $7x + 7y + 7z$

12. $4x - 4y + 4$

13. $15ax - 45ay - 15a$

14. $4a^2x^2 + 8a^3x - 4a^2$

15. $9y^4 + 9y^3 - 27y^2$

16. $150y^3 - 200y^2 - 250y$

17. $70t^4 - 105t^3 + 140t^2$

18. $a^2b^2c + ab^2c^2 + a^2bc^2$

19. $r^4s^2t^2 - r^2s^4t^2 - r^2s^2t^4$

20. $24x^3yz + 48xy^3z - 72xyz^3$

1. ______________
2. ______________
3. ______________
4. ______________
5. ______________
6. ______________
7. ______________
8. ______________
9. ______________
10. ______________
11. ______________
12. ______________
13. ______________
14. ______________
15. ______________
16. ______________
17. ______________
18. ______________
19. ______________
20. ______________

SAMPLE PROBLEM 3
Factor $-24t^5 - 36t^4 + 48t^3$.

SOLUTION

1. The greatest common numerical factor of 24, 36, and 48 is 12.
2. The highest power of t that divides t^5, t^4, and t^3 is t^3.
3. The product of all common factors of the terms of the polynomial is $12t^3$.

When the leading coefficient of a polynomial is negative, it is useful to include the minus sign in the monomial factor. This makes the polynomial factor simpler, as a rule.

Therefore, $-12t^3$ is taken as the greatest common monomial factor.

$$-24t^5 - 36t^4 + 48t^3 = (\mathbf{-12t^3})(2t^2) + (\mathbf{-12t^3})(3t) + (\mathbf{-12t^3})(-4)$$
$$= \mathbf{-12t^3}(2t^2 + 3t - 4)$$

21. $-3x^4 - 12x^2$

22. $-2y^3 + 4y^2 + 6y$

23. $9x^3 - 45x^5$

24. $10 - 5t - t^2$

25. $-4y^5 - 12y^3 - 4y^2$

26. $5n^2 + 2n^3 - n^4$

27. $20t^5 - 5t^6 - 5t^4$

28. $-18x^3 - 12x^6$

29. $n^4 - n^6 - n^2$

30. $-a^4x^4 - a^3x^3 - a^2x^2$

21. ____________________

22. ____________________

23. ____________________

24. ____________________

25. ____________________

26. ____________________

27. ____________________

28. ____________________

29. ____________________

30. ____________________

Factor completely.

1. $40x^2y - 50xy^2$

2. $25x^4 - 100x^3$

3. $4x^6 + 64x^4$

4. $-15x^2y^2 - 15y^4$

5. $42x^3 + 6x^2$

6. $-40x^2y^2 + 32xy^3$

7. $21x - 42y + 21z$

8. $x^4y^2 - x^2y^4 - x^2y^2$

1. ______
2. ______
3. ______
4. ______
5. ______
6. ______
7. ______
8. ______

NAME ______ DATE ______ COURSE ______

9. $48r - 84s + 24$

10. $6x^2 - 18x + 30$

9. ____________

10. ____________

11. $4x^3 + 16x^2 - 4x$

12. $10x^3y - 5x^2y^2 - 10xy^3$

11. ____________

12. ____________

13. $-24y^4 + 40y^3 - 32y^2$

14. $r^2st + rs^2t + rst^2$

13. ____________

14. ____________

15. $-3a^4b^2c^2 - 6a^2b^4c^2 + 3a^2b^2c^4$

15. ____________

PRODUCT OF SUM AND DIFFERENCE; DIFFERENCE OF SQUARES 4.4

The product of the sum and the difference of the same two terms is a special product, called the **difference of squares.**

Multiplying, by using the distributive axiom:

$$
\begin{aligned}
(A + B)(A - B) &= A(A - B) + B(A - B) \\
&= A^2 - AB + AB - B^2 \\
&= A^2 - B^2
\end{aligned}
$$

A polynomial having the form $A^2 - B^2$ can be expressed in factored form as the product of a sum and a difference by using the special product for a sum and a difference.

Difference of squares = Product of sum and difference

$$A^2 - B^2 = (A + B)(A - B)$$

For this type of factoring, it is necessary to recognize a square. It may be useful to refer to a Table of Squares.

N	x	x^2	x^3	1	2	3	4	5	6	7	8	9	10
N^2	x^2	x^4	x^6	1	4	9	16	25	36	49	64	81	100

EXAMPLE 1

Use the formula to obtain the product of $(7x + 2)(7x - 2)$.

Solution

Use $A = 7x$ and $B = 2$.

$$
\begin{aligned}
(A + B)(A - B) &= A^2 - B^2 \\
(7x + 2)(7x - 2) &= (7x)^2 - (2)^2 \\
&= 49x^2 - 4
\end{aligned}
$$

EXAMPLE 2

Factor $49x^2 - 4$.

Solution

$$
\begin{aligned}
A^2 - B^2 &= (A + B)(A - B) \\
(7x)^2 - (2)^2 &= (7x + 2)(7x - 2)
\end{aligned}
$$

EXERCISES 4.4

1–15. Use the formula $(A + B)(A - B) = A^2 - B^2$ to obtain each product.

SAMPLE PROBLEM 1

$(3x + 7)(3x - 7)$

SOLUTION

$(A + B)(A - B) = A^2 - B^2$

$(3x + 7)(3x - 7) = (3x)^2 - 7^2 = 9x^2 - 49$

SAMPLE PROBLEM 2

$(4x^2 - 5y)(4x^2 + 5y)$

SOLUTION

Note that $(4x^2 - 5y)(4x^2 + 5y) = (4x^2 + 5y)(4x^2 - 5y)$ by the commutative axiom for multiplication.

Use $A = 4x^2$ and $B = 5y$.

$(A + B)(A - B) = A^2 - B^2$

$(4x^2 + 5y)(4x^2 - 5y) = (4x^2)^2 - (5y)^2$

$= 16x^4 - 25y^2$

1. $(x + 3)(x - 3) = (\quad)^2 - (\quad)^2$ — 1. ______
2. $(x + 5)(x - 5) = (\quad)^2 - (\quad)^2$ — 2. ______
3. $(y - 2)(y + 2) = (\quad)^2 - (\quad)^2$ — 3. ______
4. $(y - 4)(y + 4) = (\quad)^2 - (\quad)^2$ — 4. ______
5. $(5x + 1)(5x - 1) = (\quad)^2 - (\quad)^2$ — 5. ______
6. $(4x + 7)(4x - 7) = (\quad)^2 - (\quad)^2$ — 6. ______
7. $(x + y)(x - y) = (\quad)^2 - (\quad)^2$ — 7. ______
8. $(r + s)(s - r) = (\quad)^2 - (\quad)^2$ — 8. ______
   [Note that $r + s = s + r$]
9. $(6r - s)(s + 6r)$ — 9. ______
10. $(x^2 + 6)(x^2 - 6)$ — 10. ______
11. $(x^2 - 8y^2)(x^2 + 8y^2)$ — 11. ______
12. $(3y^2 - 10)(3y^2 + 10)$ — 12. ______
13. $(a^2 + b^2)(b^2 - a^2)$ — 13. ______
14. $(7a - 2b)(2b + 7a)$ — 14. ______
15. $(1 + xy)(1 - xy)$ — 15. ______

16–31. Factor.

SAMPLE PROBLEM 3	SAMPLE PROBLEM 4
$4x^2 - 9$	$36y^2 - 1$
SOLUTION	**SOLUTION**
1. Recognize difference of squares: $(2x)^2 - 3^2$.	$A^2 - B^2 = (A + B)(A - B)$
2. Use the formula $A^2 - B^2 = (A + B)(A - B)$, replacing A by $2x$ and replacing B by 3.	$36y^2 - 1 = (6y)^2 - (1)^2 = (6y + 1)(6y - 1)$
3. $(2x)^2 - (3)^2 = (2x + 3)(2x - 3)$	

SAMPLE PROBLEM 5

$25x^6 - 16y^4$

SOLUTION

$25x^6 = (5)(5)(xxx)(xxx) = (5x^3)(5x^3) = (5x^3)^2$

$16y^4 = (4)(4)(yy)(yy) = (4y^2)(4y^2) = (4y^2)^2$

Use $A^2 - B^2 = (A + B)(A - B)$.

$$25x^6 - 16y^4 = (5x^3)^2 - (4y^2)^2 = (5x^3 + 4y^2)(5x^3 - 4y^2)$$

16. $x^2 - 25$ **17.** $x^2 - 16$

18. $y^2 - 36$ **19.** $y^2 - 49$

20. $64x^2 - 1$ **21.** $9y^2 - 1$

22. $81x^2 - 25$ **23.** $100y^2 - 49$

24. $25x^2 - 64y^2$ **25.** $4x^2 - 9y^2$

26. $16x^4 - 25y^2$ **27.** $36y^2 - x^4$

28. $x^2y^4 - z^2$ **29.** $400a^2b^2 - c^2$

30. $x^6 - y^2$ **31.** $81x^6 - 100y^4$

16. ____________
17. ____________
18. ____________
19. ____________
20. ____________
21. ____________
22. ____________
23. ____________
24. ____________
25. ____________
26. ____________
27. ____________
28. ____________
29. ____________
30. ____________
31. ____________

32–37. First multiply the binomials. Then multiply the product by the monomial.

SAMPLE PROBLEM 6	SAMPLE PROBLEM 7
$6x(x+5)(x-5)$	$-y^2(3y-1)(3y+1)$
SOLUTION	SOLUTION
$6x(x^2-25) = 6x^3 - 150x$	$-y^2(9y^2-1) = -9y^4 + y^2$

32. $5(x+4)(x-4)$

33. $x(x+3)(x-3)$

34. $3x(x-2)(x+2)$

35. $4y(y-1)(y+1)$

36. $y^2(y+7)(y-7)$

37. $-t^2(5t+1)(5t-1)$

32. ______________

33. ______________

34. ______________

35. ______________

36. ______________

37. ______________

38–45. Factor. First remove any common monomial factor. Check by multiplication, as in Exercises 32–37.

SAMPLE PROBLEM 8	SAMPLE PROBLEM 9
$6x^3 - 150x$	$9x^4 + 36x^2$
SOLUTION	SOLUTION
$(6x)(x^2) - (6x)(25)$	$(9x^2)(x^2) + (9x^2)(4)$
$6x(x^2-25)$	$9x^2(x^2+4)$
$6x(x+5)(x-5)$	x^2+4 is a sum of squares and cannot be written as a product of two binomials.
Check. See Sample Problem 6.	

38. $5x^2 - 80$

39. $x^3 - 9x$

40. $7x^3 - 7x$

41. $4x^4 - x^2$

42. $2y^4 - 200y^2$

43. $x^3y^3 - xy)$

44. $9x^2 + 81$

45. $25x^4 + 100x^2$

38. ______________

39. ______________

40. ______________

41. ______________

42. ______________

43. ______________

44. ______________

45. ______________

EXERCISES 4.4 S

1–10. Multiply by using $(A + B)(A - B) = A^2 - B^2$.

1. $(x + 6y)(x - 6y)$

2. $(2a - 7b)(2a + 7b)$

3. $(x^2 + 4)(4 - x^2)$

4. $(rs - 1)(rs + 1)$

5. $(xy - z)(z + xy)$

6. $(3x^3 - 8)(3x^3 + 8)$

7. $(9r^2 - s^2)(9r^2 + s^2)$

8. $(x + y)(y - x)$

9. $(ax + by)(by - ax)$

10. $(x^2 - y^2)(-y^2 - x^2)$

1. ______
2. ______
3. ______
4. ______
5. ______
6. ______
7. ______
8. ______
9. ______
10. ______

11–30. Factor by using $A^2 - B^2 = (A - B)(A + B)$.

11. $x^2 - 1$

12. $16x^2 - y^2$

13. $36x^2 - 49$

14. $64y^2 - 25$

15. $x^2 - 100y^2$

16. $9x^2y^2 - 4$

17. $81 - k^2$

18. $x^2y^2 - z^4$

19. $900a^4 - 1$

20. $a^6 - b^2c^2$

11. ______
12. ______
13. ______
14. ______
15. ______
16. ______
17. ______
18. ______
19. ______
20. ______

NAME ______ DATE ______ COURSE ______

21. $4 - 81y^6$

22. $c^6 - a^4$

23. $121x^6 - y^4$

24. $y^6 - 144x^4$

25. $x^6 - 169y^6$

26. $1 - 100t^2$

27. $x^4 - 49$

28. $y^4 - 64$

29. $x^4 - 9y^4$

30. $25a^4 - 16b^4$

21. ______________________

22. ______________________

23. ______________________

24. ______________________

25. ______________________

26. ______________________

27. ______________________

28. ______________________

29. ______________________

30. ______________________

31–40. Factor. First remove any common monomial factor. Check by multiplication.

31. $75x^2 - 3$

32. $4y^3 - 16y$

33. $t^4 - 36t^2$

34. $x^4y^2 - x^2y^4$

35. $64x^4 - 16x^2$

36. $64x^4 - 16x^3$

37. $64x^4 + 16x^2$

38. $9y^2 - 81$

39. $9y^2 - 81y$

40. $9y^2 + 81$

31. ______________________

32. ______________________

33. ______________________

34. ______________________

35. ______________________

36. ______________________

37. ______________________

38. ______________________

39. ______________________

40. ______________________

FACTORING SIMPLE TRINOMIALS 4.5

A **trinomial** is a polynomial having three terms.

As examples,

$x^2 - 7x + 10$ and $4x^2 + 12xy + 9y^2$ are trinomials

Multiplying $(X + A)(X + B)$, using the distributive axiom, yields the simple trinomial product formula.

$$\begin{aligned}(X + A)(X + B) &= (X + A)(X) + (X + A)(B)\\ &= X^2 + AX + BX + AB\\ &= X^2 + (A + B)X + AB\end{aligned}$$

$$\begin{aligned}&(x + 4)(x + 5)\\ &= x^2 + 4x + 5x + (4)(5)\\ &= x^2 + (4 + 5)x + (4)(5)\end{aligned}$$

Note that $(A + B)$, the coefficient of X in the trinomial, is the **sum** of the constants A and B in the related binomial factors and that AB, the constant term of the trinomial, is the **product** of the constant terms of the binomial factors.

Simple trinomial product:

$$X^2 + (A + B)X + AB = (X + A)(X + B)$$

$$\begin{aligned}&x^2 + 9x + 20\\ &= x^2 + (4 + 5)x + (4)(5)\\ &= (x + 4)(x + 5)\end{aligned}$$

To factor a simple trinomial product, it is necessary to find two integers whose product is the constant term and whose sum is the coefficient of the variable.

The following table shows the factorization of some simple trinomials. Note the sign patterns that are typical in this type of factoring: When the product AB is positive, A and B have the same sign, the sign of the coefficient of X. When the product AB is negative, A and B have opposite signs.

Trinomials and factors

$$\begin{aligned}X^2 + (A + B)X + AB &= (X + A)(X + B)\\ x^2 + 8x + 15 &= (x + 5)(x + 3)\\ x^2 - 8x + 15 &= (x - 5)(x - 3)\\ x^2 + 2x - 15 &= (x + 5)(x - 3)\\ x^2 - 2x - 15 &= (x - 5)(x + 3)\end{aligned}$$

GENERAL PROCEDURE FOR FACTORING $X^2 + SX + P$

1. **Find all possible pairs of factors of P.** $(P = ab)$
2. * **Of these pairs, select the one whose sum is S.** $(a + b = S)$
3. **Write the product: $(X + a)(X + b)$**

***If no pair has S for its sum, the given polynomial is prime.**

EXERCISES 4.5

Factor, if possible. If the polynomial is prime, write "prime." Check by multiplication.

SAMPLE PROBLEM 1

$x^2 + 9x + 20$

SOLUTION

1. Find the pairs of factors of 20: (1)(20); (2)(10); (4)(5).
2. Select those factors whose sum is 9: 4 and 5.
3. Substitute into the simple trinomial product formula, using $A = 4$ and $B = 5$.

$$X^2 + (A + B)X + AB = (X + A)(X + B)$$
$$x^2 + 9x + 20 = x^2 + (\mathbf{4} + \mathbf{5})x + (\mathbf{4})(\mathbf{5}) = (x + \mathbf{4})(x + \mathbf{5})$$

CHECK

$(x + 4)(x + 5)$
$= x^2 + 5x + 4x + 20$
$= x^2 + 9x + 20$

SAMPLE PROBLEM 2

$y^2 - 7y + 12$

SOLUTION

1. Factors of 12: (1)(12); (2)(6); (3)(4).
2. Since the product is positive and the sum is negative, the factors of 12 must both be negative.

   Factors whose sum is -7 are -3 and -4.
3. Use $A = -3$ and $B = -4$.

$$\begin{aligned} y^2 - 7y + 12 &= y^2 + ([\mathbf{-3}] + [\mathbf{-4}])y + (\mathbf{-3})(\mathbf{-4}) \\ &= (y + [\mathbf{-3}])(y + [\mathbf{-4}]) \\ &= (y - \mathbf{3})(y - \mathbf{4}) \end{aligned}$$

CHECK

$(y - 3)(y - 4)$
$= y^2 - 4y - 3y + 12$
$= y^2 - 7y + 12$

SAMPLE PROBLEM 3

$x^2 + 4x + 2$, if possible

SOLUTION

The only pairs of factors of 2 are (1)(2) and $(-1)(-2)$. Since neither combination has a sum of 4, the polynomial cannot be factored using integers for A and B. Such a polynomial is said to be **prime** with respect to the integers.

1. $x^2 + 7x + 10$

2. $y^2 + 6y + 8$

3. $x^2 - 3x + 2$

4. $t^2 - 9t + 18$

5. $x^2 - 2x + 3$

6. $n^2 - 7n + 6$

1. ______
2. ______
3. ______
4. ______
5. ______
6. ______

SAMPLE PROBLEM 4

$x^2 - 3x - 10$

SOLUTION

1. Factors of 10: (1)(10); (2)(5).
2. Since the product, -10, is negative, the factors of -10 must be selected with opposite sign and the sum must be -3. The combination desired is $+2$ and -5.
3. Use $A = -5$ and $B = 2$.

$$X^2 + (A + B)X + AB = (X + A)(X + B)$$

$$x^2 - 3x - 10 = x^2 + (\mathbf{-5} + \mathbf{2})x + (\mathbf{-5})(\mathbf{2}) = (x \mathbf{-5})(x \mathbf{+2})$$

CHECK

$(x - 5)(x + 2)$
$= x^2 + 2x - 5x - 10$
$= x^2 - 3x - 10$

SAMPLE PROBLEM 5

$x^2 + 7xy - 18y^2$

SOLUTION

1. Factors of $18y^2$: $(y)(18y)$; $(2y)(9y)$; $(3y)(6y)$.
2. Since the product, $-18y^2$, is negative, the factors of $-18y^2$ must be selected with opposite sign and the sum must be $+7y$.

   The combination desired is $-2y$ and $+9y$.
3. Use $A = -2y$ and $B = 9y$.

$$x^2 + 7xy - 18y^2 = x^2 + (\mathbf{-2y} + \mathbf{9y})x + (\mathbf{-2y})(\mathbf{9y})$$
$$= (x \mathbf{- 2y})(x \mathbf{+ 9y})$$

CHECK

$(x - 2y)(x + 9y)$
$= x^2 + 9xy - 2xy - 18y^2$
$= x^2 + 7xy - 18y^2$

7. $x^2 + 3x - 10$

8. $x^2 - 9x - 10$

9. $y^2 - 4y - 12$

10. $x^2 + xy - 12y^2$

11. $r^2 + 5rt - 14t^2$

12. $t^2 - 3t - 2$

7. ______
8. ______
9. ______
10. ______
11. ______
12. ______

13. $x^2 + 9x + 14$

14. $x^2 - 9x + 14$

15. $x^2 - 5x - 14$

16. $x^2 + 5x - 14$

17. $x^2 - 15x + 14$

18. $x^2 - 13x - 14$

19. $x^2 + 13x - 14$

20. $x^2 + 11x + 30$

21. $x^2 - 5x + 4$

22. $x^2 - 5x + 6$

23. $x^2 - 4x - 21$

24. $x^2 + 4x - 21$

25. $y^2 + y - 20$

26. $y^2 - y - 20$

27. $y^2 - 14y - 15$

28. $y^2 - 16y + 8$

29. $x^2 + 15x + 26$

30. $x^2 + 11x - 26$

31. $x^2 - 11x - 26$

32. $t^2 - 15t + 26$

33. $x^2 + 4xy - 12y^2$

34. $x^2 - 4xy - 12y^2$

35. $r^2 - 11rs - 12s^2$

36. $s^2 - 13st + 12t^2$

37. $u^2 - 10uv + 25v^2$

38. $x^2 + 12xy + 36y^2$

13. ____________________
14. ____________________
15. ____________________
16. ____________________
17. ____________________
18. ____________________
19. ____________________
20. ____________________
21. ____________________
22. ____________________
23. ____________________
24. ____________________
25. ____________________
26. ____________________
27. ____________________
28. ____________________
29. ____________________
30. ____________________
31. ____________________
32. ____________________
33. ____________________
34. ____________________
35. ____________________
36. ____________________
37. ____________________
38. ____________________

EXERCISES 4.5 S

1–34. Factor, if possible. If the polynomial is prime, write "prime." Check by multiplication.

$$X^2 + (A + B)X + AB = (X + A)(X + B)$$

1. $x^2 + 6x + 8$

2. $x^2 - 11x + 30$

3. $x^2 - 2x - 24$

4. $x^2 + 2x - 8$

5. $x^2 + 2x + 4$

6. $y^2 - 4y - 5$

7. $y^2 + 4y - 5$

8. $y^2 - 5y + 4$

9. $y^2 - 5y - 4$

10. $x^2 + 11xy + 18y^2$

11. $x^2 + 5xy - 24y^2$

12. $x^2 - 2xy - 35y^2$

13. $x^2 + 4xy - 3y^2$

14. $u^2 - uv - 2v^2$

15. $u^2 + uv + v^2$

16. $u^2 + uv - 56v^2$

17. $r^2 - 3rs - 54s^2$

18. $x^4 - 2x^2y^2 - 3y^4$

19. $x^4 - 15x^2 + 50$

20. $x^4 - 16x^2 + 64$

1. ______________
2. ______________
3. ______________
4. ______________
5. ______________
6. ______________
7. ______________
8. ______________
9. ______________
10. ______________
11. ______________
12. ______________
13. ______________
14. ______________
15. ______________
16. ______________
17. ______________
18. ______________
19. ______________
20. ______________

NAME ______________ DATE ______________ COURSE ______________

21. $x^2 + 14x + 48$

22. $x^2 - 14x + 48$

21. ______

22. ______

23. $x^2 + 2x - 48$

24. $x^2 - 2x - 48$

23. ______

24. ______

25. $y^2 + 8y + 16$

26. $y^2 - 18y + 81$

25. ______

26. ______

27. $x^4 + 2x^2 + 1$

28. $x^4 - 12x^2 + 36$

27. ______

28. ______

29. $x^4 - x^2 - 2$

30. $x^4 - 2x^2 - 1$

29. ______

30. ______

31. $x^2 - 10xy + 9y^2$

32. $x^2 - 6xy + 9y^2$

31. ______

32. ______

33. $x^4 + 10x^2y + 25y^2$

34. $x^4 + 26x^2y + 25y^2$

33. ______

34. ______

35–42. Factor by first removing the greatest common monomial factor. Then factor the trinomial, if possible.

35. $2x^2 + 10x + 12$

36. $x^3 - 10x^2 + 21x$

35. ______

36. ______

37. $5x^3 - 15x^2 - 50x$

38. $2x^3 - 28x^2 + 98x$

37. ______

38. ______

39. $6y^3 + 12y^2 - 144y$

40. $100y^4 - 200y^3 + 100y^2$

39. ______

40. ______

41. $16x^4 + 16x^3 - 16x^2$

42. $16x^4 - 32x^3 + 16x^2$

41. ______

42. ______

PERFECT SQUARE TRINOMIALS 4.6

The square of a binomial is a polynomial having three terms, called a **perfect square trinomial.**

Using the distributive axiom,

$$\begin{aligned}(A + B)^2 &= (A + B)(A + B)\\ &= (A + B)A + (A + B)B\\ &= A^2 + BA + AB + B^2\\ &= A^2 + 2AB + B^2\end{aligned}$$

The perfect square trinomial formula is

$$(A + B)^2 = A^2 + 2AB + B^2$$

This formula shows that the square of a binomial is obtained by squaring the first term, doubling the product of the two terms, and squaring the last term, and then adding these three products.

Perfect square trinomial product:

$$A^2 + 2AB + B^2 = (A + B)^2$$
$$A^2 - 2AB + B^2 = (A - B)^2$$

WARNING!
$(A + B)^2 \neq A^2 + B^2$
Do not make this common error.

To recognize a perfect square trinomial:

1. There must be three terms.
2. Two terms must be perfect square monomials: $(A)^2$ and $(B)^2$.
3. There must not be a minus sign before either A^2 or B^2.
4. The third term must be *twice* the product of A and B, where A^2 and B^2 are the perfect square monomials.

The following table may be useful. Also see the Table of Squares inside the cover.

N	x	x^2	x^3	1	2	3	4	5	6	7	8	9	10
N^2	x^2	x^4	x^6	1	4	9	16	25	36	49	64	81	100

EXERCISES 4.6

1–10. Use the perfect square trinomial formula to obtain each product.

SAMPLE PROBLEM 1

$(6x + 7)^2$

SOLUTION

Use $A = 6x$ and $B = 7$.

$(A + B)^2 = A^2 + 2AB + B^2$

$(6x + 7)^2 = (6x)^2 + 2(6x)(7) + 7^2$

$= 36x^2 + 84x + 49$

SAMPLE PROBLEM 2

$(4x^2 - 9y^3)^2$

SOLUTION

Use $A = 4x^2$ and $B = -9y^3$.

$(A + B)^2 = A^2 + 2AB + B^2$

$(4x^2 + [-9y^3])^2 = (4x^2)^2 + 2(4x^2)(-9y^3) + (-9y^3)^2$

$= 16x^4 - 72x^2y^3 + 81y^6$

1. $(x + 4)^2$

2. $(x - 3)^2$

3. $(2x - 5)^2$

4. $(5 - 2x)^2$

5. $(5x - 2)^2$

6. $(x - 7y)^2$

7. $(-y + x)^2$

8. $(x^2 + 6)^2$

9. $(4y^3 + 1)^2$

10. $(xy + 8)^2$

1. ________ **2.** ________

3. ________ **4.** ________

5. ________ **6.** ________

7. ________ **8.** ________

9. ________ **10.** ________

11–30. Express each perfect square trinomial as the square of a binomial. If the trinomial is not a perfect square, write "not a perfect square."

SAMPLE PROBLEM 3

Show that each of the following is not a perfect square trinomial:

a. $4x^2 + 9$

b. $4x^2 + 6x + 9$

c. $4x^2 + 12x - 9$

SOLUTION

Noting that $(2x + 3)^2 = 4x^2 + 2(2x)(3) + 9 = 4x^2 + 12x + 9$,

a. $4x^2 + 9$ is not a perfect square because the third term is missing.

b. $4x^2 + 6x + 9$ is not a perfect square because the term $6x$ is *not twice* the product of $2x$ and 3.

c. $4x^2 + 12x - 9$ is not a perfect square because -9 is not the square of a monomial. [Note that $(-3)(-3) = +9$.]

SAMPLE PROBLEM 4

Factor $9x^2 + 30xy + 25y^2$.

SOLUTION

1. If there are three terms, continue. If not, write "not a perfect square." — There are three terms.
2. Find two terms that are perfect squares, A^2 and B^2. If two terms are not squares, write "not a perfect square." — $9x^2 = (3x)^2$, $25y^2 = (5y)^2$
3. State A and B. — $A = 3x$ and $B = 5y$
4. Form $2AB$. — $2AB = 2(3x)(5y) = 30xy$
5. If $2AB$ is the third term, form $(A + B)^2$. — $(3x + 5y)^2$
   If $-2AB$ is the third term, form $(A - B)^2$.
   If neither $2AB$ nor $-2AB$ is the third term, write "not a perfect square."

Therefore, $9x^2 + 30xy + 25y^2 = (3x + 5y)^2$.

SAMPLE PROBLEM 5

Factor $x^4 - 14x^2 + 49$.

SOLUTION

1. Three terms? — Yes.
2. Find A^2 and B^2. — $x^4 = (x^2)^2$ and $49 = 7^2$
3. State A and B. — $A = x^2$ and $B = 7$
4. Form $2AB$. — $2AB = 2(x^2)(7) = 14x^2$
5. Middle term is $-2AB$. Form $(A - B)^2$. — $(x^2 - 7)^2$

Therefore, $x^4 - 14x^2 + 49 = (x^2 - 7)^2$.

11. $x^2 + 8x + 16$

12. $y^2 + 2y + 1$

13. $y^2 - 4y + 4$

14. $y^2 - 6y + 9$

15. $x^2 - 6x + 36$

16. $x^2 + 2x + 4$

17. $4x^2 + 20x + 25$

18. $4x^2 - 20x + 25$

19. $4x^2 - 20x - 25$

20. $4x^2 + 10x + 25$

21. $x^2 + 25$

22. $x^2 - 16xy + 64y^2$

11. ______
12. ______
13. ______
14. ______
15. ______
16. ______
17. ______
18. ______
19. ______
20. ______
21. ______
22. ______

23. $x^2 + 18xy + 81y^2$

24. $x^2 + 20xy - 100y^2$

25. $x^2 + 100y^2 - 20xy$

26. $9x^2 + 6x + 1$

27. $16y^2 - 8y + 1$

28. $25t^4 - 10t^2 + 1$

29. $49x^2 + 84xy + 36y^2$

30. $4x^2 + 1$

23. ______

24. ______

25. ______

26. ______

27. ______

28. ______

29. ______

30. ______

31–40. Fill in the missing term so that the result is a perfect square trinomial. Then write the result as the square of a binomial.

SAMPLE PROBLEM 6	SAMPLE PROBLEM 7
$x^2 - (\quad) + 900$	$25x^4 + (\quad) + 16y^4$
SOLUTION	SOLUTION
$A = x$ and $B = 30$ and $2AB = 60x$	$A = 5x^2$, $B = 4y^2$, and $2AB = 40x^2y^2$
$x^2 - 60x + 900 = (x - 30)^2$	$25x^4 + 40x^2y^2 + 16y^4 = (5x^2 + 4y^2)^2$

31. $x^2 + (\quad) + 3600$

32. $49x^2 - (\quad) + y^2$

33. $4y^4 - (\quad) + 1$

34. $64x^2 + (\quad) + 81y^2$

35. $100x^2y^2 + (\quad) + 1$

36. $36x^4 - (\quad) + 25y^4$

37. $a^2b^2 + (\quad) + c^2$

38. $t^4 - (\quad) + 4$

39. $x^2 - (\quad) + 121y^2$

40. $x^2 + (\quad) + 144$

31. ______

32. ______

33. ______

34. ______

35. ______

36. ______

37. ______

38. ______

39. ______

40. ______

EXERCISES 4.6 S

1–20. Use the perfect square trinomial formula to obtain each product.

$$(A + B)^2 = A^2 + 2AB + B^2$$

1. $(x + 5)^2$
2. $(y - 6)^2$
3. $(10x - 3y)^2$
4. $(xy + 4)^2$
5. $(8x^2 - 1)^2$
6. $(-5x - y)^2$
7. $(9a - 4b)^2$
8. $(4b - 9a)^2$
9. $(-4b - 9a)^2$
10. $(4a - 9b)^2$
11. $(9x^2 - 1)^2$
12. $(y^2 - 10)^2$
13. $(x^2 + y^2)^2$
14. $(x^2 - y^2)^2$
15. $(y^2 - x^2)^2$
16. $(6x^2 - 5y^2)^2$
17. $(x^2 + 8)^2$
18. $(x^2 - 2)^2$
19. $(4t^2 - 1)^2$
20. $(x^3 + 1)^2$

1. ______
2. ______
3. ______
4. ______
5. ______
6. ______
7. ______
8. ______
9. ______
10. ______
11. ______
12. ______
13. ______
14. ______
15. ______
16. ______
17. ______
18. ______
19. ______
20. ______

NAME ______ DATE ______ COURSE ______

21–40. Write each perfect square trinomial as the square of a binomial. If the trinomial is not a perfect square, write "not a perfect square."

21. $9x^2 + 49y^2 - 42xy$

22. $x^4 - 10x^2 + 25$

23. $y^4 + 4y^2 + 4$

24. $25x^4 + 60x^2y^2 + 36y^4$

25. $16x^4 + 20x^2y^2 + 25y^4$

26. $49x^2 + 100y^2$

27. $25x^2 + 10x + 1$

28. $49y^2 - 14y + 1$

29. $36x^2 + 25$

30. $36x^2 - 12x + 1$

31. $100t^4 + 20t^2 + 1$

32. $64t^4 - 16t^2 + 1$

33. $4x^2 + y^2 - 4xy$

34. $4x^4 + 1 - 2x^2$

35. $9x^4 + 1 - 6x^2$

36. $x^4 + 2x^2 + 1$

37. $t^4 + t^2 + 1$

38. $100c^4 - 180c^2 + 81$

39. $x^6 + 2x^3y^3 + y^6$

40. $x^2 + y^2$

21. ____________________

22. ____________________

23. ____________________

24. ____________________

25. ____________________

26. ____________________

27. ____________________

28. ____________________

29. ____________________

30. ____________________

31. ____________________

32. ____________________

33. ____________________

34. ____________________

35. ____________________

36. ____________________

37. ____________________

38. ____________________

39. ____________________

40. ____________________

FACTORING GENERAL TRINOMIALS 4.7

Using the FOIL method to multiply,

$$\begin{aligned} & \quad\quad\quad\quad\quad\quad\quad\quad\quad \text{F} \quad\quad\quad\quad \text{O} \quad\quad\quad\quad \text{I} \quad\quad\quad\quad \text{L} \\ (2x + 5)(3x + 4) &= (2x)(3x) + (2x)(4) + (5)(3x) + (5)(4) \\ &= 6x^2 + 8x + 15x + 20 \\ &= 6x^2 + 23x + 20 \end{aligned}$$

Factoring a general trinomial such as $6x^2 + 23x + 20$ is the reverse of this multiplication. In other words, we want to find the binomial factors whose product is the general trinomial. In a sense, we are "undoing" the FOIL method of multiplication. We begin by knowing that the factored form is a product of two binomials.

SAMPLE PROBLEM 1

Factor $3x^2 + 11x + 10$.

SOLUTION

1. Write the factored form format. — $(\_\_\_x + \_\_\_)(\_\_\_x + \_\_\_)$
2. State product of first terms. — 3
3. List pairs of factors of the first terms product. — $3 \cdot 1$
4. State product of last terms. — 10
5. List pairs of factors of the last terms product. — $10 \cdot 1$ and $5 \cdot 2$
6. List the possible factored forms.
   - a. $(3x + 1)(x + 10)$
   - b. $(3x + 10)(x + 1)$
   - c. $(3x + 2)(x + 5)$
   - d. **$(3x + 5)(x + 2)$**
7. Find the sum of the outer and inner products of the forms in (6).
   - a. $30x + x = 31x$
   - b. $3x + 10x = 13x$
   - c. $15x + 2x = 17x$
   - d. **$6x + 5x = 11x$**
8. Select the set of factors whose sum of inner and outer products is the middle term of the given trinomial, in this case $11x$. For polynomials in this text, if the middle term of the given polynomial does not occur in (7), the polynomial is prime.

   Therefore, $3x^2 + 11x + 10 = (3x + 5)(x + 2)$.

EXERCISES 4.7

1–10. Supply the missing factor for each of the following.

SAMPLE PROBLEM 2

$6x^2 - 7x - 20 = (3x + 4)(\quad)$

SOLUTION

1. Check first terms product. $3x(\quad) = 6x^2$
   $3x(\mathbf{2x}) = 6x^2$
   $2x$ is the first term.
2. Check last terms product. $4(\quad) = -20$
   $4(\mathbf{-5}) = -20$
   -5 is the last term.
3. Supply the missing factor. $(3x + 4)\mathbf{(2x - 5)}$

1. $6x^2 + 23x + 20 = (2x + 5)(\quad)$ **1.** ____________

2. $8x^2 + 18x + 7 = (4x + 7)(\quad)$ **2.** ____________

3. $5y^2 - 47y + 18 = (5y - 2)(\quad)$ **3.** ____________

4. $6y^2 - 13y + 6 = (3y - 2)(\quad)$ **4.** ____________

5. $10x^2 - 11x - 6 = (5x + 2)(\quad)$ **5.** ____________

6. $4x^2 + 5x - 6 = (4x - 3)(\quad)$ **6.** ____________

7. $2x^2 - 19x - 10 = (2x + 1)(\quad)$ **7.** ____________

8. $5x^2 - 48x + 27 = (x - 9)(\quad)$ **8.** ____________

9. $6t^2 - t - 15 = (3t - 5)(\quad)$ **9.** ____________

10. $6t^2 - t - 7 = (t + 1)(\quad)$ **10.** ____________

11–18. Supply the missing signs for each pair of factors.

SAMPLE PROBLEM 3

$8x^2 + 2x - 15 = (4x \quad 5)(2x \quad 3)$

SOLUTION

1. Form outer product. $4x(3) = 12x$
2. Form inner product. $5(2x) = 10x$
3. Select signs for products so sum is given middle term. $+12x$, $-10x$, sum $+ 2x$
4. Insert signs to produce products of Step 3. $8x^2 + 2x - 15 = (4x - 5)(2x + 3)$

11. $6x^2 - x - 7 = (6x \quad 7)(x \quad 1)$ **11.** ______

12. $6x^2 + x - 7 = (6x \quad 7)(x \quad 1)$ **12.** ______

13. $6x^2 - 13x + 7 = (6x \quad 7)(x \quad 1)$ **13.** ______

14. $6x^2 + 13x + 7 = (6x \quad 7)(x \quad 1)$ **14.** ______

15. $9y^2 + 9y + 2 = (3y \quad 1)(3y \quad 2)$ **15.** ______

16. $9y^2 + 3y - 2 = (3y \quad 1)(3y \quad 2)$ **16.** ______

17. $9y^2 - 3y - 2 = (3y \quad 1)(3y \quad 2)$ **17.** ______

18. $9y^2 - 9y + 2 = (3y \quad 1)(3y \quad 2)$ **18.** ______

19–40. Factor completely, if possible. If the polynomial is prime, write "prime."

SAMPLE PROBLEM 4

Factor $5x^2 + 3x - 14$.

SOLUTION

1. Write factored form format. $(\_\_x + \_\_)(\_\_x + \_\_)$
2. List pairs of factors of first terms product. $5 \cdot 1$
3. Write possible factored forms. $(5x + \_\_)(x + \_\_)$
4. List pairs of factors of last terms product. $(1)(-14)$, $(-1)(14)$; $(2)(-7)$, $(-2)(7)$
5. List the possible factored forms.

$(5x + 1)(x - 14)$	$(5x - 14)(x + 1)$
$(5x - 1)(x + 14)$	$(5x + 14)(x - 1)$
$(5x + 2)(x - 7)$	$\mathbf{(5x - 7)(x + 2)}$
$(5x - 2)(x + 7)$	$(5x + 7)(x - 2)$

6. Find the sum of the inner and outer products of the forms in Step 5. Select the factored form for which the sum is the middle term (x term) of the given trinomial. The sum should be $+3x$.

$(5x - 7)(x + 2)$

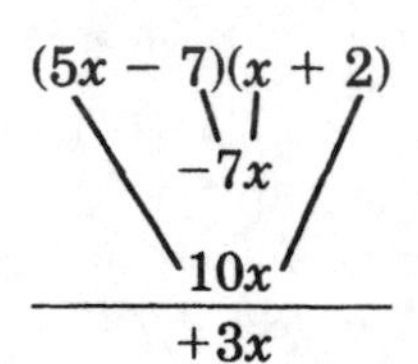

$-7x$

$10x$

$+3x$

Therefore, $5x^2 + 3x - 14 = (5x - 7)(x + 2)$.

SAMPLE PROBLEM 5

Factor $6y^2 - 7y - 5$.

SOLUTION

1. List pairs of factors of first terms product. $1 \cdot 6$ and $2 \cdot 3$
2. List possible factored forms. $(6y + \_\_)(y + \_\_)$ or $(2y + \_\_)(3y + \_\_)$
3. List pairs of factors of last terms product. $(-1)(5)$ and $(1)(-5)$
4. List possible factored forms, but omit signs. Find inner and outer products. Then determine the signs to obtain the middle term of the given trinomial.

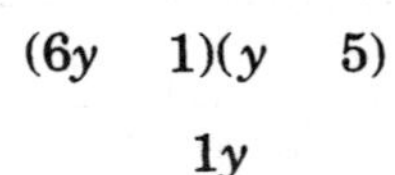

$(6y \quad 1)(y \quad 5)$	$(6y \quad 5)(y \quad 1)$	$(2y \quad 1)(3y \quad 5)$	$(2y \quad 5)(3y \quad 1)$
$1y$	$5y$	$+3y$	$15y$
$30y$	$6y$	$-10y$	$2y$
		$-7y$	

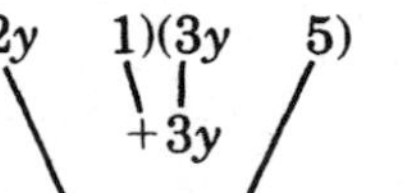

5. Therefore, $(2y + 1)(3y - 5)$ is the factored form.

19. $5x^2 + 11x + 2$ **19.** __________

20. $3x^2 - 22x + 7$ **20.** __________

21. $2x^2 - 5x - 3$ 21. ______

22. $7x^2 + 13x - 2$ 22. ______

23. $3y^2 + 2y - 5$ 23. ______

24. $2y^2 - 3y - 5$ 24. ______

25. $7t^2 - 2t - 5$ 25. ______

26. $5t^2 + 6t + 2$ 26. ______

27. $11x^2 + 9xy - 2y^2$ 27. ______

28. $33x^2 - 8xy - y^2$ 28. ______

29. $2x^3 - 7xy + 6y^2$ 29. ______

30. $2x^2 - xy - 6y^2$ 30. ______

31. $5y^2 + 6y - 8$ 31. ______

32. $5y^2 - 6y - 8$ 32. ______

33. $6x^2 - 17x + 5$ 33. ______

34. $6x^2 + 15x - 5$ 34. ______

35. $4x^2 - 8x + 3$

36. $9x^2 - 12x - 5$

37. $4x^2 - 17x + 4$

38. $9x^2 + 4x - 5$

39. $5x^2 - 4x - 1$

40. $4x^2 - 5x - 1$

35. ______________

36. ______________

37. ______________

38. ______________

39. ______________

40. ______________

EXERCISES 4.7 S

1–10. Supply the missing factor.

1. $3r^2 - 5r + 2 = (r - 1)(\quad)$ 1. ________
2. $18r^2 + 3r - 28 = (3r + 4)(\quad)$ 2. ________
3. $52x^2 + xy - 30y^2 = (4x - 3y)(\quad)$ 3. ________
4. $2x^2 - 5xy + 2y^2 = (x - 2y)(\quad)$ 4. ________
5. $15x^2 - xy - 2y^2 = (3x + y)(\quad)$ 5. ________
6. $10x^2 + 13xy - 3y^2 = (5x - y)(\quad)$ 6. ________
7. $4x^2 - 15xy - 4y^2 = (4x + y)(\quad)$ 7. ________
8. $6x^2 - 25x + 14 = (2x - 7)(\quad)$ 8. ________
9. $6x^2 - 17x - 14 = (2x - 7)(\quad)$ 9. ________
10. $6x^2 + 17x - 14 = (2x + 7)(\quad)$ 10. ________

11–18. Supply the missing signs for each pair of factors.

11. $8s^2 - 6st - 5t^2 = (4s \quad 5t)(2s \quad t)$ 11. ________
12. $40s^2 - 17st - 12t^2 = (8s \quad 3t)(5s \quad 4t)$ 12. ________
13. $12x^2 - 97xy + 8y^2 = (12x \quad y)(x \quad 8y)$ 13. ________
14. $20x^2 - 109xy - 100y^2 = (4x \quad 25y)(5x \quad 4y)$ 14. ________
15. $10a^2 + 31ab - 14b^2 = (2a \quad 7b)(5a \quad 2b)$ 15. ________
16. $30x^2 + 31xy - 21y^2 = (2x \quad 3y)(15x \quad 7y)$ 16. ________
17. $21a^2 - 26ab + 8b^2 = (7a \quad 4b)(3a \quad 2b)$ 17. ________
18. $21a^2 + 2ab - 8b^2 = (7a \quad 4b)(3a \quad 2b)$ 18. ________

NAME ________ DATE ________ COURSE ________

19–40. Factor completely. If the polynomial is prime, write "prime."

19. $15y^2 - y - 2$

20. $15y^2 + y - 2$

21. $15x^2 - 11xy + 2y^2$

22. $15x^2 - 17xy + 2y^2$

23. $15x^2 + 29x - 2$

24. $5t^2 - 8t + 7$

25. $6t^2 + t - 2$

26. $27a^2 + 48a + 5$

27. $10b^2 + b - 2$

28. $8x^2 - 37x - 15$

29. $8x^2 + 14x - 15$

30. $15x^2 - 22xy + 8y^2$

31. $6x^2 - 2xy - 5y^2$

32. $18r^2 + 3rs - 28s^2$

33. $18r^2 - 3rs - 28s^2$

34. $18r^2 + 55rs - 28s^2$

35. $10x^2 - 43x - 9$

36. $10x^2 + 13x - 9$

37. $10x^2 - 21x + 9$

38. $10x^2 + 89x - 9$

39. $10x^2 - 19x + 9$

40. $10x^2 - 27x - 9$

19. ______________

20. ______________

21. ______________

22. ______________

23. ______________

24. ______________

25. ______________

26. ______________

27. ______________

28. ______________

29. ______________

30. ______________

31. ______________

32. ______________

33. ______________

34. ______________

35. ______________

36. ______________

37. ______________

38. ______________

39. ______________

40. ______________

4.8 ONE-STEP FACTORING PROBLEMS

The factoring formulas we have studied so far are listed below for convenience.

1. Monomial factor: $AX + AY = A(X + Y)$
   $AX + AY + AZ = A(X + Y + Z)$
2. Difference of squares: $A^2 - B^2 = (A - B)(A + B)$
3. Perfect square trinomial: $A^2 + 2AB + B^2 = (A + B)^2$
   $A^2 - 2AB + B^2 = (A - B)^2$
4. Simple trinomial: $X^2 + (a + b)X + ab = (X + a)(X + b)$
5. General trinomial:

$$acX^2 + (ad + bc)X + bd = (aX + b)(cX + d)$$

NOTE TO STUDENT:
There is no form for the sum of two squares.
$A^2 + B^2$ cannot be factored as the product of two binomials.

The first step in factoring a polynomial is to recognize the polynomial as having one of the forms on the left side of the above equations. (This means that the formulas must be memorized.) Then the polynomial is written in its factored form, the form on the right side of one of the above formulas.

Unless we recognize the form immediately, we test the above five formulas in the order in which they are listed: monomial factor first, then difference of squares, then perfect square trinomial, and so on.

SAMPLE PROBLEM 1

Factor $9x^2 - 25$.

SOLUTION

We first check to see if there is a monomial factor common to each term. There is none in this case.

Next we test if the polynomial is a difference of squares:

$$9x^2 - 25 = (3x)^2 - 5^2 = A^2 - B^2$$

Recalling that $A^2 - B^2 = (A - B)(A + B)$, we write out the right side:

$$9x^2 - 25 = (3x)^2 - 5^2 = (3x - 5)(3x + 5)$$

SAMPLE PROBLEM 2

Factor $x^2 - 4x - 21$.

SOLUTION

1. Checking, we find there is no common monomial factor.
2. We do not have a difference of squares, since we have three terms and not two.
3. We do not have a perfect square trinomial, since two terms are not perfect squares.
4. So we try the simple trinomial formula:
   $X^2 + (a + b)X + ab = (X + a)(X + b)$

We need to find two numbers, a and b, whose product is -21 and whose sum is -4. The numbers are -7 and 3. Therefore,

$$x^2 - 4x - 21 = (x - 7)(x + 3)$$

SAMPLE PROBLEM 3

Factor $36x^2 + 36x + 36$.

SOLUTION

Checking for a common monomial factor first, we note that 36 divides each term. Therefore,

$$36x^2 + 36x + 36 = 36(x^2 + x + 1)$$

Since no two integers have both a sum of 1 and a product of 1, our factoring is completed.

Note how many trials it would have taken if we had tried to use the general trinomial formula first. It always pays to check for a common monomial factor first.

SAMPLE PROBLEM 4

Factor $x^2 + 25$, if possible.

SOLUTION

1. There is no common monomial factor.
2. This is a sum of squares.
   $x^2 + 25$ cannot be factored.
   $x^2 + 25$ is prime.

SAMPLE PROBLEM 5

Factor $36x^2 - 60x + 25$.

SOLUTION

Checking for the common monomial factor first, we note that there is none.

Noting that $36x^2$ and $+25$ are perfect squares, we test the perfect square trinomial formula:

If we have the form $A^2 - 2AB + B^2$, then $A = 6x$ and $B = 5$.

We form $2AB$ and check the middle term. $2AB = 2(6x)(5) = 60x$.

Therefore we form $(A - B)^2$.

$$(6x - 5)^2 = 36x^2 - 2(6x)(5) + 25 = 36x^2 - 60x + 25$$

Therefore, $36x^2 - 60x + 25 = (6x - 5)^2$.

Note how many trials were saved by not trying the general trinomial formula first.

EXERCISES 4.8

1–8. Match each polynomial in Column 1 with its factored form in Column 2.

Column 1	Column 2	
1. $4x^2 + 36$	**a.** $(2x + 3)^2$	**1.** ________
2. $4x^2 - 9$	**b.** $(2x + 3)(2x - 3)$	**2.** ________
3. $4x^2 - 9x$	**c.** $(4x + 1)(x - 3)$	**3.** ________
4. $4x^2 + 9$	**d.** $(4x + 9)(x + 1)$	**4.** ________
5. $4x^2 + 12x + 9$	**e.** $4(x^2 + 9)$	**5.** ________
6. $4x^2 + 4x + 4$	**f.** $x(4x - 9)$	**6.** ________
7. $4x^2 + 13x + 9$	**g.** $4(x^2 + x + 1)$	**7.** ________
8. $4x^2 - 11x - 3$	**h.** Not factorable	**8.** ________

9–26. Factor if possible. If not possible, write "prime."

9. $9x^2 - 25$

10. $9x^2 - 25x$

11. $9x^2 + 36$

12. $9x^2 + 25$

13. $x^2 - 12x + 36$

14. $9x^2 - 30x + 25$

15. $x^2 - 5x - 36$

16. $9x^2 + 19x + 2$

17. $9x^2 + 18x + 27$

18. $49x^2 - 1$

19. $x^2 + 4x - 32$

20. $5x^2 - 2x - 3$

21. $x^4 - 100$

22. $x^4 - 100x^3$

23. $x^4 + 20x^2 + 100$

24. $x^4 - x^3 + x^2$

25. $x^4 + x^2y - 2y^2$

26. $x^4 + 9x^2 + 20$

9. ______________________

10. ______________________

11. ______________________

12. ______________________

13. ______________________

14. ______________________

15. ______________________

16. ______________________

17. ______________________

18. ______________________

19. ______________________

20. ______________________

21. ______________________

22. ______________________

23. ______________________

24. ______________________

25. ______________________

26. ______________________

1–8. Each polynomial in Column 1 has a factor listed in Column 2. Find this factor for each polynomial.

Column 1	Column 2	
1. $25x^2 - 4$	**a.** $x + 1$	**1.** ______
2. $25x^2 + x^4$	**b.** $x + 5$	**2.** ______
3. $25x^2 + 20x$	**c.** $x - 5$	**3.** ______
4. $25x^2 + 25$	**d.** $5x + 1$	**4.** ______
5. $x^2 - 5x - 50$	**e.** $5x + 2$	**5.** ______
6. $x^2 + 26x + 25$	**f.** $5x + 4$	**6.** ______
7. $25x^2 + 10x + 1$	**g.** $x^2 + 25$	**7.** ______
8. $5x^2 - 26x + 5$	**h.** $x^2 + 1$	**8.** ______

9–40. Factor if possible. If not possible, write "prime."

9. $x^2 - 36$

10. $x^4 - 36$

11. $x^4 + 36x^2$

12. $x^2 - 14x + 49$

13. $x^4 - 14x^2 + 49$

14. $x^2 + 6x - 16$

15. $x^4 + 6x^2 - 16$

16. $7x^2 + 4x - 3$

17. $7x^4 + 4x^2 - 3$

18. $x^4 - 64y^2$

9. ______
10. ______
11. ______
12. ______
13. ______
14. ______
15. ______
16. ______
17. ______
18. ______

NAME ______ DATE ______ COURSE ______

19. $x^4 + 64x^2$

20. $36x^4 + 12x^2 + 1$

21. $4x^2 - 25$

22. $4x^2 - 100x$

23. $27y^3 - 81y^2$

24. $4x^2 - 28x + 49$

25. $t^2 - t - 30$

26. $3x^3 - 3x^2 - 30x$

27. $3x^2 - 13x - 10$

28. $x^4 - 36y^2$

29. $x^4 - 8x^2 + 12$

30. $25x^4 + 20x^2 + 4$

31. $x^3y + 4xy^3$

32. $5n^4 + 41n^2 + 8$

33. $x^4 + x^2 - 12$

34. $9y^4 - 42y^2 + 49$

35. $5x^4 + 3x^2 - 2$

36. $25x^2 + 9$

37. $4x^2 + 100$

38. $16y^2 + 1$

39. $16y^2 + 64$

40. $49t^2 + 81$

19. ______________

20. ______________

21. ______________

22. ______________

23. ______________

24. ______________

25. ______________

26. ______________

27. ______________

28. ______________

29. ______________

30. ______________

31. ______________

32. ______________

33. ______________

34. ______________

35. ______________

36. ______________

37. ______________

38. ______________

39. ______________

40. ______________

MULTIPLE-STEP FACTORING PROBLEMS 4.9

A polynomial whose coefficients are integers is said to be **completely factored** if

1. The polynomial is expressed as a product of polynomial factors whose coefficients are integers.
2. None of these factors can be expressed as a product of two different polynomials whose coefficients are integers.

After a polynomial has been factored by using one of the formulas listed in Section 4.8, each polynomial factor must be checked to see if it can be factored.

PROCEDURE FOR COMPLETE FACTORIZATION

1. **First, find the GCMF (greatest common monomial factor), if there is one.**
2. **Count the number of terms in a nonmonomial factor.**
   - **2.1 Two terms (binomial).**
     - **a. If a difference of squares, $A^2 - B^2$, factor.**
     - **b. If not, stop.**
   - **2.2 Three terms (trinomial).**
     - **a. If a perfect square trinomial, factor as $(A + B)^2$.**
     - **b. If not, try to factor as a product of binomial factors.**
     - **c. If neither Step a nor Step b is possible, stop.**
3. **Repeat Step 2 until none of the factors can be factored.**

SAMPLE PROBLEM 1

Completely factor $2x^5 - 162x$.

Check by multiplication.

SOLUTION

Remove the GCMF.	$2x^5 - 162x = \mathbf{2x}(x^4 - 81)$
Difference of squares	$= 2x\mathbf{(x^2 - 9)(x^2 + 9)}$
Difference of squares	$= 2x\mathbf{(x - 3)(x + 3)}(x^2 + 9)$

CHECK

$$2x(x^2 - 9)(x^2 + 9) = 2x(x^4 - 81)$$
$$= 2x(x^4) - 2x(81)$$
$$= 2x^5 - 162x$$

SAMPLE PROBLEM 2

Completely factor $18x^2 - 66x - 24$.

SOLUTION

Remove the GCMF. $18x^2 - 66x - 24 = 6(3x^2 - 11x - 4)$

Use the general trinomial product formula. $= 6(3x + 1)(x - 4)$

CHECK

Check by multiplication.

$$6(3x + 1)(x - 4) = 6(3x^2 - 12x + x - 4)$$
$$= 6(3x^2 - 11x - 4)$$
$$= 18x^2 - 66x - 24$$

SAMPLE PROBLEM 3

Completely factor $x^4 + 4x^2 - 32$.

SOLUTION

Simple trinomial product $x^4 + 4x^2 - 32 = (x^2 - 4)(x^2 + 8)$

Difference of squares $= (x - 2)(x + 2)(x^2 + 8)$

EXERCISES 4.9

Completely factor each of the following. Check by multiplication.

1. $5x^3 - 35x^2 + 50x$ 1. ____________

THREE TERMS METHOD

1. Remove greatest common monomial factor, if any.
   Use $AB + AC + AD = A(B + C + D)$.
2. For the form $AX^2 + BX + C$,
   a. List pairs of factors of AX^2.
   b. List pairs of factors of C.
   c. List possible factored forms: $(aX + b)(cX + d)$.
   d. Select from Step c the form whose sum of outer and inner products $= BX$.
3. Check each binomial factor for the form $A^2 - B^2$.
   Rewrite $A^2 - B^2$ as $(A - B)(A + B)$.
   Do not rewrite any other form.
4. Repeat Step 3 if necessary.

2. $16x^4 - 81y^4$

2. ______

TWO TERMS METHOD

1. Remove greatest common monomial factor, if any. Use $AB + AC = A(B + C)$.
2. Check binomial factor for form $A^2 - B^2$.
3. If binomial has form $A^2 - B^2$, rewrite it as $(A - B)(A + B)$. If not, stop.
4. Repeat Steps 2 and 3, as necessary.

3. $100x^2 - 4y^2$

4. $100x^2 - 4x$

3. ______

4. ______

5. $16y^4 + 64y^2$

6. $6x^4 - 24x^3 - 126x^2$

5. ______

6. ______

7. $9x^4 - 72x^3 + 144x^2$

8. $25x^2 - 50x + 100$

7. ______

8. ______

9. $256x^4 - 625y^4$

10. $15x^2 + 60x + 60$

9. ______

10. ______

11. $12x^2 + 60x - 72$

12. $36x^4 - 144x^2$

11. ______

12. ______

13. $21x^3 - 21x^2 - 21x$

14. $3x^3 - 8x^2 + 5x$

15. $42x^2 - 9x - 6$

16. $72x^2 - 60x - 28$

17. $x^4 - 10x^2 + 9$

18. $16x^4 - 68x^2 + 16$

19. $625x^4 + 600x^2 - 25$

20. $72x^4 + 178x^2 - 5$

13. ______________________

14. ______________________

15. ______________________

16. ______________________

17. ______________________

18. ______________________

19. ______________________

20. ______________________

EXERCISES 4.9 S

Completely factor.

1. $9x^2 - 9x - 18$
2. $9x^2y + 9xy - 18y$
3. $3x^3 - 17x^2 - 6x$
4. $144x^3y - 100xy^3$
5. $324x^4 - 4y^4$
6. $27t^3 + 36t^2 + 63t$
7. $49x - x^3$
8. $y^8 - 16y^4$
9. $5x^2 - 15x - 50$
10. $4y^2 - 28y + 49$
11. $6x^2 - 20x + 16$
12. $2c^2 - 20c^3 + 50c^4$
13. $y^5 - y$
14. $6x^2 - 54x + 84$
15. $-3x^3 - 27x$
16. $5a^5 + 60a^3 - 320a$
17. $6b^4 + b^3 - 12b^2$
18. $x^4 - 16x^2$
19. $x^4 - 13x^2 + 36$
20. $4a^4 - 16b^4$
21. $2x^4y^4 - 32$
22. $t^6 + 5t^4 - 36t^2$

1. __________
2. __________
3. __________
4. __________
5. __________
6. __________
7. __________
8. __________
9. __________
10. __________
11. __________
12. __________
13. __________
14. __________
15. __________
16. __________
17. __________
18. __________
19. __________
20. __________
21. __________
22. __________

NAME __________ DATE __________ COURSE __________

23. $4x^4 - 48x^3 + 144x^2$

24. $x^4 - 24x^2 - 25$

25. $16u^4 - 36v^4$

26. $-x^4 + 4x^3 + 5x^2$

27. $3x^3 - 15x^2 - 42x$

28. $4y^2 - 48y + 144$

29. $16n^4 - 1$

30. $x^4 - 5x^2 - 36$

31. $2x^4 - 7x^3 + 5x^2$

32. $x^4 - 81y^4$

33. $x^4 - 64x^2$

34. $x^4 - 18x^2 + 81$

35. $4y^4 + 36y^2 - 40$

36. $10x^3y + 40x^2y^2 + 40xy^3$

23. ______

24. ______

25. ______

26. ______

27. ______

28. ______

29. ______

30. ______

31. ______

32. ______

33. ______

34. ______

35. ______

36. ______

4.10 SOLUTION OF EQUATIONS BY FACTORING

The following are examples of quadratic equations in the variable x.

$$3x^2 - 5x + 2 = 0 \qquad x^2 = 8 - 2x$$
$$4x^2 - 9x = 0 \qquad x^2 = 9$$

A **quadratic equation in one variable** is a polynomial equation in which the highest power of the variable is the second.

The **standard form of a quadratic equation** in the variable x is

$$ax^2 + bx + c = 0$$

where a, b, and c are constants and a is positive.

The factoring methods studied in this chapter can be used to solve some types of quadratic equations. Also needed is a property called the **zero product property** of real numbers. This property states that if a product of two factors is zero, then one factor or the other factor must equal zero.

ZERO PRODUCT PROPERTY

If $AB = 0$, then $A = 0$ or $B = 0$

TO SOLVE A QUADRATIC EQUATION BY FACTORING:

1. **Write the equation in standard form.**
2. **Factor the quadratic polynomial.**
3. **Set each linear factor equal to zero.**
4. **Solve the resulting equations.**
5. **Check each solution in the original equation.**
6. **State the solution set.**

In general, a quadratic equation has two solutions, also called **roots** of the equation. If two linear factors of the quadratic polynomial are identical, then only one solution is obtained and it is called a **double root.**

SAMPLE PROBLEM 1

Solve by factoring:

$x^2 - 9x + 20 = 0$

SOLUTION

Step	Work	
1. Write in standard form.	$x^2 - 9x + 20 = 0$	
2. Factor.	$(x - 5)(x - 4) = 0$	
3. Set factors equal to 0.	$x - 5 = 0$ or	$x - 4 = 0$
4. Solve equations.	$x = 5$	$x = 4$
5. Check in original equation.	$5^2 - 9(5) + 20 = 0$	$4^2 - 9(4) + 20 = 0$
	$25 - 45 + 20 = 0$	$16 - 36 + 20 = 0$
	$0 = 0$	$0 = 0$
6. State solution set.	The solution set is {4, 5}.	

SAMPLE PROBLEM 2

Solve by factoring:

$(x + 3)(x + 5) = 3$

SOLUTION

Step	Work	
1. Write in standard form.	$x^2 + 8x + 15 = 3$	
	$x^2 + 8x + 12 = 0$	
2. Factor.	$(x + 6)(x + 2) = 0$	
3. Set factors equal to 0.	$x + 6 = 0$ or	$x + 2 = 0$
4. Solve equations.	$x = -6$	$x = -2$
5. Check in original equation.	$(-6 + 3)(-6 + 5) = 3$	$(-2 + 3)(-2 + 5) = 3$
	$(-3)(-1) = 3$	$(1)(3) = 3$
	$3 = 3$	$3 = 3$
6. State solution set.	The solution set is {−6, −2}.	

SAMPLE PROBLEM 3

Solve $3x^2 = 15x$.

SOLUTION

1. $3x^2 - 15x = 0$
2. $3x(x - 5) = 0$
3. $3x = 0$ or $x - 5 = 0$
4. $x = 0$ $\qquad x = 5$
5. $3(0)^2 = 15(0)$ $\qquad 3(5^2) = 15(5)$

   $0 = 0$ $\qquad 75 = 75$
6. The solution set is {0, 5}.

SAMPLE PROBLEM 4

Solve $x^2 = 25$.

SOLUTION

1. $x^2 - 25 = 0$
2. $(x + 5)(x - 5) = 0$
3. $x + 5 = 0$ or $x - 5 = 0$
4. $x = -5$ $\quad x = 5$
5. $(-5)^2 = 25$ and $(5)^2 = 25$
6. The solution set is $\{-5, 5\}$.

SAMPLE PROBLEM 5

Solve $x^2 + 9 = 6x$.

SOLUTION

1. $x^2 - 6x + 9 = 0$
2. $(x - 3)(x - 3) = 0$
3. $x - 3 = 0$ or $x - 3 = 0$
4. $x = 3$ $\quad x = 3$
5. $3^2 + 9 = 6(3)$
   $18 = 18$
6. The solution set is $\{3\}$ where 3 is a double root.

EXERCISES 4.10

Solve by factoring and check.

1. $x^2 - 2x - 8 = 0$

2. $x^2 + 5x = 14$

3. $x^2 - 25 = 0$

4. $x^2 = 49$

5. $4x(x - 6) = 0$

6. $x^2 = 12(x - 3)$

7. $x^2 + 16 = 8x$

8. $(x + 5)(x - 4) = 10$

1. __________

2. __________

3. __________

4. __________

5. __________

6. __________

7. __________

8. __________

9. $2x(2x - 25) = 50(2 - x)$

10. $6(x + 2)(x + 3) = 0$

9. ______

10. ______

11. $12x = 4x^2$

12. $y^2 = 4(y - 1)$

11. ______

12. ______

13. $100t - 25t^2 = 0$

14. $(y + 2)(y + 3) = 2$

13. ______

14. ______

15. $r^2 = 100$

16. $(u - 5)^2 = (2u + 5)^2$

15. ______

16. ______

17. $16(t + 1) = 8t(t + 1)$

18. $y(y - 2) = 5(y - 2)$

17. ______

18. ______

19. $(r - 8)^2 = 64$

20. $(4x + 7)^2 = 7(8x + 7)$

19. ______

20. ______

Solve and check.

1. $x^2 + 6x + 5 = 0$

2. $x^2 - 3x = 18$

3. $2x^2 - 18 = 0$

4. $64 = x^2$

5. $5x(x + 7) = 0$

6. $2x + 8 = x^2$

7. $(x + 5)(x + 10) - x + 1$

8. $x(x - 7) - (2x + 1)(x - 4)$

9. $(x - 4)(x + 2) = 40$

10. $(x + 3)^2 = 2(3x + 5)$

1. ____________

2. ____________

3. ____________

4. ____________

5. ____________

6. ____________

7. ____________

8. ____________

9. ____________

10. ____________

NAME ____________ DATE ____________ COURSE ____________

11. $x^2 + 15 = 8x$

12. $81y = 9y^2$

11. ________________

12. ________________

13. $t(t + 3) = t - 1$

14. $(y - 4)(y + 5) = 22$

13. ________________

14. ________________

15. $(r + 4)^2 = (3r - 4)^2$

16. $2u^2 = 22u - 56$

15. ________________

16. ________________

17. $(t + 7)(t - 7) = 3(t - 3)$

18. $5(y - 1) = y(y - 1)$

17. ________________

18. ________________

19. $(2x + 1)^2 = 3x^2 + 1$

20. $(x - 6)^2 = 36$

19. ________________

20. ________________

APPLICATIONS (OPTIONAL) 4.11

A great variety of practical problems can be solved by using quadratic equations. Some of these will be shown in this section. Others may be found in Chapter 9. Many of the problems shown here have been simplified so that the answers will be integers.

For the problems in this section, the method of short statements or the formula method, both discussed in Chapter 3, can be used to form the equation from the words of the problem.

When solving applied problems, it is very important to check each root of the equation in the statement of the problem. A quadratic equation may have two roots, but only one may meet the conditions of the problem. For example, lengths, widths, areas, and volumes must be positive numbers. Numbers of people or objects must be natural numbers. An amount to be cut from a given length of material cannot be greater than the given length.

There are many problems that require the use of the **Theorem of Pythagoras.**

Theorem of Pythagoras

In a right triangle, the sum of the squares of the legs is equal to the square of the hypotenuse. In symbols, referring to Figure 4.1, this theorem is expressed as $c^2 = a^2 + b^2$.

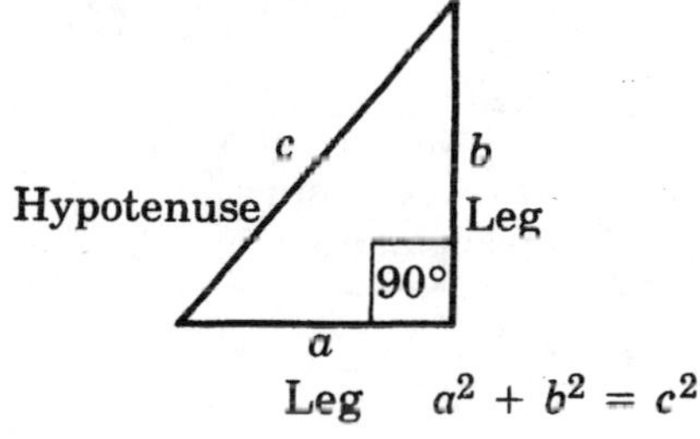

FIGURE 4.1

The **hypotenuse** is the longest side of a right triangle, the side opposite the 90-degree (right) angle. The **legs** of a right triangle are the sides that form the right angle.

For Sample Problems 1 and 2, refer to the right triangle in Figure 4.1. Use the Table of Squares on the front inside cover, if needed.

SAMPLE PROBLEM 1	SAMPLE PROBLEM 2
Find c if $a = 15$ in and $b = 20$ in. **SOLUTION** Use the Theorem of Pythagoras. $c^2 = a^2 + b^2$ $c^2 = 15^2 + 20^2 = 225 + 400$ $c^2 = 625 = (25)^2$ $c = 25$ or $c = -25$ A geometric measurement cannot be negative. $c = 25$ in	Find b if $c = 26$ meters and $a = 10$ meters. **SOLUTION** $a^2 + b^2 = c^2$ $10^2 + b^2 = 26^2$ $b^2 = 26^2 - 10^2$ $b^2 = (26 - 10)(26 + 10)$ $b^2 = 16(36) = (24)^2$ $b = 24$ or $b = -24$ $b = 24$ meters

EXERCISES 4.11

1. In a right triangle, find c for $a = 3$ meters and $b = 4$ meters.

2. In a right triangle, find b for $c = 25$ in and $a = 24$ in.

1. ____________

2. ____________

3. A builder needs to know how much lumber to cut for the roof of a certain house. To do this, he calculates the hypotenuse, c, of a right triangle whose legs are 12 ft and 16 ft. Find c.

3. ____________

4. A builder lays off a rectangular area of land that is 30 meters wide and 40 meters long. To check that the area is a true rectangle, he measures the diagonal of the area on the land. Then he calculates the diagonal by using the Theorem of Pythagoras. Find the diagonal of this rectangle.

4. ____________

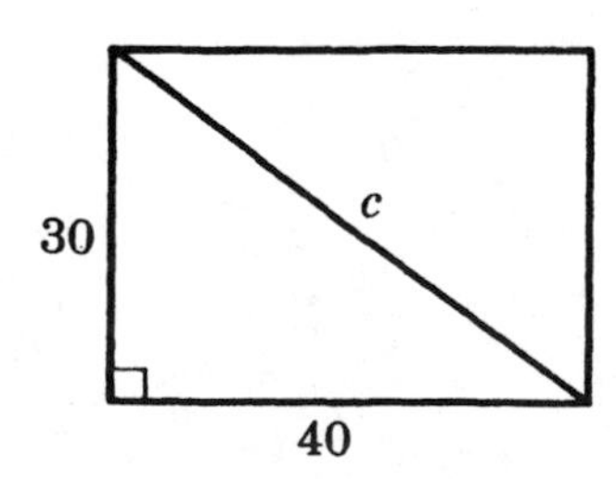

5. A 13-ft guy wire is to be used to brace a 5-ft-high television antenna on a level roof. How far from the bottom of the antenna should the wire be attached?

5. 

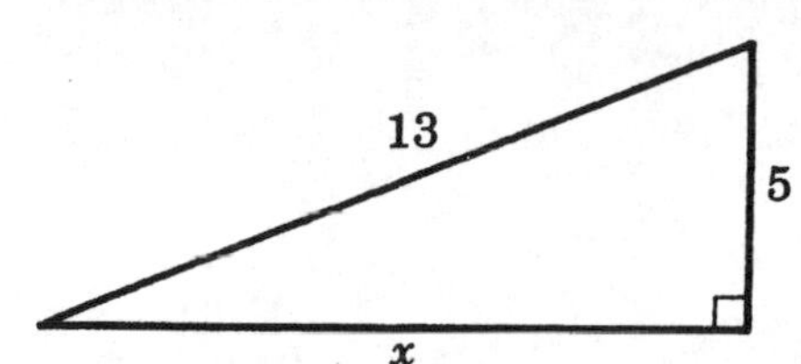

6. A 15-ft ladder leans against the side of a house. The bottom of the ladder is 9 ft from the bottom of the house. How high up will the ladder reach?

6. 

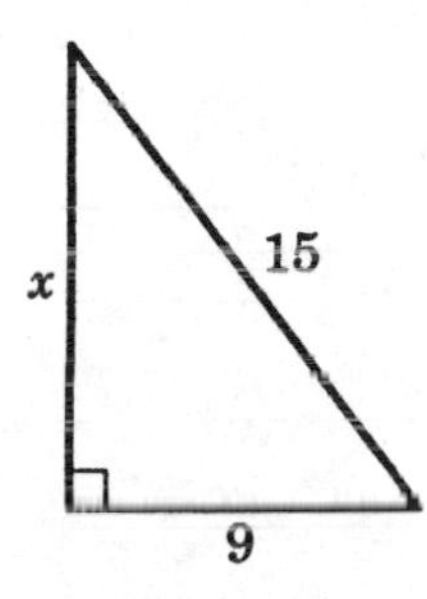

7. An architect wants to increase one side of a square room by 5 ft and decrease an adjacent side by 3 ft so that the area of the resulting rectangular room will be 240 sq ft. Find the length, x, of one side of the square room.
Use $(x - 3)(x + 5) = 240$.

7. 

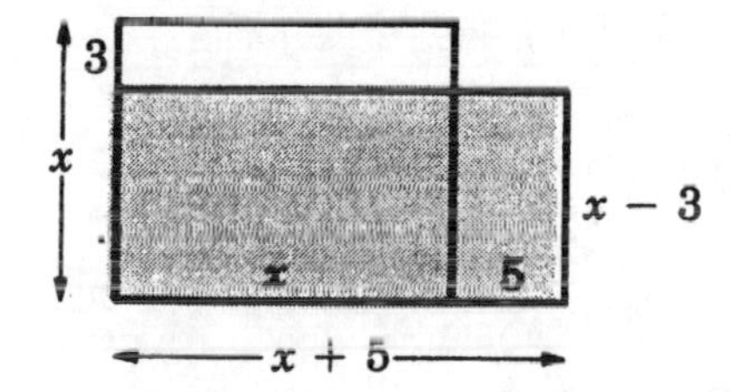

8. An architect uses the formula

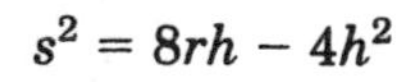

$$s^2 = 8rh - 4h^2$$

to find the height, h, of a circular arch whose radius, r, is 13 ft and whose span, s, is 24 ft. Find h. (h must be less than 13, the radius.)

8. ____________________

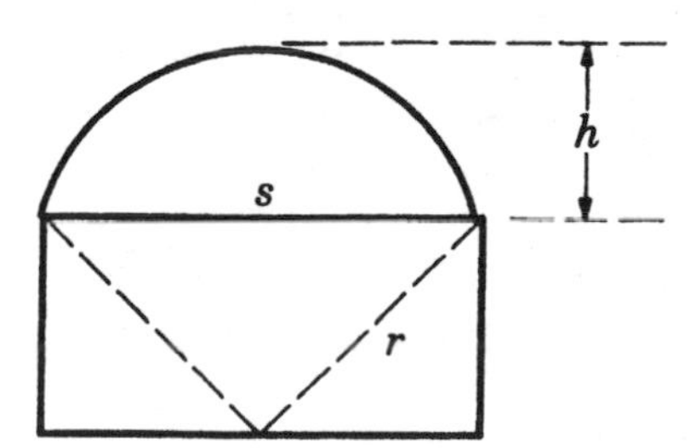

9. A gardener wants to plant a uniform width of ground cover along two widths and one length of a rectangular piece of land. The width of the rectangular area is 12 ft, and the length is 18 ft. How wide should the planted area be if the unplanted area will be 140 sq ft?
Use $(18 - 2x)(12 - x) = 140$.

9. 

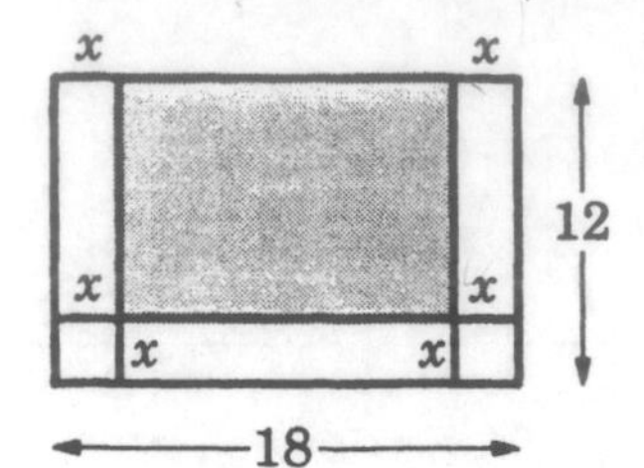

10. A landscaper wants to put a path having a uniform width of 1 meter around a rectangular area of land. The resulting area is to be 924 sq meters. Find the width and length of the original rectangular area if the length is twice the width, x.
Use $(x + 2)(2x + 2) = 924$.

10. 

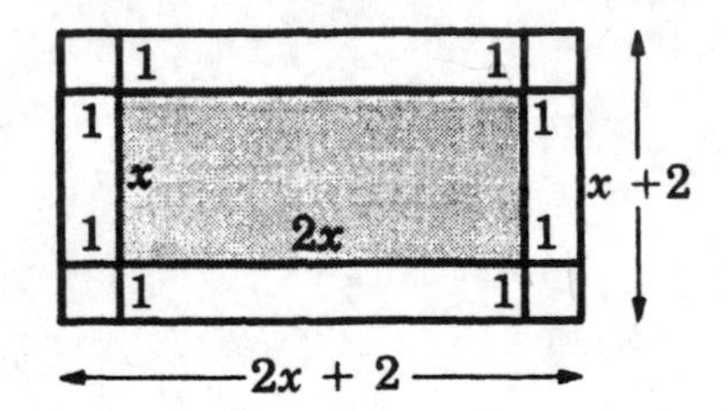

11. An artist wants to put a frame of uniform width around a rectangular painting that is 4 ft wide and 6 ft long. The artist wants the area of the framed painting to be twice the area of the painting. How wide, x, should the frame be?
Use $(4 + 2x)(6 + 2x) = 2(4)(6)$.

11. 

12. An editor wants to determine the margins for a rectangular page 15 cm wide and 20 cm long. One margin along a 20-cm length must be 2 cm wide for binding purposes. The other three margins are to have a uniform width of x cm. The area of the printed space must be 140 sq cm. How wide should each of the other three margins be?
Use $(15 - 2 - x)(20 - 2x) = 140$.

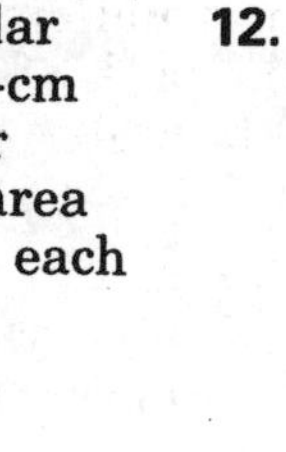

12. ____________

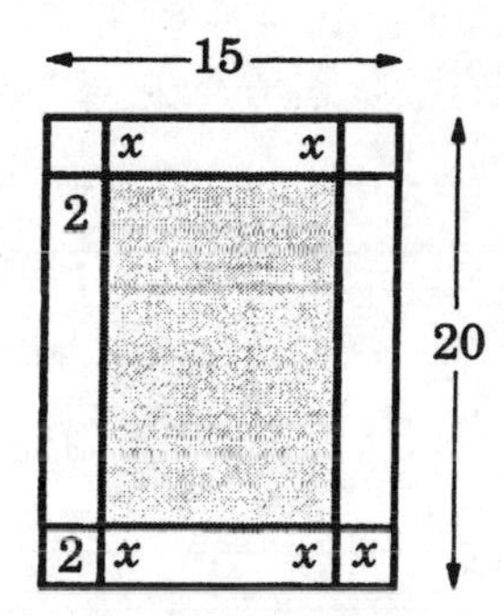

13. A chemist wants to determine the hydrogen ion concentration, h, in a certain acid solution. This is done by letting $x = 100{,}000\ h$ and then solving

$$x^2 = 9(10 - x)$$

Find x.

13. ____________

14. A chemist uses the equation

$$x^2 - 4(3 - x)^2$$

to find the number, x, of moles that react when 3 moles of pure ethyl alcohol are mixed with 3 moles of acetic acid. Find x. (By chemistry considerations, x is positive and less than 3.)

14. ____________

15. A doctor tests some medicine given to a certain patient to reduce blood pressure. The doctor finds that for x mg of medicine given to the patient 1 hour before the test is made, the blood pressure, P, is related to x by

$$P = 192 - 48x + 6x^2 \qquad \text{for } x \text{ less than or equal to } 4$$

Find x for $P = 120$.

15. ____________

16. A biologist found in a certain study of bacterial growth in an antibacterial solution that

$$N = 12 + 4t - t^2 \qquad \text{for } t \text{ less than 6 hours}$$

where t = the time in hours after the bacteria were placed in the solution and N is the number of bacteria divided by 10,000. Find t for $N = 7$.

16. ______________

17. An electrician uses the formula

$$P = EA - RA^2$$

to find the electric power, P, in watts where E is the voltage, A is the amperage, and R is the resistance in ohms. Find A for $P = 400$ watts, $E = 220$ volts, and $R = 10$ ohms.

17. ______________

18. The distance, F, in feet that are needed for a car to come safely to a stop is related to the speed, r, of the car in miles per hour by the equation

$$20F = r^2 + 22r$$

a. Find F for $r = 30$.
b. Find r for $F = 180$.

18. a. ______________

b. ______________

19. A manufacturer makes a box by cutting a square from each corner of a rectangular piece of material. The material is then folded along the dotted lines shown in the figure to form the box. The side of each square cut off is 5 cm long. The length of the rectangle is 10 cm longer than the width. The volume, V, of the box is given by

$$V = 5x(x - 10)$$

Find x for $V = 3000$ cu cm.

19. ______________

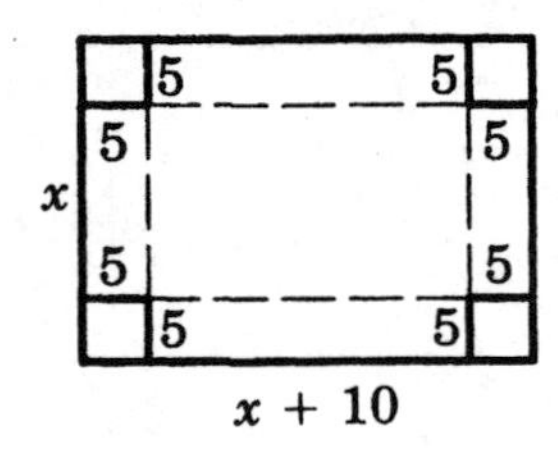

$h = 5$

$w = x - 10$

$l = x + 10 - 10$

SAMPLE TEST FOR CHAPTER 4

1. Rewrite in descending powers of x. (See Section 4.1.)

   a. $18x - x^2 - 81 + x^4$ — 1. a. ______

   b. $5x^2y^3 - 2x^3y^2 - 10xy^4$ — b. ______

2–5. Simplify. (Add or subtract as indicated.) (See Section 4.1.)

2. $(17 - x^2 - 6x) + (2x^2 - 20 + 6x)$ — 2. ______

3. $(8x - 5y - 9) + (5y - 3x + 2)$ — 3. ______

4. $(x^4 - 16x^2 + 64) - (9x^2 - 12x + 4)$ — 4. ______

5. $(5x^2 + 4y^2 - 7xy) - (3y^2 - 4x^2 - 7xy)$ — 5. ______

6–10. Multiply. (See Section 4.2.)

6. $-2x^2(7x^3 - x + 3)$ — 6. ______

7. $(4x - 3)(5x + 2)$ — 7. ______

8. $(3x - 7)(6y - 1)$ — 8. ______

9. $(6x - 7)^2$ — 9. ______

10. $(6x + 5y)(2x^2 - 3xy - y^2)$ — 10. ______

NAME ______ DATE ______ COURSE ______

11–14. Factor. (See Sections 4.3–4.8.)

11. $18ax^3 + 3ax^2$

12. $y^2 + 6y - 27$

11. ____________________

12. ____________________

13. $2x^2 - 3x + 1$

14. $y^4 + 16y^2 + 64$

13. ____________________

14. ____________________

15–18. Factor completely. (See Section 4.9.)

15. $x^4 - 16$

16. $3x^2 - 6x - 105$

15. ____________________

16. ____________________

17. $2x^4 - 7x^3 - 15x^2$

18. $y^4 - 8y^2 + 16$

17. ____________________

18. ____________________

19–20. Solve. (See Section 4.10.)

19. $4x^2 = 12x + 280$

19. ____________________

20. $(x + 2)^2 = 4(x + 10)$

20. ____________________

Algebraic Fractions

ALGEBRA IN ACTION: THE CONSUMER

A knowledge of ratio and proportion is very helpful to consumers—which is all of us—in determining quantities needed and best values for our money.

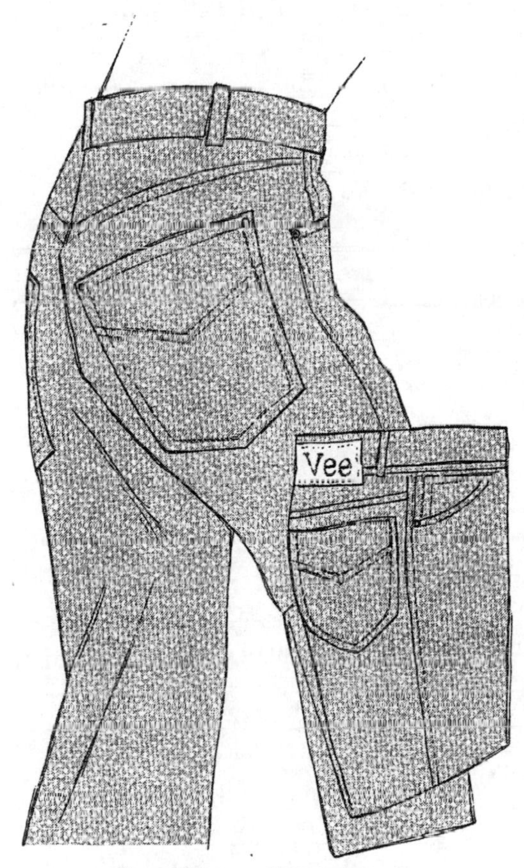

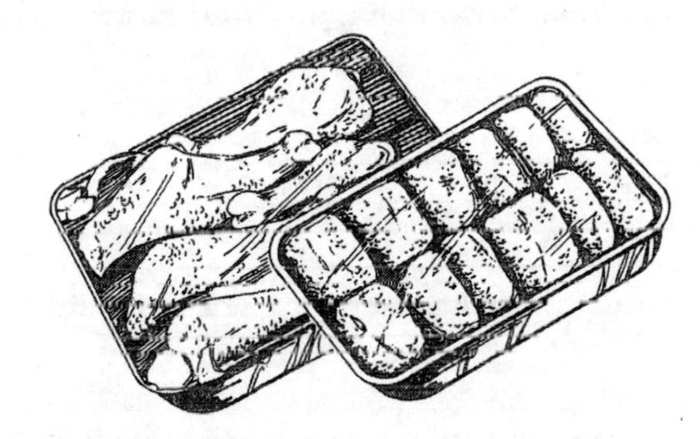

FRESH FRYER
THIGHS OR DRUMS
Fresh 1st Quality
Corn Fed
Family Pack
67c

25% OFF
Rugged jeans

A pair of jeans whose regular price is \$36 is advertised for sale at 25% off. What is the sales price?

$$\frac{x}{36} = \frac{75}{100}$$

If 20 lbs of chicken serve 25 people, how many lbs of chicken are needed to serve 15 people?

$$\frac{x}{15} = \frac{20}{25}$$

Chapter 5 Objectives

Objectives	Typical Problems
Simplify a fraction.	**1–2.** Simplify. **1.** $\dfrac{x^2 - 9x + 20}{x^2 + x - 20}$ **2.** $\dfrac{36 + 6x}{36 - x^2}$
Solve a ratio and proportion problem.	**3.** For an alloy rated 14K gold, the ratio of pure gold to the other metal in the alloy is 7 to 5. How many grams of pure gold are in 60 grams of an object rated 14K gold?
Multiply fractions.	**4.** Multiply. $\dfrac{3x^2 + 6x - 72}{2x^2 - 4x - 48} \cdot \dfrac{2x - 12}{3x - 12}$
Divide fractions.	**5.** Divide. $\dfrac{4x^2 - 8x}{2x^2 - 3x - 2} \div \dfrac{4x^2 - 2x}{4x^2 - 1}$
Raise a fraction to higher terms.	**6.** Raise to higher terms with the indicated denominator. $\dfrac{x + 1}{x + 5} = \dfrac{}{2x^2 - 50}$
Add and subtract fractions.	**7–8.** Rewrite as a single fraction. **7.** $\dfrac{x - 5}{4x - 16} + \dfrac{x - 2}{x^2 - 16}$ **8.** $\dfrac{1}{x^2 - x} - \dfrac{1}{2x^2 - 3x + 1}$
Simplify a complex fraction.	**9.** Simplify. $\dfrac{1 + \dfrac{4}{x}}{1 - \dfrac{16}{x^2}}$
Divide polynomials using long division.	**10.** Divide by using the long division algorithm. $\dfrac{x^3 - 3x^2 + 5x - 6}{x - 3}$

ANSWERS to Typical Problems

1. $\dfrac{x - 5}{x + 5}$ **2.** $\dfrac{-6}{x - 6}$ **3.** 35 grams **4.** $\dfrac{x + 6}{x + 4}$ **5.** 2 **6.** $\dfrac{2x^2 - 8x - 10}{2x^2 - 50}$ **7.** $\dfrac{x + 7}{4(x + 4)}$

8. $\dfrac{1}{x(2x - 1)}$ **9.** $\dfrac{x}{x - 4}$ **10.** $x^2 + 5 + \dfrac{9}{x - 3}$

EQUAL FRACTIONS, SIMPLIFICATION

The set of rational numbers, Q, was introduced in Chapter 2.

The set of rational numbers is the set of numbers that can be written in the form $\frac{n}{d}$, where n and d are integers and $d \neq 0$.

The number n is called the **numerator**, and the number d is called the **denominator**. For the rational number $\frac{-3}{4}$, its numerator is -3 and its denominator is 4.

This set has the important property that every sum, difference, product, or quotient (excluding zero as a divisor) of two rational numbers is a rational number.

The set of integers can be compared with the set of polynomials with integers for coefficients. If the variable or variables are replaced by integers, then the value of the polynomial is an integer.

The set of rational numbers can be compared with the set of quotients of polynomials with rational numbers for coefficients. If the variable or variables are replaced by rational numbers, then the quotient is a rational number (provided a denominator does not become zero.)

A quotient of two polynomials is called a **rational algebraic expression** or an **algebraic fraction.**

Examples of algebraic fractions are

$$\frac{1}{y-5} \quad \text{and} \quad \frac{x+2}{x^2-2x+3}$$

In this chapter we discuss the properties of algebraic fractions and their sums, differences, products, and quotients.

NOTE TO STUDENT:
For many students, this chapter is the most difficult of all. However, hard work, lots of practice, and traveling step-by-step will see you through to the top of this mountain. The climb may not be easy, but you can make it. From then on, the rest of the work is a downhill walk.

A. EQUAL FRACTIONS

In arithmetic, we learn that a fraction may have many names. For example, $\frac{1}{2} = \frac{2}{4} = \frac{3}{6}$ and so on. The same is true for algebraic fractions. We use the following definition to determine if two fractions are equal.

DEFINITION OF EQUAL FRACTIONS

Two fractions are equal if and only if the product of the numerator of one fraction and the denominator of the other fraction equals the product of the remaining numerator and denominator.

$$\frac{a}{b} = \frac{c}{d} \quad \textbf{if and only if} \quad bc = ad$$

We call bc and ad the **cross products** of the two fractions. Then we can say two fractions are equal if and only if their cross products are equal.

$\frac{a}{b} \times \frac{c}{d}$	$\frac{2}{4} \times \frac{3}{6}$	$\frac{x}{4} \times \frac{5x}{20}$
$bc = ad$	$4(3) = 2(6)$	$4(5x) = x(20)$
	$12 = 12$	$20x = 20x$

EXERCISES 5.1A

SAMPLE PROBLEM 1

Does $\frac{3}{5} = \frac{12}{20}$?

SOLUTION

Form the cross products.

$\frac{3}{5} \times \frac{12}{20}$ $\quad 5(12) = 60$ $\quad 3(20) = 60$

Since $60 = 60$,

Yes, the fractions are equal.

SAMPLE PROBLEM 2

Does $\frac{7}{9} = \frac{5}{6}$?

SOLUTION

Form the cross products.

$\frac{7}{9} \times \frac{5}{6}$ $\quad 9(5) = 45$ $\quad 7(6) = 42$

Since $45 \neq 42$,

No, the fractions are not equal.

1. Does $\frac{2}{3} = \frac{10}{15}$?

2. Does $\frac{3}{4} = \frac{9}{16}$?

3. Does $\frac{3}{5} = \frac{8}{10}$?

4. Does $\frac{5}{7} = \frac{50}{70}$?

5. Does $\frac{5}{8} = \frac{5+4}{8+4}$?

6. Does $\frac{5}{8} = \frac{5(4)}{8(4)}$?

1. ______________

2. ______________

3. ______________

4. ______________

5. ______________

6. ______________

7. Does $\frac{4k}{9k} = \frac{4}{9}$?

8. Does $\frac{26}{64} = \frac{1}{2}$?

7. ____________

8. ____________

9–20. Solve for x.

SAMPLE PROBLEM 3	**SAMPLE PROBLEM 4**
$\frac{3}{5} = \frac{3x}{20}$	$\frac{x}{x+4} = \frac{2}{3}$
SOLUTION	**SOLUTION**
Equate the cross products and solve.	Equate the cross products and solve.
$5(3x) = 3(20)$	$2(x+4) = 3x$
$15x = 60$	$2x + 8 = 3x$
$x = 4$	$x = 8$

9. $\frac{4}{7} = \frac{4x}{35}$

10. $\frac{x}{x+2} = \frac{3}{4}$

11. $\frac{x}{12} - \frac{1}{4}$

12. $\frac{1}{x} = \frac{10}{5}$

13. $\frac{x+1}{x+2} = \frac{5}{6}$

14. $\frac{x-2}{x+2} = \frac{1}{2}$

15. $\frac{x+1}{x-1} = \frac{2}{3}$

16. $\frac{x+3}{x+5} = \frac{3}{5}$

17. $\frac{1}{x} = \frac{4}{5}$

18. $\frac{2x-5}{3x-5} = \frac{2}{3}$

19. $\frac{2(x-5)}{3(x-5)} = \frac{2}{3}$

20. $\frac{6x}{7x} = \frac{6}{7}$

9. ____________

10. ____________

11. ____________

12. ____________

13. ____________

14. ____________

15. ____________

16. ____________

17. ____________

18. ____________

19. ____________

20. ____________

B. SIMPLIFICATION

Note that $\frac{3}{4} = \frac{3(5)}{4(5)}$, since $4(15) = 3(20)$ and $60 = 60$.

In general, if the numerator and denominator of a fraction are both multiplied by the same nonzero number, then the resulting fraction is equal to the original fraction.

This important property is called the **Fundamental Principle of Fractions.**

> FUNDAMENTAL PRINCIPLE OF FRACTIONS
>
> $$\frac{n}{d} = \frac{nk}{dk} \quad \text{for } d \neq 0 \quad \text{and} \quad k \neq 0$$

To **reduce an arithmetic fraction to lowest terms** means to rename the fraction so that the numerator and denominator do not have a factor in common other than the number 1.

This process is also called **simplifying the fraction.**

Thus, $\frac{3}{5}$ is the simplified form of $\frac{6}{10}$ and of $\frac{12}{20}$.

Note that $\frac{6}{10} = \frac{3(\not{2})}{5(\not{2})} = \frac{3}{5}$.

Also $\frac{12}{20} = \frac{3(\not{4})}{5(\not{4})} = \frac{3}{5}$.

The simplification of an algebraic fraction is similar to the process of reducing an arithmetic fraction to lowest terms.

An algebraic fraction of the form $\frac{P}{Q}$ where P and Q are polynomials is said to be in **simplified form** if the numerator P and the denominator Q have no common factor other than 1.

> TO SIMPLIFY AN ALGEBRAIC FRACTION
>
> **1.** Factor the numerator and the denominator.
> **2.** Use the fundamental principle of fractions
>
> $$\frac{nk}{dk} = \frac{n}{d}$$
>
> so that the resulting fraction is in simplified form.

EXERCISES 5.1B

1–18. Simplify.

SAMPLE PROBLEM 1

Reduce $\frac{30}{42}$ to lowest terms.

SOLUTION

1. Factor the numerator and the denominator: $\frac{30}{42} = \frac{(2)(3)(5)}{(2)(3)(7)}$
2. Use the fundamental principle of fractions (line out common factors of numerator and denominator): $\frac{30}{42} = \frac{(\not{2})(\not{3})(5)}{(\not{2})(\not{3})(7)} = \frac{5}{7}$

SAMPLE PROBLEM 2

Simplify $\frac{36x^3y}{60x^2y^3}$.

SOLUTION

1. Factor the numerator and the denominator: $\frac{2 \cdot 2 \cdot 3 \cdot 3xxxy}{2 \cdot 2 \cdot 3 \cdot 5xxyyy}$
2. Line out the common factors of the numerator and the denominator: $\frac{\not{2} \cdot \not{2} \cdot \not{3} \cdot 3\not{x}\not{x}x\not{y}}{\not{2} \cdot \not{2} \cdot \not{3} \cdot 5\not{x}\not{x}\not{y}yy}$
3. Write the simplified result: $\frac{3x}{5y^2}$

1. $\frac{10}{15}$

2. $\frac{56}{64}$

3. $\frac{6}{54}$

4. $\frac{4}{100}$

5. $\frac{6}{10}$

6. $\frac{125}{1000}$

7. $\frac{6x}{9x^2}$

8. $\frac{20xy}{45x^2y^2}$

9. $\frac{28x^3y^4}{7x^2y^4}$

10. $\frac{12x^2y}{18xy^2}$

1. ____________
2. ____________
3. ____________
4. ____________
5. ____________
6. ____________
7. ____________
8. ____________
9. ____________
10. ____________

11. $\dfrac{14x^3}{21x^2}$

12. $\dfrac{25x^4y^2}{125x^4y^3}$

13. $\dfrac{39a^2b^4}{65ab^5}$

14. $\dfrac{105r^4s^6}{63r^5s^3}$

15. $\dfrac{85ab^2c^3}{68ab^3c^2}$

16. $\dfrac{57x^3y^4z^5}{76x^2y^3z^4}$

17. $\dfrac{36r^2s^2t^4}{108rs^2t^2}$

18. $\dfrac{225a^4b^4t^3}{150a^2b^4t^6}$

11. ______________

12. ______________

13. ______________

14. ______________

15. ______________

16. ______________

17. ______________

18. ______________

19–28. Simplify. Check using $x = 2$ and $y = 5$.

SAMPLE PROBLEM 3

Simplify $\dfrac{5x^2 - 45}{5x^2 + 10x - 75}$.

SOLUTION

1. Factor numerator and denominator.

$$5x^2 - 45 = 5(x^2 - 9) = 5(x + 3)(x - 3)$$
$$5x^2 + 10x - 75 = 5(x^2 + 2x - 15) = 5(x + 5)(x - 3)$$

2. Line out common factors and simplify.

$$\frac{5x^2 - 45}{5x^2 + 10x - 75} = \frac{(x + 3)\cancel{(5)}\cancel{(x - 3)}}{(x + 5)\cancel{(5)}\cancel{(x - 3)}} = \frac{x + 3}{x + 5}$$

CHECK

Use $x = 2$.

1. Evaluate original fraction.

$$\frac{5(2^2) - 45}{5(2^2) + 10(2) - 75} = \frac{20 - 45}{20 + 20 - 75} = \frac{-25}{-35} = \frac{5}{7}$$

2. Evaluate final fraction.

$$\frac{2 + 3}{2 + 5} = \frac{5}{7}$$

3. Compare results.

$$\frac{5}{7} = \frac{5}{7} \quad \text{true}$$

Result checks.

If the values of Steps 1 and 2 are not equal, there is an error in the work. Check that your factoring is correct. Also check that you lined out common factors only and not terms.

SAMPLE PROBLEM 4

Simplify $\frac{x^4 - 4}{x^4 + 4x^2 + 4}$.

SOLUTION

1. Factor numerator and denominator.
2. Line out common factors and simplify.

$$\frac{x^4 - 4}{x^4 + 4x^2 + 4} = \frac{(x^2 - 2)\cancel{(x^2 + 2)}}{(x^2 + 2)\cancel{(x^2 + 2)}} = \frac{x^2 - 2}{x^2 + 2}$$

Note that $\frac{x^2 - 2}{x^2 + 2}$ is in simplified form. It would be incorrect to line out either the two x^2 terms or the 2's.

Terms of a sum cannot be lined out. Only a common *factor* of the numerator and the denominator can be omitted.

CHECK

1. Evaluate original expression. $\frac{2^4 - 4}{2^4 + 4(2^2) + 4} = \frac{16 - 4}{16 + 16 + 4} = \frac{12}{36} = \frac{1}{3}$
2. Evaluate final expression. $\frac{2^2 - 2}{2^2 + 2} = \frac{4 - 2}{4 + 2} = \frac{2}{6} = \frac{1}{3}$
3. Compare results. $\frac{1}{3} = \frac{1}{3}$ true

Result checks.

19. $\frac{5x^2 - 5}{5x^2 + 5}$ **20.** $\frac{x^2 + 5x}{x^2 - 25}$

21. $\frac{x^2 + 6x + 5}{x^2 + 7x + 10}$ **22.** $\frac{4x^2 - 100x}{x^2 - 24x - 25}$

23. $\frac{x^2 + x - 42}{x^2 - 2x - 24}$ **24.** $\frac{2x^2 - 9x - 5}{x^2 - 3x - 10}$

25. $\frac{x^2 - 49}{x^2 + 14x + 49}$ **26.** $\frac{4y^2 - 9}{4y^2 - 12y + 9}$

27. $\frac{x^2 + 9}{3x^3 + 27x}$ **28.** $\frac{3y^4 - 48}{3y^2 - 12}$

19. ______
20. ______
21. ______
22. ______
23. ______
24. ______
25. ______
26. ______
27. ______
28. ______

C. FRACTIONS AND THE NUMBER −1

Note that $\frac{3}{-4} = \frac{3(-1)}{-4(-1)} = \frac{-3}{4}$. We prefer to write the minus sign in the numerator.

$$\frac{-a}{b} \text{ is the preferred form for } \frac{a}{-b}$$

The numbers $5 - 9$ and $9 - 5$ are additive inverses ($5 - 9 = -4$ and $9 - 5 = 4$).

$$5 - 9 = (-1)(-1)(5 - 9) = (-1)(-5 + 9) = -(9 - 5)$$

In general, $b - a$ is the additive inverse of $a - b$.

$$b - a = -1(a - b)$$

Sometimes we need to use this property to simplify a fraction.

EXERCISES 5.1C

Simplify.

SAMPLE PROBLEM 1

$$\frac{4x^2}{-8x}$$

SOLUTION

$$\frac{4x^2}{-8x} = \frac{-4x^2}{8x} = \frac{(-x)(4x)}{(2)(4x)} = \frac{-x}{2}$$

SAMPLE PROBLEM 2

$$\frac{4x - 16}{4x - x^2}$$

SOLUTION

$$\frac{4(x - 4)}{x(4 - x)} = \frac{4(-1)(4 - x)}{x(4 - x)} = \frac{-4}{x}$$

1. $\frac{5x}{-15x^2}$
2. $\frac{-12x^2y}{54xy^2}$
3. $\frac{21x^2y^2}{-63x^2y^2}$
4. $\frac{x - 7}{7 - x}$
5. $\frac{4 - 2y}{y^2 - 2y}$
6. $\frac{5x^2 - 10x}{10x - 5x^2}$
7. $\frac{3x - 9}{3x - x^2}$
8. $\frac{-5 + x}{5 - x}$
9. $\frac{x^2 - x}{1 - x^2}$
10. $\frac{x^2 - 36}{6x - x^2}$

1. ______
2. ______
3. ______
4. ______
5. ______
6. ______
7. ______
8. ______
9. ______
10. ______

EXERCISES 5.1S

Simplify.

1. $\frac{21}{28}$

2. $\frac{8}{32}$

3. $\frac{36}{64}$

4. $\frac{25}{100}$

5. $\frac{8}{10}$

6. $\frac{16}{1000}$

7. $\frac{12x^2y}{18xy^2}$

8. $\frac{14x^3}{21x^2}$

9. $\frac{25x^4y^2}{125x^4y^3}$

10. $\frac{4x + 8}{x^2 - 4}$

11. $\frac{2x^2 + 18}{4x^2 - 36}$

12. $\frac{x^2 - 36}{x^2 - 6x}$

13. $\frac{x^2 - 9x + 14}{x^2 - 10x + 21}$

14. $\frac{9x^2 - 9x - 108}{9x^2 - 36x}$

1. ______
2. ______
3. ______
4. ______
5. ______
6. ______
7. ______
8. ______
9. ______
10. ______
11. ______
12. ______
13. ______
14. ______

NAME ______ DATE ______ COURSE ______

15. $\dfrac{x^2 + 3x - 40}{x^2 - 3x - 88}$

16. $\dfrac{3x^2 + 8x - 3}{3x^2 - 10x + 3}$

17. $\dfrac{y^2 - 36}{y^2 - 12y + 36}$

18. $\dfrac{25x^2 + 20xy + 4y^2}{25x^2 - 4y^2}$

19. $\dfrac{5x^2 - 10x - 40}{5x - 20}$

20. $\dfrac{2x^2 - 4}{4x^4 - 40x^2 + 64}$

21. $\dfrac{8xy}{-20x^2y}$

22. $\dfrac{-56x^3}{84x^4}$

23. $\dfrac{42xy^3}{-70x^3y}$

24. $\dfrac{y - 4}{4 - y}$

25. $\dfrac{6 - x}{x - 6}$

26. $\dfrac{x - x^2}{x^3 - x^2}$

27. $\dfrac{25 - 5x}{x^2 - 5x}$

28. $\dfrac{x^2 - 2x}{4 - x^2}$

29. $\dfrac{3 - 4y + y^2}{3 + 2y - y^2}$

30. $\dfrac{10 - 3y - y^2}{10 - 7y + y^2}$

15. ____________

16. ____________

17. ____________

18. ____________

19. ____________

20. ____________

21. ____________

22. ____________

23. ____________

24. ____________

25. ____________

26. ____________

27. ____________

28. ____________

29. ____________

30. ____________

DECIMALS, RATIO AND PROPORTION 5.2

A. DECIMALS

Common fractions are numbers such as $\frac{3}{5}$, $\frac{7}{16}$, and $\frac{25}{3}$. They are quotients of natural numbers, where any natural number may be a denominator.

A **decimal** is a common fraction whose denominator is a power of 10. Decimal notation uses a decimal point where the number of digits after the decimal point is the same as the power of 10 used for a divisor. Examples are

$$0.5 = \frac{5}{10} \qquad 0.16 = \frac{16}{100} \qquad 3.625 = \frac{3625}{1000}$$

The decimals shown as examples are called **terminating decimals.**

Now consider the common fraction $\frac{1}{3}$. Dividing,

$$\begin{array}{r} .333\ldots \\ 3\overline{)1.000} \\ \underline{9} \\ 10 \\ \underline{9} \\ 10 \end{array}$$

We see that the digit 3 repeats. When a decimal has a digit or a set of digits that repeats, the decimal is called a **repeating decimal.**

The decimal 0.333 . . . may also be written as $0.\overline{3}$ where the bar over the 3 means that the digit 3 repeats. Similarly, for $3.26\overline{14}$, the bar over 14 means that this set of digits repeats,

$$3.26\overline{14} = 3.26141414\ldots$$

All repeating and terminating decimals are rational numbers. Each rational number can be written as a repeating or terminating decimal.

There are numbers such as $\sqrt{3}$ (the positive number whose square is 3) and π (pi, the constant ratio of the circumference of any circle to its diameter) whose decimal forms are nonrepeating and nonterminating. Such numbers are called **irrational numbers.** These numbers **cannot** be written as a quotient of two integers.

$$\sqrt{3} = 1.73205\ldots \qquad \text{and} \qquad \pi = 3.14159\ldots$$

EXERCISES 5.2A

1–12. Change each common fraction to a decimal.

SAMPLE PROBLEM 1	SAMPLE PROBLEM 2	SAMPLE PROBLEM 3
$\frac{3}{4}$	$\frac{5}{6}$	$\frac{49}{11}$
SOLUTION	**SOLUTION**	**SOLUTION**
$4\overline{)3.00}$ = .75; 28; 20; 20	$6\overline{)5.000}$ = .833 . . .; 48; 20; 18; 20; 18; 2	$11\overline{)49.0000}$ = 4.4545 . . .; 44; 50; 44; 60; 55; 50
$\frac{3}{4} = 0.75$	$\frac{5}{6} = 0.8333\ldots$	$\frac{49}{11} = 4.4545\ldots = 4.\overline{45}$

1. $\frac{1}{8}$ **2.** $\frac{3}{16}$

3. $\frac{2}{25}$ **4.** $\frac{2}{3}$

5. $\frac{5}{9}$ **6.** $\frac{8}{11}$

7. $\frac{1}{7}$ **8.** $\frac{7}{4}$

9. $\frac{15}{2}$ **10.** $\frac{25}{3}$

11. $\frac{33}{32}$ **12.** $\frac{22}{7}$

1. ____________

2. ____________

3. ____________

4. ____________

5. ____________

6. ____________

7. ____________

8. ____________

9. ____________

10. ____________

11. ____________

12. ____________

13–20. Change each decimal to a common fraction.

SAMPLE PROBLEM 4

$0.\overline{5} = 0.555\ldots$

SOLUTION

Let $N = 0.555\ldots$

Multiply by 10. $10N = 5.555\ldots$

Subtract N. $N = 0.555\ldots$

$9N = 5$

$N = \frac{5}{9}$

SAMPLE PROBLEM 5

$2.\overline{18} = 2.181818\ldots$

SOLUTION

Let $N = 2.181818\ldots$

Multiply by 100 (the power of 10 necessary to place the set of repeating digits before the decimal point).

$100N = 218.181818\ldots$

Subtract N. $N = 2.181818\ldots$

$99N = 216$

$N = \frac{216}{99} = \frac{24}{11}$

13. $0.\overline{6}$

14. $0.\overline{25}$

15. $0.2\overline{8}$

16. $5.\overline{9}$

17. $5.\overline{04}$

18. $2.\overline{36}$

19. $4.\overline{123}$

20. $3.8\overline{24}$

13. ____________

14. ____________

15. ____________

16. ____________

17. ____________

18. ____________

19. ____________

20. ____________

B. RATIO AND PROPORTION

A **ratio** is a quotient used to compare two numbers.

A **proportion** is an equation stating that two ratios are equal. For example,

$$\frac{6}{2} = \frac{9}{3}$$

is a proportion stating that the ratio of 6 to 2 is the same as the ratio of 9 to 3. In general, a proportion has the form

$$\frac{a}{b} = \frac{c}{d}$$

When three numbers of a proportion are known, the fourth number can be found by using the definition of equal fractions and setting the cross products equal. Many word problems involve solving a proportion.

EXAMPLE

If 5 pounds of apples cost $2, what is the cost for 8 pounds?

Solution

Let x = the cost of 8 pounds of apples.
Compare the ratio of dollars to pounds.

$$\frac{\text{Dollars}}{\text{Pounds}} \qquad \frac{\overset{\text{dollars}}{2}}{\underset{\text{pounds}}{5}} = \frac{\overset{\text{dollars}}{x}}{\underset{\text{pounds}}{8}}$$

Solving, $5x = 16$ and $x = \frac{16}{5} = 3.2$.

The cost is $3.20.

In writing a proportion, one must be careful that the ratios are set up in the same way on each side of the equation. In the example, both ratios should be dollars to pounds. The problem could also be solved by comparing pounds to dollars. In this case, the ratio on each side of the equation should be pounds to dollars.

EXERCISES 5.2B

SAMPLE PROBLEM 1

A car used 16 gal of gasoline for a trip of 448 mi. How many gallons of gasoline would be needed for a trip of 672 mi?

SOLUTION

1. Let N = unknown number of gallons. N gal
2. Form a ratio of given numbers. $\frac{16 \text{ gal}}{448 \text{ mi}}$
3. Form the same type of ratio using N. $\frac{N \text{ gal}}{672 \text{ mi}}$
4. Set the ratios equal. $\frac{16}{448} = \frac{N}{672}$
5. Solve the proportion. (Set the cross products equal.)

$$448N = 16(672)$$

$$N = \frac{16(672)}{448}$$

$$N = 24$$

24 gal would be needed.

SAMPLE PROBLEM 2

A car used 16 gal of gasoline for a trip of 448 mi. How many miles could the car travel on 10 gal of gasoline?

SOLUTION

1. Let N = unknown number of miles. N mi
2. Form a ratio of given numbers. $\frac{16 \text{ gal}}{448 \text{ mi}}$
3. Form the same type of ratio using N. $\frac{10 \text{ gal}}{N \text{ mi}}$
4. Set the ratios equal. $\frac{10}{N} = \frac{16}{448}$
5. Solve the proportion.

$$16N = 10(448)$$

$$N = \frac{4480}{16}$$

$$N = 280$$

The answer is 280 mi.

1–2. A car uses 8 gal of gasoline for a trip of 192 mi.

1. How many gallons of gasoline would be needed for a trip of 264 mi?

2. How many miles could the car travel on 20 gal of gasoline?

1. ____________________

2. ____________________

3–4. Coffee mugs are advertised as 4 for $3.

3. How much would 12 mugs cost?

4. How many mugs can be bought for $15?

3. ____________________

4. ____________________

5–6. Color film is advertised as 2 rolls for $5.

5. How much film can be bought for $30?

6. What is the cost for 5 rolls of film?

5. ____________________

6. ____________________

7–8. For an item marked 40% off, the ratio of the sales price to the regular price is 3 to 5.

7. A vacuum cleaner is marked 40% off. What is the sales price if the regular price is $140?

8. A lawn mower is marked 40% off. What was the regular price if the sales price was $180?

7. ____________________

8. ____________________

9–10. For an item marked 25% off, the ratio of the sales price to the regular price is 3 to 4.

9. Wallpaper is marked 25% off. If the regular price is $10 a roll, what is the sales price?

10. Mosaic tile is marked 25% off. If the sales price is 45¢ per tile, what is the regular price?

9. ____________________

10. ____________________

11–12. For an item marked $\frac{1}{3}$ off, the ratio of the sales price to the regular price is 2 to 3.

11. Jeans are marked $\frac{1}{3}$ off. What is the sales price if the regular price is $27?

12. A sewing machine is marked $\frac{1}{3}$ off. What was the regular price if the sales price is $120?

11. ______________

12. ______________

13–14. It takes 5 oz of yarn to knit 2 slippers.

13. How many ounces are needed to knit 8 slippers?

14. How many slippers would 15 ounces of yarn make?

13. ______________

14. ______________

15–16. It takes 3 oz of yarn to crochet 5 coat hanger covers.

15. How many covers would 24 oz make?

16. How many ounces of yarn are needed to make 60 covers?

15. ______________

16. ______________

17–18. Yarn is sold both by the ounce and by the gram. Use 200 grams = 7 oz.

17. How many grams are there in a $3\frac{1}{2}$-oz skein of yarn?

18. How many ounces are in a 50-gram ball of yarn?

17. ______________

18. ______________

19–20. 5 yd of pillow tubing make 6 pillowcases.

19. How many cases can be made from 10 yd of tubing?

20. How many yards of tubing are needed to make 2 cases?

19. ______________

20. ______________

21–22. 5 meters of material make 4 blouses. (1 meter = 1.09 yd, approximately.)

21. How many meters are needed to make 6 blouses?

22. How many blouses can be made from 2.5 meters?

21. ______________

22. ______________

23–24. Certain fabrics are on sale at 2 yd for $1.

23. How many yards can be bought for $4.50?

24. What is the cost of 5 yd of this fabric?

23. ______

24. ______

25. Corn is advertised at 6 ears for 99¢.

a. How many ears can be bought for $1.65?

b. What is the cost for 8 ears?

c. What is the cost for 1 ear?

d. Is this a better buy than corn sold for 15¢ each?

25. a. ______

b. ______

c. ______

d. ______

26. Ham is advertised at 5 lb for $14.50?

a. How many pounds can be bought for $58?

b. What is the cost for 10 lb?

c. What is the cost for 1 lb?

d. Is this a better buy than ham sold at $2.95 a pound?

26. a. ______

b. ______

c. ______

d. ______

27–28. 20 lb of roast chicken serve 25 people.

27. How many pounds are needed to serve 10 people?

28. How many people will 36 lb serve?

27. ______

28. ______

29–30. 8 lb of carrots serve 25 people.

29. How many people will 40 lb serve?

30. How many pounds are needed to serve 60 people?

29. ______

30. ______

31–32. 3 oz of fish contain 135 calories and 22 grams of protein.

31. How many calories do 12 oz of fish contain?

32. How many ounces should be eaten to get 55 grams of protein?

31. ______

32. ______

EXERCISES 5.2S

1–6. Change each common fraction to a decimal.

1. $\frac{5}{8}$ **2.** $\frac{17}{4}$

3. $\frac{1}{6}$ **4.** $\frac{5}{9}$

5. $\frac{25}{11}$ **6.** $\frac{5}{12}$

1. ______
2. ______
3. ______
4. ______
5. ______
6. ______

7–12. Change each decimal to a common fraction.

7. $0.777\ldots = 0.\overline{7}$ **8.** 0.875

9. $0.1515\ldots = 0.\overline{15}$ **10.** 5.25

11. $7.0808\ldots = 7.\overline{08}$ **12.** $0.135135\ldots = 0.\overline{135}$

7. ______
8. ______
9. ______
10. ______
11. ______
12. ______

13–24. Answer each question.

13. If 4 ounces of hamburger contain 300 calories, how many calories are in 6 ounces of hamburger?

14. If 3 ounces of chicken contain 20 grams of protein, how many ounces of chicken contain 50 grams of protein?

13. ______
14. ______

15–16. A person uses up 35 calories by walking for 10 min.

15. How long should the person walk to use up 140 calories?

16. How many calories would the person use up by walking for 60 min?

15. ______
16. ______

NAME ______ DATE ______ COURSE ______

17–18. The price for 5 tulip bulbs was $4.

17. Find the price for 20 tulip bulbs.

18. How many tulip bulbs can be bought for $20?

17. ______________

18. ______________

19. 8 oz of orange juice contain 124 mg of vitamin C. How many ounces of orange juice contain 310 mg of vitamin C?

20. 8 oz of whole milk contain 290 mg of calcium. How much calcium is in 20 oz of whole milk?

19. ______________

20. ______________

21–22. 3 lb of ground beef serve 10 people.

21. How many pounds are needed to serve 25 people?

22. How many people will 45 lb serve?

21. ______________

22. ______________

23–24. A 12-oz package of frozen fish fillets is advertised for $1.29.

23. What would be the cost of 16 oz of this fish? (Assume the same unit price.)

24. Is this fish cheaper than fresh fish sold at $1.69 a pound? (16 oz = 1 lb)

23. ______________

24. ______________

MULTIPLICATION AND DIVISION OF FRACTIONS 5.3

A. MULTIPLICATION

The product of two fractions is defined as a single fraction whose numerator is the product of the numerators of the two fractions being multiplied and whose denominator is the product of the denominators of the two fractions being multiplied.

MULTIPLICATION OF FRACTIONS

$$\frac{a}{b} \cdot \frac{c}{d} = \frac{ac}{bd} \quad \text{where } bd \neq 0$$

PROCEDURE FOR MULTIPLYING FRACTIONS

1. **Write the product of the numerators over the product of the denominators. Leave these products in factored form. Do not do the multiplication.**
2. **Factor the resulting numerator and factor the resulting denominator.**
3. **Simplify, by lining out the common factors of the numerator and denominator.**

EXERCISES 5.3A

Simplify. (Multiply and reduce to lowest terms.)

SAMPLE PROBLEM 1

Simplify $\left(\frac{5}{12}\right)\left(\frac{-18}{35}\right)$.

SOLUTION

1. $\left(\frac{a}{b}\right)\left(\frac{c}{d}\right) = \frac{ac}{bd}$ $\qquad \left(\frac{5}{12}\right)\left(\frac{-18}{35}\right) = \frac{(-1)(5)(18)}{(12)(35)}$ Write product of numerators over product of denominators.
2. Factor. $\qquad = \frac{(-1)(\not{5})(\not{2})(\not{3})(3)}{(\not{2})(2)(\not{3})(\not{5})(7)}$ Line out common factors.
3. Simplify. $\qquad = \frac{-3}{14}$

SAMPLE PROBLEM 2

Multiply $\frac{15}{x^2} \cdot \frac{x^3}{20} \cdot \frac{12}{x^4}$.

SOLUTION

1. $\frac{a}{b} \cdot \frac{c}{d} \cdot \frac{e}{f} = \frac{ace}{bdf}$ $\qquad \frac{(15)(12)x^3}{(20)x^2x^4}$ Write product of numerators over product of denominators.
2. Factor. $\qquad = \frac{(3)(\not{5})(3)(\not{4})\not{x}^3}{(\not{5})(\not{4})x^2x\not{x}^3}$ Line out common factors.
3. Simplify. $\qquad = \frac{9}{x^3}$

1. $\left(\frac{1}{3}\right)\left(\frac{3}{5}\right)$
2. $\left(\frac{5}{16}\right)\left(\frac{-32}{35}\right)$
3. $\left(\frac{2}{9}\right)(54)$
4. $\left(\frac{-5}{14}\right)\left(\frac{-21}{15}\right)\left(\frac{-4}{13}\right)$
5. $(-12)\left(\frac{5}{6}\right)\left(\frac{1}{4}\right)$
6. $\left(\frac{-7}{9}\right)(0)$
7. $\frac{36x^2}{60x^3} \cdot \frac{10x^4}{21x}$
8. $\frac{3x}{2y^2} \cdot \frac{8y^3}{9x}$
9. $30x^2\left(\frac{-7}{5x}\right)\left(\frac{-1}{21}\right)$
10. $\left(\frac{2a}{3b}\right)\left(\frac{-5b}{4c}\right)\left(\frac{-6c}{35a^2}\right)$

1. ______
2. ______
3. ______
4. ______
5. ______
6. ______
7. ______
8. ______
9. ______
10. ______

SAMPLE PROBLEM 3

Multiply $\frac{2x+6}{3x} \cdot \frac{15x^3}{3x+9}$.

SOLUTION

1. Factor. $\frac{2(x+3)}{3x} \cdot \frac{\not{3}(5)xxx}{\not{3}(x+3)}$ Line out common factors.
2. Use $\frac{a}{b} \cdot \frac{c}{d} = \frac{ac}{bd}$. $= \frac{2\cancel{(x+3)}5\not{x}xx}{3\not{x}\cancel{(x+3)}}$ $\frac{\text{Product of numerators}}{\text{Product of denominators}}$
3. Simplify. $= \frac{10x^2}{3}$ Eliminate lined out common factors.

SAMPLE PROBLEM 4

Multiply $\frac{x^2+3x-10}{2x^2-x-10} \cdot \frac{2x^2-5x}{x^3-4x^2+4x}$.

SOLUTION

1. Factor. $\frac{(x+5)(x-2)}{(2x-5)(x+2)} \cdot \frac{\not{x}(2x-5)}{\not{x}(x-2)(x-2)}$ Line out common factors.
2. Use $\frac{a}{b} \cdot \frac{c}{d} = \frac{ac}{bd}$. $= \frac{(x+5)\cancel{(x-2)}\cancel{(2x-5)}}{\cancel{(2x-5)}(x+2)\cancel{(x-2)}(x-2)}$ $\frac{\text{Product of numerators}}{\text{Product of denominators}}$
3. Simplify. $= \frac{x+5}{(x+2)(x-2)}$ Eliminate common factors.

$= \frac{x+5}{x^2-4}$

SAMPLE PROBLEM 5

Multiply $\frac{t^2-9}{2t^2+2t} \cdot \frac{t^2+4t+3}{t^2+6t+9}$.

SOLUTION

1. Factor. $\frac{(t+3)(t-3)}{2t(t+1)} \cdot \frac{(t+1)\cancel{(t+3)}}{(t+3)\cancel{(t+3)}}$ Line out common factors.
2. Use $\frac{a}{b} \cdot \frac{c}{d} = \frac{ac}{bd}$. $= \frac{\cancel{(t+3)}(t-3)\cancel{(t+1)}}{2t\cancel{(t+1)}\cancel{(t+3)}}$ $\frac{\text{Product of numerators}}{\text{Product of denominators}}$
3. Simplify. $= \frac{t-3}{2t}$ Eliminate common factors.

11. $\left(\frac{9}{t^2-9}\right)\left(\frac{t-3}{3}\right)$

12. $\frac{4}{3x+3y} \cdot \frac{6x+6y}{2x+3y}$

11. ______________________

12. ______________________

13. $\dfrac{6x^2 - 42x}{45x - 90} \cdot \dfrac{5x^2 - 10x}{2x^2 - 28x + 98}$

14. $\dfrac{x^2 - x}{x^2 + 2x} \cdot \dfrac{8x + 8}{4x^3 - 4x^2}$

15. $\dfrac{ab + a}{b - ab} \cdot \dfrac{a^2b - ab}{a^2b + a^2}$

16. $\dfrac{2t^2 + 3t - 2}{2t^2 - t - 1} \cdot \dfrac{2t^2 - 3t - 2}{2t^2 + t - 1}$

17. $\dfrac{2 + x}{2 - x} \cdot \dfrac{x - 2}{x + 2}$

18. $\dfrac{3 - x}{x - 5} \cdot \dfrac{5 - x}{x - 3}$

19. $\dfrac{-x - 4}{x - 4} \cdot \dfrac{4 - x}{4 + x}$

20. $\dfrac{3x - x^2}{3x - 9} \cdot \dfrac{9}{x^2}$

13. ____________

14. ____________

15. ____________

16. ____________

17. ____________

18. ____________

19. ____________

20. ____________

B. DIVISION

Since every nonzero rational number r has a reciprocal $\dfrac{1}{r}$ with the property that $r\left(\dfrac{1}{r}\right) = 1$, we can use the reciprocal to define division as a multiplication. See Chapter 2, Section 2.C.

Note that multiplying the numerator and denominator of $\dfrac{3}{4}$ by the reciprocal of the denominator, $\dfrac{1}{4}$,

$$\frac{3}{4} = \frac{3\left(\frac{1}{4}\right)}{4\left(\frac{1}{4}\right)} = \frac{3\left(\frac{1}{4}\right)}{1} = 3\left(\frac{1}{4}\right)$$

changes the division into a multiplication.

DEFINITION OF DIVISION

$$\frac{a}{b} = (a)\left(\frac{1}{b}\right), \text{ where } b \neq 0$$

Note that the reciprocal of $\frac{3}{4}$ is $\frac{4}{3}$, since $\frac{3}{4} \cdot \frac{4}{3} = \frac{12}{12} = 1$.

To divide $\frac{2}{5}$ by $\frac{3}{4}$ we can again use the fundamental principle of fractions and multiply numerator and denominator by the reciprocal of the denominator.

$$\frac{\frac{2}{5}}{\frac{3}{4}} = \frac{\frac{2}{5} \cdot \frac{4}{3}}{\frac{3}{4} \cdot \frac{4}{3}} = \frac{\frac{2}{5} \cdot \frac{4}{3}}{1} = \frac{2}{5} \cdot \frac{4}{3}$$

This agrees with the definition of division as multiplication by the reciprocal of the divisor.

DIVISION OF FRACTIONS

$$\frac{\frac{a}{b}}{\frac{c}{d}} = \frac{a}{b} \div \frac{c}{d} = \frac{a}{b} \cdot \frac{d}{c} \quad \text{where } bcd \neq 0$$

The same rule used for arithmetic fractions also applies; that is, "invert the divisor and multiply."

SAMPLE PROBLEM 1

Simplify $\frac{\left(\frac{2}{3}\right)}{\left(\frac{4}{15}\right)}$.

SOLUTION

$$\frac{\frac{2}{3}}{\frac{4}{15}} = \left(\frac{2}{3}\right)\left(\frac{15}{4}\right) = \frac{(2)(3)(5)}{(2)(3)(2)} = \frac{5}{2}$$

SAMPLE PROBLEM 2

Simplify $\frac{12}{\frac{3}{4}}$.

SOLUTION

$$\frac{12}{\frac{3}{4}} = \left(\frac{12}{1}\right)\left(\frac{4}{3}\right) = \frac{(4)(4)(3)}{(1)(3)} = \frac{16}{1} = 16$$

EXERCISES 5.3B

Simplify. (Divide and reduce to lowest terms.)

SAMPLE PROBLEM 3

$\frac{15}{x^2} \div \frac{10}{x^3}$

SOLUTION

1. Invert divisor and multiply. $\frac{15}{x^2} \div \frac{10}{x^3} = \frac{15}{x^2} \cdot \frac{x^3}{10}$
2. Factor. $= \frac{(3)(\cancel{5})(\cancel{x})(\cancel{x})(x)}{(2)(\cancel{5})(\cancel{x})(\cancel{x})}$
3. Simplify. $= \frac{3x}{2}$

SAMPLE PROBLEM 4

$\frac{\frac{6x}{7y^2}}{\frac{3x^2}{14y}}$

SOLUTION

1. Invert divisor and multiply. $\frac{6x}{7y^2} \div \frac{3x^2}{14y} = \frac{6x}{7y^2} \cdot \frac{14y}{3x^2}$
2. Factor. $= \frac{2(\cancel{3})\cancel{x}(2)(\cancel{7})\cancel{y}}{\cancel{7}\cancel{y}y\cancel{3}\cancel{x}x}$
3. Simplify. $= \frac{4}{xy}$

1. $\frac{2}{x} \div \frac{5}{y}$

2. $\frac{\frac{24}{x^2}}{\frac{18}{x^3}}$

3. $6t \div \frac{18}{t^2}$

4. $\frac{\frac{18a^2b}{5c^2}}{60a^2b^2c^2}$

1. ____________________

2. ____________________

3. ____________________

4. ____________________

SAMPLE PROBLEM 5

$$\frac{\frac{x+y}{x-y}}{x^2-y^2}$$

SOLUTION

1. Use $N = \frac{N}{1}$. $\quad \frac{x+y}{x-y} \div (x^2 - y^2) = \frac{x+y}{x-y} \div \frac{x^2-y^2}{1}$
2. Invert divisor and multiply. $\quad = \frac{x+y}{x-y} \cdot \frac{1}{x^2-y^2}$
3. Factor and simplify. $\quad = \frac{(x+y)(1)}{(x-y)(x+y)(x-y)} = \frac{1}{(x-y)^2}$

SAMPLE PROBLEM 6

$$\frac{r^3-36r}{12r^2+2r} \div \frac{r^3-6r^2}{36r+6}$$

SOLUTION

1. Invert divisor and multiply. $\quad \frac{r^3-36r}{12r^2+2r} \cdot \frac{36r+6}{r^3-6r^2}$
2. Factor. $\quad = \frac{r(r+6)(r-6)6(6r+1)}{2r(6r+1)r^2(r-6)}$
3. Simplify. $\quad = \frac{3(r+6)}{r^2}$

$$= \frac{3r+18}{r^2}$$

5. $\frac{3x+3y}{10x-10y} \div \frac{6x-6y}{5x+5y}$

6. $\frac{-5x}{x^2+5} \div \frac{-5x}{x^2+5}$

7. $(25a^2 - 36b^2) \div \frac{5a-6b}{5a+6b}$

8. $\frac{\frac{(x-4)^2}{3x^2-12x}}{\frac{20-5x}{10}}$

5. ____________________

6. ____________________

7. ____________________

8. ____________________

9. $\dfrac{4a^2 - 4a}{10a - 20} \div \dfrac{2a^2 + 2a}{5a - 10}$

9. ______________

10. $\dfrac{15x^2y}{x^2 - 8xy + 12y^2} \div \dfrac{75xy^2}{x^2 - 4xy - 12y^2}$

10. ______________

11. $(x^3 - 25x) \div \dfrac{x^2 - 5x}{5x + 25}$

11. ______________

12. $\dfrac{2 - 2x}{x^2 - 2x} \div (2x^2 + 2x - 4)$

12. ______________

Simplify.

1. $\left(\frac{4}{7}\right)\left(\frac{14}{15}\right)$

2. $\left(\frac{35}{36}\right)\left(\frac{-12}{55}\right)$

3. $\left(\frac{3}{5}\right)(45)$

4. $\left(-\frac{7}{9}\right)\left(-\frac{36}{14}\right)$

5. $\left(-\frac{3}{18}\right)(27)$

6. $\left(\frac{8}{3}\right)\left(\frac{3}{8}\right)$

7. $\frac{xy}{xz} \cdot \frac{yz}{4}$

8. $\left(\frac{-2x}{y}\right) \cdot \left(\frac{-y}{2x}\right)$

9. $14xy\left(\frac{2}{7y}\right)$

10. $(x^2 - y^2)\left(\frac{1}{x + y}\right)$

11. $\left(\frac{6}{(x + 8)^2}\right)\left(\frac{x + 8}{12}\right)$

12. $\left(\frac{35}{x - 10}\right)\left(\frac{(x - 10)^2}{45}\right)$

13. $\frac{6x + 12y}{2x - 6y} \cdot \frac{4x^2 - 36y^2}{9x^2 - 36y^2}$

14. $\frac{xy - x}{xy + y} \cdot \frac{x^2y^2 - y^2}{x^2y^2 - x^2}$

15. $\frac{b^2 - 3b + 2}{b^2 - b - 2} \cdot \frac{b^2 - 2b - 3}{b^2 + b - 2}$

16. $\frac{9r^4s^2 - r^2s^4}{36r^4 - 108r^2s} \cdot \frac{9r^2 - 27s}{3r^2s^2 - 10rs^3 + 3s^4}$

1. ______
2. ______
3. ______
4. ______
5. ______
6. ______
7. ______
8. ______
9. ______
10. ______
11. ______
12. ______
13. ______
14. ______
15. ______
16. ______

NAME ______ DATE ______ COURSE ______

17. $\dfrac{x+1}{x-2} \cdot \dfrac{2-x}{1+x}$

18. $\dfrac{7-x}{x-9} \cdot \dfrac{9-x}{7+x}$

17. ____________

18. ____________

19. $\dfrac{6+x}{6-x} \cdot \dfrac{x-6}{-x-6}$

20. $\dfrac{x^2}{16} \cdot \dfrac{4x-16}{4x-x^2}$

19. ____________

20. ____________

21. $\dfrac{6}{x} \div \dfrac{18}{xy}$

22. $\dfrac{\dfrac{25x^2}{y^2}}{\dfrac{5x}{y^2}}$

21. ____________

22. ____________

23. $35r^2s \div \dfrac{42r^2}{s^2}$

24. $\dfrac{\dfrac{4a}{5t}}{20at}$

23. ____________

24. ____________

25. $\dfrac{a^2-4b^2}{4a+2b} \div \dfrac{a^2-4ab+4b^2}{8a+4b}$

26. $\dfrac{y^2-36}{y^2+36} \div \dfrac{36-y^2}{36+y^2}$

25. ____________

26. ____________

27. $\dfrac{7r+2s}{7r-2s} \div (49r^2-4s^2)$

28. $\dfrac{\dfrac{10y}{y^2-36}}{\dfrac{5y^2}{(6-y)^2}}$

27. ____________

28. ____________

29. $\dfrac{10t^2}{t^2-7t+10} \div \dfrac{25t^2}{t^2+3t-10}$

29. ____________

30. $\dfrac{b^2x^2-b^4}{a^2x^2+2a^2bx+a^2b^2} \div \dfrac{ax-ab}{bx+b^2}$

30. ____________

5.4 BUILDING UP FRACTIONS

To add or subtract certain fractions, the fractions must be renamed so that they have the same denominator, called a **common denominator.** The process of renaming a fraction by multiplying the numerator and denominator by the same number is called **raising the fraction to higher terms.** This process is justified by the fundamental principle of fractions.

$$\frac{n}{d} = \frac{nk}{dk}$$

In practice, it is useful to consider the renaming of a fraction as a multiplication of the original fraction by the number 1, with 1 renamed as $\frac{k}{k}$. Thus

$$\frac{n}{d} = \frac{n}{d} \cdot 1 = \frac{n}{d} \cdot \frac{k}{k} = \frac{nk}{dk}$$

EXERCISES 5.4

Build up as indicated.

SAMPLE PROBLEM 1

$$\frac{5}{x} = \frac{?}{4x^3}$$

SOLUTION

1. Divide new denominator by old denominator. The simplified quotient is the build-up factor. $\quad \frac{4x^3}{x} = 4x^2$
2. Multiply numerator and denominator of the original fraction by the build-up factor. $\quad \frac{5}{x} = \frac{5(4x^2)}{x(4x^2)}$
3. Do the multiplications. $\quad = \frac{20x^2}{4x^3}$

SAMPLE PROBLEM 2

$$\frac{x}{x+7} = \frac{\quad}{x^2+2x-35}$$

SOLUTION

1. Divide new denominator by old denominator. $\dfrac{x^2+2x-35}{x+7}$
2. Factor the numerator.
   Factor the denominator. $= \dfrac{(x+7)(x-5)}{1(x+7)}$
3. Simplify. (Line out common factors of numerator and denominator.) $= x - 5$
   The simplified quotient is the build-up factor. Build-up factor $= x - 5$
4. Multiply numerator and denominator of original fraction by the build-up factor. $\dfrac{x}{x+7} = \dfrac{x(x-5)}{(x+7)(x-5)}$
5. Do the multiplications. $= \dfrac{x^2-5x}{x^2+2x-35}$

SAMPLE PROBLEM 3

$$\frac{3}{4-y} = \frac{\quad}{y-4}$$

SOLUTION

1. $y - 4 = (-1)(4-y)$ or $\dfrac{y-4}{4-y} = -1$
2. Build-up factor is (-1).
3. $\dfrac{3}{4-y} = \dfrac{(-1)3}{(-1)(4-y)} = \dfrac{-3}{y-4}$

SAMPLE PROBLEM 4

$$x - 3 = \frac{\quad}{x+5}$$

SOLUTION

1. $x - 3 = \dfrac{x-3}{1}$
2. Build-up factor is $x + 5$.
3. $\dfrac{x-3}{1} = \dfrac{(x-3)(x+5)}{1(x+5)} = \dfrac{x^2+2x-15}{x+5}$

1. $\dfrac{3}{4} = \dfrac{\quad}{20}$

2. $\dfrac{2}{3} = \dfrac{\quad}{18}$

3. $4 = \dfrac{\quad}{15}$

4. $\dfrac{5}{2} = \dfrac{\quad}{16}$

5. $\dfrac{2}{x} = \dfrac{\quad}{5x^2}$

6. $\dfrac{x}{-5} = \dfrac{\quad}{5}$

7. $\dfrac{4}{-x} = \dfrac{\quad}{x}$

8. $\dfrac{3}{-x} = \dfrac{\quad}{2x^3}$

1. ________ **2.** ________

3. ________ **4.** ________

5. ________ **6.** ________

7. ________ **8.** ________

9. $\dfrac{1}{5-x} = \dfrac{\quad}{x-5}$

10. $\dfrac{4}{y-7} = \dfrac{\quad}{7-y}$

9. ________ 10. ________

11. $x + 2 = \dfrac{\quad}{x+4}$

12. $y - 4 = \dfrac{\quad}{y+4}$

11. ________ 12. ________

13. $\dfrac{3}{x-3} = \dfrac{\quad}{x^2-9}$

14. $\dfrac{3}{x-3} = \dfrac{\quad}{(x-3)^2}$

13. ________ 14. ________

15. $\dfrac{x}{1-x} = \dfrac{\quad}{(x-1)^2}$

16. $\dfrac{-2}{4-y} = \dfrac{\quad}{y^2-16}$

15. ________ 16. ________

17. $\dfrac{x}{x+1} = \dfrac{\quad}{5x+5}$

18. $\dfrac{6}{t-6} = \dfrac{\quad}{2t^2-12t}$

17. ________ 18. ________

19. $1 = \dfrac{\quad}{x^2-3x+2}$

20. $\dfrac{5}{x+2} = \dfrac{\quad}{x^2-3x-10}$

19. ________ 20. ________

21. $\dfrac{t}{2t-2} = \dfrac{\quad}{6t^2-6}$

22. $\dfrac{y}{y-5} = \dfrac{\quad}{5y^2-5y-100}$

21. ________ 22. ________

23. $\dfrac{t}{2t-2} = \dfrac{\quad}{4t^2-4t}$

24. $4 = \dfrac{\quad}{2x^2-x-1}$

23. ________ 24. ________

25. $\dfrac{5}{x+4} = \dfrac{\quad}{2x^2 - 32}$

25. ______________________

26. $\dfrac{x+1}{x-1} = \dfrac{\quad}{x^2 - 1}$

26. ______________________

27. $\dfrac{x-4}{4x} = \dfrac{\quad}{4x^2 + 16x}$

27. ______________________

28. $\dfrac{x+4}{x+6} = \dfrac{\quad}{x^2 + 5x - 6}$

28. ______________________

29. $\dfrac{5x+1}{5x-1} = \dfrac{\quad}{5x^2 + 24x - 5}$

29. ______________________

30. $\dfrac{x}{(x-8)^2} = \dfrac{\quad}{(x^2 - 64)^2}$

30. ______________________

EXERCISES 5.4S

Build up as indicated.

1. $\frac{7}{8} = \frac{}{32}$

2. $\frac{5}{9} = \frac{}{36}$

3. $6 = \frac{}{8}$

4. $\frac{7}{5} = \frac{}{25}$

5. $\frac{5}{4} = \frac{}{100}$

6. $\frac{1}{7} = \frac{}{56}$

7. $1 = \frac{}{16}$

8. $\frac{7}{x} = \frac{}{9x^3}$

9. $\frac{5}{x-5} = \frac{}{2x-10}$

10. $\frac{8}{-y} = \frac{}{y^2}$

11. $\frac{y}{-7} = \frac{}{7}$

12. $\frac{2}{y-6} = \frac{}{6-y}$

13. $\frac{1}{8-5x} = \frac{}{5x-8}$

14. $\frac{-5}{6-x} = \frac{}{x^2-36}$

15. $\frac{3}{x-4} = \frac{}{x^2-16}$

16. $\frac{2}{x+7} = \frac{}{(x+7)^2}$

1. ______
2. ______
3. ______
4. ______
5. ______
6. ______
7. ______
8. ______
9. ______
10. ______
11. ______
12. ______
13. ______
14. ______
15. ______
16. ______

NAME ______ DATE ______ COURSE ______

17. $\dfrac{8}{x-2} = \dfrac{\quad}{5x^2 - 20}$

18. $\dfrac{x-5}{x+5} = \dfrac{\quad}{x^2 - 25}$

19. $\dfrac{x-3}{9x} = \dfrac{\quad}{9x^2 + 27x}$

20. $\dfrac{x-2}{x+4} = \dfrac{\quad}{x^2 + 2x - 8}$

21. $\dfrac{4x-3}{4x-5} = \dfrac{\quad}{4x^2 - x - 5}$

22. $\dfrac{5}{(x+9)^2} = \dfrac{\quad}{(x^2 - 81)^2}$

23. $1 = \dfrac{\quad}{t^2 - 16}$

24. $2 = \dfrac{\quad}{3t^2 + 6t}$

25. $x = \dfrac{\quad}{5x^2}$

26. $3x = \dfrac{\quad}{x^2 - 1}$

27. $y + 1 = \dfrac{\quad}{y - 1}$

28. $y - 3 = \dfrac{\quad}{y + 3}$

29. $t - 4 = \dfrac{\quad}{t^2 + 4t}$

30. $t + 2 = \dfrac{\quad}{5t^2 - 10t}$

17. ______________________

18. ______________________

19. ______________________

20. ______________________

21. ______________________

22. ______________________

23. ______________________

24. ______________________

25. ______________________

26. ______________________

27. ______________________

28. ______________________

29. ______________________

30. ______________________

ADDITION AND SUBTRACTION 5.5

A. SAME DENOMINATOR

BASIC IDEA

Addition of fractions:

$$\frac{a}{d} + \frac{b}{d} = \frac{a+b}{d} \quad \text{where} \quad d \neq 0$$

Subtraction of fractions:

$$\frac{a}{d} - \frac{b}{d} = \frac{a-b}{d} \quad \text{where} \quad d \neq 0$$

The basic idea states that an indicated sum (or difference) of two fractions having the **same** denominator can be expressed as a single fraction whose numerator is the sum (or difference) of the numerators of the terms and whose denominator is the **common denominator.**

SAMPLE PROBLEM 1

$$\frac{1}{3x} + \frac{5}{3x}$$

SOLUTION

1. Add the numerators and place the sum over the common denominator. $\frac{1+5}{3x} = \frac{6}{3x}$
2. Factor the numerator and the denominator. $\frac{2(3)}{x(3)}$
3. Line out the factor common to the numerator and denominator.
   Write the result as a single simplified fraction. $\frac{2}{x}$

SAMPLE PROBLEM 2

$$\frac{2x}{x-4} - \frac{8}{x-4}$$

SOLUTION

1. Subtract numerators and place over the common denominator. $\dfrac{2x-8}{x-4}$
2. Factor numerator and denominator. $\dfrac{2(x-4)}{1(x-4)}$
3. Simplify (line out common factors). $\dfrac{2}{1} = 2$

SAMPLE PROBLEM 3

$$\frac{x}{2x+6} - \frac{x-2}{2x+6}$$

SOLUTION

1. Put a binomial numerator in parentheses. Use the definition of subtraction. $\dfrac{x+(-1)(x-2)}{2x+6}$

$$a - b = a + (-b) = a + (-1)b$$

2. Carefully simplify the numerator. $\dfrac{x+(-x+2)}{2x+6} = \dfrac{2}{2x+6}$
3. Simplify the fraction, if possible. (Factor numerator and denominator and line out common factors.) $\dfrac{2(1)}{2(x+3)} = \dfrac{1}{x+3}$

EXERCISES 5.5A

Write as a polynomial or as a single simplified fraction.

1. $\dfrac{3}{16} + \dfrac{5}{16}$ **2.** $\dfrac{1}{10} + \dfrac{7}{10}$

3. $\dfrac{5}{8} - \dfrac{3}{8}$ **4.** $\dfrac{13}{12} - \dfrac{5}{12}$

5. $\dfrac{3}{2x} + \dfrac{5}{2x}$ **6.** $\dfrac{7}{6y} - \dfrac{5}{6y}$

1. ____________
2. ____________
3. ____________
4. ____________
5. ____________
6. ____________

7. $\dfrac{x}{x-2}+\dfrac{2}{x-2}$

8. $\dfrac{x}{x-2}-\dfrac{2}{x-2}$

9. $\dfrac{2x}{x+3}+\dfrac{6}{x+3}$

10. $\dfrac{3x}{2x-8}-\dfrac{12}{2x-8}$

11. $\dfrac{x-1}{x+1}+\dfrac{x-3}{x+1}$

12. $\dfrac{x+1}{x-2}-\dfrac{x-3}{x-2}$

13. $\dfrac{2x}{x-4}-\dfrac{x+4}{x-4}$

14. $\dfrac{2}{x-5}-\dfrac{x-3}{x-5}$

15. $\dfrac{x^2+x-2}{2x}+\dfrac{x^2+3x+2}{2x}$

16. $\dfrac{1}{6}+\dfrac{7}{6}-\dfrac{5}{6}$

17. $\dfrac{1}{x}+\dfrac{7}{x}-\dfrac{5}{x}$

18. $\dfrac{x}{5}-\dfrac{3}{5}-\dfrac{2}{5}$

19. $\dfrac{x+1}{x+5}-\dfrac{x+2}{x+5}+\dfrac{x+3}{x+5}$

20. $\dfrac{x+6}{2x-6}-\dfrac{2x+1}{2x-6}-\dfrac{8-2x}{2x-6}$

7. ______

8. ______

9. ______

10. ______

11. ______

12. ______

13. ______

14. ______

15. ______

16. ______

17. ______

18. ______

19. ______

20. ______

B. FINDING THE LCM

To add or subtract fractions that have different denominators, we must first rename the fractions so they have the same denominator. The work is easiest when the **least common denominator, LCD,** is used. The LCD is the **least common multiple, LCM,** of the denominators. So first we consider how to find the LCM of two or more natural numbers and of two or more polynomials.

The least common multiple (LCM) of two or more natural numbers is the smallest natural number that is divisible by each of the given numbers.

The least common multiple (LCM) of two or more given polynomials is the polynomial of least degree that is divisible by each of the given polynomials.

EXERCISES 5.5B

1–20. Find the LCM of the given natural numbers.

To find the LCM
1. **Factor each given number into prime factors.**
2. **Use, as a factor of the LCM, each prime factor the greatest number of times it occurs in any one of the given numbers.**

SAMPLE PROBLEM 1	SAMPLE PROBLEM 2
Find the LCM of 120 and 126.	Find the LCM of 10, 12, and 15.
SOLUTION	**SOLUTION**
1. $120 = (2)(2)(2)(3)(5)$ $126 = (2)(3)(3)(7)$	1. $10 = (2)(5)$ $12 = (2)(2)(3)$ $15 = (3)(5)$
2. $\text{LCM} = (2)(2)(2)(3)(3)(5)(7)$ $= 2520$	2. $\text{LCM} = (2)(2)(3)(5)$ $= 60$

1. 12 and 18

2. 25 and 30

3. 6 and 8

4. 9 and 12

5. 42 and 70

6. 24 and 144

7. 4 and 100

8. 6 and 42

9. 75 and 125

10. 48 and 54

11. 16, 18, and 24

12. 3, 4, and 6

13. 4, 8, and 16

14. 5, 10, and 25

15. 9, 16, and 72

16. 6, 12, and 36

17. 2, 5, and 25

18. 6, 7, and 8

19. 9, 10, and 12

20. 15, 25, and 30

1. ________ **2.** ________

3. ________ **4.** ________

5. ________ **6.** ________

7. ________ **8.** ________

9. ________ **10.** ________

11. ________ **12.** ________

13. ________ **14.** ________

15. ________ **16.** ________

17. ________ **18.** ________

19. ________ **20.** ________

21–40. Find the LCM of the given polynomials.

To find the LCM
1. Completely factor each given polynomial.
2. Use as a factor of the LCM each factor of a given polynomial the greatest number of times it occurs in any of the given polynomials.

SAMPLE PROBLEM 3	SAMPLE PROBLEM 4
$12x^2y$ and $30xy^3$	$4x^2 - 25$ and $4x^2 + 20x + 25$
SOLUTION	SOLUTION
1. $12x^2y = (2)(2)(3)(x)(x)(y)$ $30xy^3 = (2)(3)(5)(x)(y)(y)(y)$	1. $4x^2 - 25 = (2x + 5)(2x - 5)$ $4x^2 + 20x + 25 = (2x + 5)(2x + 5)$
2. $\text{LCM} = (2)(2)(3)(5)(x)(x)(y)(y)(y)$ $= 60x^2y^3$	2. $\text{LCM} = (2x + 5)(2x + 5)(2x - 5)$ $= (2x + 5)^2(2x - 5)$

21. $5x$ and x^2

22. $4y$ and $6y^3$

23. $45xy^2$ and $75x^3y$

24. $10x$ and $15x^3$

25. $4(x + y)$ and $6(x + y)^2$

26. $10x + 50$ and $15x + 75$

27. $2x$ and $x + 2$

28. $4y$ and $y - 4$

29. y and $3y^2 - 6y$

30. x^2 and $4x^3 + 8x^2$

31. $x^2 - 4$ and $x^2 + 2x$

32. $y^2 - 3y$ and $y^2 - 9$

33. $y^2 - 25$ and $y^2 + 10y + 25$

34. $36x^2 - 49$ and $36x^2 - 84x + 49$

35. $x^2 + 5x + 6$ and $x^2 + 6x + 8$

36. $y^2 - 9y + 20$ and $y^2 - 11y + 30$

21. ______________

22. ______________

23. ______________

24. ______________

25. ______________

26. ______________

27. ______________

28. ______________

29. ______________

30. ______________

31. ______________

32. ______________

33. ______________

34. ______________

35. ______________

36. ______________

SAMPLE PROBLEM 5

Find the LCM of $y^2 - 12y + 36$, $y^2 + 12y + 36$, and $y^2 - 36$.

SOLUTION

1. $y^2 - 12y + 36 = (y - 6)(y - 6)$
   $y^2 + 12y + 36 = (y + 6)(y + 6)$
   $y^2 - 36 = (y - 6)(y + 6)$
2. LCM $= (y - 6)^2(y + 6)^2$

37. $x - 4, x + 4, x^2 - 16$

38. $(x - 5)^2, (x + 5)^2, x^2 - 25$

39. $2x + 6, 3x + 9, 6x + 18$

40. $4x - 20, 2x - 10, 5x - 25$

37. ______________

38. ______________

39. ______________

40. ______________

C. ADDITION AND SUBTRACTION, DIFFERENT DENOMINATORS

An indicated sum (or difference) of two fractions having **different denominators** can be expressed as a single fraction by first renaming the given fractions so that the resulting fractions have a common denominator.

Although any common denominator can be used, the resulting fraction is the simplest when the least common denominator (LCD) is used.

The LCD (least common denominator) of two or more given fractions is the LCM (least common multiple) of their denominators.

To add (or subtract) fractions with different denominators

1. Factor each denominator.
2. Find the LCD.
3. Rename each fraction so the new denominator is the LCD.
4. Add (or subtract) the new numerators and place the result over the LCD.
5. Simplify the resulting fraction, if possible.

EXERCISES 5.5C

Write each of the following as a single simplified fraction.

ARITHMETIC FRACTIONS

METHOD	SAMPLE PROBLEM 1	SAMPLE PROBLEM 2
	$\frac{5}{18} + \frac{7}{12}$	$\frac{3}{10} - \frac{2}{15}$
	SOLUTION	SOLUTION
1. Factor denominators.	$18 = (2)(3)(3)$ $12 = (2)(2)(3)$	$10 = (2)(5)$ $15 = (3)(5)$
2. Find LCD.	$\text{LCD} = (2)(2)(3)(3) = 36$	$\text{LCD} = (2)(3)(5) = 30$
3. Rename fractions.	$\frac{5}{18} = \frac{10}{36}$ and $\frac{7}{12} = \frac{21}{36}$	$\frac{3}{10} = \frac{9}{30}$ and $\frac{2}{15} = \frac{4}{30}$
4. Combine numerators.	$\frac{10}{36} + \frac{21}{36} = \frac{31}{36}$	$\frac{9}{30} - \frac{4}{30} = \frac{5}{30}$
5. Simplify.	Not possible	$\frac{5}{30} = \frac{(1)(5)}{(6)(5)} = \frac{1}{6}$

1. $\frac{1}{3} + \frac{1}{9}$

2. $\frac{1}{4} + \frac{5}{12}$

3. $\frac{2}{9} - \frac{1}{18}$

4. $\frac{3}{14} - \frac{2}{7}$

5. $\frac{5}{6} + \frac{7}{8}$

6. $\frac{7}{30} - \frac{5}{24}$

7. $\frac{1}{2} - \frac{1}{3}$

8. $\frac{1}{14} + \frac{1}{35}$

9. $2 - \frac{1}{3}$

10. $1 + \frac{2}{5}$

1. ____________________

2. ____________________

3. ____________________

4. ____________________

5. ____________________

6. ____________________

7. ____________________

8. ____________________

9. ____________________

10. ____________________

MONOMIAL DENOMINATORS

METHOD	SAMPLE PROBLEM 3	SAMPLE PROBLEM 4
	$\dfrac{x+1}{36x} + \dfrac{x+1}{45x}$	$\dfrac{x+2}{4x} - \dfrac{3x-1}{6x^2}$
	SOLUTION	SOLUTION
1. Factor denominators.	$36x = (2)(2)(3)(3)x$ $45x = (3)(3)(5)x$	$4x = (2)(2)x$ $6x^2 = (2)(3)xx$
2. Find LCD.	$\text{LCD} = 180x$	$\text{LCD} = 12x^2$
3. Rename fractions.	$\dfrac{x+1}{36x} = \dfrac{5(x+1)}{180x}$ $\dfrac{x+1}{45x} = \dfrac{4(x+1)}{180x}$	$\dfrac{x+2}{4x} = \dfrac{3x(x+2)}{12x^2}$ $\dfrac{3x-1}{6x^2} = \dfrac{2(3x-1)}{12x^2}$
4. Combine numerators.	$\dfrac{5x+5+(4x+4)}{180x}$	$\dfrac{3x^2+6x-(6x-2)^*}{12x^2}$
5. Simplify.	$\dfrac{9x+9}{180x} = \dfrac{\cancel{9}(x+1)}{\cancel{9}(20x)}$ $= \dfrac{x+1}{20x}$	$\dfrac{3x^2+6x-6x+2}{12x^2}$ $\dfrac{3x^2+2}{12x^2}$

11. $\dfrac{x}{14} + \dfrac{x}{35}$

12. $\dfrac{x}{5} - \dfrac{x}{30}$

13. $\dfrac{2}{x^2} + \dfrac{5}{3x}$

14. $\dfrac{7}{3y^2} - \dfrac{2}{9y}$

15. $\dfrac{x+1}{6x} - \dfrac{x+1}{10x}$

16. $\dfrac{t+4}{36t} + \dfrac{t-5}{45t}$

17. $\dfrac{1}{y} - \dfrac{y-1}{y^2}$

18. $x - 2 + \dfrac{4x-1}{2x}$

19. $\dfrac{1}{x} + \dfrac{1}{y}$

20. $\dfrac{1}{12xy} - \dfrac{1}{18y^2}$

11. ______________

12. ______________

13. ______________

14. ______________

15. ______________

16. ______________

17. ______________

18. ______________

19. ______________

20. ______________

*WARNING! Always enclose a polynomial subtrahend in parentheses. Remember that each term of the polynomial must have its sign changed. This is a common source of error.

BINOMIAL DENOMINATORS

METHOD	SAMPLE PROBLEM 5	SAMPLE PROBLEM 6
	$\dfrac{y+2}{2y+10} - \dfrac{y+3}{3y+15}$	$\dfrac{4}{x^2+4x} + \dfrac{4}{x^2-4x}$
	SOLUTION	**SOLUTION**
1. Factor denominators.	$2y + 10 = 2(y+5)$ $3y + 15 = 3(y+5)$	$x^2 + 4x = x(x+4)$ $x^2 - 4x = x(x-4)$
2. Find LCD.	LCD $= 6(y+5)$	LCD $= x(x+4)(x-4)$
3. Rename fractions.	$\dfrac{y+2}{2(y+5)} = \dfrac{3(y+2)}{6(y+5)}$ $\dfrac{y+3}{3(y+5)} = \dfrac{2(y+3)}{6(y+5)}$	$\dfrac{4}{x(x+4)} = \dfrac{4(x-4)}{4(x+4)(x-4)}$ $\dfrac{4}{x(x-4)} = \dfrac{4(x+4)}{x(x-4)(x+4)}$
4. Combine numerators.	$\dfrac{3y+6-(2y+6)^*}{6(y+5)}$	$\dfrac{4x-16+(4x+16)}{x(x+4)(x-4)}$
5. Simplify.	$\dfrac{3y+6-2y-6}{6(y+5)}$ $\dfrac{y}{6(y+5)}$ or $\dfrac{y}{6y+30}$	$\dfrac{8\cancel{x}}{\cancel{x}(x+4)(x-4)}$ $\dfrac{8}{(x+4)(x-4)}$ or $\dfrac{8}{x^2-16}$

21. $\dfrac{2}{x+3} + \dfrac{5}{2x+6}$

22. $\dfrac{1}{6y-18} + \dfrac{1}{10y-30}$

23. $\dfrac{1}{7x-49} - \dfrac{1}{x^2-7x}$

24. $\dfrac{2}{y^2+5y} + \dfrac{2}{y^2-5y}$

25. $\dfrac{1}{y} - \dfrac{1}{y+2}$

26. $\dfrac{1}{2x-12} - \dfrac{3}{x^2-6x}$

27. $\dfrac{n+1}{2n+10} - \dfrac{n-1}{3n+15}$

28. $\dfrac{x+2}{4x^2+16x} - \dfrac{x-2}{4x^2}$

29. $\dfrac{t+3}{t-3} - \dfrac{t-3}{t+3}$

30. $\dfrac{y+2}{y+1} + \dfrac{y-2}{y-1}$

21. ______________

22. ______________

23. ______________

24. ______________

25. ______________

26. ______________

27. ______________

28. ______________

29. ______________

30. ______________

*See note to Sample Problem 4.

GENERAL POLYNOMIAL DENOMINATORS

SAMPLE PROBLEM 7

$$\frac{2}{x^2 + 10x + 25} + \frac{3}{x^2 - 25}$$

SOLUTION

1. Factor denominators. $x^2 + 10x + 25 = (x + 5)(x + 5)$
   $x^2 - 25 = (x - 5)(x + 5)$
2. Find LCD. $\text{LCD} = (x - 5)(x + 5)(x + 5)$
3. Rename fractions. $$\frac{2}{(x + 5)(x + 5)} = \frac{2(x - 5)}{(x + 5)(x + 5)(x - 5)}$$
   $$\frac{3}{(x - 5)(x + 5)} = \frac{3(x + 5)}{(x - 5)(x + 5)(x + 5)}$$
4. Combine numerators. $$\frac{2x - 10 + (3x + 15)}{(x - 5)(x + 5)(x + 5)}$$
5. Simplify. $$\frac{5x + 5}{(x - 5)(x + 5)(x + 5)}$$

SAMPLE PROBLEM 8

$$\frac{x - 4}{x^2 - 2x - 3} - \frac{x - 5}{x^2 - x - 6}$$

SOLUTION

1. Factor denominators. $x^2 - 2x - 3 = (x - 3)(x + 1)$
   $x^2 - x - 6 = (x - 3)(x + 2)$
2. Find LCD. $\text{LCD} = (x - 3)(x + 1)(x + 2)$
3. Rename fractions. $$\frac{x - 4}{(x - 3)(x + 1)} = \frac{(x - 4)(x + 2)}{(x - 3)(x + 1)(x + 2)}$$
   $$\frac{x - 5}{(x - 3)(x + 2)} = \frac{(x - 5)(x + 1)}{(x - 3)(x + 2)(x + 1)}$$
4. Combine numerators. $$\frac{(x^2 - 2x - 8) - (x^2 - 4x - 5)^*}{(x - 3)(x + 1)(x + 2)}$$
5. Simplify. $$\frac{x^2 - 2x - 8 - x^2 + 4x + 5}{(x - 3)(x + 1)(x + 2)}$$
   $$\frac{2x - 3}{(x - 3)(x + 1)(x + 2)}$$

*See note to Sample Problem 4.

31. $\dfrac{3}{x^2 - 12x + 36} + \dfrac{4}{x^2 - 36}$

31. ____________________

32. $\dfrac{1}{x^2 + 8x + 16} - \dfrac{1}{x^2 - 16}$

32. ____________________

33. $\dfrac{4}{x - 7} + \dfrac{x - 20}{x^2 - 49}$

33. ____________________

34. $\dfrac{3}{y^2 + 6y + 9} - \dfrac{1}{y + 3}$

34. ____________________

35. $\dfrac{4}{y^2 - 4} - \dfrac{3}{y^2 - y - 2}$

35. ____________________

36. $\dfrac{4}{4t^2 - 1} - \dfrac{5}{6t^2 - t - 1}$

36. ____________________

37. $\dfrac{1}{x^2 + 3x + 2} + \dfrac{1}{x^2 + 5x + 6}$

37. ____________________

38. $\dfrac{1}{3t^2 - 7t + 2} - \dfrac{1}{2t^2 - 3t - 2}$

38. ____________________

39. $\dfrac{x + 16}{x^2 + 2x - 8} + \dfrac{x - 10}{x^2 + x - 12}$

39. ____________________

40. $\dfrac{5}{2y^2 - 3y - 2} - \dfrac{3}{2y^2 - y - 1}$

40. ____________________

THREE TERMS

SAMPLE PROBLEM 9

$$\frac{x}{x-4}+\frac{4}{x+4}+\frac{32}{x^2-16}$$

SOLUTION

1. Find LCD. $\text{LCD} = (x-4)(x+4)$

2. Rename fractions. $$\frac{x(x+4)}{(x-4)(x+4)}+\frac{4(x-4)}{(x-4)(x+4)}+\frac{32}{(x-4)(x+4)}$$

3. Combine numerators. $$=\frac{(x^2+4x)+(4x-16)+32}{(x-4)(x+4)}$$

4. Simplify. $$=\frac{x^2+8x+16}{(x-4)(x+4)}$$

$$=\frac{(x+4)\cancel{(x+4)}}{(x-4)\cancel{(x+4)}}$$

$$=\frac{x+4}{x-4}$$

41. $\dfrac{2}{3x}+\dfrac{3}{4x^2}-\dfrac{5}{6x^3}$

42. $\dfrac{x-6}{6x}+\dfrac{1}{12}-\dfrac{x-4}{4x}$

43. $\dfrac{1}{3t+3}+\dfrac{1}{6t+6}+\dfrac{1}{2t+2}$

44. $\dfrac{1}{y^2-4y}+\dfrac{1}{4y-16}-\dfrac{1}{y-4}$

45. $\dfrac{1}{x+3}-\dfrac{1}{x-3}+\dfrac{2x}{x^2-9}$

46. $\dfrac{t+2}{t-2}-\dfrac{t-2}{t+2}-\dfrac{8t}{(t+2)^2}$

47. $\dfrac{1}{x}+\dfrac{2}{x-5}-\dfrac{3}{x+5}$

48. $\dfrac{1}{(x-3)^2}+\dfrac{1}{(x+3)^2}-\dfrac{2}{x^2-9}$

49. $\dfrac{2}{x^2+3x+2}-\dfrac{3}{x^2+x-2}+\dfrac{2}{x^2-1}$

50. $\dfrac{x}{x^2-8x+15}+\dfrac{3}{2x-6}-\dfrac{5}{2x-10}$

41. ______

42. ______

43. ______

44. ______

45. ______

46. ______

47. ______

48. ______

49. ______

50. ______

EXERCISES 5.5S

1. $\frac{1}{5} + \frac{1}{12} + \frac{1}{20}$

2. $\frac{1}{6} + \frac{1}{4} - \frac{1}{2}$

3. $\frac{1}{5} - \frac{1}{15} - \frac{1}{30}$

4. $\frac{5}{12x} - \frac{1}{12x}$

5. $\frac{x}{x-5} + \frac{5}{x-5}$

6. $\frac{x}{x-7} - \frac{7}{x-7}$

7. $\frac{y+2}{y-6} - \frac{y-4}{y-6}$

8. $\frac{1}{x-10} - \frac{1}{10-x}$

9. $2 + \frac{4}{y}$

10. $\frac{4}{5x} - \frac{3}{2x^2}$

11. $\frac{5}{6x^2} - \frac{7}{8x^2}$

12. $\frac{10}{x^2-25} + \frac{2}{x+5}$

13. $\frac{3}{(x-6)^2} - \frac{2}{(x-6)}$

14. $\frac{5}{x-5} - \frac{5}{x+5}$

15. $\frac{6}{25x^2-9} + \frac{6}{(5x-3)^2}$

16. $\frac{1}{5} + \frac{1}{y+5}$

1. ______________
2. ______________
3. ______________
4. ______________
5. ______________
6. ______________
7. ______________
8. ______________
9. ______________
10. ______________
11. ______________
12. ______________
13. ______________
14. ______________
15. ______________
16. ______________

NAME ______________ DATE ______________ COURSE ______________

17. $\dfrac{1}{y^2 + 3y} + \dfrac{1}{3y + 9}$

18. $\dfrac{2}{2k - 1} - \dfrac{2}{2k + 1}$

19. $\dfrac{n - 2}{n - 3} - \dfrac{n + 2}{2n - 6}$

20. $\dfrac{4}{4x^2 - 1} - \dfrac{4}{4x^2 - 4x + 1}$

21. $1 + \dfrac{1}{x + 1} + \dfrac{1}{x - 1}$

22. $5 + \dfrac{1}{5x} + \dfrac{1}{x^2}$

23. $\dfrac{1}{3x^2y} + \dfrac{1}{5xy^2} - \dfrac{1}{15xy}$

24. $\dfrac{1}{4} - \dfrac{1}{x} + \dfrac{1}{x - 4}$

25. $\dfrac{2}{3x + 9} - \dfrac{2}{2x + 6} - \dfrac{1}{x^2 + 3x}$

26. $\dfrac{2}{x^2 - 1} + \dfrac{1}{x^2 - 2x + 1} + \dfrac{1}{x^2 + 2x + 1}$

27. $x - 3 + \dfrac{9}{x + 3} - \dfrac{x^2}{x - 3}$

28. $2 + \dfrac{2}{x - 2} + \dfrac{x}{(x - 2)^2}$

29. $\dfrac{x - 1}{2} - \dfrac{3x}{2x + 8} + \dfrac{2}{x^2 + 4}$

30. $\dfrac{3}{x^2 - x - 2} - \dfrac{1}{x^2 + x - 6} + \dfrac{2}{x^2 + 4x + 3}$

17. ____________________

18. ____________________

19. ____________________

20. ____________________

21. ____________________

22. ____________________

23. ____________________

24. ____________________

25. ____________________

26. ____________________

27. ____________________

28. ____________________

29. ____________________

30. ____________________

COMBINED OPERATIONS, EVALUATION 5.6

A. COMBINED OPERATIONS

If a fraction contains more than one division bar, it is called a **complex fraction.**

PROCEDURE FOR SIMPLIFYING A COMPLEX FRACTION

1. Replace the longest bar with the division symbol ÷ , and put the main numerator and the main denominator inside parentheses.
2. Do the operations inside the parentheses, following the Order Rules (see Section 1.4).
3. Invert the denominator when it is a single fraction.
4. Multiply the resulting simple fractions (see Section 5.3).
5. Simplify (see Section 5.1).

SAMPLE PROBLEM 1

Simplify $\dfrac{\frac{1}{2} - \frac{1}{3}}{1 + \frac{1}{6}}$.

SOLUTION

1. Replace bar with ÷ and put numerator and denominator inside parentheses. $\left(\frac{1}{2} - \frac{1}{3}\right) \div \left(1 + \frac{1}{6}\right)$

2. Do operations inside parentheses. $\left(\frac{3}{6} - \frac{2}{6}\right) \div \left(\frac{6}{6} + \frac{1}{6}\right)$

$\frac{1}{6} \div \frac{7}{6}$

3. Invert divisor and multiply. $\frac{1}{6} \cdot \frac{6}{7}$

4–5. Multiply and simplify. $\frac{1(\not{6})}{\not{6}(7)} = \frac{1}{7}$

SAMPLE PROBLEM 2

Simplify $\dfrac{\frac{1}{2} - \frac{1}{x}}{\frac{1}{4} - \frac{1}{x^2}}$.

SOLUTION

1. Replace bar with ÷ and insert parentheses. $\left(\frac{1}{2} - \frac{1}{x}\right) \div \left(\frac{1}{4} - \frac{1}{x^2}\right)$

2. Do operations inside parentheses. $\left(\frac{x}{2x} - \frac{2}{2x}\right) \div \left(\frac{x^2}{4x^2} - \frac{4}{4x^2}\right)$

   $\left(\frac{x-2}{2x}\right) \div \left(\frac{x^2-4}{4x^2}\right)$

3. Invert divisor and multiply. $\frac{x-2}{2x} \cdot \frac{4x^2}{x^2-4}$

4. Multiply and factor. $\frac{\cancel{(x-2)}\cancel{2}(2)\cancel{x}x}{\cancel{2}\cancel{x}\cancel{(x-2)}(x+2)}$

5. Simplify. $\frac{2x}{x+2}$

Another way to simplify a complex fraction involves the fundamental principle of fractions. The main numerator and the main denominator of the complex fraction are each multiplied by the LCD of all fractions in the main numerator and main denominator. This method is shown in Sample Problem 3.

SAMPLE PROBLEM 3

Simplify $\dfrac{\frac{1}{x} + \frac{1}{2}}{\frac{1}{4}}$.

SOLUTION (Alternate method)

1. The LCD of $\frac{1}{x}$ and $\frac{1}{2}$ and $\frac{1}{4}$ is $4x$.
2. Multiply the main numerator and main denominator by the LCD,

$$\frac{4x\left(\frac{1}{x} + \frac{1}{2}\right)}{4x\left(\frac{1}{4}\right)} = \frac{4 + 2x}{x} = \frac{2(x+2)}{x}$$

EXERCISES 5.6A

Simplify.

1. $\dfrac{\frac{3}{5}}{\frac{6}{20}}$

2. $\dfrac{5 + \frac{1}{2}}{4}$

3. $\dfrac{1 - \frac{1}{x}}{x}$

4. $\dfrac{x - 5}{\frac{1}{2}}$

5. $\dfrac{\frac{1}{6} + \frac{1}{9}}{\frac{2}{3}}$

6. $\dfrac{2x}{\frac{1}{x} + \frac{1}{5}}$

7. $\dfrac{\frac{1}{3} - \frac{1}{x}}{\frac{1}{9} - \frac{1}{x^2}}$

8. $\dfrac{\frac{1}{4x} + \frac{1}{6x}}{\frac{5}{3x}}$

9. $\dfrac{1 + \frac{5}{t}}{1 + \frac{25}{t^2}}$

10. $\dfrac{1 - \frac{6}{x} - \frac{7}{x^2}}{1 - \frac{8}{x} + \frac{7}{x^2}}$

1. ____________

2. ____________

3. ____________

4. ____________

5. ____________

6. ____________

7. ____________

8. ____________

9. ____________

10. ____________

B. EVALUATION

In some applications you will have to substitute one or more rational numbers in a fractional expression. A formula to be evaluated may already be fractional, or it may become fractional after the replacement of a letter by a fraction. In checking fractional equations, you will have to evaluate rational expressions for rational values of one or more variables.

EXERCISES 5.6B

1–4. Evaluate for the given values.

SAMPLE PROBLEM 1

$$m = \frac{a - b}{1 + ab} \text{ for } a = \frac{1}{4} \text{ and } b = \frac{1}{6}$$

SOLUTION

$$m = \frac{\frac{1}{4} - \frac{1}{6}}{1 + \left(\frac{1}{4}\right)\left(\frac{1}{6}\right)} = \left(\frac{(1)(3)}{2(2)(3)} - \frac{(1)(2)}{2(3)(2)}\right) \div \left(\frac{1(24)}{1(24)} + \frac{1}{24}\right)$$

$$= \frac{1}{12} \div \frac{25}{24}$$

$$= \frac{1}{12} \cdot \frac{24}{25} = \frac{2}{25}$$

ALTERNATE SOLUTION

$$\frac{24\left(\frac{1}{4} - \frac{1}{6}\right)}{24\left[1 + \left(\frac{1}{4}\right)\left(\frac{1}{6}\right)\right]} = \frac{6 - 4}{24 + 1}$$

$$= \frac{2}{25}$$

1. $\dfrac{m - n}{1 + mn}$ for $m = 1$ and $n = \dfrac{1}{3}$

1. ________________

2. $\dfrac{1}{\frac{1}{a} + \frac{1}{b}}$ for $a = 10$ and $b = 15$

2. ________________

3. $\dfrac{200}{\frac{100}{x} + \frac{100}{y}}$ for $x = 40$ and $y = 60$

3. ________________

4. $\dfrac{h(a + b)}{2}$ for $h = \dfrac{5}{2}$ and $a = \dfrac{11}{5}$ and $b = \dfrac{9}{5}$

4. ________________

5–10. Check that the given value of the variable is a solution of the given equation.

METHOD	SAMPLE PROBLEM 2	SAMPLE PROBLEM 3
	$\frac{3x-2}{3x-1} = \frac{3x-1}{3x+2}$ and $x = \frac{5}{6}$	$2 - \frac{8x}{4x-7} = \frac{7}{4x^2-3x-7}$ and $x = \frac{-3}{2}$
	SOLUTION	**SOLUTION**
1. Evaluate left side.	$\frac{3x-2}{3x-1} = \frac{3\left(\frac{5}{6}\right)-2}{3\left(\frac{5}{6}\right)-1}$ $= \left(\frac{5}{2} - \frac{2}{1}\right) \div \left(\frac{5}{2} - \frac{1}{1}\right)$ $= \left(\frac{5}{2} - \frac{4}{2}\right) \div \left(\frac{5}{2} - \frac{2}{2}\right)$ $= \frac{1}{2} \div \frac{3}{2}$ $= \frac{1}{2} \cdot \frac{2}{3} = \frac{1}{3}$	$2 - \frac{8x}{4x-7} = 2 - \frac{8\left(\frac{-3}{2}\right)}{4\left(\frac{-3}{2}\right)-7}$ $= 2 - \frac{-12}{-6-7}$ $= \frac{2(13)}{1(13)} - \frac{12}{13}$ $= \frac{14}{13}$
2. Evaluate right side.	$\frac{3x-1}{3x+2} = \frac{3\left(\frac{5}{6}\right)-1}{3\left(\frac{5}{6}\right)+2}$ $= \left(\frac{5}{2} - \frac{2}{2}\right) \div \left(\frac{5}{2} + \frac{4}{2}\right)$ $= \frac{3}{2} \cdot \frac{2}{9} = \frac{1}{3}$	$\frac{7}{4x^2-3x-7} = \frac{7}{4\left(\frac{-3}{2}\right)^2 - 3\left(\frac{-3}{2}\right) - 7}$ $= 7 \div \left(4\left(\frac{9}{4}\right) + \frac{9}{2} - 7\right)$ $= 7 \div \left(2 + \frac{9}{2}\right)$ $= \frac{7}{1} \div \left(\frac{2(2)}{1(2)} + \frac{9}{2}\right)$ $= \frac{7}{1} \cdot \frac{2}{13} = \frac{14}{13}$
3. Compare values.	$\frac{1}{3} = \frac{1}{3}$ true $\frac{5}{6}$ is a solution.	$\frac{14}{13} = \frac{14}{13}$ true $\frac{-3}{2}$ is a solution.

If the value of the left side does not equal the value of the right side, then either there is an error in the calculations or the value of x is not a solution.

5. $\frac{8}{2x+3} = \frac{6}{5-4x}$ and $x = \frac{1}{2}$

5. ____________

6. $\frac{1}{x} = 12 - \frac{1}{5x}$ and $x = \frac{1}{10}$

6. ______________________

7. $\frac{3}{2t+1} + \frac{6}{2t-1} = \frac{2t+27}{4t^2-1}$ and $t = \frac{3}{2}$

7. ______________________

8. $1 + \frac{1}{x} = \frac{x+5}{x+1}$ and $x = \frac{1}{3}$

8. ______________________

9. $\frac{t+2}{t-3} = 1 - \frac{5}{t+2}$ and $t = \frac{1}{2}$

9. ______________________

10. $\frac{6}{y+2} + \frac{4}{y-2} = \frac{9}{4-y^2}$ and $y = \frac{-1}{2}$

10. ______________________

1–10. Simplify.

1. $\dfrac{\dfrac{4x^2}{9y^2}}{\dfrac{6x}{y}}$

2. $\dfrac{1-\dfrac{x}{y}}{2}$

3. $\dfrac{2x+14}{\dfrac{2}{7}}$

4. $\dfrac{1-\dfrac{1}{x}}{1+\dfrac{1}{x}}$

5. $\dfrac{\dfrac{1}{2}+\dfrac{1}{3}}{1-\dfrac{1}{3}}$

6. $\dfrac{\dfrac{8}{y}}{\dfrac{1}{4}-\dfrac{1}{y}}$

7. $\dfrac{\dfrac{x}{5}+\dfrac{5}{x}}{\dfrac{x}{5}-\dfrac{5}{x}}$

8. $\dfrac{2d}{\dfrac{d}{x}+\dfrac{d}{y}}$

9. $\dfrac{\dfrac{7}{t}+1}{\dfrac{49}{t^2}-1}$

10. $\dfrac{1+\dfrac{1}{x}-\dfrac{20}{x^2}}{1+\dfrac{9}{x}+\dfrac{20}{x^2}}$

1. ______________
2. ______________
3. ______________
4. ______________
5. ______________
6. ______________
7. ______________
8. ______________
9. ______________
10. ______________

NAME ______________ DATE ______________ COURSE ______________

11–14. Evaluate.

11. $P(1 + r)^2$ for $P = 2500$ and $r = \frac{8}{100}$ **11.** ____________

12. $\frac{4\pi r^3}{3}$ for $\pi = \frac{22}{7}$ and $r = \frac{3}{2}$ **12.** ____________

13. $\frac{1 - r^n}{1 - r}$ for $r = \frac{1}{2}$ and $n = 5$ **13.** ____________

14. $gt - \frac{1}{2}gt^2$ for $g = 32$ and $t = \frac{1}{4}$ **14.** ____________

15–20. Check that the given value of the variable is a solution of the given equation.

15. $\frac{2y}{2y - 1} = \frac{y + 1}{y - 1}$ and $y = \frac{1}{3}$ **15.** ____________

16. $3 - \frac{1}{t} = \frac{1}{5t^2 + 3t}$ and $t = \frac{2}{5}$ **16.** ____________

17. $\frac{1}{2} - \frac{2x}{x + 3} = \frac{5}{2x + 6}$ and $x = \frac{-2}{3}$ **17.** ____________

18. $\frac{1 - y}{5y} - \frac{3}{10} = \frac{y + 2}{4y}$ and $y = \frac{-2}{5}$ **18.** ____________

19. $\frac{2}{x^2 + 5x} + \frac{1}{x^2 - 5x} = \frac{1}{x^2 - 25}$ and $x = \frac{5}{2}$ **19.** ____________

20. $\frac{1}{2x + 5} + \frac{1}{2x - 5} = \frac{6}{4x^2 - 25}$ and $x = \frac{3}{2}$ **20.** ____________

DIVISION OF POLYNOMIAL BY POLYNOMIAL 5.7

TO DIVIDE A POLYNOMIAL BY A POLYNOMIAL

1. Arrange each polynomial in descending powers of a variable, if necessary.
2. Write the polynomials in the long division format used in arithmetic, leaving space for any missing powers of a variable.
3. Obtain the first term of the quotient by dividing the first term of the dividend by the first term of the divisor.
4. Multiply the term of the quotient by each term of the divisor.
5. Subtract this product from the dividend.
6. Repeat Steps 3 through 5 until the degree of the remainder is less than the degree of the divisor.
7. Add $\left(\frac{\text{remainder}}{\text{divisor}}\right)$ to the quotient.

To check, multiply the quotient polynomial by the divisor and add the remainder to this product. The result should be the dividend.

SAMPLE PROBLEM 1

Divide and check $\frac{6x^2 - 10x - 15}{3x + 4}$.

SOLUTION

1–2. $$3x + 4\overline{)6x^2 - 10x - 15}\quad \text{quotient: } 2x - 6$$

3–4. $\frac{6x^2}{3x} = 2x$ **and** $2x(3x + 4) = 6x^2 + 8x$ **Subtract**

5. $-18x - 15$

6. $\frac{-18x}{3x} = -6$ **and** $-6(3x + 4) = -18x - 24$ **Subtract**

$+\ 9$

7. Therefore, $\frac{6x^2 - 10x - 15}{3x + 4} = \mathbf{2x - 6 + \frac{9}{3x + 4}}$ (Answer)

CHECK

$$(3x + 4)(2x - 6) + 9 = 6x^2 - 18x + 8x - 24 + 9$$
$$= 6x^2 - 10x - 15$$

SAMPLE PROBLEM 2

Divide and check $\dfrac{2x - 66 - 23x^2 + 5x^3}{x - 5}$.

SOLUTION

1. Arrange each polynomial is descending powers of x. $\qquad \dfrac{5x^3 - 23x^2 + 2x - 66}{x - 5}$

2. Use the long division format.

$$\begin{array}{r} 5x^2 + 2x + 12 \\ x - 5\overline{)5x^3 - 23x^2 + 2x - 66} \\ \underline{5x^3 - 25x^2 \qquad\qquad} \\ 2x^2 + 2x - 66 \\ \underline{2x^2 - 10x \qquad} \\ 12x - 66 \\ \underline{12x - 60} \\ -6 \end{array}$$

3–4. $\dfrac{5x^3}{x} = 5x^2$ and $5x^2(x - 5) = 5x^3 - 25x^2$

5. Subtract

6. $\dfrac{2x^2}{x} = 2x$ and $2x(x - 5) = 2x^2 - 10x$

$\dfrac{12x}{x} = 12$ and $12(x - 5) = 12x - 60$

7. The quotient is $5x^2 + 2x + 12 + \dfrac{-6}{x - 5}$.

CHECK

$$\begin{aligned}(x - 5)(5x^2 + 2x + 12) + (-6) &= 5x^3 - 25x^2 + 2x^2 - 10x + 12x - 60 - 6 \\ &= 5x^3 - 23x^2 + 2x - 66\end{aligned}$$

SAMPLE PROBLEM 3

Divide and check $\dfrac{x^3 + 8}{x + 2}$.

SOLUTION

$$\begin{array}{r} x^2 - 2x + 4 \\ x + 2\overline{)x^3 + ox^2 + ox + 8} \\ \underline{x^3 + 2x^2 \qquad\qquad} \\ -2x^2 \qquad + 8 \\ \underline{-2x^2 - 4x \qquad} \\ 4x + 8 \\ \underline{4x + 8} \end{array}$$

Note the insertion of ox^2 and ox as place holders for the missing x and x^2 terms.

CHECK

$$\begin{aligned}(x + 2)(x^2 - 2x + 4) &= x(x^2 - 2x + 4) + 2(x^2 - 2x + 4) \\ &= (x^3 - 2x^2 + 4x) + (2x^2 - 4x + 8) \\ &= x^3 + (-2x^2 + 2x^2) + (4x - 4x) + 8 \\ &= x^3 + 8\end{aligned}$$

SAMPLE PROBLEM 4

Divide and check $\dfrac{x^3 - 4 - 5x}{x - 2}$.

SOLUTION

Rewrite polynomials in descending powers of a variable.

$$\begin{array}{r} x^2 + 2x \quad - 1 \\ x - 2\overline{)x^3 \qquad\quad - 5x - 4} \\ x^3 - 2x^2 \qquad\qquad \\ \hline 2x^2 - 5x - 4 \\ 2x^2 - 4x \qquad \\ \hline -x - 4 \\ -x + 2 \\ \hline -6 \end{array}$$

Leave space for missing powers

$\dfrac{x^3}{x} = x^2$ and $x^2(x - 2) = x^3 - 2x^2$

$\dfrac{2x^2}{x} = 2x$ and $2x(x - 2) = 2x^2 - 4x$

Subtract

$\dfrac{-x}{x} = -1$ and $-1(x - 2) = -x + 2$

$$\frac{x^3 - 5x - 4}{x - 2} = x^2 + 2x - 1 + \frac{-6}{x - 2}$$

Answer

CHECK

$(x - 2)(x^2 + 2x - 1) + (-6) = (x^3 + 2x^2 - x) + (-2x^2 - 4x + 2) - 6$
$= x^3 - 5x - 4$

EXERCISES 5.7

Divide, using the long division format, and check.

1. $\dfrac{x^2 - 7x + 10}{x + 2}$

$x + 2\overline{)\qquad\qquad}$

2. $\dfrac{y^2 - 6y + 5}{y - 5}$

$y - 5\overline{)\qquad\qquad}$

3. $\dfrac{6x^2 - 5x - 4}{2x - 1}$

$2x - 1\overline{)\qquad\qquad}$

4. $\dfrac{12x^2 + 11x - 15}{3x + 5}$

$3x + 5\overline{)\qquad\qquad}$

1. ____________________

2. ____________________

3. ____________________

4. ____________________

5. $\dfrac{4y^2 - y + 2y^3 - 5}{1 + y}$

$y + 1\overline{)}$

6. $\dfrac{x^3 - 10x - 24}{x - 4}$

$x - 4\overline{)}$

5. ______________________

6. ______________________

7. $\dfrac{y^3 + 125}{y + 5}$

$y + 5\overline{)}$

8. $\dfrac{4x^2 + 9}{2x + 3}$

$2x + 3\overline{)}$

7. ______________________

8. ______________________

9. $\dfrac{5x^2 - 20 + x^3 - 4x}{x + 5}$

$x + 5\overline{)}$

10. $\dfrac{24 + y^4 - 10y^2}{y - 2}$

$y - 2\overline{)}$

9. ______________________

10. ______________________

Divide, using the long division format. Check.

1. $\frac{x^2 + 4x - 8}{x + 3}$

$x + 3\overline{)}$

2. $\frac{y^2 - 7y + 10}{y - 2}$

$y - 2\overline{)}$

3. $\frac{8x^2 + 6x - 9}{4x + 3}$

$4x + 3\overline{)}$

4. $\frac{5x^2 - 22x + 8}{5x - 2}$

$5x - 2\overline{)}$

1. ______________

2. ______________

3. ______________

4. ______________

NAME ______________ DATE ______________ COURSE ______________

5. $\dfrac{5y + 7 - 8y^3 - 2y^2}{1 - 2y}$

$-2y + 1\overline{)\qquad\qquad}$

6. $\dfrac{x^3 + 5x^2 - 12}{x + 2}$

$x + 2\overline{)\qquad\qquad}$

5. ______________

6. ______________

7. $\dfrac{y^3 - 27}{y - 3}$

$y - 3\overline{)\qquad\qquad}$

8. $\dfrac{25x^2 + 16y^2}{4y + 5x}$

$5x + 4y\overline{)\qquad\qquad}$

7. ______________

8. ______________

9. $\dfrac{x^3 + 11x - 6x^2 - 6}{x - 2}$

$x - 2\overline{)\qquad\qquad}$

10. $\dfrac{45 - 14y^2 + y^4}{y + 3}$

$y + 3\overline{)\qquad\qquad}$

9. ______________

10. ______________

SAMPLE TEST FOR CHAPTER 

1–4. Simplify. (See Section 5.1.)

1. $\dfrac{x^3 - 4x}{x^2 - 2x}$

2. $\dfrac{12x^2y}{-8xy^2}$

3. $\dfrac{2x^2 - 3x - 35}{2x^2 - 10x}$

4. $\dfrac{4 - x^2}{4x - 8}$

1. ____________

2. ____________

3. ____________

4. ____________

5–6. Answer each question. (See Section 5.2.) The ratio of quarts to liters is 20 to 19.

5. How many quarts are equal to 95 liters?

6. How many liters are equal to 4 qt?

5. ____________

6. ____________

7–8. Write as a single simplified fraction. (See Section 5.3.)

7. $\dfrac{x^2 - 6x + 9}{x^2 - 4x + 3} \cdot \dfrac{2x^2 - 2x}{x^2 - 9}$

7. ____________

8. $\dfrac{x^4y - 4x^3y}{9x^2 - 4} \div \dfrac{x^3y - 4x^2y}{3x^2 + 10x - 8}$

8. ____________

NAME ____________ DATE ____________ COURSE ____________

9–10. Find the missing numerator. (See Section 5.4.)

9. $\dfrac{6t}{t-6} = \dfrac{\quad}{t^3 - 36t}$

10. $\dfrac{x+4}{x+6} = \dfrac{\quad}{x^2 + 3x - 18}$

9. ______________

10. ______________

11–12. Write each sum as a single simplified fraction. (See Section 5.5.)

11. $1 - \dfrac{6}{x+3} - \dfrac{1}{(x+3)^2}$

11. ______________

12. $\dfrac{2}{4x^2 - 1} - \dfrac{2}{2x - 1} + \dfrac{1}{2x + 1}$

12. ______________

13. Simplify. (See Section 5.6.)

13. ______________

$$\frac{1 - \dfrac{1}{x+1}}{1 + \dfrac{1}{x-1}}$$

14. Divide and check. (See Section 5.7.)

14. ______________

$$\frac{8x^3 - 60x^2 + 150x - 125}{2x - 5}$$

$2x - 5\overline{\big)\qquad\qquad}$

Fractional Equations, Applications

ALGEBRA IN ACTION: BUSINESS

Many types of mixture problems and work problems occur in business.
An equation for a mixture problem may be useful to determine the best way to invest money.
A contractor may use a work problem equation to find the time needed for a certain job.

Mortgage rates

Lending Institutions	Rate	Loan	Orig. Fee
Banks			
Brown*	$14\frac{3}{8}\%$	$2\frac{1}{2}\%$	+ $250
	15%	2%	+ $200
	$15\frac{1}{4}\%$		2%
		2%	+ $200

U.S. Treasury bonds

COUPON RATE	MATURITY	PRICE	YIELD TO MATURITY
13%	6/15/86	100 16/32	12.67
$13\frac{1}{8}\%$	5/15/94	99 10/32	12
$8\frac{3}{8}\%$	8/15/00	68 1/	
12%	8/15/		

A corporation invested \$200,000 in bonds earning 9% interest per year. How much should it invest in mortgages earning 14% per year so the yearly income from the investments will be 12% of the total amount invested?

$$0.09\,(200{,}000) + 0.14x = 0.12\,(x + 200{,}000)$$

A painter with a power machine takes 30 minutes to paint a room. Another painter using a roller takes 60 minutes to paint this room. How long would it take both painters working together to paint the room?

$$\frac{x}{30} + \frac{x}{60} = 1$$

Chapter 6 Objectives

Objectives	Typical Problems
Solve a linear equation with rational coefficients.	1–2. Solve. 1. $\frac{x}{24} + \frac{x}{40} = \frac{3}{5}$ 2. $0.70x + 1.10(200 - x) = 160$
Solve a stated mixture problem.	3. Walnuts worth 70¢ per pound are mixed with almonds worth \$1.10 per pound to produce 200 lb of nuts worth 80¢ per pound. How many pounds of walnuts and how many pounds of almonds are used?
Solve a fractional equation with monomial denominators.	4. Solve. $\frac{2}{3x} - \frac{1}{x^2} = \frac{1}{2x}$
Solve a stated work problem.	5. An old machine can process a certain amount of data in 40 min. It takes a new machine 24 min to do the same job. How long would it take both machines, working together, to process these data?
Solve a general fractional equation.	6. Solve. $\frac{3}{x-4} - \frac{2}{x+6} = \frac{24}{x^2 + 2x - 24}$

ANSWERS to Typical Problems

1. 9 **2.** 150 **3.** 150 lb walnuts, 50 lb almonds **4.** 6 **5.** $\frac{x}{40} + \frac{x}{24} = 1$; 15 min **6.** -2

EQUATIONS, FRACTIONAL COEFFICIENTS

A **fractional equation** is an equation that has one or more terms that are algebraic fractions.

The basic idea in solving a fractional equation is to eliminate the fractions. This is done by multiplying each side of the equation by the same number, namely the LCD (least common denominator).

> THEOREM
>
> $\frac{A}{C} = \frac{B}{C}$ has the same solution set as $A = B$ and $C \neq 0$.

For example, $\frac{6}{x} = \frac{3x}{x}$ and $6 = 3x$ have the same solution set, since $6 = 3x$ is obtained by multiplying each side of $\frac{6}{x} = \frac{3x}{x}$ by x and since $x \neq 0$ ($x = 2$ in this case).

Division by 0 is undefined, and a variable may not equal a value that makes a denominator 0. For example, consider

$$\frac{x-2}{x-5} + \frac{5x}{x+4} = 7$$

NOTE TO STUDENT:
There are restrictions on fractional equations.
No number can be a solution of a fractional equation if it causes a denominator to become zero.

If $x = 5$, then $x - 5 = 5 - 5 = 0$.
If $x = -4$, then $x + 4 = -4 + 4 = 0$.
Therefore, for this equation, neither 5 nor -4 can be a solution.

> PROCEDURE FOR SOLVING A FRACTIONAL EQUATION
>
> 1. Factor each denominator to find the LCD (least common denominator).
> 2. Multiply each side of the equation by the LCD. (Each term of the equation must be multiplied by the LCD.)
> 3. Solve the resulting equation.
> 4. Check the solution of Step 3 in the original equation. Evaluate the left side and the right side separately, doing the operations according to the Rules of Order.
> 5. Reject any value of the variable that produces a zero denominator, and state the solution.

EXERCISES 6.1

Solve and check.

FRACTIONAL COEFFICIENTS

METHOD	SAMPLE PROBLEM 1	SAMPLE PROBLEM 2
	$\frac{x}{36} + \frac{x}{45} = \frac{3}{4}$	$\frac{t}{10} + \frac{t}{15} = 1 - \frac{t}{3}$
	SOLUTION	**SOLUTION**
1. Factor each denominator to find the LCD.	$36 = 2 \cdot 2 \cdot 3 \cdot 3$ $45 = 3 \cdot 3 \cdot 5$ $4 = 2 \cdot 2$ LCD = 180	$10 = 2 \cdot 5$ $15 = 3 \cdot 5$ $3 = 3 \cdot 1$ LCD = 30
2. Multiply each side by the LCD.	$180\left(\frac{x}{36} + \frac{x}{45}\right) = 180\left(\frac{3}{4}\right)$	$30\left(\frac{t}{10} + \frac{t}{15}\right) = 30\left(1 - \frac{t}{3}\right)$
3. Solve the equation.	$5x + 4x = 135$ $9x = 135$ $x = 15$	$3t + 2t = 30 - 10t$ $15t = 30$ $t = 2$
4. Check in the original equation.	$\frac{15}{36} + \frac{15}{45} = \frac{5}{12} + \frac{1}{3} \cdot \frac{4}{4}$ $= \frac{9}{12} = \frac{3}{4}$ $\frac{3}{4} = \frac{3}{4}$	$\frac{2}{10} + \frac{2}{15} = \frac{1(3)}{5(3)} + \frac{2}{15}$ $= \frac{5}{15} = \frac{1}{3}$ $1 - \frac{2}{3} = \frac{3}{3} - \frac{2}{3} = \frac{1}{3}$
5. State the solution.	The solution is 15.	The solution is 2.

1. $\frac{x}{3} + 6 = x$

2. $5 - \frac{x}{2} = 1$

3. $\frac{y}{6} + \frac{y}{3} = 12$

4. $\frac{t}{15} + 1 = \frac{t}{10}$

5. $\frac{x}{4} + \frac{x}{12} = 1$

6. $\frac{y}{6} + \frac{y}{30} = 1$

1. ______________

2. ______________

3. ______________

4. ______________

5. ______________

6. ______________

7. $\frac{x}{24} + \frac{x}{40} = \frac{3}{5}$

8. $\frac{y}{21} + \frac{y}{28} = \frac{2}{3}$

7. ____________________

8. ____________________

9. $\frac{x}{7} + \frac{x}{42} = 1 - \frac{x}{3}$

10. $\frac{t}{3} + \frac{t}{24} = 1 - \frac{t}{8}$

9. ____________________

10. ____________________

DECIMAL COEFFICIENTS

SAMPLE PROBLEM 3

$1.65x + 0.05(160 - x) = 40$

SOLUTION

1. Multiply each side by a multiple of 10 to remove the decimal point.

   $100[1.65x + 0.05(160 - x)] = 100(40)$

   $165x + 5(160 - x) = 4000$

2. Solve the equation.

   $165x + 800 - 5x = 4000$

   $160x = 3200$

   $x = 20$

3. Check in the original equation.

   $1.65(20) + 0.05(140) = 33 + 7 = 40$

4. State the solution.

   The solution is 20.

11. $1.50x + 0.50(450 - x) = 545$

11. ____________________

12. $0.10x + 0.14(x + 3000) = 1260$

12. ____________________

13. $y + 0.16(18 - y) = 5.4$

13. ____________________

14. $0.03(60 - y) = 0.01(60)$

14. ____________________

15. $0.45x + 0.12x = 3.42$ 15. ____________

16. $t + 0.75(30) = 0.90(t + 30)$ 16. ____________

17. $0.35x + 0.04(115) = 0.12(x + 115)$ 17. ____________

18. $0.75t + 6t = 2430$ 18. ____________

19. $0.72x + 0.12(120 - x) = 43.2$ 19. ____________

20. $0.15d + 0.12(10{,}000 - d) = 1320$ 20. ____________

Solve and check.

1. $x + \dfrac{x}{7} = 16$

2. $\dfrac{3x}{5} - 2 = \dfrac{2x}{5}$

3. $\dfrac{y}{4} - \dfrac{y}{8} = 5$

4. $\dfrac{t}{12} + 2 = \dfrac{t}{9}$

5. $\dfrac{3}{4} + \dfrac{x}{8} = \dfrac{1}{2}$

6. $\dfrac{t}{14} + \dfrac{t}{35} = 1$

7. $\dfrac{x}{2} + \dfrac{x}{3} = \dfrac{1}{2} - \dfrac{x}{6}$

8. $\dfrac{y}{12} + \dfrac{y}{20} = 1 - \dfrac{y}{5}$

1. ______________
2. ______________
3. ______________
4. ______________
5. ______________
6. ______________
7. ______________
8. ______________

NAME ______________ DATE ______________ COURSE ______________

9. $\frac{x}{36} + \frac{x}{45} + \frac{x}{5} = 1$

10. $\frac{x}{18} + \frac{x}{36} + \frac{x}{12} = \frac{2}{3}$

9. ____________

10. ____________

11. $0.60x + 0.80(40 - x) = 30$

11. ____________

12. $40 + 0.30x = 0.50(x + 60)$

12. ____________

13. $0.15y + 0.10(10 - y) = 1.20$

13. ____________

14. $1.20y + 60 = 1.36(y + 40)$

14. ____________

15. $0.2x + 0.3(x - 10) + 0.4(x + 10) = 28$

15. ____________

16. $x + 0.5(20 - x) + 0.3(20 - x) = 40$

16. ____________

17. $0.06x + 0.09x + 0.07(50{,}000 - 2x) = 3650$

17. ____________

18. $0.075t + 0.12(5000 - t) = 510$

18. ____________

19. $0.25x + 0.60x + 0.20x = 42$

19. ____________

20. $0.125y + 1.35y + 0.675y = 17.2$

20. ____________

MIXTURE PROBLEMS 

A. THE PRODUCT RULE

Many practical problems can be classified as mixture problems. The product rule plays an important role in the solution of these problems. Several examples of this rule are shown in the Table of Special Cases, below.

> BASIC IDEA 1
> THE PRODUCT RULE
>
> (Unit value) · (Amount) = Value

Table of special cases

Subject matter	*Unit value* ·	*Amount* =	*Value*
Coins:	No. of cents in one coin	No. of coins	Total value of coins
Nickels	5	n	$5n$
Dimes	10	n	$10n$
Quarters	25	n	$25n$
Dollars	100	n	$100n$
Money invested:	Percent as decimal	Investment in dollars	Income (interest) in dollars
$500, at 6%	0.06	500	0.06(500)
Sales:	Price per unit weight (or volume)	Weight (or volume)	Cost
Walnuts	40¢ per lb	3 lb	40(3) cents
Solutions:	Percent of component as decimal	Volume of solution (or weight)	Volume (or weight) of component
3% salt solution	0.03	20 cc	0.03(20)
Salary:	Daily (hourly) wage	Number of days (hours)	Salary received
	$2.15 per hr	50 hr	2.15(50) dollars

EXERCISES 6.2A

Basic idea 1: (Unit value) · (Amount) = Value

Express each of the following in symbols.

1. The value in cents of 4 quarters. 1. ______
2. The value in cents of x quarters. 2. ______
3. The income in dollars from \$2000 invested at 5%. 3. ______
4. The income in dollars from \$5000 invested at $6\frac{1}{2}$%. 4. ______
5. The income in dollars from d dollars invested at 5%. 5. ______
6. The income in dollars from $x + 500$ dollars invested at 5%. 6. ______
7. The cost in cents of 5 lb of walnuts if the price is 15¢ per lb. 7. ______
8. The cost in cents of y lb of walnuts if the price is 15¢ per lb. 8. ______
9. The cost in cents of $20 - x$ lb of walnuts if the price is 15¢ per lb. 9. ______
10. The cost in cents of 5 lb of walnuts if the price is c cents per lb. 10. ______
11. The volume of acid in 40 cc of a 5% acid solution. 11. ______
12. The volume of acid in x cc of a 5% acid solution. 12. ______
13. The volume of acid in $x + 15$ cc of a 5% acid solution. 13. ______
14. The volume of acid in $50 - x$ cc of a 5% acid solution. 14. ______
15. The weekly salary in dollars of a man who works 40 hr at \$3.10 per hour. 15. ______
16. The weekly salary in dollars of a man who works h hr at \$3.10 per hour. 16. ______
17. The weekly salary in dollars of a man who works 40 hr at x dollars per hour. 17. ______
18. The weight of copper in 60% copper ore weighing 20 lb. 18. ______
19. The weight of copper in 60% copper ore weighing y lb. 19. ______
20. The weight of copper in 60% copper ore weighing $x + 20$ lb. 20. ______

B. THE SUM RULE

A second important idea involved in a mixture problem is that the sum of the amounts of the components of a mixture is equal to the amount of the mixture. If 30 lb of almonds are combined with 50 lb of walnuts, then there are 30 + 50, or 80, lb of nuts in the mixture. In general, if a lb of one ingredient are mixed with b lb of another ingredient, then $a + b$ lb are in the mixture.

> BASIC IDEA 2
> THE SUM RULE
>
> Sum of the amounts of the components equals the amount of the mixture.

SAMPLE PROBLEM 1

Express in symbols the number of pounds of tea in a mixture of 8 lb of domestic tea and x lb of imported tea.

SOLUTION

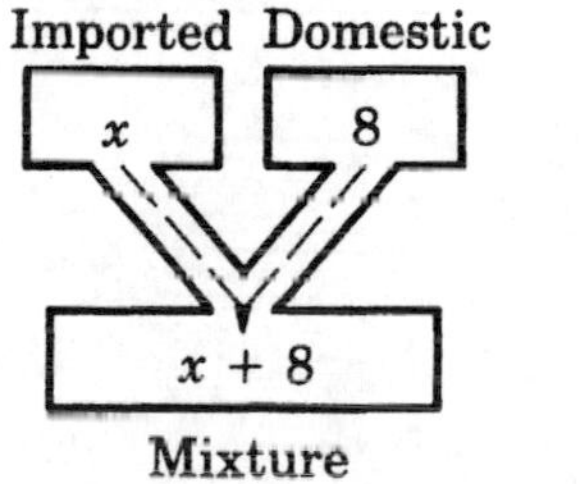

Answer

$x + 8$

SAMPLE PROBLEM 2

Express in symbols the number of pounds of imported tea in a 20-lb mixture consisting of imported tea and x lb of domestic tea.

SOLUTION

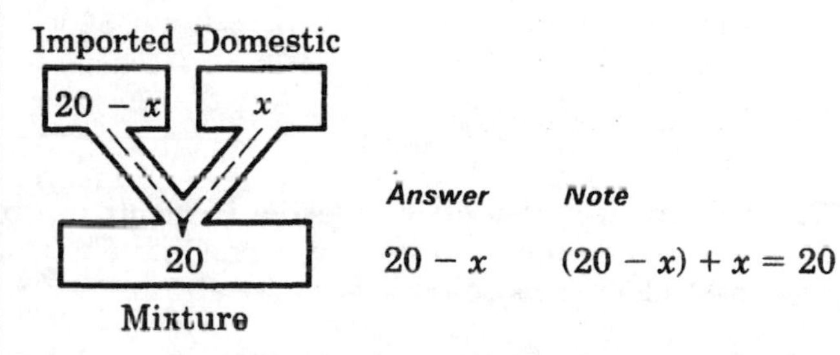

Answer

$20 - x$

Note

$(20 - x) + x = 20$

EXERCISES 6.2B

Basic idea 2: The sum of the amounts of the components is equal to the amount of the mixture.

Express each of the following in symbols. If necessary, draw a diagram like the ones shown in the first four problems.

1. The number of pounds in a mixture of 5 lb of walnuts and 3 lb of peanuts.

1. 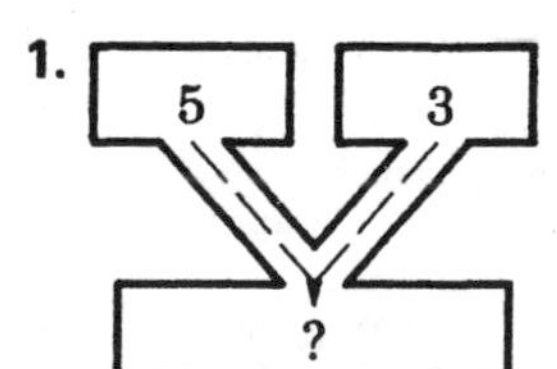

2. The number of pounds in a mixture of x lb of walnuts and 3 lb of peanuts.

2.

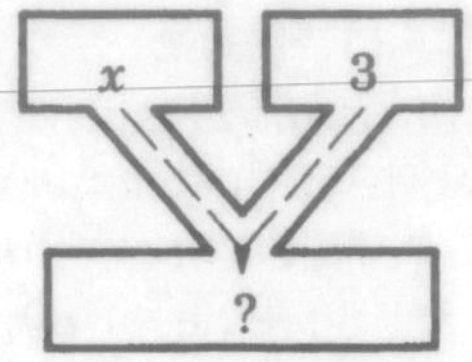

3. The number of pounds of walnuts in a 12-lb mixture consisting of walnuts and 3 lb of peanuts.

3.

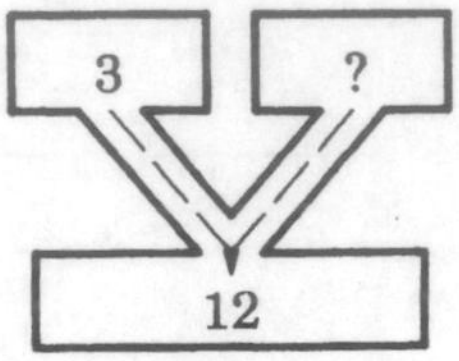

4. The number of pounds of walnuts in a 12-lb mixture consisting of walnuts and x lb of peanuts.

4.

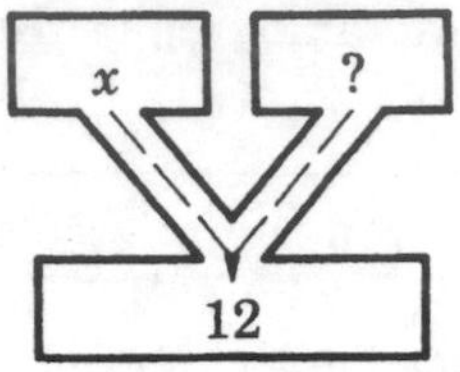

5. The total amount of money invested if \$2000 is invested in bonds and \$3000 is invested in stocks.

5. ________________

6. The total amount of money invested if \$2000 is invested in bonds and d dollars is invested in stocks.

6. ________________

7. The amount of money invested in bonds if a total of \$5000 is invested in both stocks and bonds and \$3000 of this is invested in stocks.

7. ________________

8. The amount of money invested in bonds if a total of \$5000 is invested in both stocks and bonds and x dollars of this is invested in stocks.

8. ________________

9. The volume of the mixture obtained when 20 cc of water is added to 40 cc of a certain acid solution.

9. ________________

10. The volume of the mixture obtained when x cc of water is added to 40 cc of a certain acid solution.

10. ________________

11. The volume of water added to 50 cc of a certain acid solution to obtain a mixture whose volume is 75 cc.

11. ________________

12. The volume of water added to 50 cc of a certain acid solution to obtain a mixture whose volume is x cc.

12. ________________

C. COMPLETE SOLUTION

A table such as that shown in Sample Problem 1 is a convenient way to summarize the information stated in a mixture problem.

The equation for a mixture problem is obtained by using the following rule:

> BASIC IDEA 3
>
> **The sum of the values of the components equals the value of the mixture.**

Using the table, the equation is obtained by setting the sum of the entries in the last column above the double line equal to the entry in the last column below the double line.

EXERCISES 6.2C

SAMPLE PROBLEM 1

How many pounds of dried prunes costing 70¢ per pound should be mixed with 30 lb of dried apricots costing \$1.20 per pound to obtain a mixture costing \$1.00 per pound?

SOLUTION

First, complete the amount column. Using the sum rule, the amount of the mixture is $x + 30$.

Then, complete the last column, using the product rule. The table should look as follows:

Component	*Unit value* · *(¢ per lb)*	*Amount* = *(in lb)*	*Value* *(in ¢)*
Prunes	70	x	
Apricots	120	30	
Mixture	100		

Prunes	70	x	$70x$
Apricots	120	30	$120(30)$
Mixture	100	$x + 30$	$100(x + 30)$

Equation

$$\begin{aligned} 100(x + 30) &= 70x + 120(30) \\ 100x + 3000 &= 70x + 3600 \\ 30x &= 600 \\ x &= 20 \end{aligned}$$

20 lb of dried prunes should be used.

SAMPLE PROBLEM 2

A corporation invested \$200,000 in bonds earning 9% interest per year. How much should it invest in mortgages earning 14% per year so the yearly income from the investments will be 12% of the total amount invested?

SOLUTION

Component	*Unit value (% as decimal)*	· *Amount invested*	= *Value (in \$)*
Bonds	0.09	200,000	0.09(200,000)
Mortgages	0.14	x	$0.14x$
Mixture	0.12	$x + 200{,}000$	$0.12(x + 200{,}000)$

Equation

$$0.09(200{,}000) + 0.14x = 0.12(x + 200{,}000)$$
$$9(200{,}000) + 14x = 12(x + 200{,}000)$$
$$2x = 600{,}000$$
$$x = 300{,}000$$

\$300,000 should be invested in 14% mortgages.

1. A dealer wants to mix 30 lb of rice costing 40¢ per pound with macaroni worth 55¢ per pound to produce a mixture worth 45¢ per pound. How many pounds of macaroni should he use?

1. ____________________

Component	*Unit value (in ¢)*	· *Amount (in lb)*	= *Value (in ¢)*
Rice			
Macaroni			
Mixture			

2. \$12,000 is invested in mortgages earning 15%. How much should be invested in bonds earning 8% so the income from the investments is 10% of the total amount invested?

2. ____________________

Component	*Unit value (% as decimal)*	· *Amount invested*	= *Value (in \$)*
Bonds			
Mortgages			
Mixture			

3. How many grams of an alloy containing 80% gold should be combined with 60 grams of an alloy containing 50% gold to produce an alloy containing 60% gold?

4. 5 liters of a 20% acid solution are mixed with a 35% acid solution to produce a 25% acid solution. How many liters of the 35% acid solution are used?

3. ______________

4. ______________

SAMPLE PROBLEM 3

A manufacturer combines wild rice costing $1.65 an ounce with a long-grain rice costing 5¢ an ounce to obtain a 160-oz mixture costing 25¢ an ounce. How many ounces of wild rice does he use?

SOLUTION

Component	*Unit value (in $)* ·	*Amount (in oz)* =	*Value (in $)*
Wild	1.65	x	$1.65x$
Long grain	0.05	$160 - x$	$0.05(160 - x)$
Mixture	0.25	160	40

Equation

$$1.65x + 0.05(160 - x) = 0.25(160)$$
$$165x + 5(160 - x) = 4000$$
$$165x - 5x = 4000 - 800$$
$$160x = 3200$$
$$x = 20$$

20 oz of wild rice are used.

SAMPLE PROBLEM 4

An advertising manager wants to place ads in magazines, costing \$1200 an ad, and in newspapers, costing \$800 an ad. He wants to place a total of 120 ads. His budget allowance for the total cost of the ads is \$124,000. How many magazine ads should he place?

SOLUTION

Component	*Unit value* ·	*Amount* =	*Value*
Magazines	1200	x	$1200x$
Newspapers	800	$120 - x$	$800(120 - x)$
Mixture		120	124,000

Equation

$$1200x + 800(120 - x) = 124{,}000$$
$$1200x - 800x = 124{,}000 - 96{,}000$$
$$400x = 28{,}000$$
$$x = 70$$

He should place 70 magazine ads.

5. Tickets to a certain event cost \$1.50 for each adult and 50¢ for each child. A total of \$545 was paid by 450 persons who attended the event. How many adult tickets were sold?

5. ____________

Component	*Unit value* · *(in ¢)*	*Amount* = *(no. of people)*	*Value (in ¢)*
Adults			
Children			
Mixture			

6. How many quarts of pure methanol (100%) should be added to an antifreeze containing 16% methanol to obtain 18 qt of an antifreeze containing 30% methanol?

6. ____________

Components	*Unit value* · *(% as decimal)*	*Amount* = *(qt)*	*Value (amount of methanol)*
Pure			
16%			
30%			

7. How many pounds of caramels worth 45¢ per pound should be combined with chocolates worth \$1.20 per pound to obtain a 10-lb mixture of chocolates and caramels worth 60¢ per pound?

7. ____________

8. A total of $50,000 is invested in savings certificates earning 11% and in a money market deposit account earning 9%. The income from the two investments is $5300. How much is invested in savings certificates?

8. ______________

SAMPLE PROBLEM 5

How much distilled water should be added to a 12% acid solution to obtain 100 cc of a solution that is 6% acid?

SOLUTION

(There is 0% acid in distilled water.)

Component	*Unit value (% as decimal)*	· *Amount (in cc)*	= *Value (in cc)*
Water	0	x	0
12% acid	0.12	$100 - x$	$0.12(100 - x)$
6% acid	0.06	100	0.06(100)

Equation

$$0 + 0.12(100 - x) = 0.06(100)$$
$$12(100 - x) = 6(100)$$
$$100 - x = 50$$
$$x = 50$$

9. How much water should be added to a 3% salt solution to obtain 60 gal of a 1% salt solution? (The water contains 0% salt.)

9. ______________

Components	*Unit value* ·	*Amount* =	*Value*
Water			
3% solution			
1% solution			

10. How much water should be added to a 60% alcohol solution to obtain 30 oz of a 40% alcohol solution?

10. ______________

Components	*Unit value* ·	*Amount* =	*Value*
Water			
60% solution			
40% solution			

SAMPLE PROBLEM 6

A corporation has \$100,000 invested in 6% bonds, 9% stocks, and 7% mortgages. If the amount invested in bonds is the same as the amount invested in stocks and if the total income from the bonds, stocks, and mortgages is \$7300, how much money is invested in mortgages?

SOLUTION

Component	*Unit value* ·	*Amount*	= *Value*
Bonds	0.06	x	$0.06x$
Stocks	0.09	x	$0.09x$
Mortgages	0.07	$100{,}000 - 2x$	$0.07(100{,}000 - 2x)$
Mixture		100,000	7300

Equation

$$0.06x + 0.09x + 0.07(100{,}000 - 2x) = 7300$$
$$6x + 9x + 7(100{,}000 - 2x) = 730{,}000$$
$$x = 30{,}000$$
$$100{,}000 - 2x = 40{,}000$$

\$40,000 is invested in mortgages.

11. A 60-lb mixture of nuts contains walnuts worth 50¢ per pound, cashews worth \$1.00 per pound, and almonds worth 75¢ per pound. If there are twice as many pounds of walnuts as cashews, how many pounds of cashews are in the mixture that is worth 70¢ per pound?

11. ______________

Component	*Unit value* ·	*Amount* =	*Value*
Cashews			
Walnuts			
Almonds			
Mixture			

12. Equal amounts of a 70% copper alloy and an 80% copper alloy are combined with 30 lb of a 40% copper alloy to produce an alloy containing 60% copper. How many pounds of the 70% alloy are used?

12. ______________

Component	*Unit value* ·	*Amount* =	*Value*
Mixture			

1. How much pure copper should be combined with 30 lb of an alloy containing 75% copper to obtain another alloy containing 90% copper?

1. ______

2. How much cream containing 35% butterfat should be added to milk containing 4% butterfat to produce 155 qt of half-and-half containing 12% butterfat?

2. ______

3. How much water (distilled water containing no salt) should be added to brine containing 3% salt to obtain 900 gal of a brine containing 2% salt?

3. ______

4. Twice as much money is invested in 14% stocks as in 10% bonds. The total income from these stocks and bonds is $1520. How much money is invested in bonds and how much money is invested in stocks?

4. ______

5. Maple syrup worth 20¢ an ounce is combined with corn syrup worth 8¢ an ounce to produce 480 oz worth 12¢ an ounce. How much maple syrup is used?

5. ______

NAME ______ DATE ______ COURSE ______

6. How many cubic centimeters of a 5% glucose solution should be combined with 20 cc of a 25% glucose solution to produce a 10% solution?

6. ____________________

7. $20,000 is invested in 6% tax-free bonds. How much should be invested in 9% certificates so that the total income is 8% of the total amount invested?

7. ____________________

8. A trucking company charges 6¢ a pound for boxed produce and 8¢ a pound for unboxed produce. A bill of $46 was received for a shipment of 700 lb. How much of the order was boxed?

8. ____________________

9. Equal amounts of onions and carrots are combined with 12 kg of peas to produce a mixture costing 60¢ a kilogram. The onions cost 80¢ a kilogram, the carrots 70¢ a kilogram, and the peas 50¢ a kilogram. How many kilograms of onions should be used?

9. ____________________

10. A baseball stadium has a total seating capacity of 50,000 seats. The ticket prices for a game are $4 for general admission, $8 for reserved seats, and $12 for box seats. The number of general admission seats is twice that of the reserved seats. When all the seats are sold, the receipts are $300,000. How many general admission seats are there?

10. ____________________

GENERAL FRACTIONAL EQUATIONS 6.3

The procedure stated in Section 6.1 is used to solve a general fractional equation.
This procedure is also shown in the following Sample Problems.

A. MONOMIAL DENOMINATORS

EXERCISES 6.3A

Solve and check.

SAMPLE PROBLEM 1

$$\frac{2}{3x} - \frac{1}{x^2} = \frac{1}{2x}$$

SOLUTION

1. Factor each denominator to find the LCD. $3x,\ 2x,\ x^2 = x(x)$; $\text{LCD} = 6x^2$

2. Multiply each side by the LCD. $6x^2\left(\frac{2}{3x} - \frac{1}{x^2}\right) = 6x^2\left(\frac{1}{2x}\right)$

$$4x - 6 = 3x$$

3. Solve the equation. $x = 6$

4. Check in the original equation.

Left side	Right side
$\frac{2}{3(6)} - \frac{1}{6(6)}$	$\frac{1}{2(6)} = \frac{1}{12}$
$= \frac{4}{6(6)} - \frac{1}{6(6)}$	
$= \frac{3}{36} = \frac{1}{12}$	

$$\frac{1}{12} = \frac{1}{12}$$

5. State the solution. The solution is 6.

SAMPLE PROBLEM 2

$$\frac{15-x}{15x}-\frac{5-x}{10x}=\frac{1}{5}$$

SOLUTION

1. Factor denominators to find LCD.

$15x = 3(5)x$ and $10x = 2(5)x$
LCD $= 30x$

2. Multiply each side by the LCD. (See Special Reminders, below.)

$$\mathbf{30x}\left(\frac{15-x}{15x}-\frac{5-x}{10x}\right)=\mathbf{30x}\left(\frac{1}{5}\right)$$

3. Solve the equation.

$$2(15-x)\mathbf{-3(5-x)}=6x$$
$$30-2x\mathbf{-15+3x}=6x$$
$$6x=x+15$$
$5x = 15$ and $x = 3$

4. Check in the original equation.

$$\frac{15-3}{45}-\frac{5-3}{30}$$
$$=\frac{12}{45}-\frac{2}{30}$$
$$=\frac{4}{15}-\frac{1}{15}=\frac{3}{15}=\frac{1}{5}$$

5. State the solution.

The solution is 3.

SPECIAL REMINDERS (SOURCES OF COMMON ERRORS)

1. Be sure to multiply the right side of the equation by the LCD.
2. Remember that each term of the equation must be multiplied by the LCD.
3. Enclose a binomial numerator of a subtrahend in parentheses. Be very careful when removing the parentheses. Each term inside the parentheses must be multiplied by the outside factor and must have its sign changed.

1. $\frac{1}{2x} + \frac{1}{5x} = \frac{1}{10}$

2. $\frac{4}{5x} - \frac{8}{x} = \frac{9}{5}$

3. $\frac{1}{6} + \frac{2}{x} = \frac{x^2 + 12}{6x^2}$

4. $\frac{1}{2x^2} + \frac{1}{8x} = \frac{1}{x^2}$

5. $\frac{2}{x^2} + \frac{5}{3x} = \frac{16}{3x^2}$

6. $\frac{7}{3y^2} - \frac{2}{9y} = \frac{y - 5}{y^2}$

7. $\frac{x + 1}{6x} - \frac{x + 1}{10x} = \frac{2x - 3}{15x}$

8. $\frac{t + 4}{36t} - \frac{t - 5}{45t} = \frac{1}{3t}$

9. $\frac{1}{y} - \frac{y - 1}{y^2} = \frac{1}{2y}$

10. $\frac{5}{2x} - \frac{x + 4}{x^2} = \frac{1}{x}$

1. ____________________

2. ____________________

3. ____________________

4. ____________________

5. ____________________

6. ____________________

7. ____________________

8. ____________________

9. ____________________

10. ____________________

EXERCISES 6.3B

Solve and check.

SAMPLE PROBLEM 1

$$\frac{5}{x+1} - \frac{3}{x-1} = \frac{8}{x^2-1}$$

SOLUTION

1. Factor each denominator to find the LCD.

$$\frac{5}{x+1} - \frac{3}{x-1} = \frac{8}{(x+1)(x-1)}$$

LCD of all denominators is $(x + 1)(x - 1)$

2. Multiply each side by the LCD.

$$\frac{(x+1)(x-1)}{1}\left[\frac{5}{x+1} - \frac{3}{x-1}\right] = \frac{(x+1)(x-1)}{1}\left[\frac{8}{(x+1)(x-1)}\right]$$

$$5(x-1) \mathbf{- 3(x+1)} = 8$$

3. Remove parentheses and solve.

$$5x - 5 \mathbf{- 3x - 3} = 8$$
$$2x - 8 = 8$$
$$2x = 16$$
$$x = 8$$

4. Check in the original equation.

Left side	Right side
$\frac{5}{8+1} - \frac{3}{8-1}$	$\frac{8}{8^2-1}$
$\frac{5}{9} - \frac{3}{7}$	$\frac{8}{64-1}$
$\frac{5(7)}{9(7)} - \frac{3(9)}{7(9)}$	$\frac{8}{63}$
$\frac{35-27}{63}$	$\frac{8}{63}$

$$\frac{8}{63} = \frac{8}{63} \quad \text{(check)}$$

5. State the solution. The solution is 8.

1. $\frac{2}{x-1} + \frac{3}{x+1} = \frac{24}{x^2-1}$

2. $\frac{3}{y^2-4} - \frac{5}{y-2} = \frac{12}{y+2}$

1. ____________________

2. ____________________

SAMPLE PROBLEM 2

Solve and check $\dfrac{x+6}{4x+20} - \dfrac{x-5}{6x+30} = 2.$

SOLUTION

1. Factor each denominator to find the LCD.

$$4x + 20 = 4(x+5) = 2(2)(x+5)$$
$$6x + 30 = 6(x+5) = 2(3)(x+5)$$
$$\text{LCD} = 12(x+5)$$

2. Multiply by the LCD.

$$\frac{12(x+5)}{1}\left(\frac{x+6}{4(x+5)}\right) - \frac{12(x+5)}{1}\left(\frac{x-5}{6(x+5)}\right) = 12(x+5)(2)$$

3. Solve.

$$3(x+6) - 2(x-5) = 24x + 120$$
$$3x + 18 - 2x + 10 = 24x + 120$$
$$x + 28 = 24x + 120$$
$$24x + 120 = x + 28$$
$$23x = -92$$
$$x = -4$$

4. Check in the original equation.

Left side:

$$\frac{-4+6}{4(-4)+20} - \frac{-4-5}{6(-4)+30}$$
$$\frac{2}{4} - \frac{-9}{6}$$
$$\frac{1}{2} + \frac{3}{2}$$
$$\frac{4}{2}$$
$$2 = 2 \quad \text{(check)}$$

5. State the solution.

The solution is -4.

3. $\dfrac{x+4}{2x+2} + \dfrac{x+1}{5x+5} = 1$

4. $\dfrac{y+8}{6y-12} - \dfrac{y-4}{10y-20} = \dfrac{8}{15}$

5. $\dfrac{5}{x-3} - \dfrac{3}{x} = \dfrac{3}{x^2-3x}$

6. $\dfrac{2}{t+2} - \dfrac{1}{2t} = \dfrac{t+1}{t^2+2t}$

3. ____________

4. ____________

5. ____________

6. ____________

7. $\dfrac{1}{5t+3}+\dfrac{2t}{25t^2-9}=\dfrac{1}{5t-3}$

7. ____________

8. $\dfrac{5}{x^2-25}-\dfrac{4}{x-5}=\dfrac{1}{x+5}$

8. ____________

9. $\dfrac{x+1}{x-2}+\dfrac{x+2}{x-1}=\dfrac{2x^2-x+1}{x^2-3x+2}$

9. ____________

10. $\dfrac{5}{x^2-2x-8}-\dfrac{3}{x^2-6x+8}=0$

10. ____________

11. $\dfrac{1}{3x^2+x-2}=\dfrac{1}{5x^2+7x+2}$

11. ____________

12. $\dfrac{1}{(x-6)^2}=\dfrac{2}{(x+6)^2}-\dfrac{1}{x^2-36}$

12. ____________

13. $1+\dfrac{1}{3x}=\dfrac{3x+3}{3x+1}$

13. ____________

14. $\dfrac{1}{4x-8}+\dfrac{1}{4x+8}=\dfrac{1}{4x^2-16}$

14. ____________

C. RESTRICTIONS

There are restrictions on a fractional equation.

No number can be a solution of a fractional equation if it causes a denominator to become zero. Remember that division by zero is undefined.

Checking is a very important step in the procedure for solving a fractional equation. We must reject any value that produces a zero denominator.

EXERCISES 6.3C

Solve and check.

METHOD	SAMPLE PROBLEM 1	SAMPLE PROBLEM 2
	$\frac{x+2}{x-6} = \frac{8}{x-6}$	$1 + \frac{1}{x+2} = \frac{x+3}{x+2}$
	SOLUTION	**SOLUTION**
1. Find the LCD.	LCD $= x - 6$	LCD $= x + 2$
2. Multiply each side by the LCD.	$x + 2 = 8$	$x + 2 + 1 = x + 3$
3. Solve.	$x = 6$	$3 = 3$ This is true for all x.
4. Check in the original equation.	$\frac{x+2}{x-6} = \frac{6+2}{6-6} = \frac{8}{0}$ $\frac{8}{0}$ is undefined. 6 cannot be a solution. There is no solution.	If the denominator $x + 2 = 0$, then $x = -2$. Check with any other value. For $x = 2$, $1 + \frac{1}{2+2} = 1 + \frac{1}{4} = \frac{5}{4}$ $\frac{2+3}{2+2} = \frac{5}{4}$
5. State the solution set.	$\varnothing$, the empty set	all x, $x \neq -2$ or $Q - \{-2\}$

1. $\frac{x}{x-4} = \frac{4}{x-4}$

2. $\frac{1}{3x} + \frac{1}{6x} = \frac{1}{2x}$

1. ______________________

2. ______________________

3. $\frac{1}{2t-2} + \frac{1}{4t-4} = \frac{3}{4t-4}$ 3. ______

4. $\frac{3}{t} + \frac{5}{t} = \frac{8}{2t}$ 4. ______

5. $\frac{y+2}{3y+3} = \frac{y+3}{6y+6}$ 5. ______

6. $\frac{y+3}{5y+10} = \frac{y+6}{10y+20}$ 6. ______

7. $\frac{x}{x-6} = 1 + \frac{6}{x-6}$ 7. ______

8. $\frac{2x-1}{x-1} = 2 + \frac{x}{x-1}$ 8. ______

9. $\frac{x-2}{x^2-3x} = \frac{2}{3x}$ 9. ______

10. $\frac{2x}{x-5} - \frac{x+5}{x-5} = 2$ 10. ______

EXERCISES 6.3 S

Solve and check.

1. $\dfrac{x+4}{6x-18} + \dfrac{x}{10x-30} = 1$

2. $\dfrac{t}{10t+40} - \dfrac{3}{2t+8} = 2$

3. $\dfrac{3}{2x+4} - \dfrac{x-2}{2x+x^2} = \dfrac{3}{4x+8}$

4. $\dfrac{5}{6x-6} - \dfrac{x+3}{3x^2-3x} = \dfrac{2}{5x}$

5. $\dfrac{7}{t+5} + \dfrac{20}{t^2+5t} = \dfrac{6}{t}$

6. $\dfrac{1}{t-6} - \dfrac{t+6}{t^2} = \dfrac{4}{t^2-6t}$

7. $\dfrac{6}{3+y} - \dfrac{5}{3-y} = \dfrac{25}{9-y^2}$

8. $\dfrac{y}{y^2-4} + \dfrac{1}{y+2} = \dfrac{1}{y-2}$

9. $\dfrac{x}{7-x} + \dfrac{2x}{x-7} = \dfrac{3}{10}$

10. $\dfrac{3}{4-x} + \dfrac{8}{4x-16} = 1$

11. $\dfrac{8}{5y+15} - \dfrac{7}{5y-15} = \dfrac{1}{5y}$

12. $\dfrac{3}{2y+5} = \dfrac{1}{2y} + \dfrac{2}{2y-5}$

13. $\dfrac{x}{x-4} = 2 + \dfrac{4}{x-4}$

14. $\dfrac{x-1}{x+5} = 1 - \dfrac{6}{x+5}$

15. $\dfrac{x+3}{x^2-2x} + \dfrac{3}{2x} = \dfrac{5}{2x-4}$

16. $\dfrac{x-2}{x+3} - \dfrac{x-7}{2x+6} = 1$

1. ______
2. ______
3. ______
4. ______
5. ______
6. ______
7. ______
8. ______
9. ______
10. ______
11. ______
12. ______
13. ______
14. ______
15. ______
16. ______

NAME ______ DATE ______ COURSE ______

17. $\frac{x-4}{x+6} - \frac{x-6}{x+4} = \frac{4x-8}{x^2+10x+24}$ 17. ______

18. $\frac{x+2}{x+5} - \frac{x-5}{x-2} = \frac{3x+24}{x^2+3x-10}$ 18. ______

19. $\frac{5}{x^2-7x+10} = \frac{6}{x^2-9x+20}$ 19. ______

20. $\frac{2}{2x^2+5x-3} = \frac{3}{2x^2+3x-2}$ 20. ______

21. $\frac{1}{(x-2)^2} + \frac{2}{(x+2)^2} = \frac{3}{x^2-4}$ 21. ______

22. $\frac{4}{y^2-1} - \frac{1}{(y-1)^2} = \frac{3}{(y+1)^2}$ 22. ______

23. $\frac{1}{x^2-8x} + \frac{2}{x^2+8x} = \frac{11}{x^2-64}$ 23. ______

24. $\frac{1}{5x-1} - \frac{2}{5x+1} = \frac{1}{25x^2-1}$ 24. ______

6.4 WORK PROBLEMS

If a bricklayer can build a certain fireplace in 4 hr, then he builds $\frac{1}{4}$ of the fireplace in 1 hr. The fraction $\frac{1}{4}$ is the hourly rate at which the bricklayer works. The rate of work plays an important role in work problems.

Basic idea 1. Let T = the time to do a certain job. Then $r = \frac{1}{T}$ is the rate at which the job is done.

EXAMPLE 1	EXAMPLE 2
A painter paints a certain room in 50 min. What is his rate of working?	A machine takes x hr to process certain data. What is its rate of working?
Solution	**Solution**
Rate $= \frac{1}{50}$ of the room per minute.	Rate $= \frac{1}{x}$ of the data per hour.

PRACTICE EXERCISES

1. Bill takes 40 min to wash his car. What is his rate of working? 1. ____________

2. A pipe takes 5 hr to fill a certain tank with water. What is the rate at which the pipe fills the tank? 2. ____________

3. A man requires x hr to brick a certain fireplace. What is the rate at which he works? 3. ____________

4. A machine takes $x + 20$ min to make a certain article. What is the rate at which the machine works? 4. ____________

5. A carpenter takes $2x$ hr to make a set of cabinets. Find the rate of the carpenter. 5. ____________

6. A computer takes $5x$ sec to run a certain program. Find the rate of the computer. 6. ____________

Basic idea 2. The amount of work (W) done is the product of the rate (r) and the time (t) spent working: $W = rt$.

EXAMPLE 3

John can build a certain type of fence in 5 hr. If he works for 3 hr, how much of the fence does he build?

Solution

Using $r = \frac{1}{T}$, where T = time for the complete job, $r = \frac{1}{5}$.

Using $W = rt$, $W = \frac{1}{5}(3) = \frac{3}{5}$ of the fence.

7. A roofing crew takes 10 days to roof a certain building. How much of the work does the crew do in 6 days? 7. ______
8. A mechanic takes 10 hr to do a certain job. How much of the job is done in x hr? 8. ______
9. A card sorter takes 25 min to sort a set of cards. How much of the job does it do in $x + 2$ min? 9. ______
10. A company takes x mo to build a bridge. How much of the bridge is built in 2 mo? 10. ______

Basic idea 3. The total work done is the sum of the amounts of work done by those persons (or machines) working.

EXAMPLE 4

In 1 hr, machine A can process $\frac{1}{10}$ while machine B can process $\frac{1}{15}$ of a certain amount of data. How much data is processed if A works for 3 hr and B works for 4 hr?

Solution

Total work = (work of A) + (work of B).

$W = 3\left(\frac{1}{10}\right) + 4\left(\frac{1}{15}\right) = \frac{3}{10} + \frac{4}{15} = \frac{17}{30}$

Therefore, $\frac{17}{30}$ of the data is processed.

11. In 1 min, one typist does $\frac{1}{50}$ of a job, while another does $\frac{3}{50}$ of the job. How much of the job is done if they both work for 10 min? 11. ______
12. In reference to Question 11, how much of the job is done if they both work for x min? 12. ______
13. In 1 hr, one machine does $\frac{1}{x}$ of a job, while another does $\frac{1}{3x}$ of the job. How much of the job is done in 2 hr? 13. ______
14. In reference to Question 13, how much of the job is done if the first machine works for 3 hr and the second for 5 hr? 14. ______

ANSWERS

1. $\frac{1}{40}$ 2. $\frac{1}{5}$ 3. $\frac{1}{x}$ 4. $\frac{1}{x+20}$ 5. $\frac{1}{2x}$ 6. $\frac{1}{5x}$ 7. $\frac{6}{10}$ 8. $\frac{x}{10}$ 9. $\frac{x+2}{25}$ 10. $\frac{2}{x}$
11. $\frac{10}{50} + \frac{30}{50}$ 12. $\frac{x}{50} + \frac{3x}{50}$ 13. $\frac{2}{x} + \frac{2}{3x}$ 14. $\frac{3}{x} + \frac{5}{3x}$

Basic idea 4. If the whole job is completed, the amount of work done is 1; therefore, $W = 1$.

EXERCISES 6.4

SAMPLE PROBLEM 1

A painter with a power machine takes 30 min to paint a room. Another painter using a roller takes 60 min to paint this room. How long would it take both painters working together to paint the room?

SOLUTION

Lot x = the time for both working together.

Working together

Formula:	t	$\cdot\ r$	$= W$
Power painter	x	$\frac{1}{30}$	$\frac{x}{30}$
Roller painter	x	$\frac{1}{60}$	$\frac{x}{60}$

Equation

Work of one + Work of other = Total work done

$$\frac{x}{30} + \frac{x}{60} = 1 \qquad \text{(The whole job is done.)}$$

$$60\left(\frac{x}{30} + \frac{x}{60}\right) = 60$$

$$2x + x = 60$$

$$3x = 60$$

$$x = 20$$

They take 20 min to paint the room.

1. One machine takes 20 min to do a certain job. A slower machine takes 30 min to do the same job. Working together, how many minutes would it take them to do the job? 1. ____________

2. A roofer takes 14 hr to shingle a certain roof. A helper takes 35 hr to do the same work. Working together, how long does it take them to shingle the roof? 2. ____________

items	R	T together	W
Roofer	$\frac{1}{14}$	X	$\frac{1}{14}X$
helper	$\frac{1}{35}$	X	$\frac{1}{35}X$

$\frac{1}{14}X + \frac{1}{35}X = 1$

$\frac{5}{70}X + \frac{2}{70}X = 70$

$\frac{7}{70}X = 70$

SAMPLE PROBLEM 2

A man can build a certain type of fireplace in 4 hr. If his son helps him, then working together they can build the fireplace in 3 hr. How long would it take the son working alone to build the fireplace?

SOLUTION

Let $x =$ the time of the son.

Working together

Formula:	t	$\cdot\ r$	$= w$
Father	3	$\frac{1}{4}$	$\frac{3}{4}$
Son	3	$\frac{1}{x}$	$\frac{3}{x}$

Equation

Father's work + Son's work = Total work

$$\frac{3}{4} + \frac{3}{x} = 1$$

$$3x + 12 = 4x$$

$$x = 12 \text{ hr}$$

3. An old machine takes 40 min to process a certain amount of data. When the old machine and a new machine work together, they can process this amount of data in 15 min. How long would it take the new machine working alone to process the same data?

3. ______________

4. A farmer can plant a certain area in 6 hr. The farmer and a friend, working together, plant the same area in 4 hr. How long would it take the friend working alone to plant this area?

4. ______________

SAMPLE PROBLEM 3

An old machine takes 5 times as long to manufacture a certain number of parts as a newer one. Working together, they can make this same number of parts in 3 hr. How long does it take each machine, working alone, to make this number of parts?

SOLUTION

Let x = time of new machine, working alone.
$5x$ = time of old machine, working alone.

Working together

Formula:	t	$\cdot\ r$	$= w$
New machine	3	$\frac{1}{x}$	$\frac{3}{x}$
Old machine	3	$\frac{1}{5x}$	$\frac{3}{5x}$

Equation

Work of new + Work of old = Total work

$$\frac{3}{x} + \frac{3}{5x} = 1$$

$$5x\left(\frac{3}{x} + \frac{3}{5x}\right) = 5x(1)$$

$$15 + 3 = 5x$$

$$5x = 18$$

$$x = \frac{18}{5} = 3\frac{3}{5}$$

It takes the new machine $3\frac{3}{5}$ hr.

It takes the old machine 18 hr.

5. One crew takes twice as long as another crew to pave a certain length of a highway. When they work together, they can pave this length in 10 days. How long does it take each crew working alone to pave this length?

5. ______________

6. A small pipe takes 3 times as long to fill a tank as a large one. Working together, they take 20 min to fill the tank. How long does it take each to fill the tank alone?

6. ______________

SAMPLE PROBLEM 4

One pipe takes 10 min to fill a certain tank full of water. A second pipe takes 15 min to fill the tank, and a third pipe takes 30 min to fill the tank. If all three pipes are open, how long would it take to fill $\frac{3}{4}$ of the tank?

SOLUTION

Let x = the time to fill $\frac{3}{4}$ of the tank. Then $\frac{x}{10} + \frac{x}{15} + \frac{x}{30} = \frac{3}{4}$

$$60\left(\frac{x}{10} + \frac{x}{15} + \frac{x}{30}\right) = 60\left(\frac{3}{4}\right)$$

$$6x + 4x + 2x = 45$$

$$12x = 45$$

$$x = \frac{15}{4}$$

$$x = 3\frac{3}{4} \text{ min}$$

7. Three chutes take 20 min, 30 min, and 60 min, respectively, to fill a granary. How long would it take them, working together, to fill $\frac{4}{5}$ of the granary?

7. ______________

8. Three transcribers take 10 hr, 15 hr, and 6 hr, respectively, to type a certain amount of information. How long does it take them, working together, to type this same amount of information?

8. ______________

SAMPLE PROBLEM 5

A man can mow his lawn in 35 min, and his son can mow the lawn in 50 min. One day the man started to mow the lawn alone. At the end of 18 min, his son joined him, and they completed the job. How long did the son work?

SOLUTION

Let $x =$ the time the son worked.

Formula:	*rate*	$\cdot$ *time*	$=$ *work*
Father	$\frac{1}{35}$	$x + 18$	$\frac{x+18}{35}$
Son	$\frac{1}{50}$	x	$\frac{x}{50}$

Equation

$$\frac{x+18}{35} + \frac{x}{50} = 1$$

$$350\left(\frac{x+18}{35} + \frac{x}{50}\right) = 350(1)$$

$$10(x + 18) + 7x = 350$$

$$17x = 170$$

$$x = 10 \text{ min}$$

9. One electrician takes 10 hr to wire a certain building. The partner of the electrician takes 15 hr to do the same job. On a similar job, the electrician worked for 5 hr. Then the partner joined him, and they completed the job. How long did the partner work?

9. ____________________

10. Working alone, an old copier takes 66 min to do a certain job, while a new copier takes 55 min to do the same job. On a similar job, both copiers worked together until the old one broke down. It took the new one 22 min to complete the job. How long did the copiers work together?

10. ____________________

11. A painter working alone could paint a certain house in 30 hr, while it would take his helper 70 hr working alone. How long would it take them to paint the house if they worked together?

11. ____________________

12. One company could build a certain bridge in 24 days. To rush the work, another company is called in; and both companies working together build the bridge in 15 days. How long would it have taken the second company working alone to build the bridge?

12. ____________________

13. Three machines, each working alone, can do a certain job in 10 min, 15 min, and 30 min, respectively. How long would it take to complete the job if all three machines were in operation at the same time?

13. ____________________

14. An outlet pipe can empty a full tank in 105 min. Another outlet pipe can empty the tank in 120 min. If the tank is $\frac{3}{4}$ full, how long would it take to empty the tank if both pipes were open?

14. ____________________

15. A cabinet maker can make a certain set of cabinets in 21 days. His assistant working alone would take 28 days to make the same set of cabinets. The cabinet maker started to make the cabinets. After he had worked for 7 days alone, his assistant then helped him to complete the job. How long did his assistant work?

15. ____________________

1. One machine can produce 100 articles in 20 min, while it takes another machine 1 hr to produce 100 of the same type of article. How long would it take to produce 100 articles of the same type if both machines were in operation at the same time?

1. ______________________

2. One inlet pipe can fill a certain tank in 75 min. It is desired to install a second inlet pipe so that when both pipes are open, the tank will be filled in 30 min. How many minutes should it take the second inlet pipe alone to fill the tank?

2. ______________________

3. When three machines are in operation together, they can do a certain job in 12 hr. The second machine alone takes twice as long as the first machine alone to do the job. The third machine alone takes 3 times as long as the first machine alone to do the job. Find the time that it takes each of the three machines alone to do the job.

3. ______________________

NAME ______________________ DATE ______________________ COURSE ______________________

4. It takes an old machine 5 times as long to disc a certain field as it does a new one. Both machines started to disc the field. At the end of $2\frac{1}{2}$ hr, the old machine broke down. It took $1\frac{1}{2}$ hr more for the new machine to complete the job. Find the time it would have taken the new machine if it had done the whole job alone.

4. ______________________

5. A mechanic can do a certain job in 12 hr. He works alone on the job for 8 hr but then is called away on an emergency. His helper completes the job in 10 more hr. How long would it have taken the helper to do the entire job alone?

5. ______________________

6. An experienced crew can do a job in 14 days. A new crew takes 35 days to do the same job. Both crews work together on a similar job until the experienced crew is called away. The new crew completes the job in 7 days. How long did both crews work together?

6. ______________________

SAMPLE TEST FOR CHAPTER 

1–2. Solve and check. (See Section 6.1.)

1. $\frac{x}{30} + \frac{x}{45} = \frac{2}{3}$

2. $0.12x + 0.08(200 - x) = 18$

1. ______________

2. ______________

3–4. Solve. (See Section 6.2.)

3. How many liters of water should be mixed with a 75% alkali solution to obtain 60 liters of a 50% alkali solution?

3. ______________

4. 15 lb of apples worth 60¢ a pound are combined with oranges worth 45¢ a pound. How many pounds of oranges should be used to obtain a mixture worth 50¢ a pound?

4. ______________

5–10. Solve and check. (See Section 6.3.)

5. $\frac{1}{36x} + \frac{1}{45x} = \frac{1}{x^2}$

6. $\frac{x+3}{x-3} = \frac{2x}{x-3}$

5. ______________

6. ______________

7. $\frac{1}{x-4} + \frac{1}{x+4} = \frac{2x}{x^2-16}$

8. $\frac{1}{x-3} - \frac{1}{x+3} = \frac{3x}{x^2-9}$

7. ______________

8. ______________

NAME ______________ DATE ______________ COURSE ______________

9. $\frac{3}{x-4} - \frac{2}{x+3} = \frac{22}{x^2 - x - 12}$

9. ____________________

10. $\frac{2}{x^2 - 2x} + \frac{7}{x^2 + 2x} = \frac{8}{x^2 - 4}$

10. ____________________

11–12. Solve. (See Section 6.4.)

11. A new machine can manufacture certain parts 5 times as fast as an old one. With both machines operating, 500 parts can be made in 30 min. How long would it take each machine alone to make 500 parts?

11. ____________________

12. One mason takes 10 hr to brick a patio. An apprentice takes 15 hr to do the same job. How long does it take them, working together, to brick the patio?

12. ____________________

SAMPLE TEST FOR CHAPTERS  AND 6

1–3. Simplify

1. $\dfrac{x^2 - 4x}{16 - x^2}$

1. ____________________

2. $\dfrac{2n^2 - 3n - 2}{n^2 + 2n - 3} \div \dfrac{6n + 3}{6n - 6}$

2. ____________________

3. $\left(6 - \dfrac{x^2}{6}\right) \div \left(1 - \dfrac{x}{6}\right)$

3. ____________________

4. Divide, using long division

$\dfrac{x^3 - 18x + 8}{x - 4}$

$x - 4\overline{)}$

4. ____________________

NAME ____________________ DATE ____________________ COURSE ____________________

5. Solve $\frac{2}{x^2 + 5x} + \frac{1}{x^2 - 5x} = \frac{1}{x^2 - 25}$ 5. ____________

6. Check Problem 5 6. ____________

7. Solve $\frac{3}{2x - 1} - \frac{7}{3x + 1} = \frac{x - 8}{6x^2 - x - 1}$ 7. ____________

8. Check Problem 7 8. ____________

9. Solve $\frac{x + 1}{x - 4} = \frac{3 - (2 - x)}{x - 4}$ 9. ____________

10. Solve $\frac{x - 2}{x - 4} = \frac{2}{x - 4}$ 10. ____________

Graphing Points and Lines

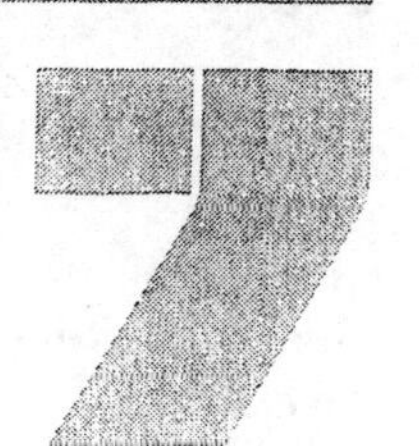

ALGEBRA IN ACTION: SLOPES

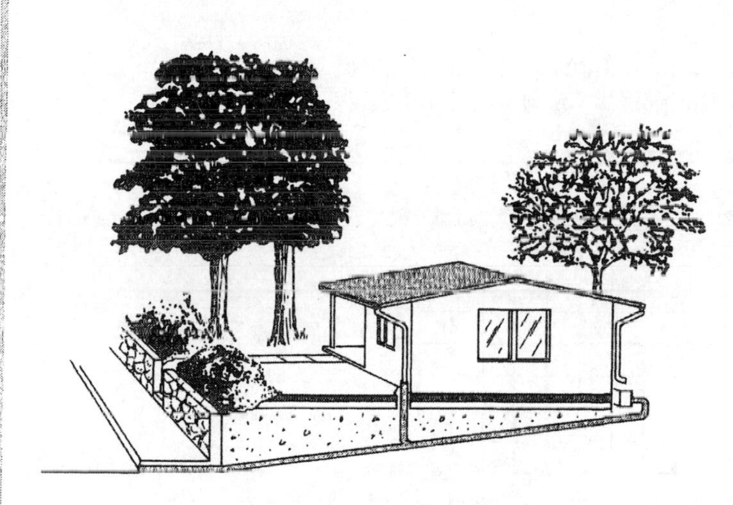

Building contractors use the slope concept to construct roof frames and to properly lay drainage pipes.

Highway engineers use the slope concept for road grades and for back slopes of roads in hilly country.

Pitch of Roof $= \dfrac{\text{rise}}{\text{half-span}}$

Common slopes are $\frac{1}{3}$ and $\frac{1}{4}$

Drainage Pipes

Slope $\geq \frac{1}{4}$

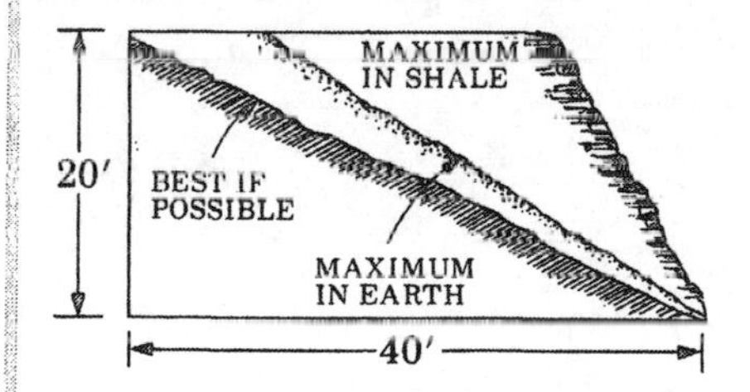

8% GRADE

2

25

Back Slopes for Roads

Earth, best if possible $m = \frac{1}{2}$

Earth, maximum $m = \frac{2}{3}$

Shale, maximum $m = 2$

Road Grades

1% gentle grade, $m = \frac{1}{100}$

8% maximum for cars in high gear, $m = \frac{2}{25}$

20% steep grade, $m = \frac{1}{5}$

Chapter 7 Objectives

Objectives	Typical Problems
Graph ordered pairs of real numbers.	1. Graph $A(6, 2)$ and $B(-1, 9)$. Join A to B with a straight line. 2. Graph $C(4, 8)$ and $D(-1, -7)$. Join C to D with a straight line. 3. Name the point of intersection of line AB and line CD.
Graph a linear equation.	4. Graph $3x - 4y = 12$ on a number plane.
Find the slope of a line, given two points on the line.	5. Find the slope of the line joining $(5, -3)$ and $(-2, 4)$.
Find the slope and y-intercept of a line using the equation $y = mx + b$.	6. Write in the form $y = mx + b$ and state the slope and y-intercept. $5y - 4x = 20$
Find the equation of a line.	7. Write in standard form the equation of the line that passes through the points $(5, 2)$ and $(-3, 6)$.

ANSWERS to Typical Problems

1–3.

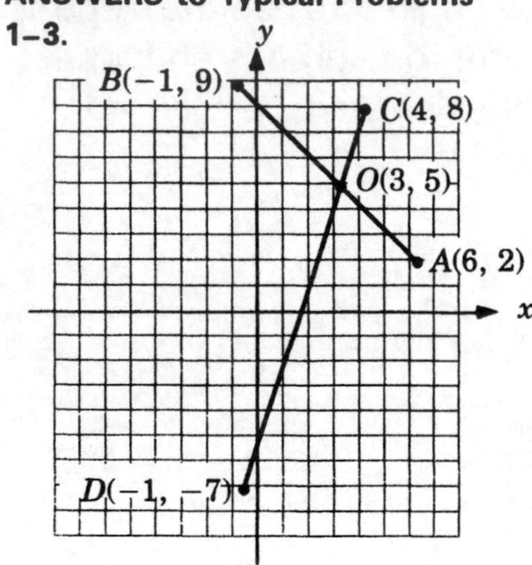

4.

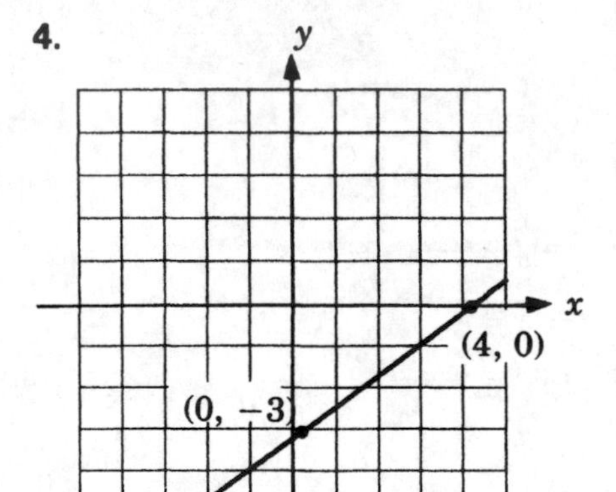

5. -1 6. $y = \frac{4}{5}x + 4$; $m = \frac{4}{5}$, $b = 4$ 7. $x + 2y - 9 = 0$

REAL NUMBERS AND REAL NUMBER LINES

Is there a number x such that $x^2 = 2$? There is no integer whose square is 2. Moreover, there is no rational number whose square is 2.

The symbol $\sqrt{2}$ (read "positive square root of 2") is used to name a positive number whose square is 2; that is, $(\sqrt{2})^2 = 2$.

We know there is such a number as $\sqrt{2}$ because it can be identified as the diagonal of a unit square. See Figure 7.1. (Also see Section 4.11.)

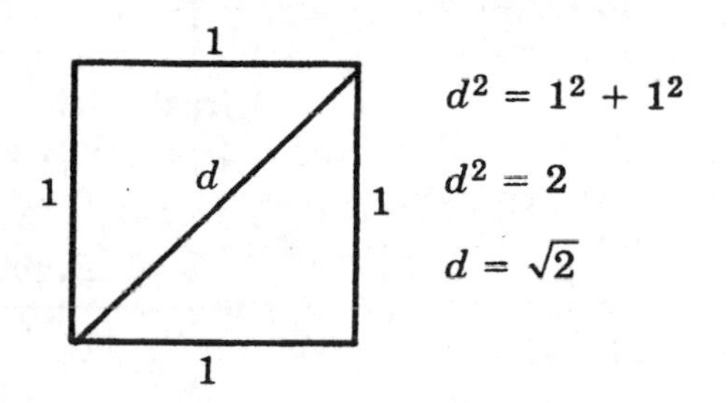

FIGURE 7.1

In geometry we learn that the ratio of the circumference of any circle to its diameter, d, is a constant called π (the Greek letter pi). See Figure 7.2. For a circle whose radius is 1, $d = 2r = 2(1) = 2$. For this case, one half the circumference, the length of the semicircle, is π. π is another number that cannot be written as a quotient of two integers.

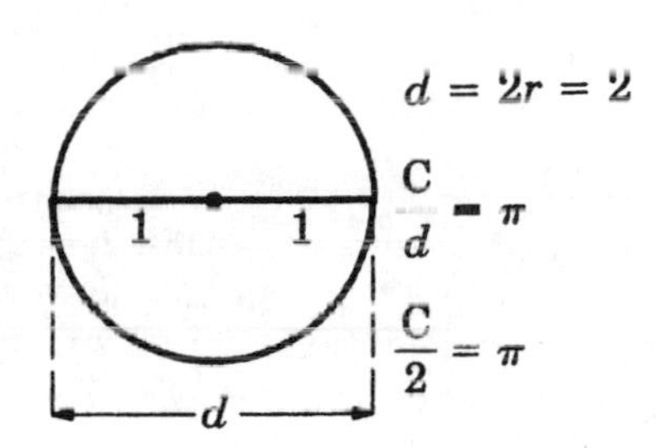

FIGURE 7.2

In Chapter 5, Section 5.2, we saw that every rational number (a quotient of two integers) can be written as either a terminating or a repeating decimal. A decimal that did not terminate and did not have a repeating set of digits was identified as an **irrational number.** (Irrational means not rational.)

Numbers such as $\sqrt{2}$, $\sqrt{5}$, $-\sqrt{2}$, and π are irrational numbers.

$$\sqrt{2} = 1.41421\ldots \quad \text{and} \quad \pi = 3.14159\ldots$$

An irrational number is a number that can be written as a decimal whose digits are nonterminating and nonrepeating.

Irrational numbers have positions on the number line just as rational numbers do. See Figure 7.3.

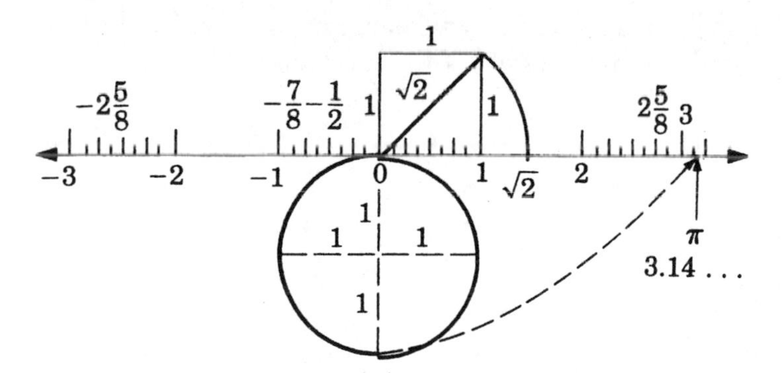

FIGURE 7.3
The real number line.

The **set of real numbers** is the set consisting of all rational numbers and all irrational numbers. See Figure 7.4.

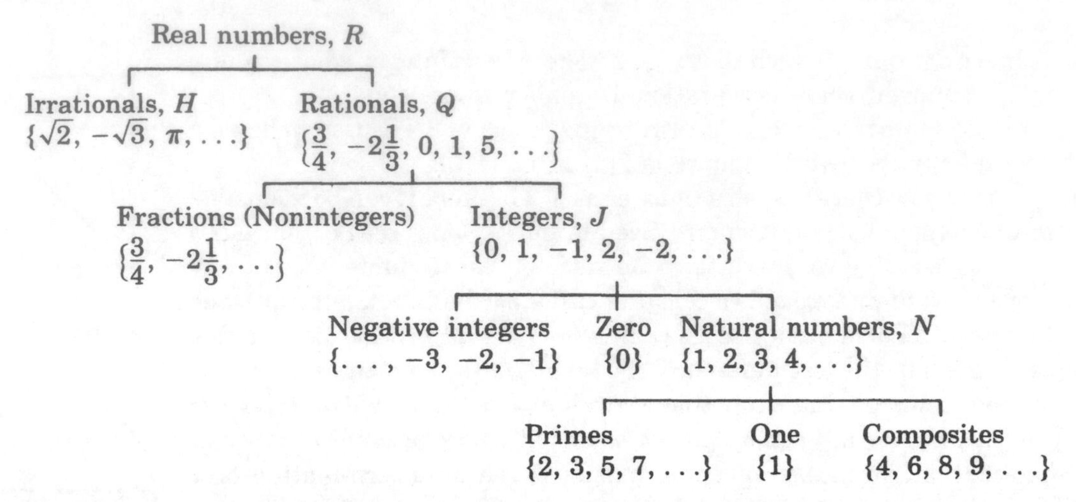

FIGURE 7.4
Structure of the real numbers.

The set of real numbers has the important property that it is **complete**. This means that **every real number has exactly one position on the number line and every point on the number line has exactly one real number name.** For this reason, a number line is called a **real number line.**

Irrational numbers and operations involving irrational numbers are discussed in Chapter 9. This chapter and Chapter 8 deal with equations in two variables, such as $3x - 4y = 12$, and their geometric models, straight lines.

The purpose of this discussion is to provide insight into the algebraic nature of the geometric straight line.

It helps us to know that the straight line is completely filled up with points having real number names. There are no "holes" in the line needing some other kinds of numbers to name them.

ORDERED PAIRS, SOLUTIONS, RECTANGULAR COORDINATES 7.1

A. ORDERED PAIRS, SOLUTIONS

Expressions such as (4, 7) and (−3, 8) are called **ordered pairs.** For the ordered pair (4, 7), 4 is the first number and 7 is the second number.

An **ordered pair** is an expression having the form (a, b), where a is called the **first component** (or first member) of the ordered pair and b is called the **second component** (or second member) of the ordered pair.

The order in which the components of an ordered pair are written is important. For example, the ordered pair (3, 5) is not the same as the ordered pair (5, 3).

> A **solution of an equation in two variables,** x and y, is an ordered pair (a, b) such that the equation becomes a true statement when x is replaced by a and y is replaced by b.

EXERCISES 7.1A

SAMPLE PROBLEM

Determine if the given ordered pair is a solution of $2x + 3y = 12$.

a. (−6, 8)

b. (8, −6)

SOLUTION

a. For (−6, 8), $x = -6$ and $y = 8$.

$$2x + 3y = 12$$
$$2(-6) + 3(8) = 12$$
$$-12 + 24 = 12 \text{ (true)}$$

(−6, 8) is a solution.

b. For (8, −6), $x = 8$ and $y = -6$.

$$2(8) + 3(-6) = 12$$
$$16 - 18 = 12$$
$$-2 = 12 \text{ (false)}$$

(8, −6) is not a solution.

1–8. Determine whether the given ordered pair is a solution of $2x + y = 6$.

1. (1, 4)

2. (4, 1)

3. (2, 2)

4. (0, 6)

5. (3, 0)

6. (0, 3)

7. (−4, 10)

8. (5, −4)

1. ________ **2.** ________

3. ________ **4.** ________

5. ________ **6.** ________

7. ________ **8.** ________

9–14. Determine whether the given ordered pair is a solution of $3x - y = 15$.

9. (6, 3)

10. (3, 6)

11. (3, −6)

12. (5, 0)

13. (0, 15)

14. (0, −15)

9. ________

10. ________

11. ________

12. ________

13. ________

14. ________

15–30. Determine whether the given ordered pair is a solution of the given equation.

15. $5x + 2y = 10$; $(4, -5)$

16. $5x + 2y = 10$; $(5, 2)$

17. $4x - 3y = 12$; $(0, 3)$

18. $4x - 3y = 12$; $(3, 4)$

19. $4x - 3y = 12$; $\left(\frac{3}{2}, -2\right)$

20. $y = 5 - x$; $(0, 5)$

21. $y = 5 - x$; $(-2, 3)$

22. $y = 5 - x$; $(-3, 8)$

23. $y = 2x - 3$; $(0, -3)$

24. $y = 2x - 3$; $\left(\frac{5}{2}, 2\right)$

25. $xy = 6$; $(2, 3)$

26. $xy = 6$; $(-6, -1)$

27. $xy = 6$; $\left(-12, -\frac{1}{2}\right)$

28. $xy = 6$; $(12, -6)$

29. $y = x^2$; $(3, 9)$

30. $y = x^2$; $(-3, 9)$

15. ________

16. ________

17. ________

18. ________

19. ________

20. ________

21. ________

22. ________

23. ________

24. ________

25. ________

26. ________

27. ________

28. ________

29. ________

30. ________

B. FINDING SOLUTIONS OF $Ax + By + C = 0$

A **linear equation in two variables,** x and y, is an equation of the form $Ax + By + C = 0$, where A, B, and C are constants and A is not zero or B is not zero.

To find a solution (a, b) of $Ax + By + C = 0$:

1. Select any value for one of the variables.
2. Replace the variable by this value.
3. Solve the resulting equation for the other variable.
4. State the solution as an ordered pair.

SAMPLE PROBLEM 1

Find the solution of $5x - 3y + 15 = 0$ for which $x = 0$.

SOLUTION

1. Replace x by 0 and solve for y.
2. Substitute. $5x - 3y + 15 = 0$

   $5(0) - 3y + 15 = 0$
3. Solve for y. $0 - 3y + 15 = 0$

   $-3y = -15$

   $y = 5$
4. State solution as an ordered pair. The solution is $(0, 5)$.

SAMPLE PROBLEM 2

Find the solution of $5x - 3y + 15 = 0$ for which $y = 0$.

SOLUTION

1. Replace y by 0 and solve for x.
2. Substitute. $5x - 3y + 15 = 0$

   $5x - 3(0) + 15 = 0$
3. Solve for x. $5x = -15$

   $x = -3$
4. State solution as an ordered pair. The solution is $(-3, 0)$.

SAMPLE PROBLEM 3

Find the solution of $2x + y = 5$ for which $x = 4$.

SOLUTION

1. Use $x = 4$ in the equation.
2. Substitute. $2x + y = 5$
   $2(4) + y = 5$
3. Solve for y. $y = 5 - 8$
   $y = -3$
4. State solution as an ordered pair. The solution is $(4, -3)$.

SAMPLE PROBLEM 4

Find the solution of $2x - 3y = 7$ for which $y = 3$.

SOLUTION

1. Use $y = 3$ in the equation.
2. Substitute. $2x - 3y = 7$
   $2x - 3(3) = 7$
3. Solve for x. $2x = 7 + 9$
   $2x = 16$
   $x = 8$
4. State solution as an ordered pair. The solution is $(8, 3)$.

EXERCISES 7.1B

Find the solution of each given equation corresponding to the given value for one of the variables.

1. $x + y = 12$ and $x = 7$

2. $x + y = 12$ and $y = -2$

3. $x - y = 5$ and $y = -3$

4. $x - y = 5$ and $x = -4$

5. $2x + y = 8$ and $x = 1$

6. $3x + y = 7$ and $x = -2$

7. $3x - y = 5$ and $x = -1$

8. $2x - y = 6$ and $y = -2$

9. $2x - 5y = 10$ and $x = 0$

10. $2x - 5y = 10$ and $y = 0$

1. ____________

2. ____________

3. ____________

4. ____________

5. ____________

6. ____________

7. ____________

8. ____________

9. ____________

10. ____________

C. GRAPHING POINTS ON A NUMBER PLANE

A **rectangular coordinate system** (or rectangular number plane) is formed by a vertical number line and a horizontal number line intersecting at their origins. This point of intersection is called the **origin** of the coordinate system.

To be a rectangular coordinate system, the two number lines must form right angles at their point of intersection.

The plane is divided into four regions called **quadrants.** The quadrants are numbered by Roman numerals in counterclockwise order, beginning with quadrant I in the upper right region.

The horizontal number line is called the ***x*-axis,** and its positive direction is conventionally taken to the right.

The vertical number line is called the ***y*-axis,** and its positive direction is conventionally taken upward.

If vertical lines are drawn through every point that corresponds to an integer on the x-axis and horizontal lines are drawn through every point that corresponds to an integer on the y-axis, the intersections of these lines will represent a one-to-one correspondence between the points of intersection of these lines and the set of ordered pairs of integers. The lines form a grid, as shown in Figure 7.5.

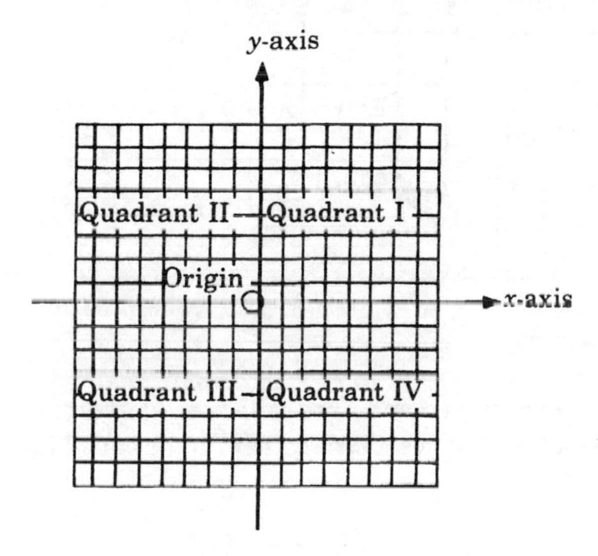

FIGURE 7.5
Rectangular coordinate system.

There are many more points on a number plane other than those corresponding to ordered pairs of integers, just as there are many more points on a number line other than those corresponding to integers.

There is a **one-to-one correspondence** between the set of points on the plane and the set of ordered pairs of real numbers. A point P is assigned the ordered pair (x, y) if and only if the vertical line through P intersects the x-axis at x and the horizontal line through P intersects the y-axis at y.

A point a on the x-axis is assigned the ordered pair $(a, 0)$, and a point b on the y-axis is assigned the ordered pair $(0, b)$. The origin is assigned the ordered pair $(0, 0)$.

On the figure, four points are graphed, one in each quadrant, namely, $A(3, 5)$, $B(-2, 3)$, $C(-4, -3)$, and $D(2, -4)$.

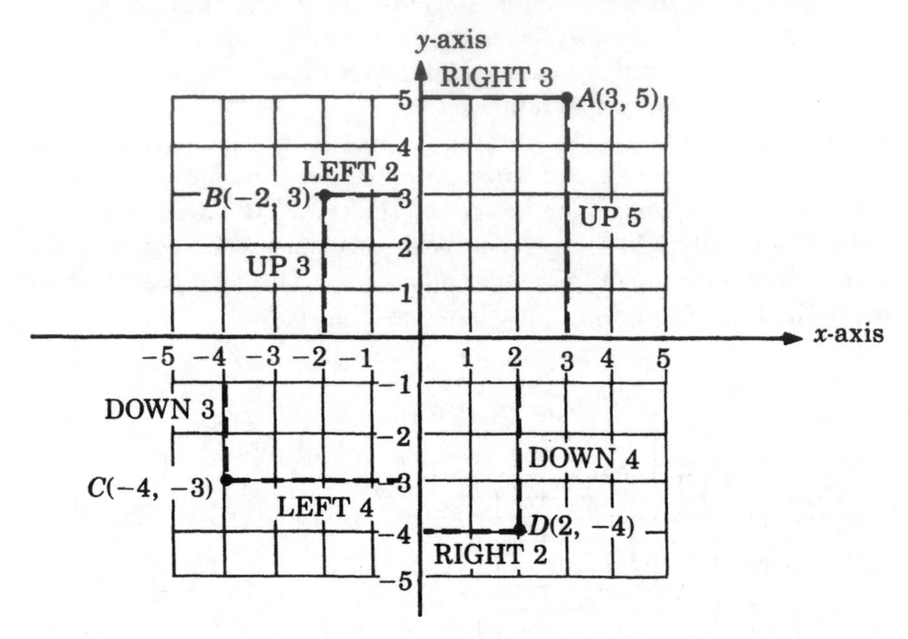

FIGURE 7.6
A rectangular coordinate system

For a point $P(x, y)$, the numbers x and y are called the **coordinates** of P, with x called the **x-coordinate,** or **abscissa,** of P and y called the **y-coordinate,** or **ordinate,** of P.

For the point $A(3, 5)$

the number 3 is the x-coordinate, or abscissa, of A.

the number 5 is the y-coordinate, or ordinate, of A.

EXERCISES 7.1C

1–8. Graph each of the following points on the number plane at the right.

1. $A(5, 2)$ **2.** $B(2, 5)$

3. $C(-3, -4)$ **4.** $D(1, -5)$

5. $E(-4, 3)$ **6.** $F(-5, 0)$

7. $G(0, 2)$ **8.** $H(0, -2)$

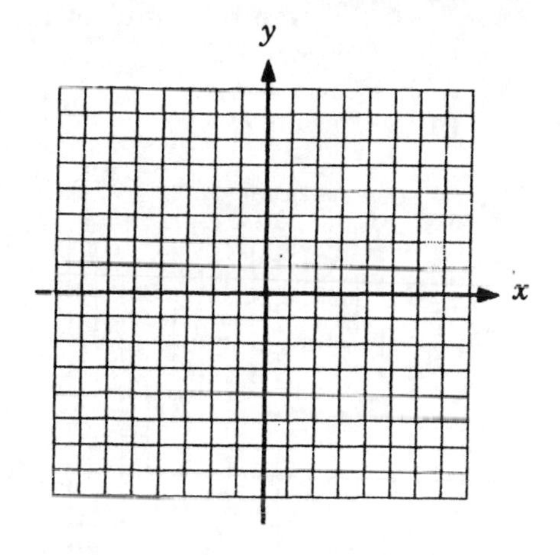

9–14. State the coordinates of each point whose graph is shown in the figure below

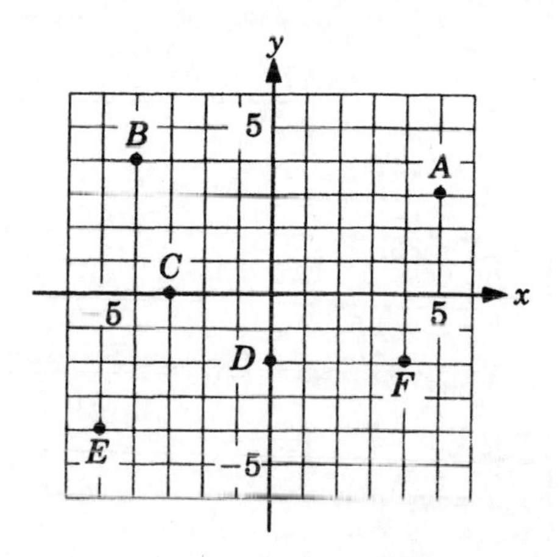

9. A________ **10.** B________

11. C________ **12.** D________

13. E________ **14.** F________

15. Graph each of the following points on the number plane shown: $A(-4, -5)$, $B(0, 7)$, $C(4, -5)$, $D(-6, 2)$, $E(6, 2)$. Join A to B to C to D to E to A. Name the figure so formed.

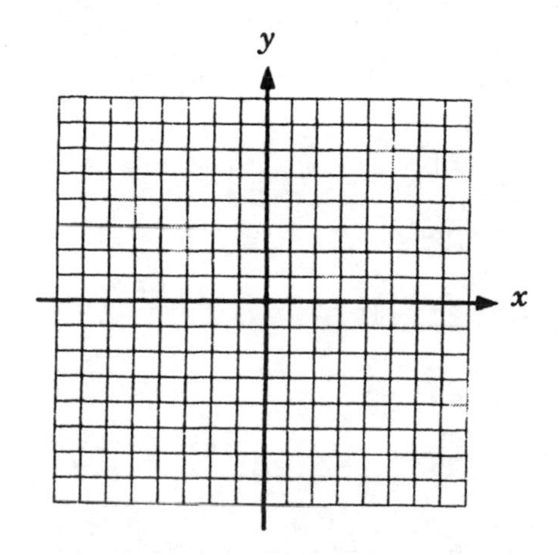

16. Graph each of the following points on the number plane shown: $A(5, 3)$, $B(-5, 3)$, $C(0, -6)$, $P(-5, -3)$, $Q(0, 6)$, $R(5, -3)$. Join A to B to C to A. Then join P to Q to R to P. Name the figure so formed.

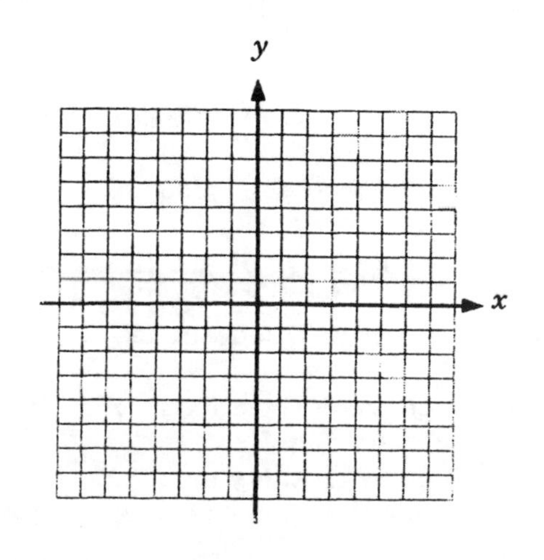

17–20. For the given points,

a. Graph A and B and join with a straight line.
b. Graph C and D and join with a straight line.
c. Name the point of intersection of the two lines.

17. $A(0, 4)$ and $B(8, 0)$
$C(0, -2)$ and $D(2, 0)$

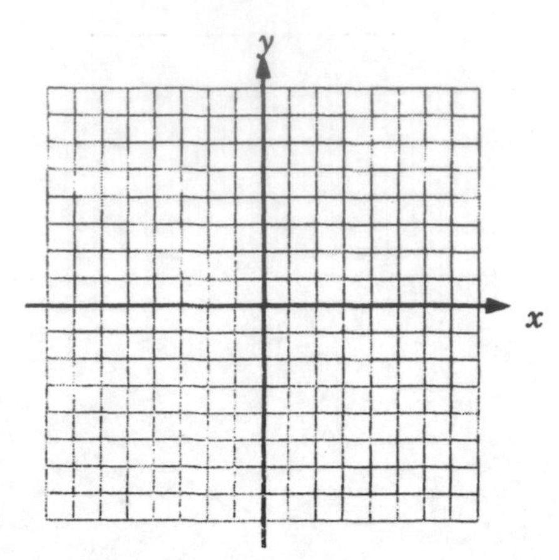

18. $A(-5, 0)$ and $B(2, 7)$
$C(6, -1)$ and $D(-6, 5)$

17. ____________

18. ____________

19. $A(5, 1)$ and $B(-1, -5)$
$C(0, 0)$ and $D(6, -2)$

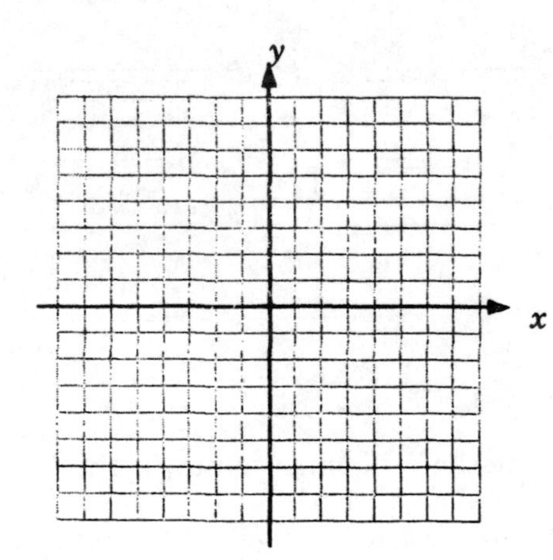

20. $A(3, 2)$ and $B(-5, -6)$
$C(-8, 1)$ and $D(0, -7)$

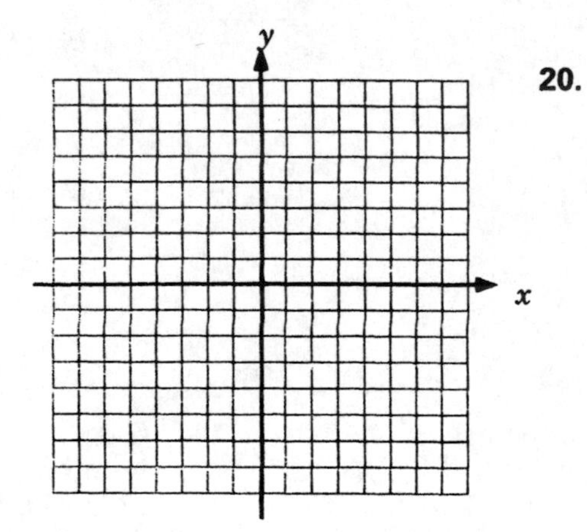

19. ____________

20. ____________

21–26. Select a point from the column on the left as the answer to each question.

$(-5, 0)$	**21.** A point whose abscissa is 3.	**21.** ____________
$(-3, 5)$	**22.** A point whose ordinate is -5.	**22.** ____________
$(3, 5)$	**23.** A point in Quadrant II.	**23.** ____________
$(0, 3)$	**24.** A point in Quadrant IV.	**24.** ____________
$(-3, -5)$	**25.** A point on the x-axis.	**25.** ____________
$(5, -3)$	**26.** A point on the y-axis.	**26.** ____________

EXERCISES 7.1S

1–40. Determine whether each given ordered pair is a solution of the given equation.

1–4. $x + y = 10$

1. $(2, 8)$ **2.** $(-3, -7)$

3. $(12, -2)$ **4.** $(5, -15)$

5–8. $x - y = 5$

5. $(7, 2)$ **6.** $(-3, -8)$

7. $(5, 0)$ **8.** $(0, 5)$

9–12. $2x + y = 8$

9. $(4, 2)$ **10.** $(2, 4)$

11. $(0, 8)$ **12.** $(-4, 0)$

13–16. $x - 2y = 7$

13. $(5, -1)$ **14.** $(3, 2)$

15. $(0, 9)$ **16.** $(-3, -5)$

17–20. $3x - 2y = 12$

17. $(3, 6)$ **18.** $(6, 3)$

19. $(-2, -3)$ **20.** $(-2, -9)$

21–24. $4x + 3y = 24$

21. $(-3, 4)$ **22.** $(0, 8)$

23. $(6, 0)$ **24.** $(-6, 16)$

25–28. $y = x^2 - 1$

25. $(-5, 24)$ **26.** $(0, -1)$

27. $(-1, 0)$ **28.** $(4, 3)$

1. ________ 2. ________
3. ________ 4. ________
5. ________ 6. ________
7. ________ 8. ________
9. ________ 10. ________
11. ________ 12. ________
13. ________ 14. ________
15. ________ 16. ________
17. ________ 18. ________
19. ________ 20. ________
21. ________ 22. ________
23. ________ 24. ________
25. ________ 26. ________
27. ________ 28. ________

NAME ________ DATE ________ COURSE ________

29–32. $y = 9 - x^2$

29. $(-3, 0)$ **30.** $(-2, 5)$ **29.** ________ **30.** ________

31. $(4, 5)$ **32.** $(4, -7)$ **31.** ________ **32.** ________

33–36. $x^2 + y^2 = 25$

33. $(3, -4)$ **34.** $(4, -3)$ **33.** ________ **34.** ________

35. $(0, -5)$ **36.** $(5, 0)$ **35.** ________ **36.** ________

37–40. $x^2 - y^2 = 16$

37. $(-5, -3)$ **38.** $(-3, -5)$ **37.** ________ **38.** ________

39. $(0, -4)$ **40.** $(-4, 0)$ **39.** ________ **40.** ________

41–50. **Find the solution of each given equation corresponding to the given value for one of the variables.**

41. $x - 4y = 12$ and $y = -3$ **42.** $x + 5y = 10$ and $y = -2$

41. ________________

42. ________________

43. $2x + y - 9 = 0$ and $x = 4$ **44.** $2x + y - 9 = 0$ and $y = 3$

43. ________________

44. ________________

45. $3x - y + 9 = 0$ and $x = 5$ **46.** $3x - y + 9 = 0$ and $y = 5$

45. ________________

46. ________________

47. $2x - 3y - 8 = 0$ and $y = -2$ **48.** $4x + 5y + 10 = 0$ and $y = 0$

47. ________________

48. ________________

49. $4x + 5y + 10 = 0$ and $x = 0$ **50.** $6y - 2x + 5 = 0$ and $y = 1$

49. ________________

50. ________________

51–58. **Graph each of the following points on the number plane at the right.**

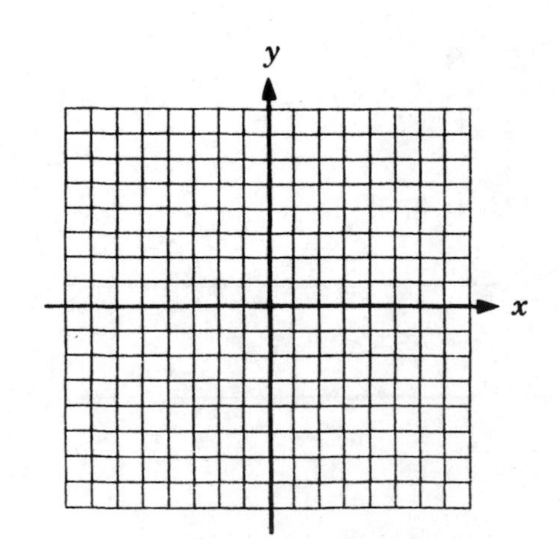

51. $A(6, 3)$ **52.** $B(-3, -6)$

53. $C(-6, -3)$ **54.** $D(2, -4)$

55. $E(4, -2)$ **56.** $F(1, 0)$

57. $G(-4, 0)$ **58.** $H(0, 5)$

59–64. State the coordinates of each point whose graph is shown in the figure below.

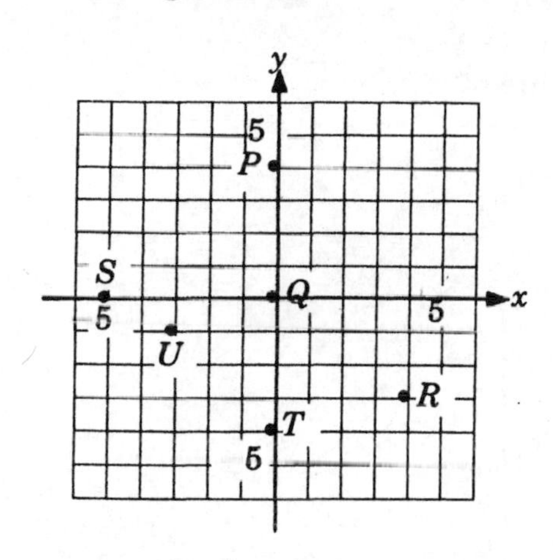

59. P__________ 60. Q__________

61. R__________ 62. S__________

63. T__________ 64. U__________

65–68. For the given points,

a. Graph A and B and join with a straight line.
b. Graph C and D and join with a straight line.
c. State the coordinates of the point of intersection.

65. $A(-2, 1)$ and $B(4, 7)$
$C(8, -1)$ and $D(0, 7)$

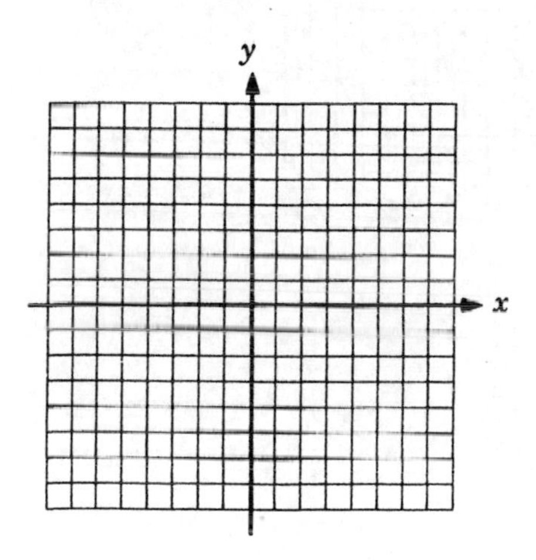

66. $A(-6, -4)$ and $B(3, 5)$
$C(-6, 0)$ and $D(1, -7)$

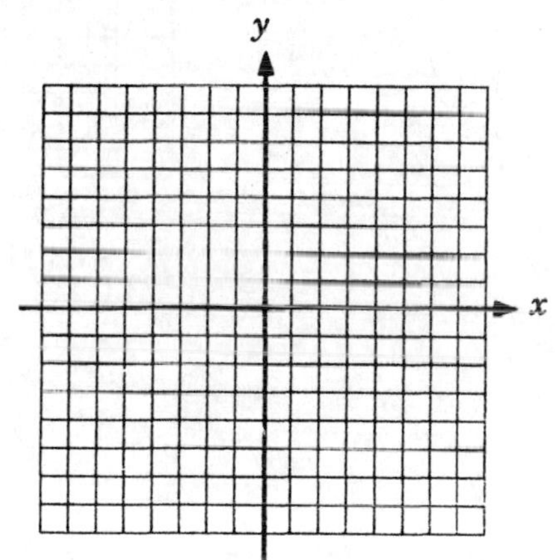

65. ________________

66. ________________

67. $A(-3, 2)$ and $B(5, -2)$
$C(5, 0)$ and $D(-3, -4)$

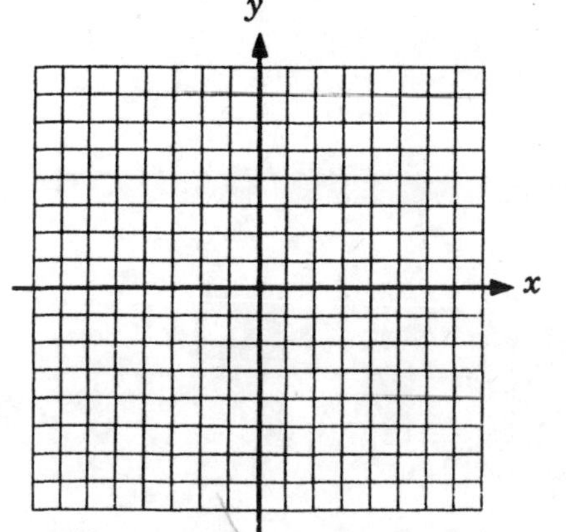

68. $A(-4, 4)$ and $B(6, -1)$
$C(-6, 1)$ and $D(0, 4)$

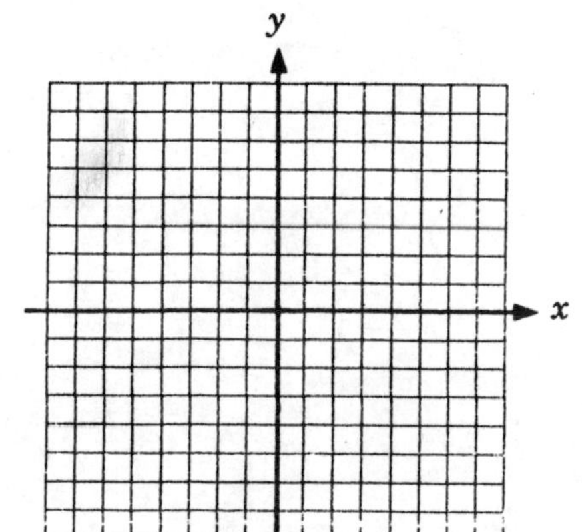

67. ________________

68. ________________

69–70. **a.** Find the solutions (ordered pairs) of the given equation corresponding to the given values of the variable in the table.

b. Graph the points corresponding to the ordered pairs of Part **a**.

c. If the points were connected, what geometric term would describe the graph?

69. $x + 2y = 6$

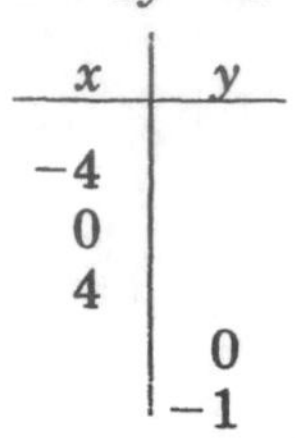

x	y
-4	
0	
4	
	0
	-1

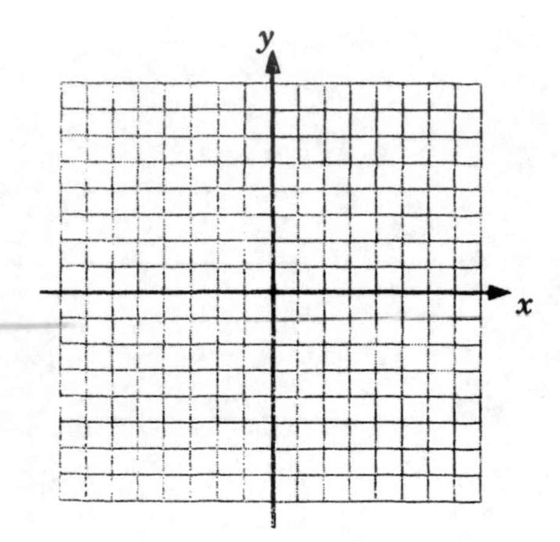

70. $2x - y = 4$

x	y
-1	
0	
	0
	2
	6

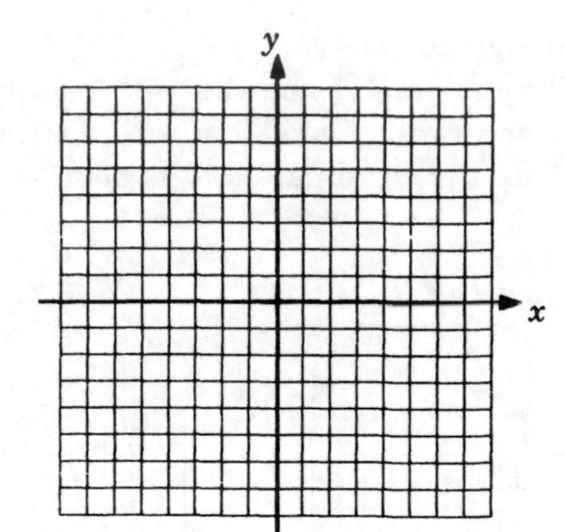

GRAPHING LINEAR EQUATIONS 7.2

A. GRAPHING THE LINEAR FUNCTION

A **linear function** is a set of ordered pairs (x, y) defined by the rule $y = mx + b$.

The **graph of a linear function** is the graph of all ordered pairs for which $y = mx + b$.

There are infinitely many ordered pairs that are solutions of $y = mx + b$, and thus there are infinitely many points on the graph of $y = mx + b$.

The graph of a linear function is a straight line.

PROCEDURE FOR GRAPHING A LINEAR FUNCTION

1. Select any three values for x.
2. Use the given equation to find the corresponding y value.
3. Make a table for the three ordered pairs.
4. Graph the three ordered pairs on a rectangular coordinate system.
5. Draw a straight line joining the three plotted points.

(Note: While only two points are needed to graph a line, the third point serves as a check point in case an error was made in the calculations.)

SAMPLE PROBLEM 1

Graph $y = 2x - 6$.

SOLUTION

1. Select any three values for x; say, $-1, 0, 3$.
2. Find each corresponding y value.
3. Make a table for the three ordered pairs.

$(x$	$y)$	$y = 2x - 6$
$(-1$	$-8)$	$2(-1) - 6 = -8$
$(0$	$-6)$	$2(0) - 6 = -6$
$(3$	$0)$	$2(3) - 6 = 0$

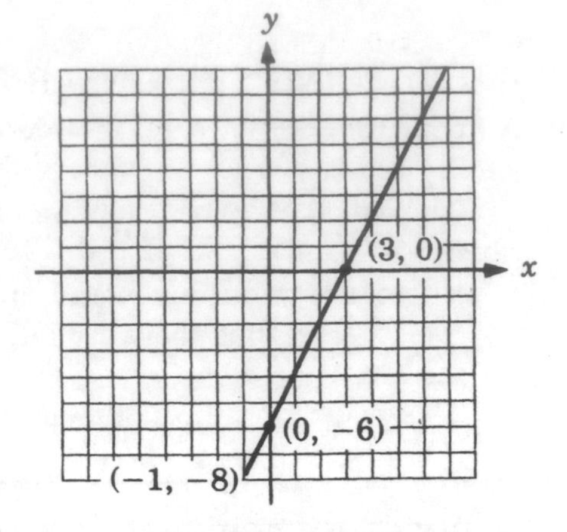

4. Plot the three ordered pairs on a rectangular coordinate system.
5. Draw a straight line joining the three plotted points.

SAMPLE PROBLEM 2

Graph $y = 8 - 4x$.

SOLUTION

1. Select any three values for x; say, $0, 2, 4$.
2. Find each corresponding y value.
3. Make a table for the three ordered pairs.

$(x$	$y)$	$8 - 4x \quad = y$
$(0$	$8)$	$8 - 4(0) = 8 - 0 = 8$
$(2$	$0)$	$8 - 4(2) = 8 - 8 = 0$
$(4$	$-8)$	$8 - 4(4) = 8 - 16 = -8$

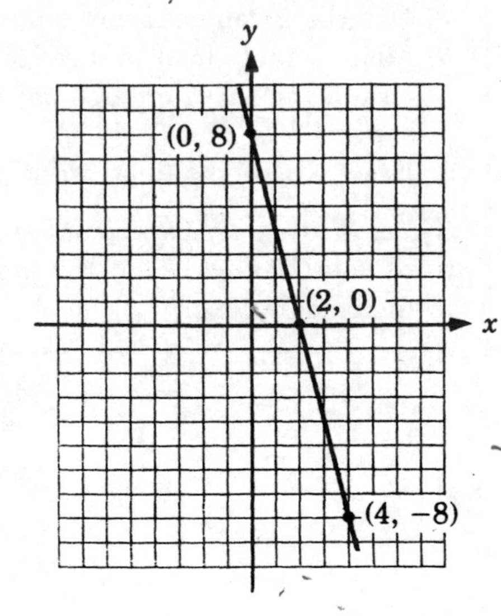

4. Plot the three ordered pairs on a rectangular coordinate system.
5. Draw a straight line joining the three plotted points.

EXERCISES 7.2A

1–6. Complete each table by finding the corresponding y value for each given x value. Plot the three points represented by these ordered pairs on the coordinate system provided and draw a straight line through the plotted points.

1. $y = x + 4$

$y = x + 4$

x	y
5	
0	
3	

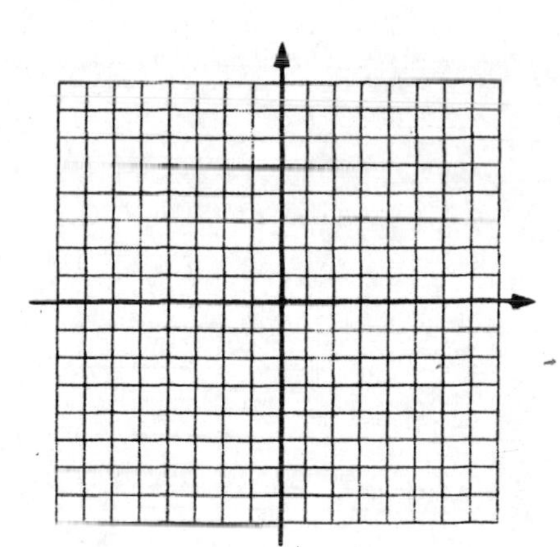

2. $y = 2x + 3$

$y = 2x + 3$

x	y
−4	
0	
2	

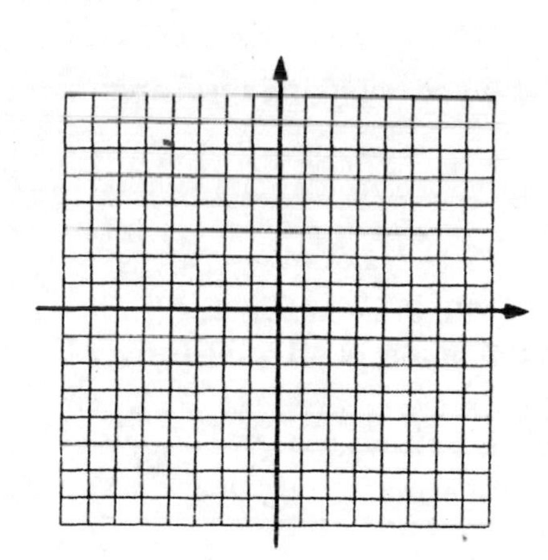

3. $y = 5 - x$

$y = 5 - x$

x	y
−2	
0	
6	

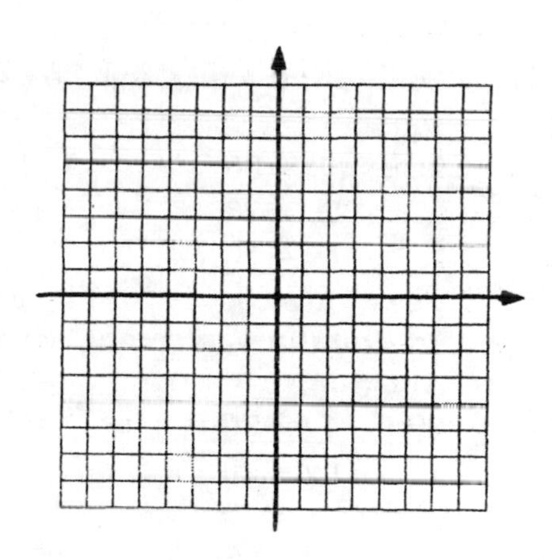

4. $y = 3x - 6$

$y = 3x - 6$

x	y
−1	
0	
4	

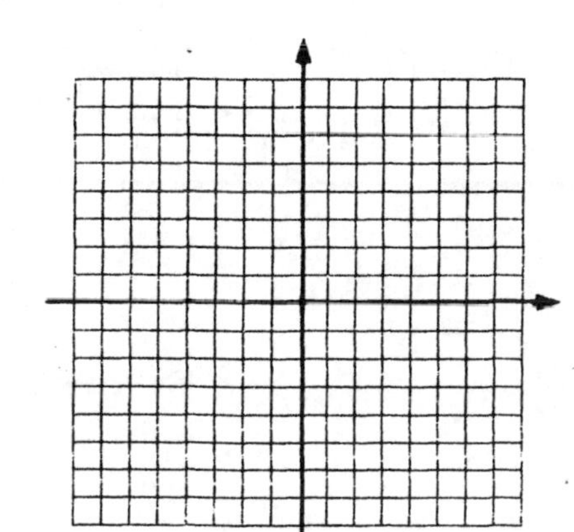

5. $y = 4 - 2x$

$y = 4 - 2x$

x	y
−2	
0	
4	

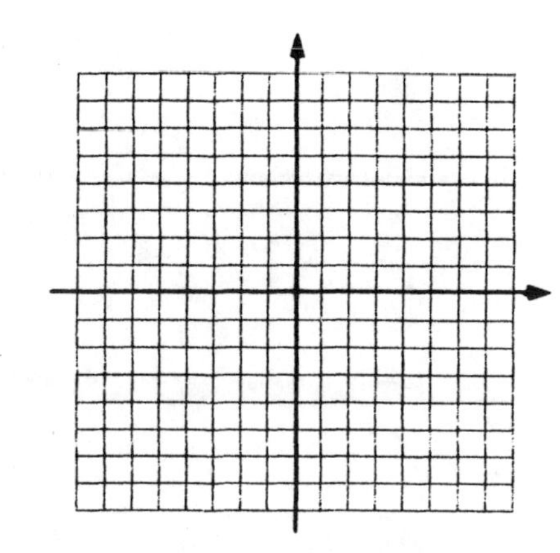

6. $y = \frac{-x}{3}$

x	y
−3	
0	
6	

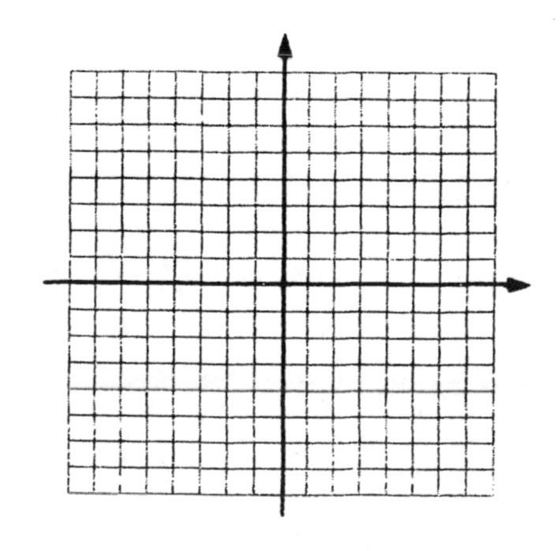

B. GRAPHING $Ax + By + C = 0$, INTERCEPTS

An equation of the form $Ax + By + C = 0$, where x and y are variables and A, B, and C are constants such that A and B are not both zero, is called a **linear equation.**

The **graph of a linear equation** is a straight line passing through the points whose coordinates satisfy the equation.

A line that crosses the x-axis has an **x-intercept.** If a is the x-intercept of a line, then the point $(a, 0)$ is on the line.

A line that crosses the y-axis has a **y-intercept.** If b is the y-intercept of a line, then the point $(0, b)$ is on the line.

SAMPLE PROBLEM 1

Find the x-intercept and the y-intercept of the line whose equation is $2x - 5y = 20$.

SOLUTION

1. **To find the x-intercept,** set $\mathbf{y = 0}$ and solve the resulting equation.

   $2x - 5(\mathbf{0}) = 20$

   $2x = 20$

   $x = 10$

   The x-intercept is 10. The point $(10, 0)$ is on the line.

2. **To find the y-intercept,** set $\mathbf{x = 0}$ and solve the resulting equation.

   $2(\mathbf{0}) - 5y = 20$

   $-5y = 20$

   $y = -4$

   The y-intercept is -4. The point $(0, -4)$ is on the line.

SAMPLE PROBLEM 2

Sketch the graph of $2x - 5y = 20$ by plotting the two points corresponding to the x- and y-intercepts.

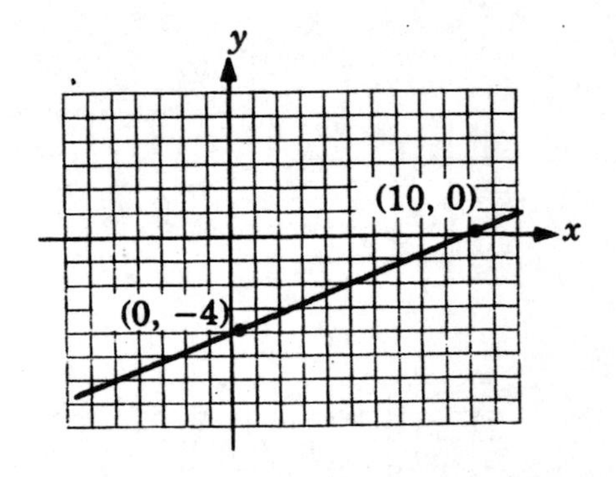

SOLUTION

Referring to Sample Problem 1, the points are $(10, 0)$ and $(0, -4)$.

EXERCISES 7.2B

Find the x- and y-intercepts and graph the equation.

1. $2x + y = 4$
x-intercept =
y-intercept =

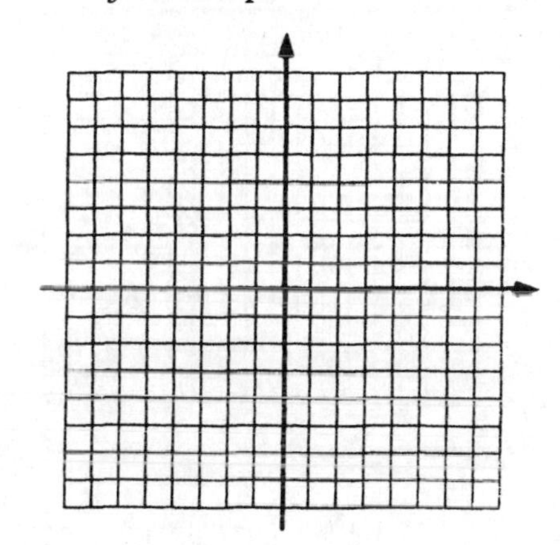

2. $y - 3x = 9$
x-intercept =
y-intercept =

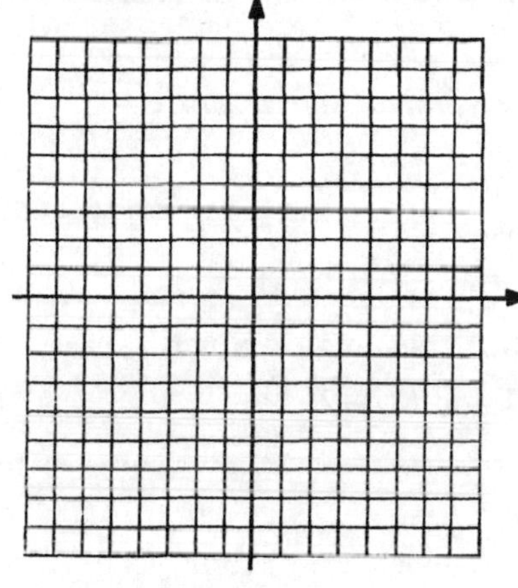

3. $x - y = 5$
x-intercept =
y-intercept =

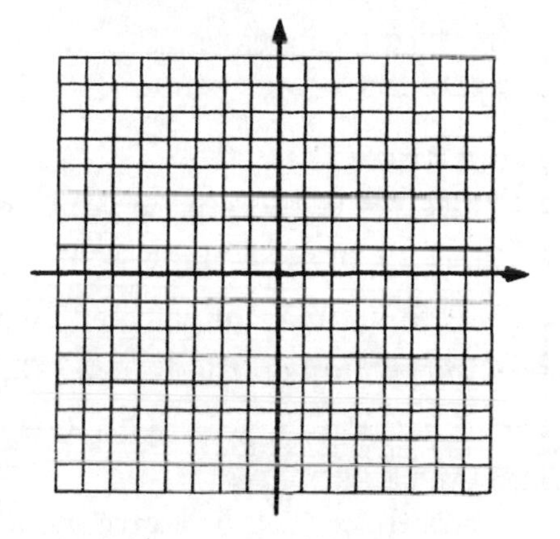

4. $5x + 2y - 10 = 0$
x-intercept =
y-intercept =

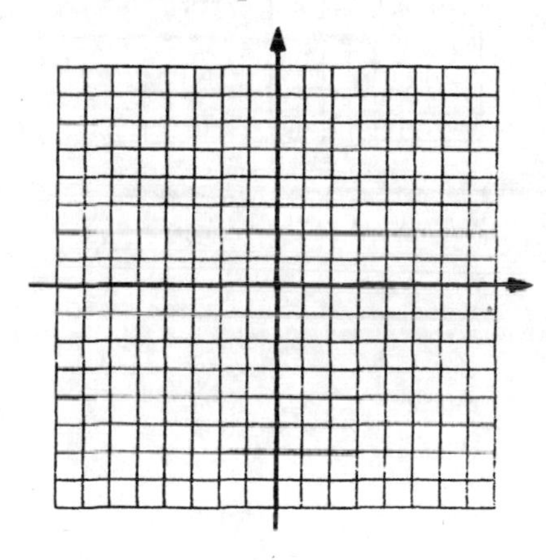

5. $3x + 4y + 12 = 0$
x-intercept =
y-intercept =

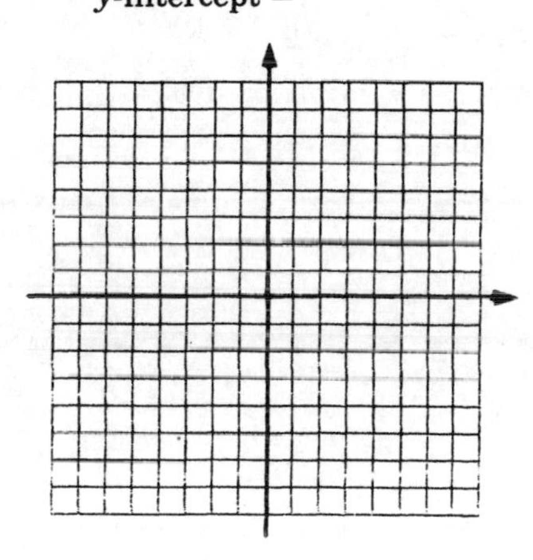

6. $2x - 5y - 10 = 0$
x-intercept =
y-intercept =

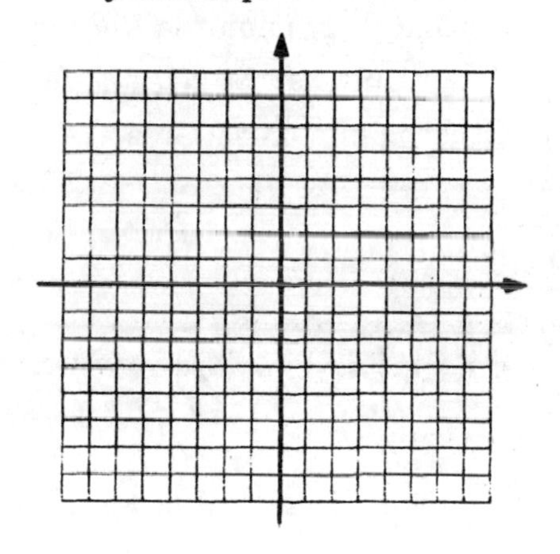

7. $4y - x = 8$
x-intercept =
y-intercept =

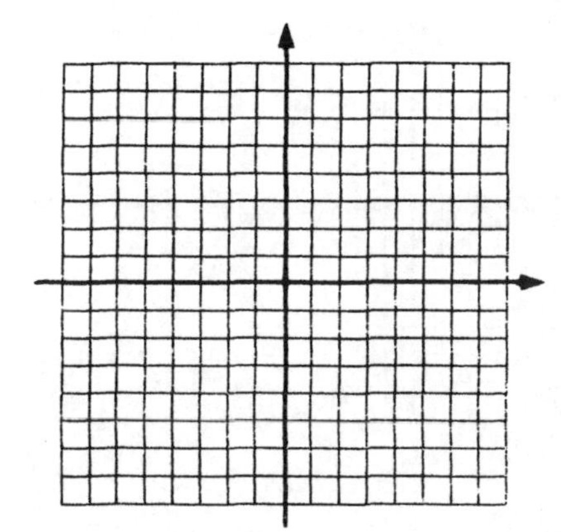

8. $x + y + 4 = 0$
x-intercept =
y-intercept =

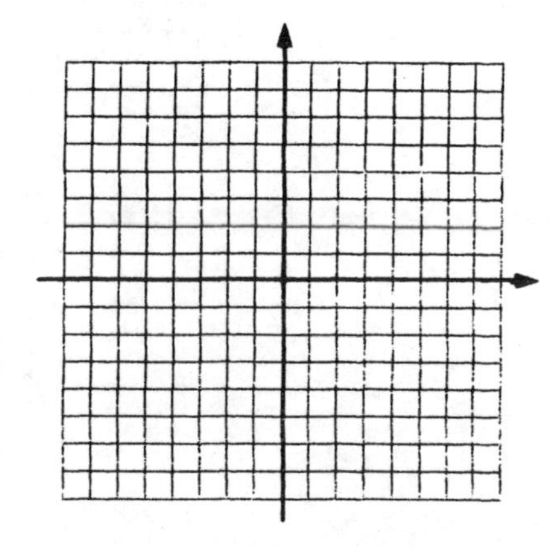

9. $y = 4 - x$
x-intercept =
y-intercept =

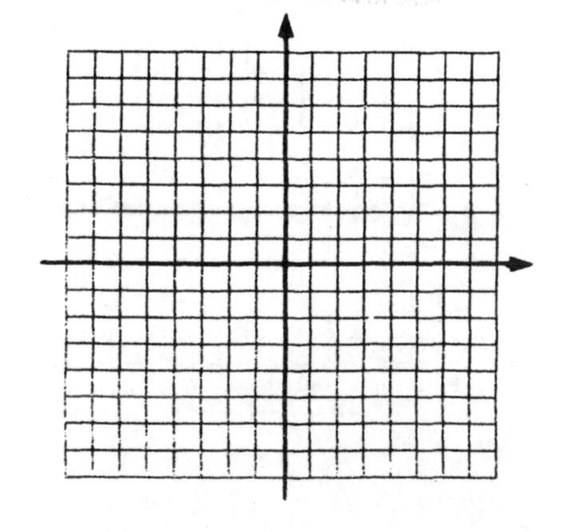

C. VERTICAL AND HORIZONTAL LINES

The graph of an equation having the form **$x = a$ is a vertical line** passing through the point $(a, 0)$ on the x-axis. Its x-intercept is a, and it has no y-intercept for $a \neq 0$.

The graph of an equation having the form **$y = b$ is a horizontal line** passing through the point $(0, b)$ on the y-axis. Its y-intercept is b, and it has no x-intercept for $b \neq 0$.

SAMPLE PROBLEM 1

State the x- and y-intercepts, if they exist, and graph $x = 3$.

SOLUTION

Since $x \neq 0$, there is no y-intercept.

Since $(3, 0)$ is on the line, the x-intercept is 3.

Any value may be selected for y but the x value must be 3.

Other points on the line are $(3, 4)$ and $(3, -2)$.

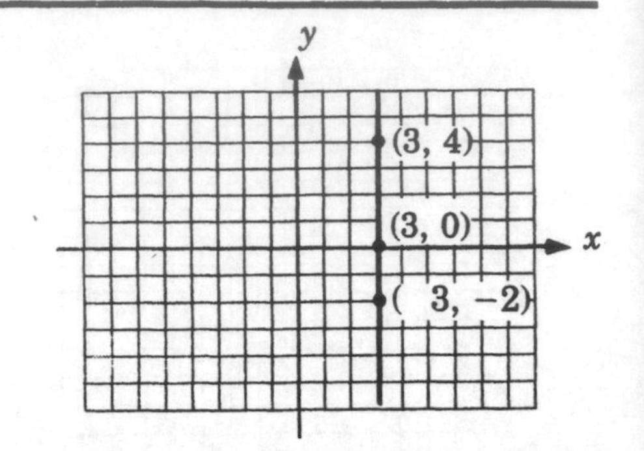

SAMPLE PROBLEM 2

State the x- and y-intercepts, if they exist, and graph $y = -2$.

SOLUTION

Since $y \neq 0$, there is no x-intercept.

Since $(0, -2)$ is on the line, the y-intercept is -2.

Any value may be selected for x but the y value must be -2.

Other points on the line are $(-3, -2)$ and $(4, -2)$.

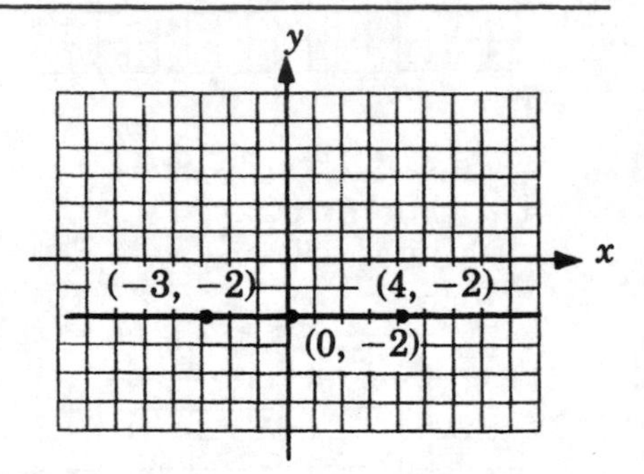

EXERCISES 7.2C

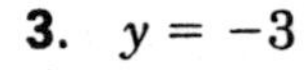

Graph each vertical or horizontal line.

1. $x = 2$
2. $x = -4$
3. $y = -3$
4. $y = 5$

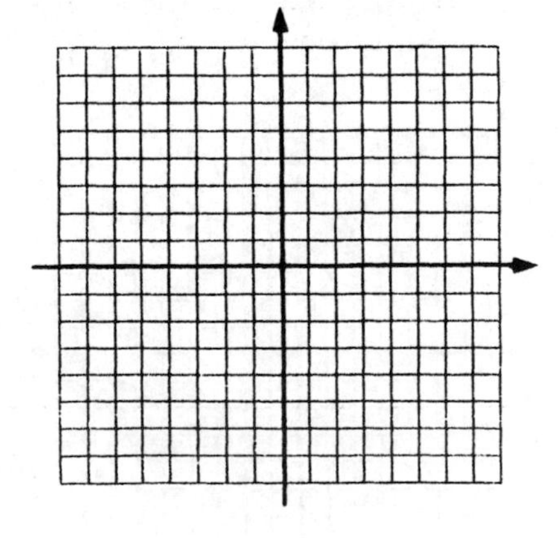

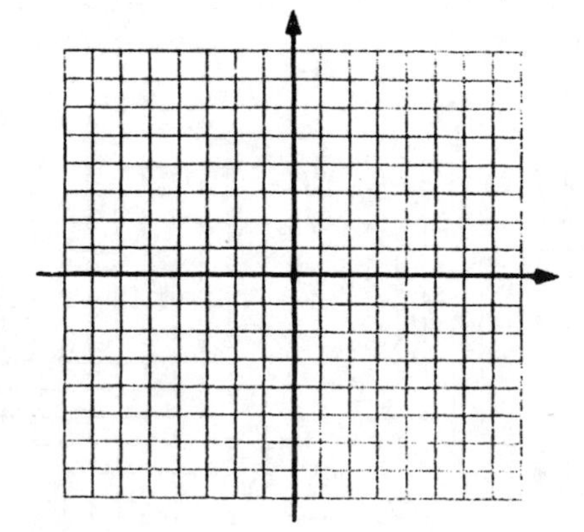

1–21. Graph each of the following.

You may use any two points on the line to graph the line.

The intercept points are usually the easiest to calculate.

If the graph passes through the origin, another point must be used.

If the intercept points (or any other two points being used) are too close together, another point should be used for accuracy.

1. $y = x$

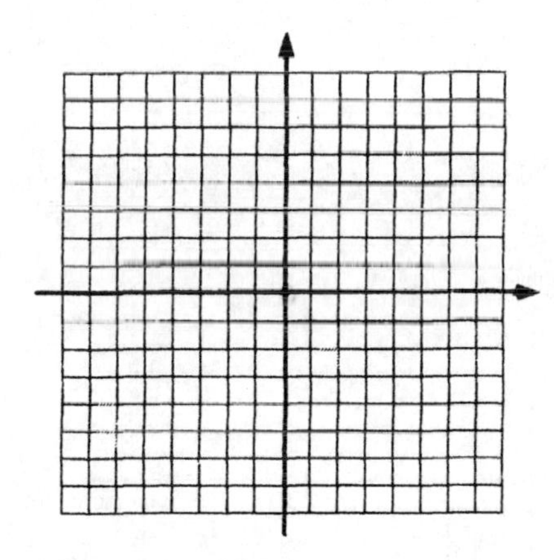

2. $y = x + 5$

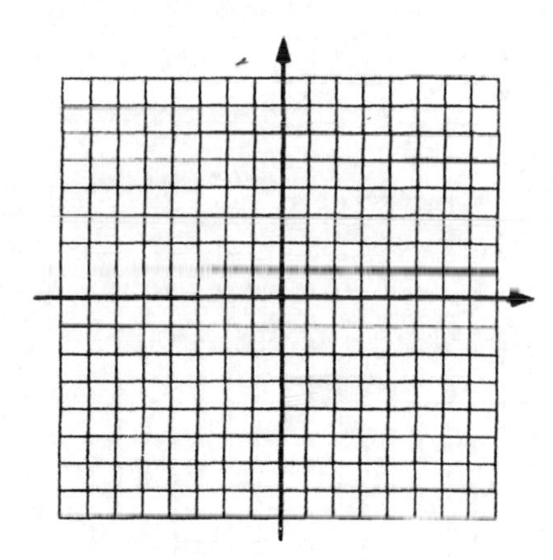

3. $y = x - 3$

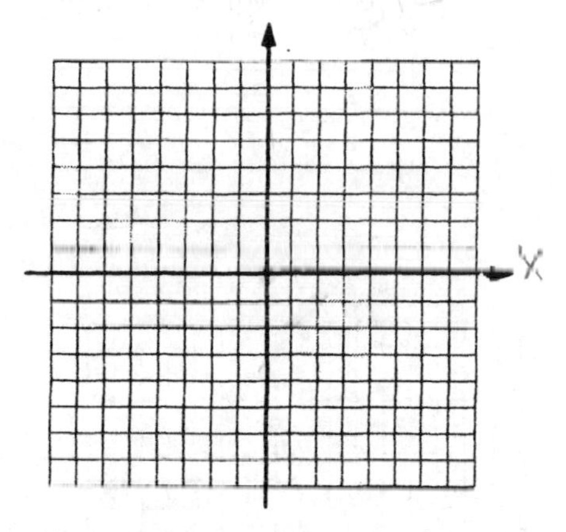

4. $y = 6 - x$

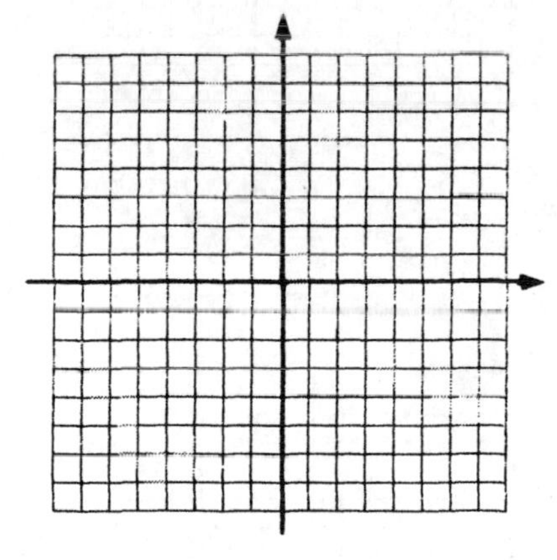

5. $y = -x - 8$

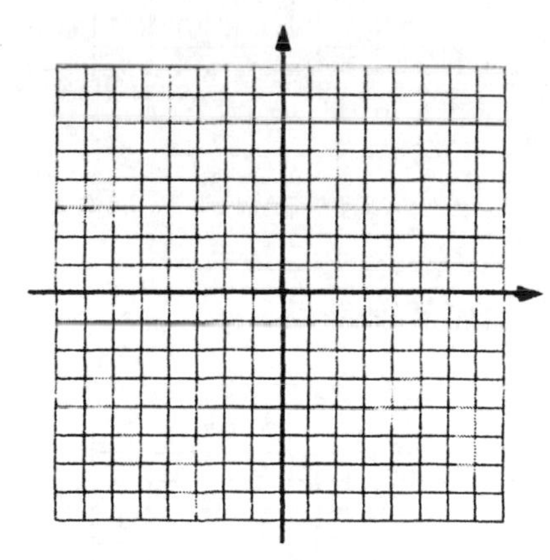

6. $y = 2x - 6$

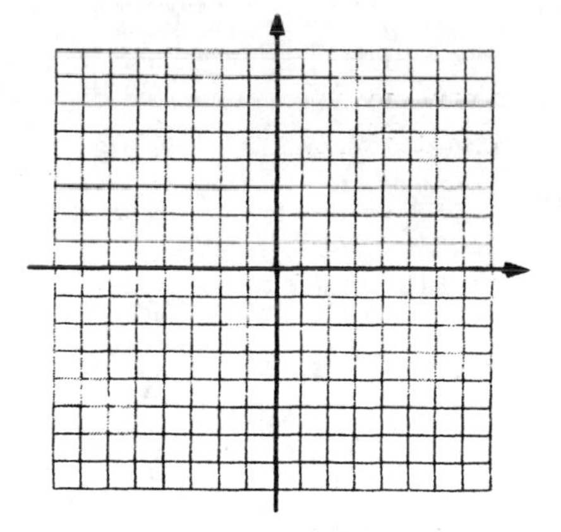

NAME ______ DATE ______ COURSE ______

7. $y = 3x + 12$

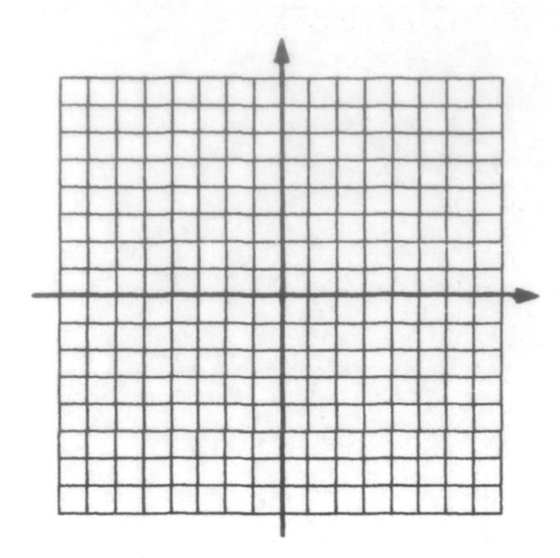

8. $y = 8 - 4x$

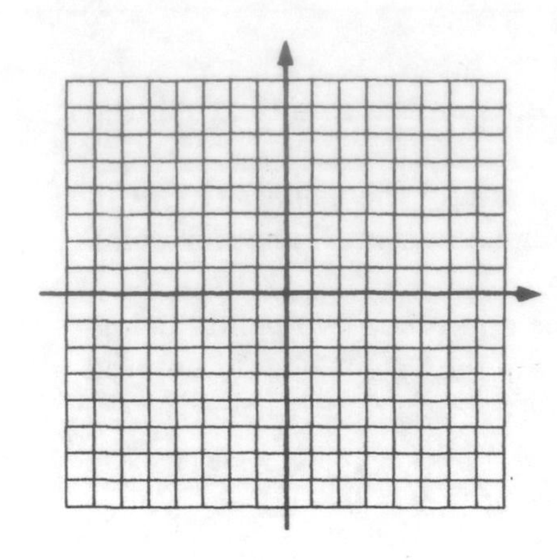

9. $y = 6 - 3x$

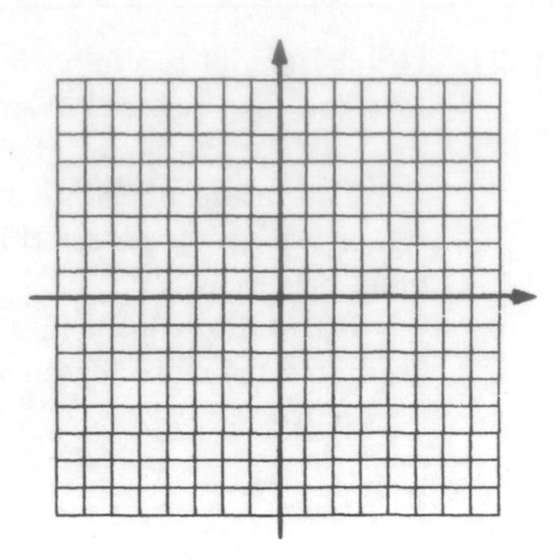

10. $y = \frac{x}{2} + 4$

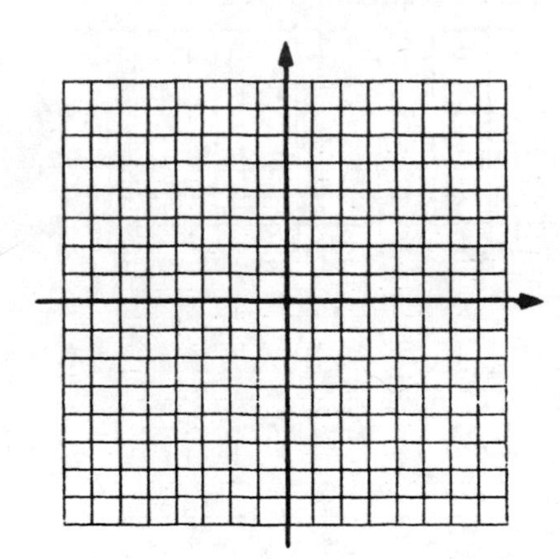

11. $x = -5$
x-intercept =
y-intercept =

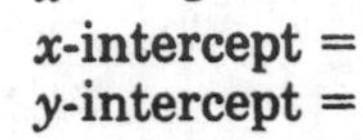

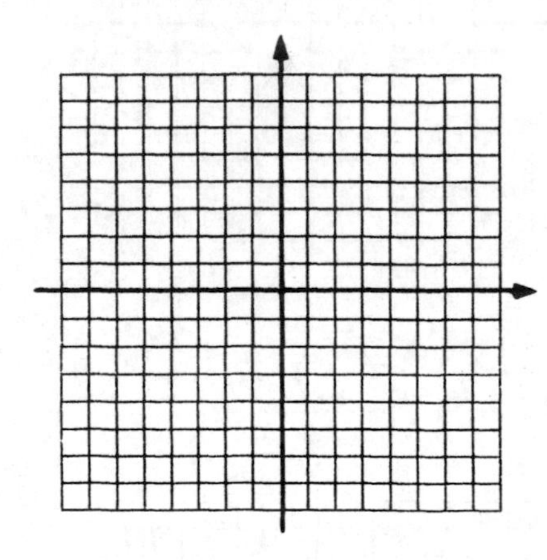

12. $y = 4$
x-intercept =
y-intercept =

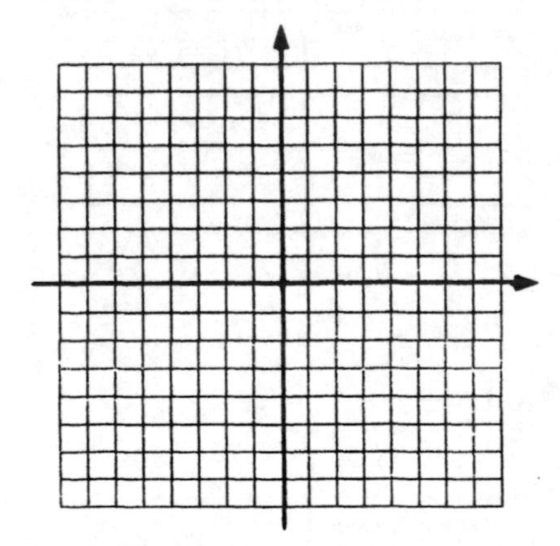

13. $x = 4$
x-intercept =
y-intercept =

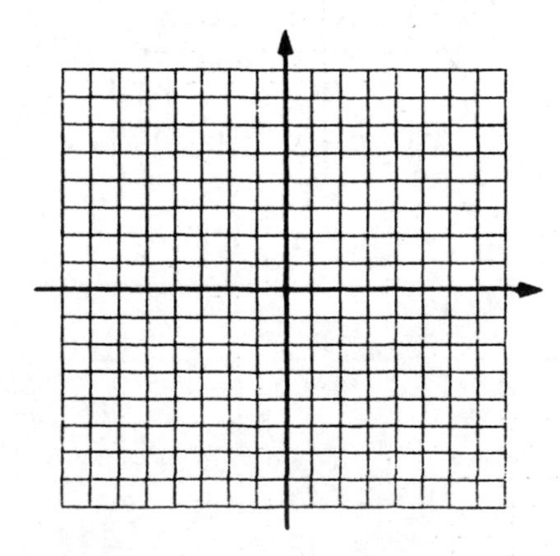

14. $y = -1$
x-intercept =
y-intercept =

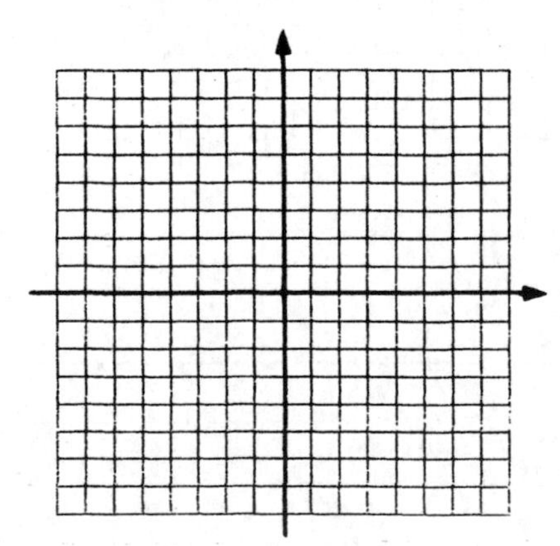

15. $3x - y = 6$
x-intercept =
y-intercept =

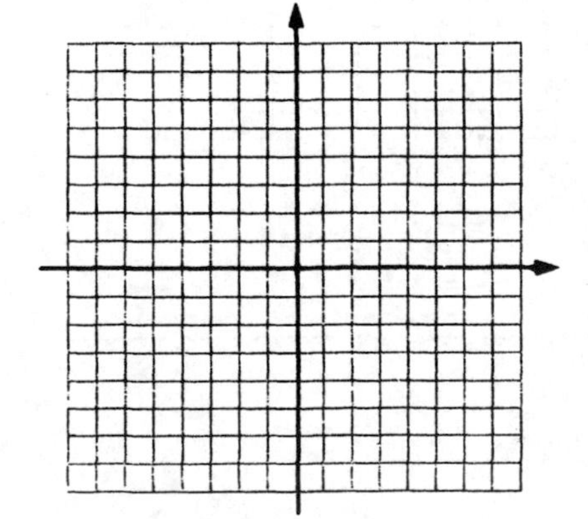

16. $5x - 2y + 10 = 0$
x-intercept =
y-intercept =

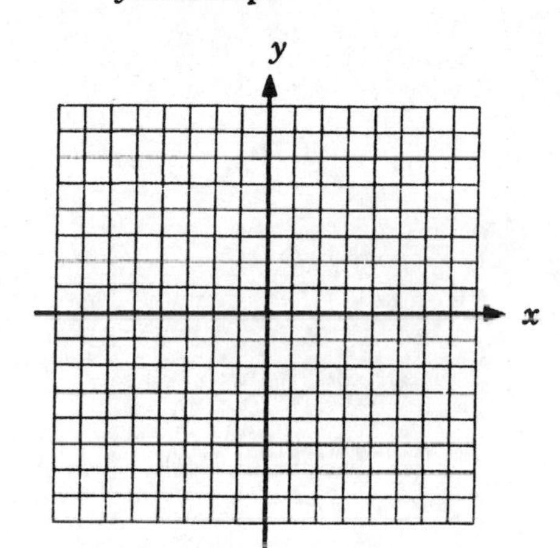

17. $5x - 6y - 30 = 0$
x-intercept =
y-intercept =

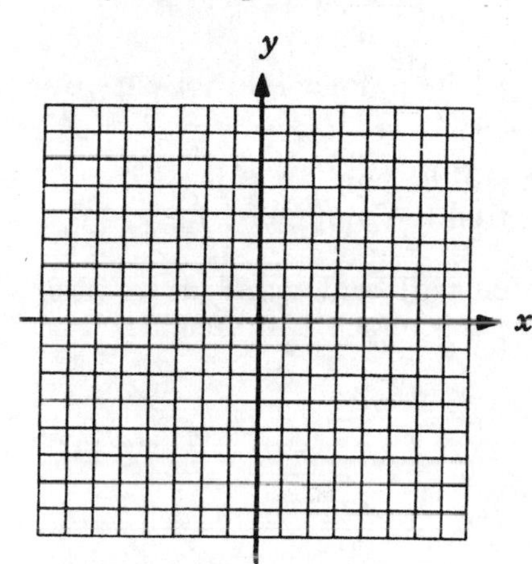

18. $2x + 7y + 14 = 0$
x-intercept =
y-intercept =

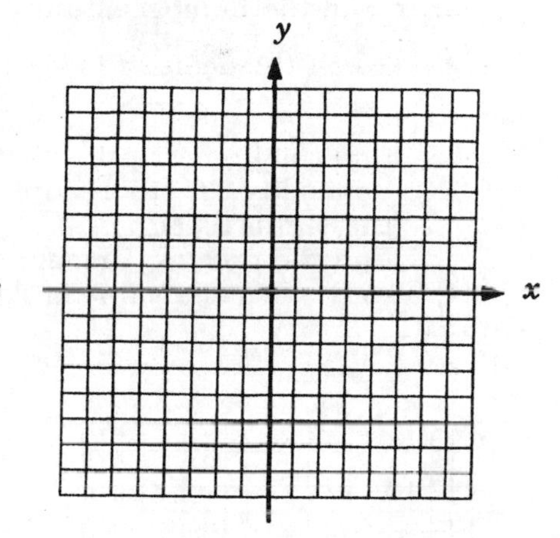

19. $x + 2y = 6$
x-intercept =
y-intercept =

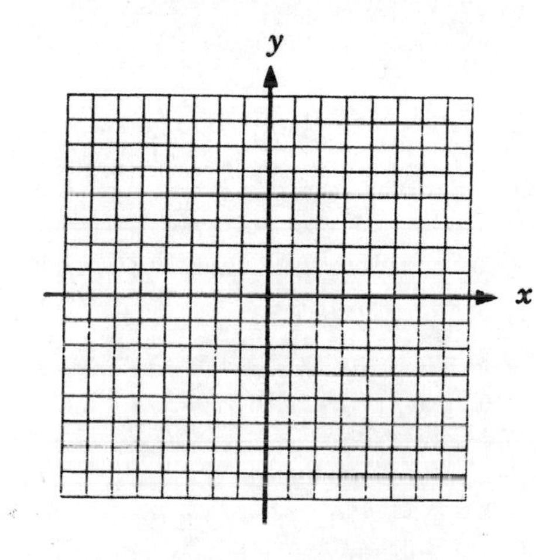

20. 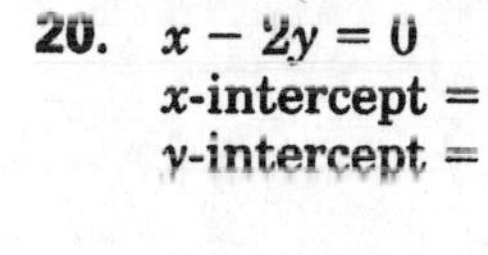
$x - 2y = 0$
x-intercept =
y-intercept =

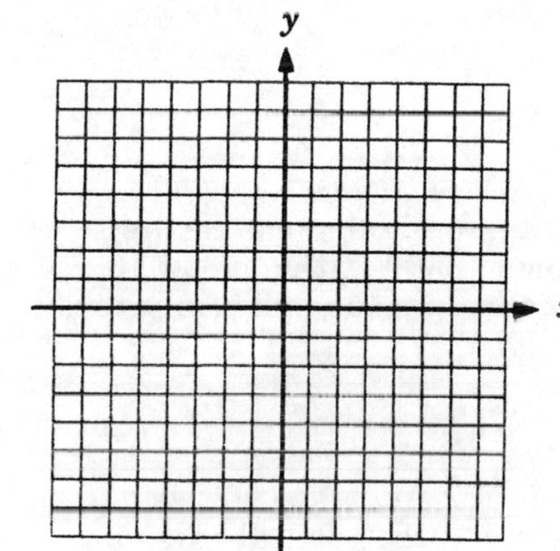

21. 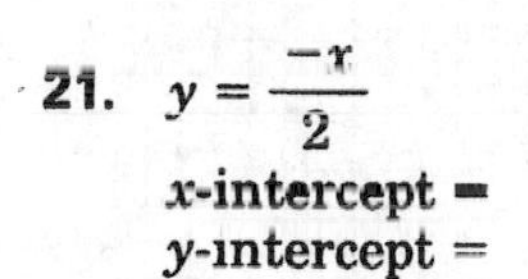
$y = \frac{-x}{2}$
x-intercept =
y-intercept =

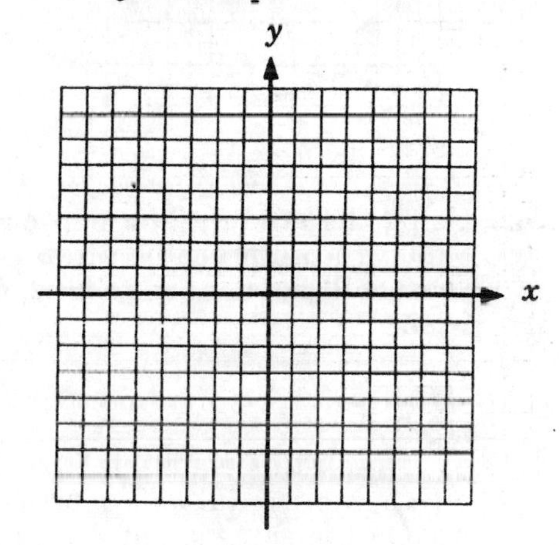

22. Graph $C = \frac{5(F - 32)}{9}$, where F = degrees Fahrenheit and C = degrees Celsius. Let the horizontal axis represent F and the vertical axis, C. Graph for F between −40 and 230, using 1 square = 10°.

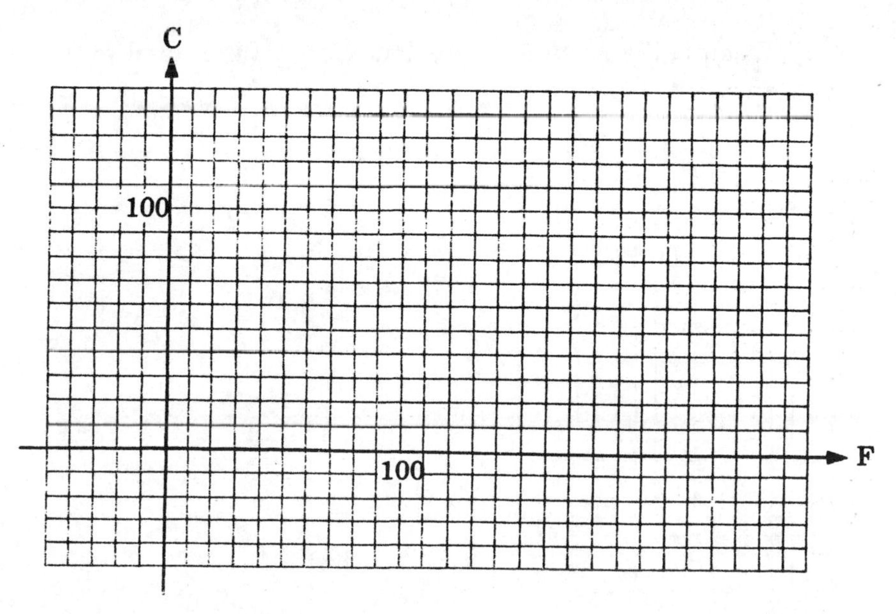

23. In **competitive business,** the price, y, for each unit of a commodity depends on the number of units, x, demanded by the consumers. A **demand law** is an equation relating the price and the number of units.

a. Graph the demand law $y = 12 - \frac{3x}{4}$ for x and y positive or 0.

b. Find the highest price that will be paid for this commodity. In other words, find the value of y for $x = 0$ (the y-intercept).

c. Find the greatest amount that will be demanded. In other words, find the value of x for $y = 0$ (the x-intercept).

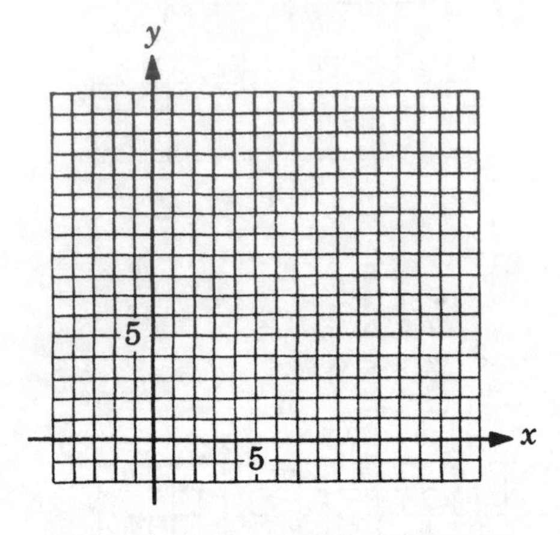

24. Using the **straight-line depreciation method,** a car costing \$3000 with a probable scrap value of \$500 at the end of 10 years will have a book value, b dollars, at the end of n years, where

$$b = 3000 - 250n$$

a. Graph this equation for values of n from 0 to 10, letting the horizontal axis represent n and the vertical axis, b.

b. On the same set of axes, graph $a = 250n$, the amount in the depreciation fund.

c. When is the amount in the depreciation fund equal to the book value?

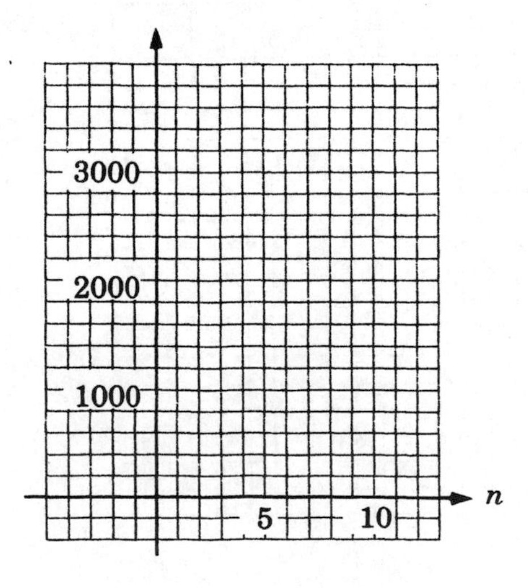

SLOPE 7.3

A. DEFINITION OF SLOPE

The steepness or slope of a road, called the grade of the road, is the ratio of the number of feet in the change of elevation (the *rise*) to the horizontal distance along which that change was measured (the *run*).

If a road has a 5% grade, then the road rises 5 ft for every run of 100 ft (the horizontal change).

$$\text{grade} = \frac{\text{rise}}{\text{run}} = \frac{5}{100}$$

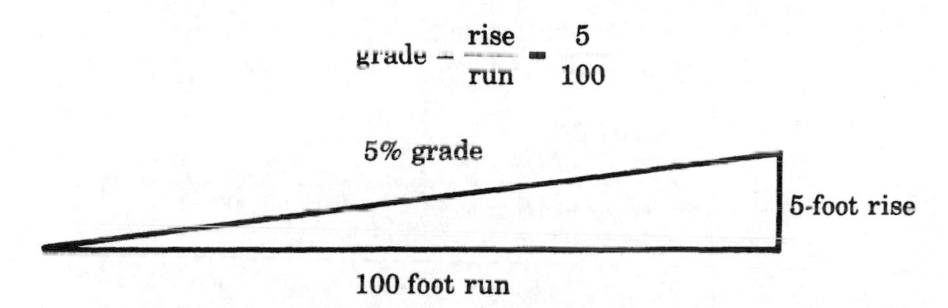

The slope of a line measures the steepness of the line, and it is also the ratio of the rise to the run of the line.

The slope m of a line is a constant. This means that the slope is the same no matter what two points on the line are used to calculate its value.

DEFINITION OF SLOPE

The slope, m, of the line through points A and B having different x-coordinates is defined as follows:

$$m = \frac{(y\text{-coordinate of } B) - (y\text{-coordinate of } A)}{(x\text{-coordinate of } B) - (x\text{-coordinate of } A)}$$

Let $A(2, 3)$ and $B(10, 8)$ be two points on a line, shown in the figure below.

$$\text{slope} = m = \frac{\text{rise}}{\text{run}} = \frac{8 - 3}{10 - 2} = \frac{5}{8}$$

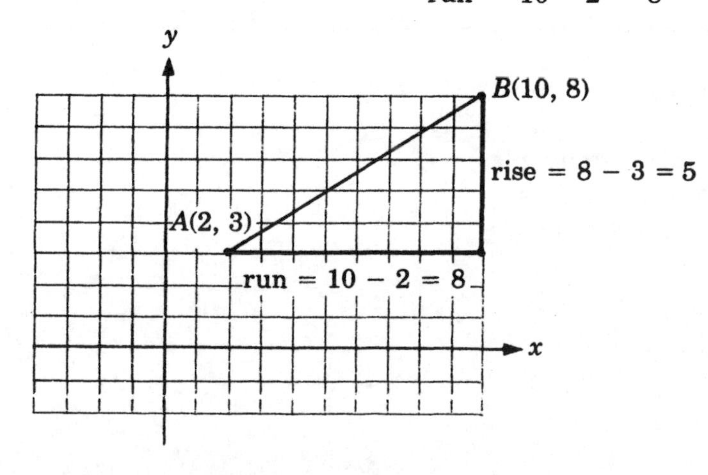

EXERCISES 7.3A

For each given pair of points, find the slope, m, by using the definition. Check by graphing the line and finding the slope graphically.

$$m = \text{slope of } AB = \frac{(y \text{ of } B) - (y \text{ of } A)}{(x \text{ of } B) - (x \text{ of } A)}$$

SAMPLE PROBLEM 1

$A(1, 2)$ and $B(3, 8)$

SOLUTION

$$m = \frac{(y \text{ of } B) - (y \text{ of } A)}{(x \text{ of } B) - (x \text{ of } A)} = \frac{8-2}{3-1} = \frac{6}{2} = 3$$

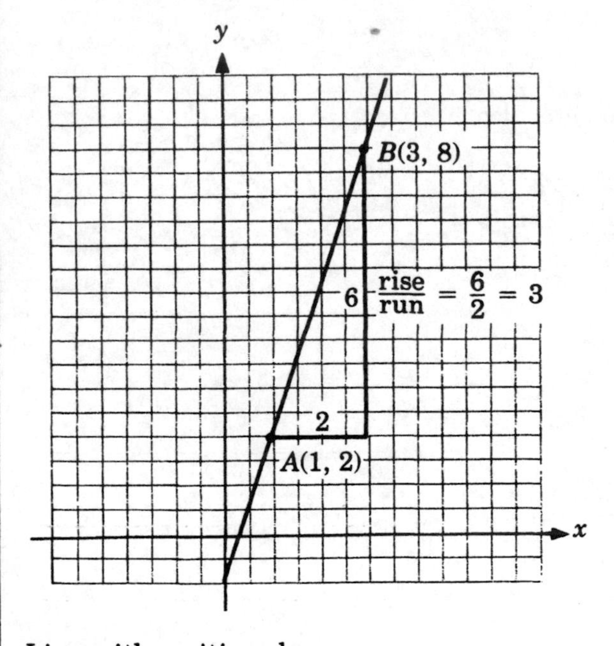

Line with positive slope.

ALTERNATE SOLUTION

$$m = \frac{(y \text{ of } A) - (y \text{ of } B)}{(x \text{ of } A) - (x \text{ of } B)} = \frac{2-8}{1-3} = \frac{-6}{-2} = 3$$

SAMPLE PROBLEM 2

$P(3, 4)$ and $Q(4, 2)$

SOLUTION

$$m = \frac{(y \text{ of } Q) - (y \text{ of } P)}{(x \text{ of } Q) - (x \text{ of } P)} = \frac{2-4}{4-3} = \frac{-2}{1} = -2$$

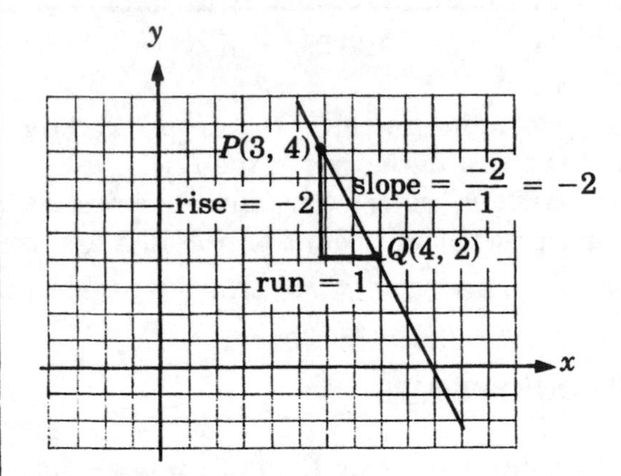

Line with negative slope.

Note in Sample Problems 1 and 2 that a line with positive slope "leans" to the right and a line with negative slope "leans" to the left. Alternatively, you can think of the line as a road. As you go from left to right (a positive run), a line with **positive slope** goes **upward** (a positive rise). As you go from left to right, a line with **negative slope** goes **downward** (a negative rise).

SAMPLE PROBLEM 3

$C(3, 2)$ and $D(-5, 2)$

SOLUTION

$$m = \frac{(y \text{ of } D) - (y \text{ of } C)}{(x \text{ of } D) - (x \text{ of } C)} = \frac{2 - 2}{-5 - 3} = \frac{0}{-8} = 0$$

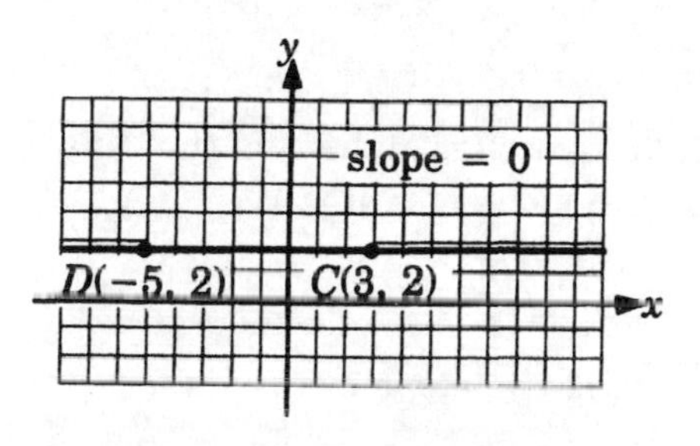

Horizontal line, slope = 0

Note that this line is horizontal. In general, a horizontal line has a slope of 0.

SAMPLE PROBLEM 4

$R(4, 6)$ and $S(4, 2)$

SOLUTION

$$m = \frac{(y \text{ of } S) - (y \text{ of } R)}{(x \text{ of } S) - (x \text{ of } R)} = \frac{2 - 6}{4 - 4} = \frac{-4}{0}$$

(undefined)

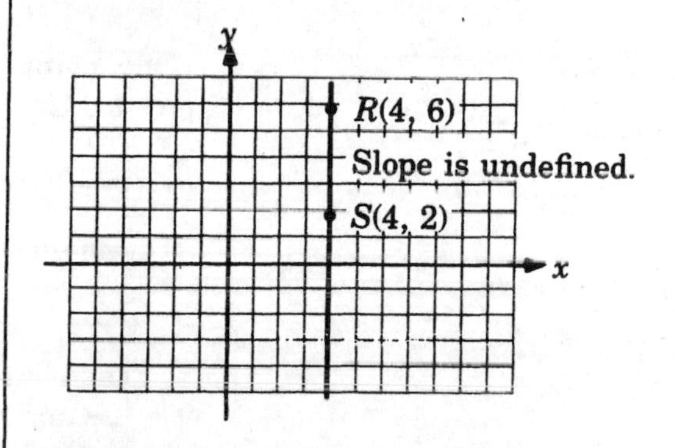

Vertical line, slope is undefined.

Since division by zero is undefined, this line has no slope.

Note that this line is vertical.

In general, the slope of a vertical line is undefined.

1. $A(2, 7)$ and $B(3, 9)$
2. $A(1, 5)$ and $B(2, 1)$
3. $A(3, -1)$ and $B(6, 0)$
4. $P(-2, 9)$ and $Q(0, 6)$
5. $P(0, -5)$ and $Q(4, 0)$
6. $C(5, 3)$ and $D(-4, 3)$
7. $C(0, 0)$ and $D(5, -2)$
8. $R(-2, 4)$ and $S(-2, 9)$
9. $A(4, 2)$ and $B(4, -2)$
10. $C(5, 6)$ and $D(-5, 6)$

1. ________________
2. ________________
3. ________________
4. ________________
5. ________________
6. ________________
7. ________________
8. ________________
9. ________________
10. ________________

B. SLOPE-INTERCEPT EQUATION

Every equation of the form $Ax + By + C = 0$, where $B \neq 0$, can be written in the form $y = mx + b$ by solving for y.

$$Ax + By + C = 0$$

$$By = -Ax - C$$

$$y = \frac{-A}{B}x + \frac{-C}{B}$$

$$y = mx + b$$

One point on the line is $(0, b)$. Another point on the line is $(1, m + b)$. Using the definition for the slope of a line,

$$\text{slope} = \frac{(m + b) - b}{1 - 0} = \frac{m}{1} = m$$

Therefore, m, the coefficient of x, is the slope of the line.

> The equation $y = mx + b$ is called the **slope-intercept form** of a linear equation.
> The equation $Ax + By + C = 0$ is called the **standard form** of a linear equation.

The slope-intercept form is useful because the slope and y-intercept of a line can be read directly from this equation.

$$y = \mathbf{m}x + \mathbf{b}$$

↑ slope ↑ y-intercept

EXERCISES 7.3B

State the slope, m, and the y-intercept, b, for each of the following, if they exist. If they do not exist, write "undefined."

SAMPLE PROBLEM 1

$y = 2x + 4$

SOLUTION

Comparing $y = \mathbf{m}x + \mathbf{b}$

↓ ↓

with $y = \mathbf{2}x + \mathbf{4}$

$m = 2$ and $b = 4$.

The slope is 2. The y-intercept is 4.

SAMPLE PROBLEM 2

$y = 3x$

SOLUTION

Using $y = \mathbf{m}x + \mathbf{b}$

$y = \mathbf{3}x + \mathbf{0}$

$m = 3$ and $b = 0$.

The slope is 3. The y-intercept is 0.

SAMPLE PROBLEM 3

$y = 3$

SOLUTION

Using $y = mx + b$

$y = 0x + 3$

$m = 0$ and $b = 3$.

The slope is 0. The y-intercept is 3.

SAMPLE PROBLEM 4

$x = 4$

SOLUTION

The graph of $x = 4$ is a vertical line.

The slope is undefined (does not exist).

The line has no y-intercept.

(The line does not intersect the y-axis.)

SAMPLE PROBLEM 5

$5x - 2y - 10 = 0$

SOLUTION

Solve for y.

$5x - 10 = 2y$

$2y = 5x - 10$

$y = \frac{5}{2}x - 5$

Using $y = mx + b$, $m = \frac{5}{2}$ and $b = -5$.

The slope is $\frac{5}{2}$. The y-intercept is -5.

SAMPLE PROBLEM 6

$y = 4 - x$

SOLUTION

Since $4 - x = -x + 4$

$y = -x + 4$

$y = (-1)x + 4.$

Using $y = mx + b$, $m = -1$ and $b = 4$.

The slope is -1.

The y-intercept is 4.

1. $y = 2x + 6$

2. $y = -2x + 6$

3. $y = 2x$

4. $y = 2$

5. $x = 2$

6. $y = \frac{1}{2}x + 4$

7. $2x + 3y = 6$

8. $2x - 3y + 6 = 0$

9. $2x + 3y + 6 = 0$

10. $3y - 2x = 12$

1. $m =$ ________, $b =$ ________

2. $m =$ ________, $b =$ ________

3. $m =$ ________, $b =$ ________

4. $m =$ ________, $b =$ ________

5. $m =$ ________, $b =$ ________

6. $m =$ ________, $b =$ ________

7. $m =$ ________, $b =$ ________

8. $m =$ ________, $b =$ ________

9. $m =$ ________, $b =$ ________

10. $m =$ ________, $b =$ ________

C. APPLIED PROBLEMS (OPTIONAL)

EXERCISES 7.3C

1. A road has an 8% grade. How many feet does it rise in a 2500-ft (horizontal) run?

1. ____________

2. A road has a -20% grade. How many feet does it drop in a 5000-ft (horizontal) run?

2. ____________

3. The back slope in earth for a road is to be $\frac{1}{2}$. What should the horizontal run be for a rise of 45 ft?

3. ____________

4. The back slope in shale for a road is to be $\frac{3}{2}$. What should the horizontal run be for a rise of 45 ft?

4. ____________

5. What is the slope of a ski slope if the change in elevation is 250 meters over a horizontal run of 1000 meters?

5. ____________

6. What is the slope of a stairway if the rise is 4 meters and the run is 4 meters?

6. ____________

7. A slide in a playground has a 16-ft rise for a run of 10 ft. What is the slope of the slide?

7. ____________

8. The slope of a stairway is to be $\frac{3}{4}$. If the rise is to be 15 ft, what should the run be?

8. ____________

EXERCISES 7.3 S

1–8. For each given pair of points, find the slope, m, by using the definition. Check by graphing the line and finding the slope graphically.

1. $A(1, -2)$ and $B(3, 4)$
2. $A(4, 2)$ and $B(1, 8)$
3. $A(4, 5)$ and $B(-2, -2)$
4. $P(0, -2)$ and $Q(-6, 8)$
5. $P(-3, 0)$ and $Q(0, -2)$
6. $C(-6, -1)$ and $D(0, 0)$
7. $R(-6, -1)$ and $S(-6, 4)$
8. $R(-7, -4)$ and $S(-2, -4)$

1. ____________________
2. ____________________
3. ____________________
4. ____________________
5. ____________________
6. ____________________
7. ____________________
8. ____________________

9. A highway has a $3\frac{1}{2}\%$ grade. How many feet does it rise in a $\frac{1}{2}$-mi run? (1 mi = 5280 ft)

9. ____________________

10. A highway has a $-2\frac{1}{4}\%$ grade. How many feet does it drop in a 2-mi run?

10. ____________________

11. A certain county specification requires that an inclined water pipe must have a slope greater than or equal to $\frac{1}{4}$. Which of the water pipes whose rises and runs are given below meets this specification?
    - **a.** Rise = 20 ft, run = 64 ft
    - **b.** Rise = 125 ft, run = 500 ft
    - **c.** Rise = 60 ft, run = 250 ft

11. **a.** ____________________
    b. ____________________
    c. ____________________

NAME ____________________ DATE ____________________ COURSE ____________________

12. The pitch (slope) of a roof is the ratio of its vertical rise to its horizontal half-span.
   a. If the pitch of a roof is 3 to 12, find the rise for a half-span of 18 ft.
   b. If the pitch of a roof is 4 to 12, find the half-span for a rise of 6 ft.

12. a. ________

b. ________

13–22. Find the slope and y-intercept, if they exist.

13. $y = 3x - 6$

14. $y = 8 - 2x$

13. $m =$ ________, $b =$ ________

14. $m =$ ________, $b =$ ________

15. $y = x - 6$

16. $y = 12 - 2x$

15. $m =$ ________, $b =$ ________

16. $m =$ ________, $b =$ ________

17. $2y + 5 = 0$

18. $x - 2y + 8 = 0$

17. $m =$ ________, $b =$ ________

18. $m =$ ________, $b =$ ________

19. $y - 2x + 4 = 0$

20. $2y - 2x = 5$

19. $m =$ ________, $b =$ ________

20. $m =$ ________, $b =$ ________

21. $x + 2y = 0$

22. $3y - 2x = 0$

21. $m =$ ________, $b =$ ________

22. $m =$ ________, $b =$ ________

7.4 FAMILIES OF LINES, EQUATIONS OF LINES (OPTIONAL)

A. FAMILIES OF LINES

In geometry, we learn that two lines in the same plane either intersect in exactly one point or are parallel and never intersect no matter how far the lines are extended.

> Any two horizontal lines are parallel.
> Any two vertical lines are parallel.
> Two distinct nonvertical lines are parallel if and only if they have the same slope.

EXERCISES 7.4A

Graph the given equations on the same rectangular coordinate system.
State whether the lines intersect or are parallel.
If the lines are parallel, state the slope.
If the lines intersect, state the point of intersection.

SAMPLE PROBLEM 1

$x - 2y + 6 = 0$ and $x = 2y$ and $2y = x - 6$

SOLUTION

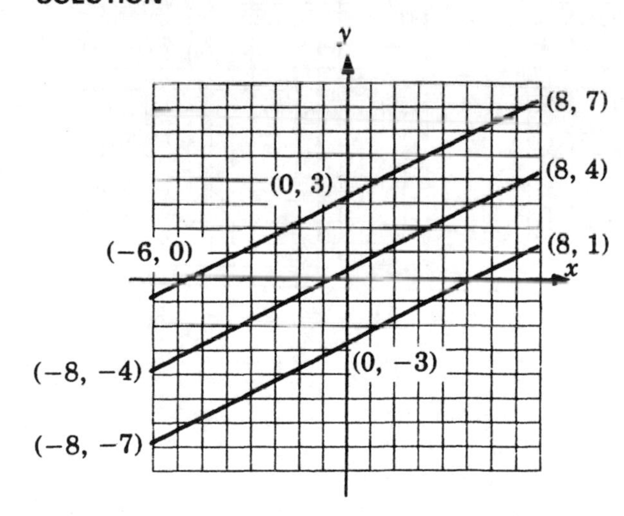

From the equations, we see that the slope of each line is $\frac{1}{2}$.

From the graphs, we see that the lines are parallel.

Note that the y-intercepts are different.

SAMPLE PROBLEM 2

$2y = x + 6$ and $x + 2y = 6$ and $y = -2x + 3$

SOLUTION

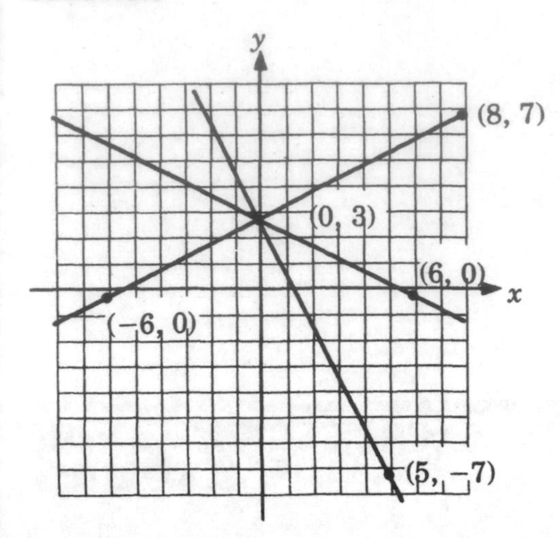

From the equations, we note that each has the same y-intercept point, (0, 3).

From the graphs, we see that the graphs intersect at the point (0, 3).

Note that the slopes are different.

1. $y = x$
$y = x + 4$
$y = x - 4$

2. $y = x + 4$
$y = 2x + 4$
$y = 4 - 2x$

3.
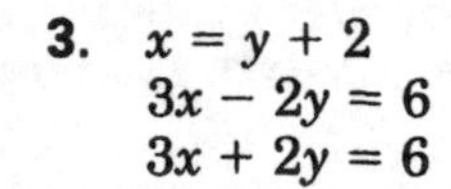
$x = y + 2$
$3x - 2y = 6$
$3x + 2y = 6$

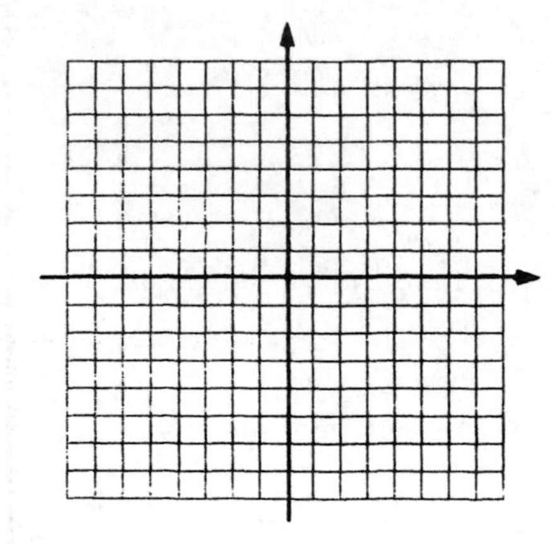

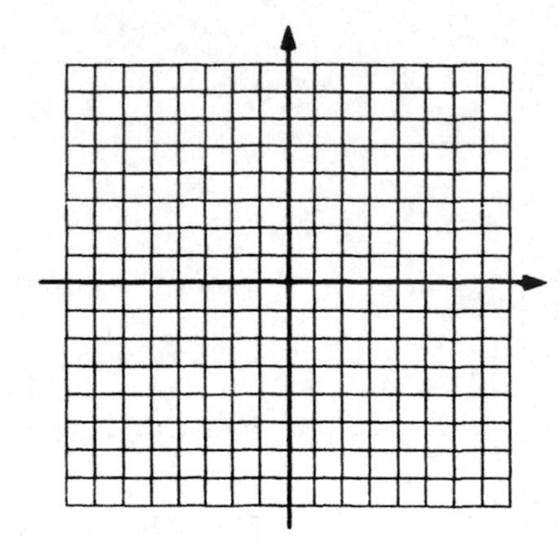

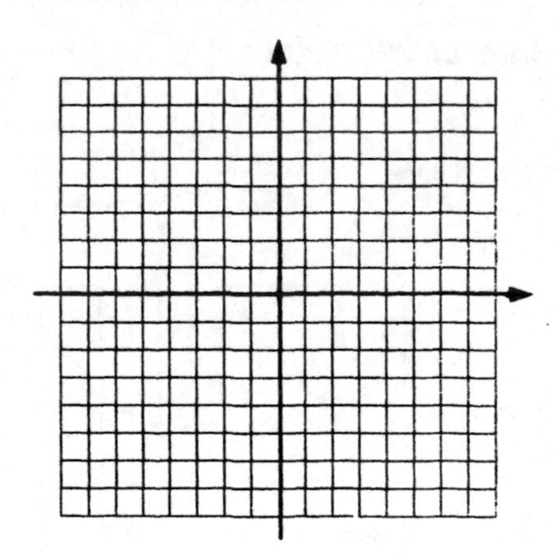

1. ______________

2. ______________

3. ______________

4. $2x + y = 0$
$2x + y = 4$
$2x + y + 4 = 0$

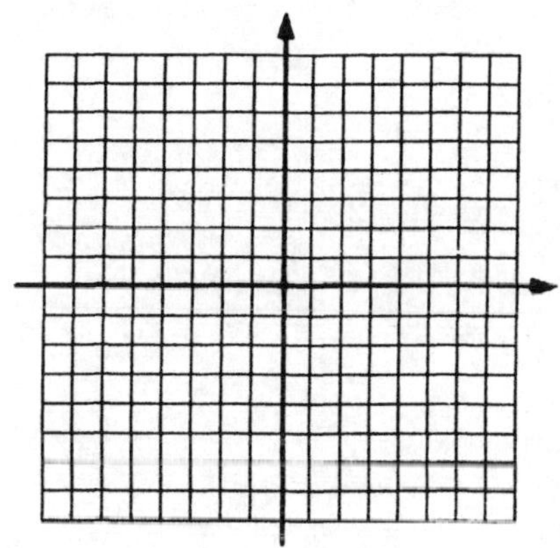

4. ______________

5. $x + y = 8$
$x - y = 4$
$x = 6$

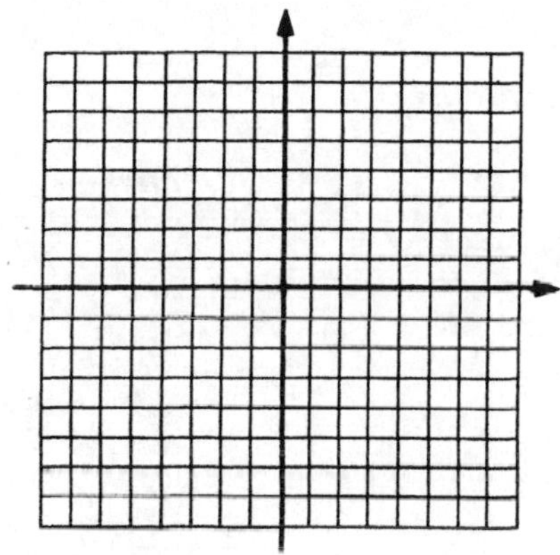

5. ______________

6. $x = 3y$
$x - 3y = 3$
$x - 3y = 6$

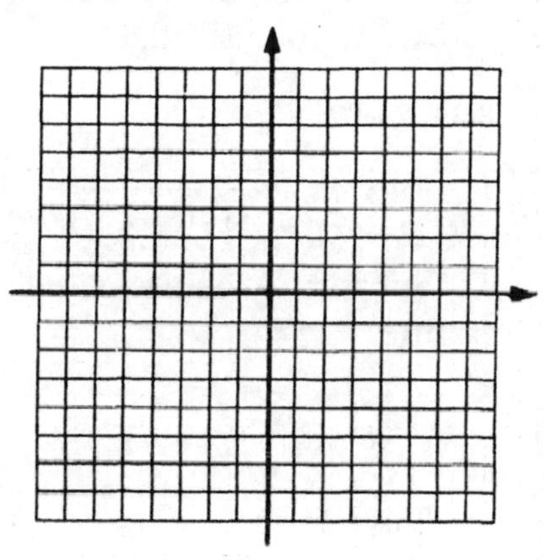

6. ______________

B. FINDING EQUATIONS

The equation of a line can be determined if

1. two points of the line are given
or **2.** one point and the slope of the line are given

PROCEDURE FOR FINDING THE EQUATION OF A LINE

1. If m is not given, use $m = \dfrac{(y \text{ of } B) - (y \text{ of } A)}{(x \text{ of } B) - (x \text{ of } A)}$ to find m.
2. Substitute known values in $y = mx + b$ to find b.
3. Substitute values for m and b in $y = mx + b$.
4. Rewrite in form $Ax + By + C = 0$, the standard form.

SAMPLE PROBLEM 1

Find an equation of the form $Ax + By + C = 0$ for the line whose slope is $\frac{2}{3}$ and that passes through the point $(3, -2)$.

SOLUTION

1. $m = \frac{2}{3}$

2. Substitute known values in $y = mx + b$. $\quad m = \frac{2}{3}$ and $x = 3$ and $y = -2$

$$y = mx + b$$
$$-2 = \frac{2}{3}(3) + b$$

Solve for b. $\quad b = -2 - 2 = -4$

3. Substitute values for m and b in $y = mx + b$.

$$y = mx + b$$
$$y = \frac{2}{3}x - 4$$

4. Multiply both sides by LCD. Write in form $Ax + By + C = 0$.

$3y = 2x - 12$ Multiply each term by 3.

$2x - 3y - 12 = 0$ Subtract $3y$ from each side.

SAMPLE PROBLEM 2

Find an equation of the form $Ax + By + C = 0$ passing through the points (4, 2) and (−2, 5).

SOLUTION

1. Find $m = \dfrac{(y \text{ of } B) - (y \text{ of } A)}{(x \text{ of } B) - (x \text{ of } A)}$ $\qquad m = \dfrac{5-2}{-2-4} = \dfrac{3}{-6} = \dfrac{-1}{2}$

2. Substitute values for m, x, and y in $y = mx + b$. $\qquad m = \dfrac{-1}{2}$ and $x = 4$ and $y = 2$

   $y = mx \qquad + b$

   (Use either point.) $\qquad 2 = \left(\dfrac{-1}{2}\right)(4) + b$

   Solve for b. $\qquad b = 2 + 2 = 4$

3. Substitute values for m and b in $y = mx + b$. $\qquad y = \dfrac{-1}{2}x + 4$

4. Write in form $Ax + By + C = 0$. $\qquad 2y = -x + 8$

   $x + 2y - 8 = 0$

EXERCISES 7.4B

Find an equation of the form $Ax + By + C = 0$ for the line satisfying the given conditions.

1. Slope = −1 and passes through (2, 3). 1. ____________
2. Slope = $\dfrac{1}{2}$ and passes through (6, −1). 2. ____________
3. Slope = $\dfrac{1}{3}$ and passes through the origin. 3. ____________
4. Slope = $\dfrac{-4}{5}$ and passes through the origin. 4. ____________
5. Slope = $\dfrac{2}{5}$ and y-intercept = −2. 5. ____________
6. Slope = $\dfrac{-5}{2}$ and x-intercept = 2. 6. ____________
7. Passes through (0, 2) and (−3, 4). 7. ____________
8. Passes through (6, 3) and (2, 0). 8. ____________
9. Passes through (4, 6) and (−2, −3). 9. ____________
10. Passes through (3, −4) and (−3, 4). 10. ____________

1–9. Graph each set of equations on the same graph.
If the lines are parallel, state the common slope.
If the lines intersect, state the point of intersection.

1. $x + y = 0$
$x + y = 8$
$x + y + 8 = 0$

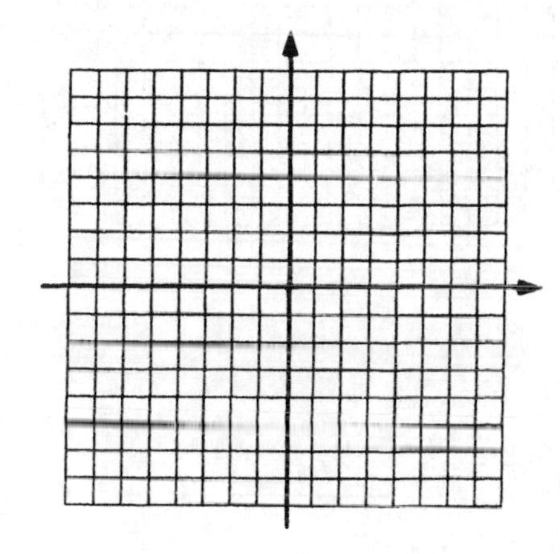

2. $x + y = 6$
$y - x = 6$
$y = 2x + 6$

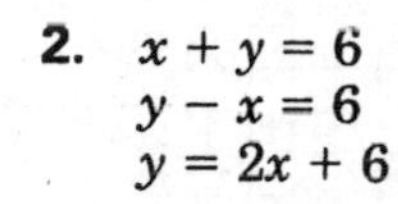

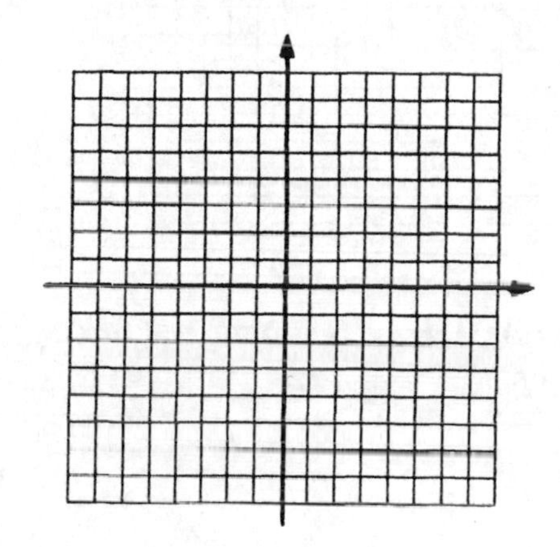

3. $x + 2y = 8$
$2x - y = 6$
$x = 2y$

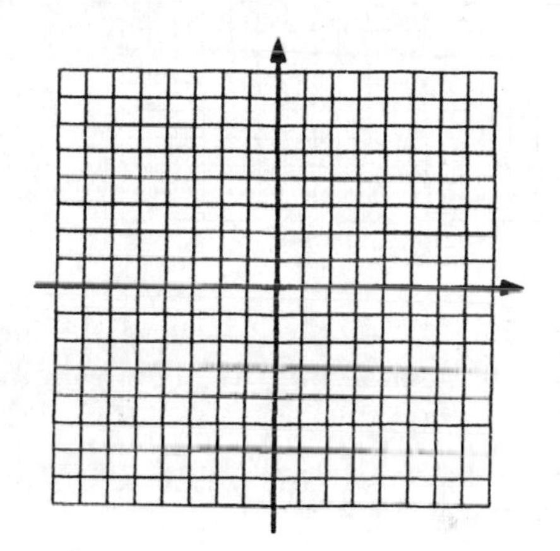

4. $3x - y = 6$
$y = 3x + 6$
$y - 3x = 3$

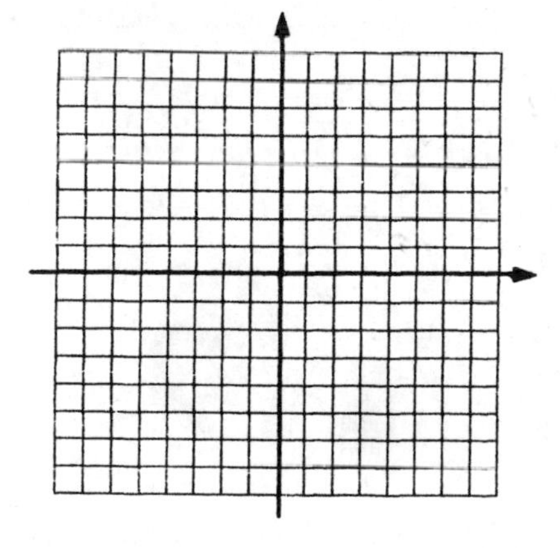

5. $2x + 3y = 0$
$2x + y + 4 = 0$
$y - x = 5$

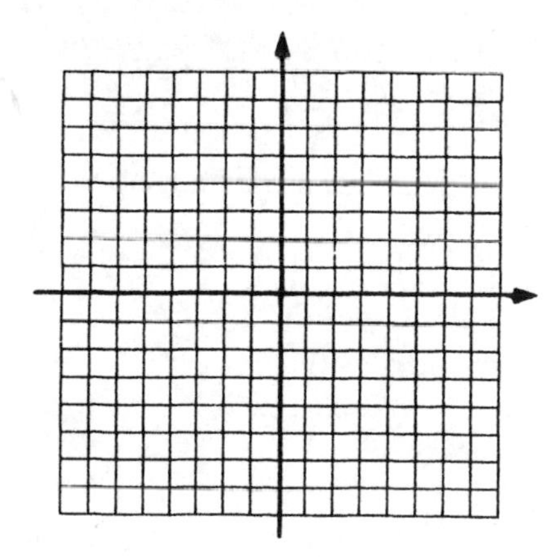

6. $x = 3y - 3$
$x = 3y + 6$
$x = 3y - 6$

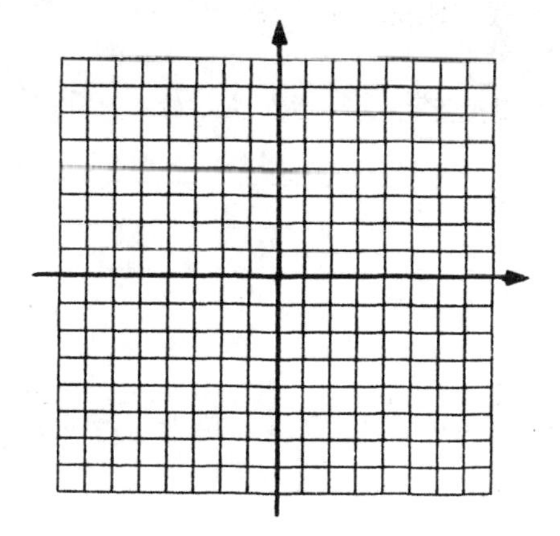

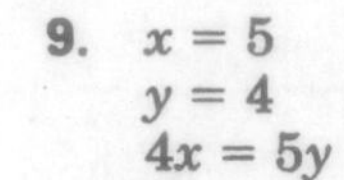

7. $2x - y = 8$
$2x + y = 4$
$x + y = 1$

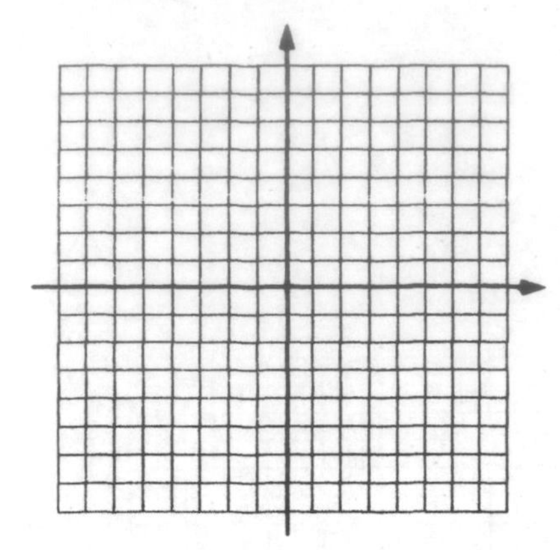

8. $x + 4y = 8$
$x = 4 - 4y$
$x + 4y + 8 = 0$

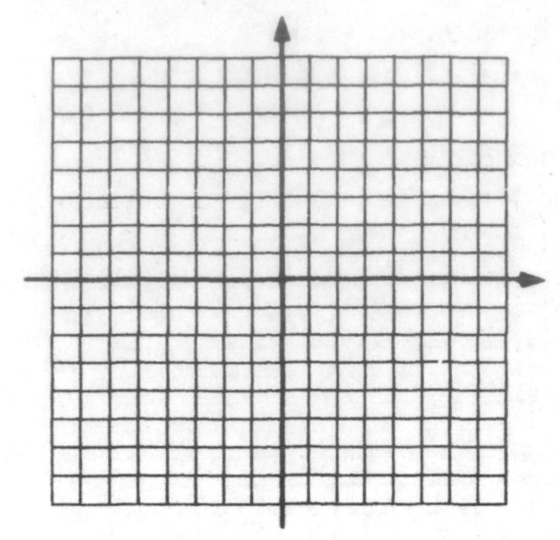

9. $x = 5$
$y = 4$
$4x = 5y$

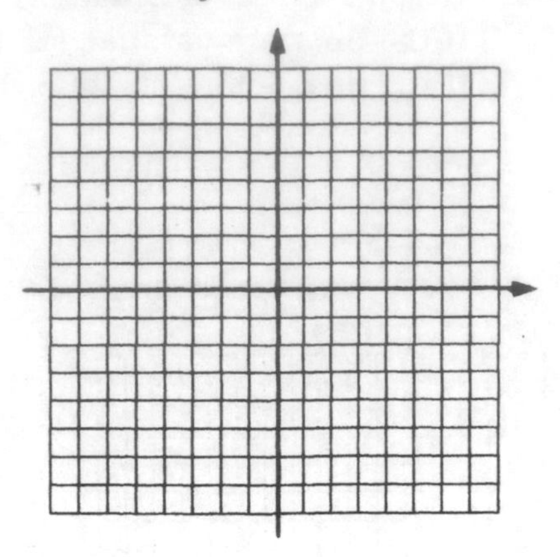

10–20. **Find an equation of the form $Ax + By + C = 0$ for the line satisfying the given conditions.**

10. Slope $= \frac{3}{4}$ and y-intercept $= 5$. **10.** __________

11. Slope $= -5$ and passes through $(4, 0)$. **11.** __________

12. Slope $= \frac{1}{3}$ and passes through $(2, 5)$. **12.** __________

13. Slope $= \frac{-5}{2}$ and passes through $(3, -1)$. **13.** __________

14. Passes through $(-5, 4)$ and $(10, -2)$. **14.** __________

15. Passes through $(6, 3)$ and $(2, -3)$. **15.** __________

16. x-intercept $= 2$ and y-intercept $= 5$. **16.** __________

17. x-intercept $= 4$ and y-intercept $= -3$. **17.** __________

18. Parallel to $x + y = 10$, passes through $(9, -2)$. **18.** __________

SAMPLE TEST FOR CHAPTER 7

1–5. Graph each of the following on the number plane below. (See Section 7.1.)

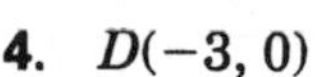

1. $A(4, 2)$
2. $B(2, -4)$
3. $C(-5, 1)$
4. $D(-3, 0)$
5. $E(-2, -5)$

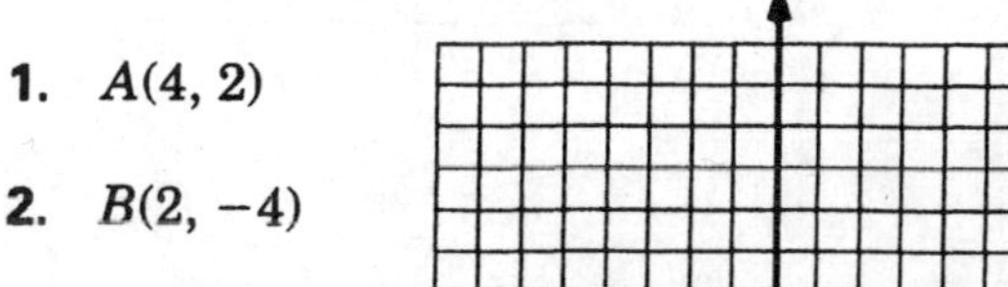

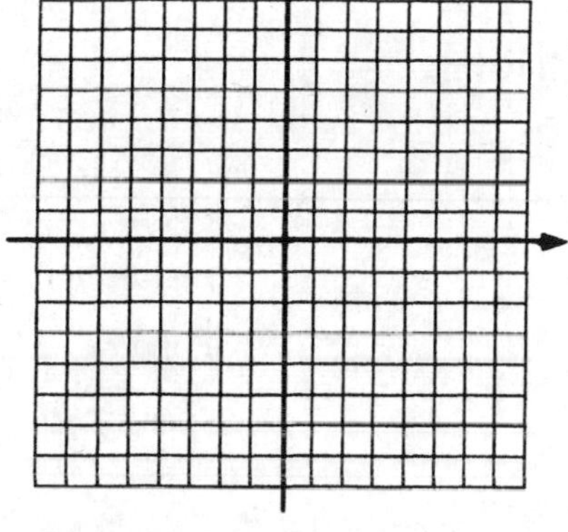

6–10. State the coordinates of each point whose graph is shown on the figure. (See Section 7.1.)

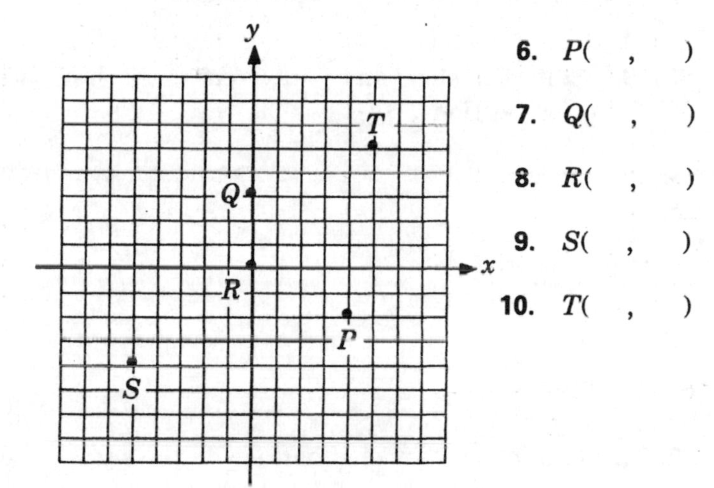

6. $P($, $)$
7. $Q($, $)$
8. $R($, $)$
9. $S($, $)$
10. $T($, $)$

11. Find the solution of $2x + y = 4$ for which $x = 5$. (See Section 7.1.) **11.** ____________

12. Find the solution $4x - 3y = 10$ for which $y = 2$. (See Section 7.1.) **12.** ____________

13–15. State the x- and y-intercepts and graph. (See Section 7.2.)

13. $y = 3x - 2$
x-intercept =
y-intercept =

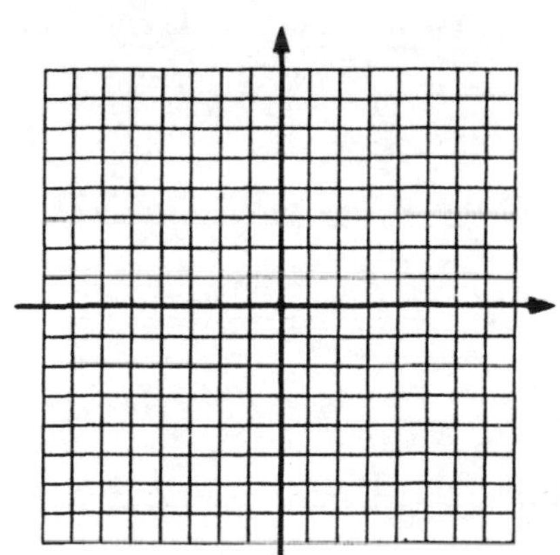

14. $x - 4y = 8$
x-intercept =
y-intercept =

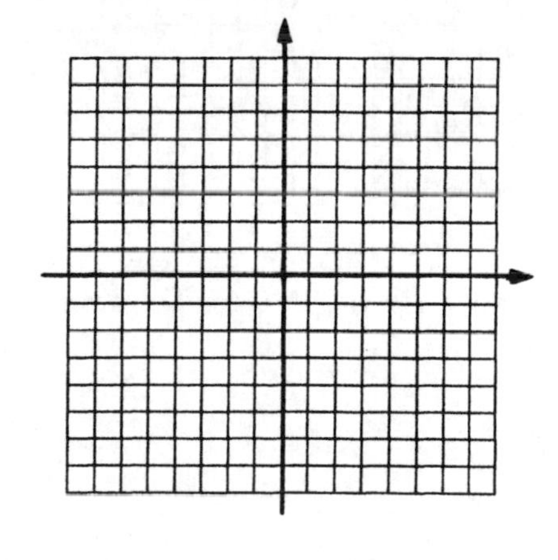

15. $5x + y + 10 = 0$
x-intercept =
y-intercept =

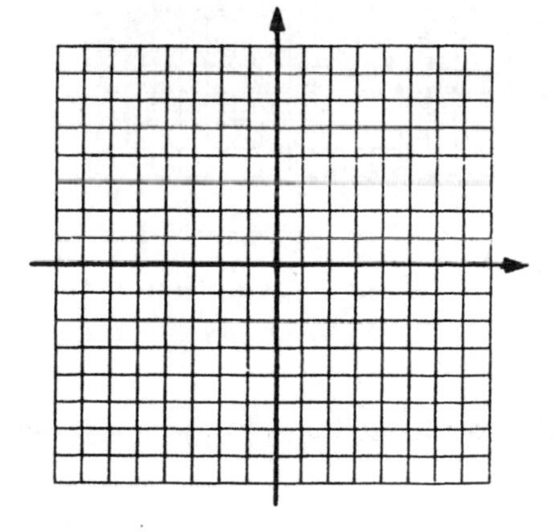

NAME ____________ DATE ____________ COURSE ____________

16–17. Find the slope of the line joining the two given points. Check by graphing the line. (See Section 7.3.)

16. $P(8, -5)$ and $Q(4, -3)$

17. $A(-7, 2)$ and $B(2, 5)$

16. ____________________

17. ____________________

18–19. State the slope and y-intercept for each of the following. (See Section 7.3.)

18. $y = 2x + 6$

19. $2x + 4y - 16 = 0$

18. slope = ____________________

y-intercept = ____________________

19. slope = ____________________

y-intercept = ____________________

20–24. Optional

20–21. Graph each set of equations on the same set of axes. If the lines are parallel, state the common slope. If the lines intersect, state the point of intersection.

20. $3x - y = 12$
$x - 3y = 12$
$3x + y = 6$

21. $y = 3x + 6$
$3x - y = 6$
$y - 3x = 9$

20. ____________________

21. ____________________

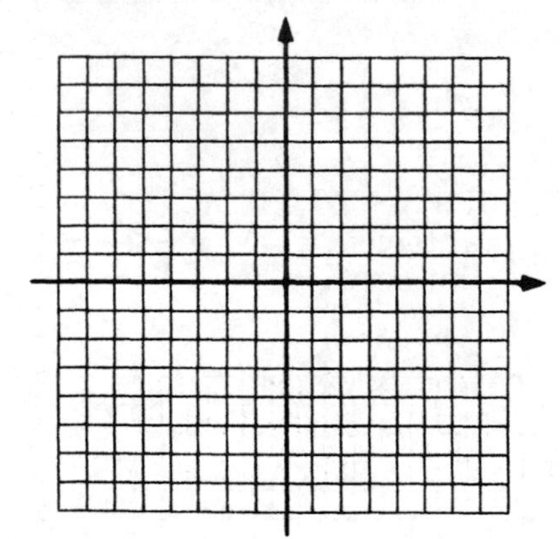

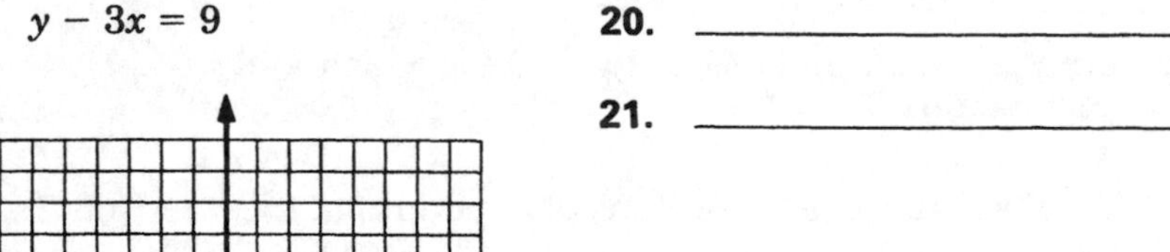

22–24. Find an equation of the form $Ax + By + C = 0$ for the line satisfying the given conditions.

22. Slope $= \frac{3}{5}$ and y-intercept $= -4$.

23. Slope $= -2$ and passes through $(5, -1)$.

24. Passes through $(4, -3)$ and $(2, 5)$.

22. ____________________

23. ____________________

24. ____________________

Linear Systems, Two Variables

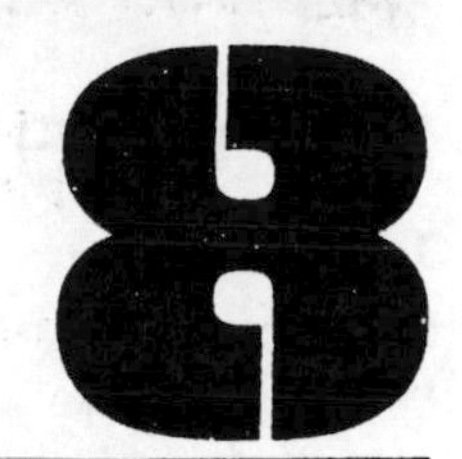

ALGEBRA IN ACTION: TRANSPORTATION

Linear systems are useful for finding times, rates, or distances involved in the transportation of goods and people.

A boat traveling downstream (with a current) goes 36 mi in 2 hr. Traveling upstream (against the same current), the same boat takes 3 hr to go 36 mi. Find the rate of the current.

$$2(x + y) = 36$$
$$3(x - y) = 36$$

A helicopter sets out to overtake a train that is 35 mi ahead. If the rate of the train is 90 m.p.h. and the rate of the helicopter is 160 m.p.h., how long does it take the helicopter to overtake the train?

$$90y + 35 = 160x$$
$$y = x$$

Chapter 8 Objectives

Objectives	Typical Problems
Determine if the graphs of two equations are parallel, intersecting, or coincident lines.	**1–3.** State whether the graphs of each pair of equations are parallel, intersecting, or coincident lines. **1.** $x = y + 2$ and $y = x + 2$ **2.** $x - y = 2$ and $x + y = 4$ **3.** $x - y = 2$ and $2x = 4 + 2y$
Solve a system of two linear equations in two variables by the graphic method.	**4.** Solve by the graphic method. $2x + 3y = 6$ $x - 2y = 10$
Solve a system of two linear equations in two variables by the substitution method.	**5.** Solve by the substitution method. $2x - y = 10$ $3x - 4y = 5$
Solve a system of two linear equations in two variables by the addition method.	**6.** Solve by the addition method. $2x + 5y = 30$ $4x - 3y = 8$
Solve a general system of two linear equations in two variables.	**7–8.** Solve each system. **7.** $x = y + 2$ and $y = x + 2$ **8.** $x - y = 2$ and $2x = 4 + 2y$
Solve a verbal problem using two variables.	**9.** A certain city charged \$320 for the inspection of 5 single homes and 2 duplexes. It charged \$340 for the inspection of 4 single homes and 3 duplexes. Find the inspection fee for 1 single home and 1 duplex.

ANSWERS to Typical Problems

1. Parallel **2.** Intersecting **3.** Coincident **4.** $(6, -2)$ **5.** $(7, 4)$ **6.** $(5, 4)$ **7.** $\varnothing$
8. $\{(x, y) | x - y = 2\}$ **9.** \$40, single home; \$60, duplex

8.1 LINE RELATIONS, SOLUTIONS

A. INTERSECTING, PARALLEL, AND COINCIDENT LINES

Two straight lines in a plane either intersect in exactly one point or are parallel and never intersect no matter how far the lines are extended.

Given two linear equations in two variables, one of three things can happen.

1. The graphs are **two intersecting lines.** The lines do not have the same slope.
2. The graphs are **two parallel lines.** The lines have the same slope (or both lines are vertical) and have different intercepts.
3. The graphs are **coincident lines** (the two equations have the same line for their graphs). The lines have the same slope and the same y-intercept, or both lines are vertical with the same x-intercept.

These three possibilities are illustrated by the graphs in Figure 8.1.

(1)

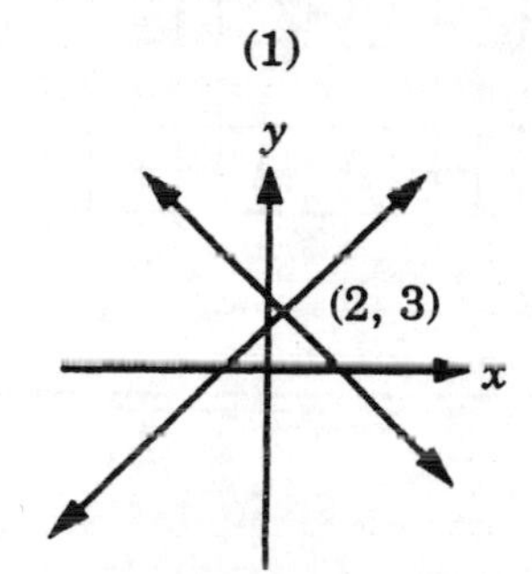

Intersecting Lines
$x + y = 5$
$y = x + 1$
Exactly one solution

(2)

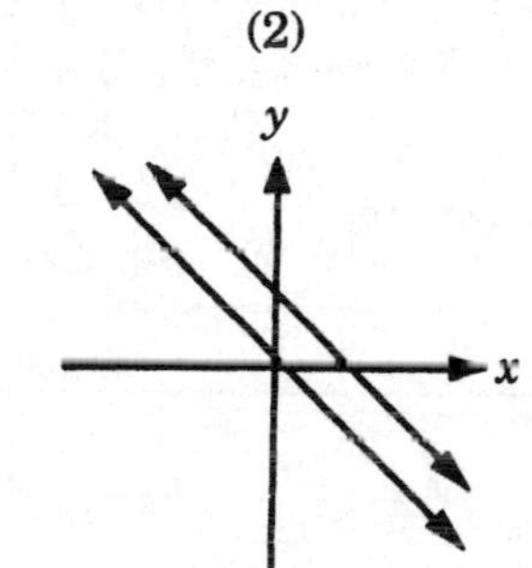

Parallel lines
$x + y = 5$
$x + y = 2$
No solution

(3)

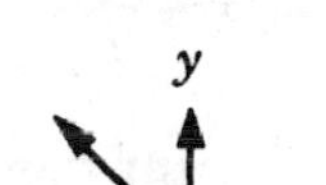
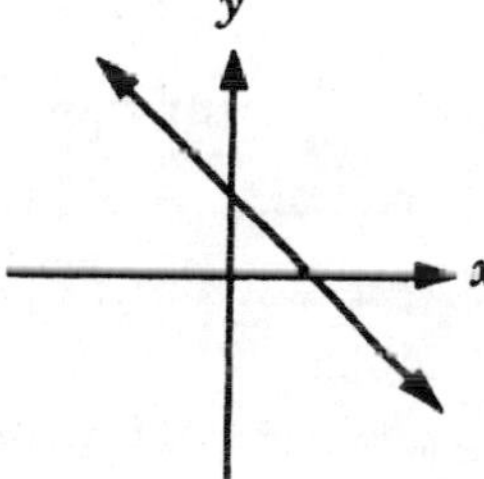

Coincident lines
$x + y = 5$
$3x + 3y = 15$
Infinitely many solutions

FIGURE 8.1

EXERCISES 8.1A

Find the slopes and y-intercepts, and determine whether the graphs of the two given equations are intersecting lines, parallel lines, or coincident lines.

Check the result by graphing each pair of equations on the same rectangular coordinate system.

SAMPLE PROBLEM 1

$2x - y - 5 = 0$ and $x + 2y - 5 = 0$

SOLUTION

Write each equation in the $y = mx + b$ form and compare the slopes and y-intercepts.

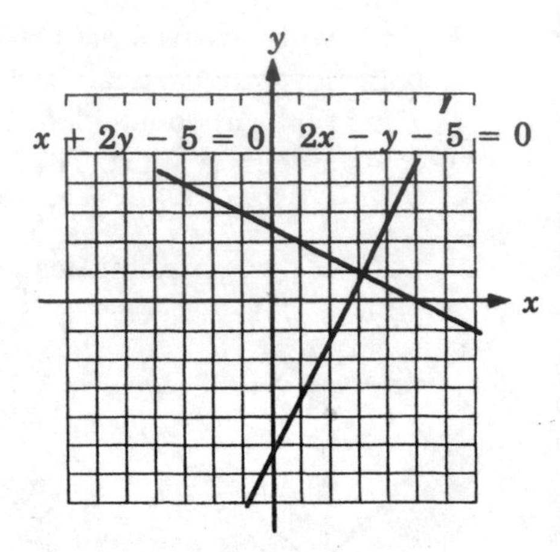

$2x - y - 5 = 0$ $\qquad$ $x + 2y - 5 = 0$

$y = 2x - 5$ $\qquad$ $y = -\frac{1}{2}x + \frac{5}{2}$

$m = 2$ and $b = -5$ $\qquad$ $m = -\frac{1}{2}$ and $b = \frac{5}{2}$

Since the **slopes are different,** $2 \neq -\frac{1}{2}$, the **lines intersect in exactly one point.**

SAMPLE PROBLEM 2

$2x + 4y = 2$ and $3x + 6y = 3$

SOLUTION

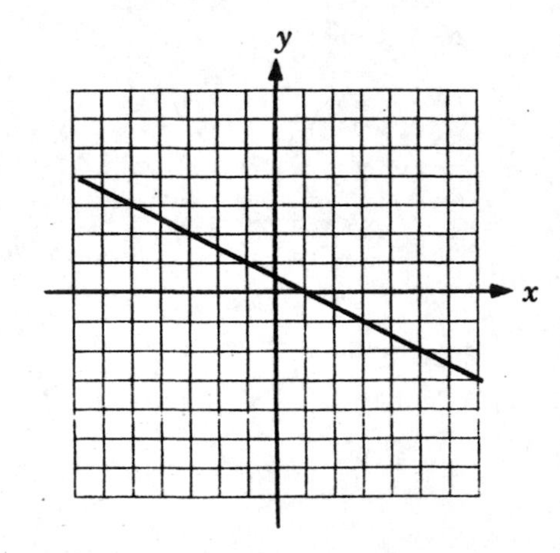

$2x + 4y = 2$ $\qquad$ $3x + 6y = 3$

$y = -\frac{1}{2}x + \frac{1}{2}$ $\qquad$ $y = -\frac{1}{2}x + \frac{1}{2}$

$m = -\frac{1}{2}$ and $b = \frac{1}{2}$ $\qquad$ $m = -\frac{1}{2}$ and $b = \frac{1}{2}$

Since the **slopes and the y-intercepts are equal,** the **lines are coincident.**

The graph of the first equation is the same as the graph of the second equation.

SAMPLE PROBLEM 3

$x + 2y = 5$ and $2x + 4y = 3$

SOLUTION

$x + 2y = 5$ | $2x + 4y = 3$

$y = -\frac{1}{2}x + \frac{5}{2}$ | $y = -\frac{1}{2}x + \frac{3}{4}$

$m = -\frac{1}{2}$ and $b = \frac{5}{2}$ | $m = -\frac{1}{2}$ and $b = \frac{3}{4}$

Since the **slopes are equal and the *y*-intercepts are different,** the **lines are parallel.**

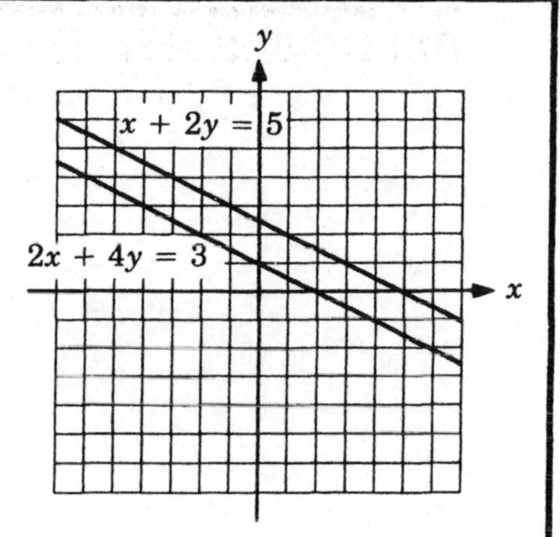

1. $y = 2x - 3$ and $y = \frac{1}{2}x - 3$

2. $3x - 2y + 6 = 0$ and $2y - 3x = 6$

3. $x - y = 2$ and $2x - 2y = 5$

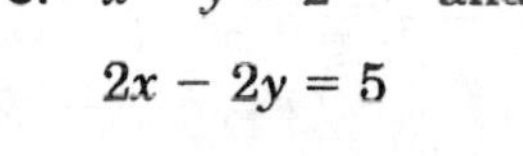

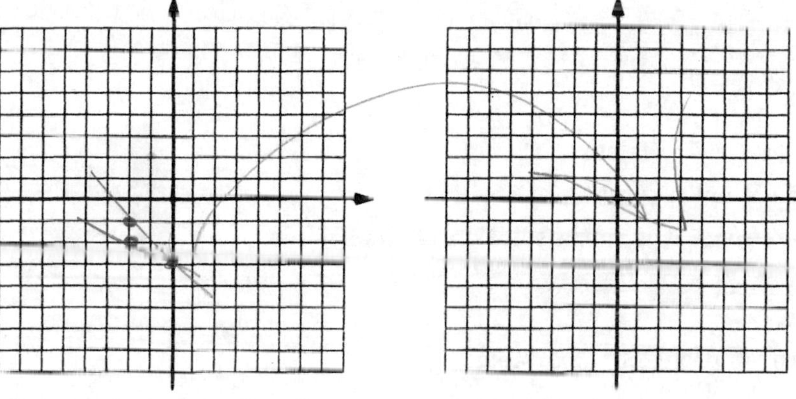

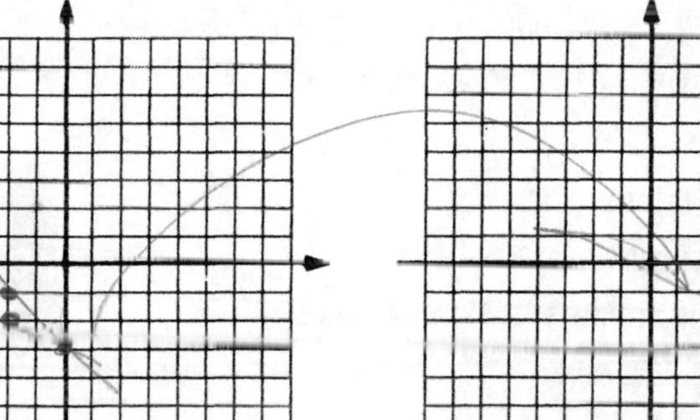

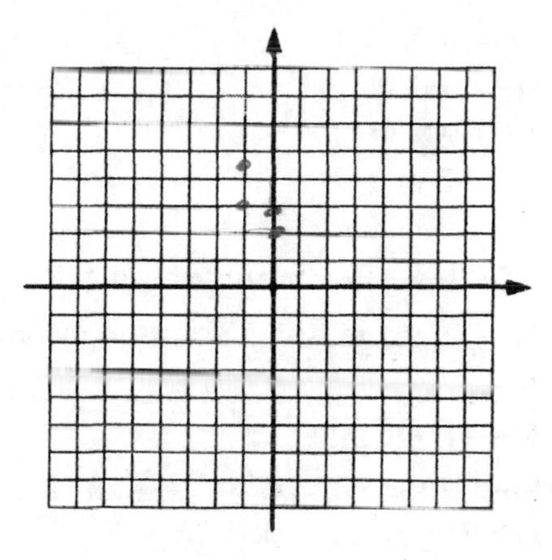

1. ______________ **2.** ______________ **3.** ______________

4. $x + y = 8$ and $x - y = 2$

5. $x + 2y = 3$ and $2x + 4y = 6$

6. $4x - y = 4$ and $y = 4x + 8$

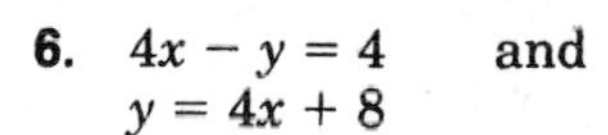

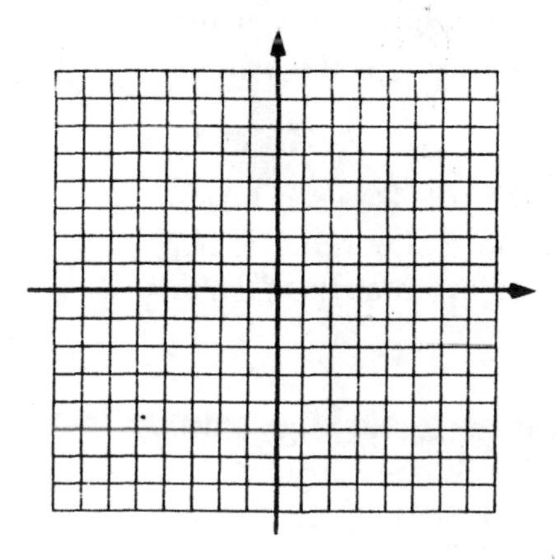

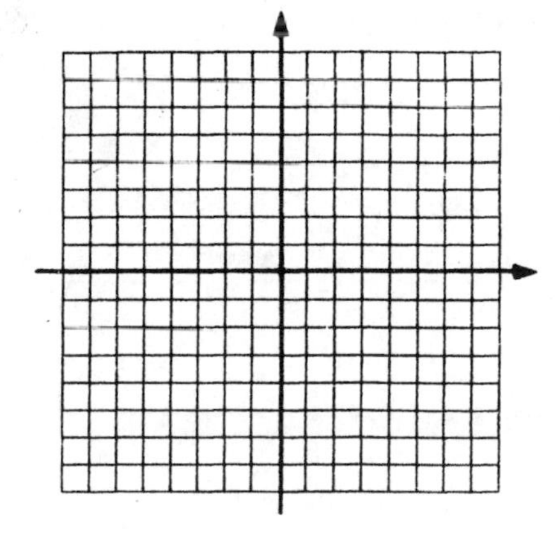

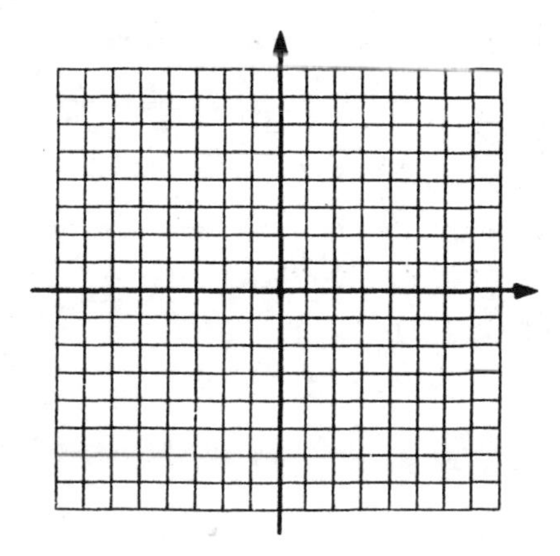

4. ______________ **5.** ______________ **6.** ______________

B. SOLUTIONS OF LINEAR SYSTEMS

A **solution of a system of two linear equations in two variables,** x and y, is an ordered pair (a, b) such that each equation becomes true when x is replaced by a and y by b.

The **solution set of a system of two linear equations in two variables** is the set of all ordered pairs that are solutions of both equations.

SAMPLE PROBLEM 1

Show that (4, 5) is not a solution of the system
$2x - y = 3$ and $x + 2y = 7$.

SOLUTION

$2x - y = 3$	$x + 2y = 7$
$2(4) - 5 = 3$	$4 + 2(5) = 7$
$8 - 5 = 3$	$4 + 10 = 7$
$3 = 3$ (true)	$14 = 7$ (false)
(4, 5) is a solution of $2x - y = 3$.	(4, 5) is not a solution of $x + 2y = 7$.

Since (4, 5) is **not** a solution of both equations, it is **not** a solution of the system.

SAMPLE PROBLEM 2

Show that (−3, 7), (0, 1), and (2, −3) are solutions of the system
$4x + 2y = 2$ and $6x + 3y = 3$.

SOLUTION

For (−3, 7),

$4x + 2y = 2$	$6x + 3y = 3$	
$4(-3) + 2(7) = 2$	$6(-3) + 3(7) = 3$	Since (−3, 7) is a solution
$-12 + 14 = 2$	$-18 + 21 = 3$	of both equations, it is
$2 = 2$ (true)	$3 = 3$ (true)	a solution of the system.

For (0, 1),

$4x + 2y = 2$	$6x + 3y = 3$	Since (0, 1) is a solution
$4(0) + 2(1) = 2$	$6(0) + 3(1) = 3$	of both equations, it is
$2 = 2$ (true)	$3 = 3$ (true)	a solution of the system.

For (2, −3),

$4x + 2y = 2$	$6x + 3y = 3$	
$4(2) + 2(-3) = 2$	$6(2) + 3(-3) = 3$	Since (2, −3) is a solution
$8 - 6 = 2$	$12 - 9 = 3$	of both equations, it is
$2 = 2$ (true)	$3 = 3$ (true)	a solution of the system.

EXERCISES 8.1B

Determine whether or not each given ordered pair is a solution of the given system.

1. $x + y = 8$ and $x - y = 2$
a. (5, 3) **b.** (3, 5) **c.** (4, 2)

1. a. ____________
b. ____________
c. ____________

2. $2x - y - 9$ and $x + 2y = 2$
a. (5, 1) **b.** (4, −1) **c.** (−2, 2)

2. a. ____________
b. ____________
c. ____________

3. $x = 4$ and $y = -1$
a. (4, 1) **b.** (−1, 4) **c.** (4, −1)

3. a. ____________
b. ____________
c. ____________

4. $x = -5$ and $y = 2$
a. (−5, 0) **b.** (0, 2) **c.** (−5, 2)

4. a. ____________
b. ____________
c. ____________

5. $y = 2x$ and $3x - y = 3$
a. (−3, −6) **b.** (3, 6) **c.** (−6, −12)

5. a. ____________
b. ____________
c. ____________

6. $x + y = 5$ and $y = 8 - x$
a. (3, 2) **b.** (3, 5) **c.** (0, 0)

6. a. ____________
b. ____________
c. ____________

7. $x + y = 5$ and $y = 5 - x$
a. (3, 2) **b.** (−1, 6) **c.** (0, 5)

7. a. ____________
b. ____________
c. ____________

8. $x + y = 5$ and $y = 5 + x$
a. (3, 2) **b.** (−1, 4) **c.** (0, 5)

8. a. ____________
b. ____________
c. ____________

9. $y = 2x - 1$ and $2x - y = 1$
a. (3, 5) **b.** (−4, −9) **c.** $\left(\frac{1}{2}, 0\right)$

9. a. ____________
b. ____________
c. ____________

10. $x = y - 6$ and $y = x - 6$
a. (0, 6) **b.** (6, 0) **c.** (12, 6)

10. a. ____________
b. ____________
c. ____________

1–10. **Find the slopes and y-intercepts and determine whether the graphs of the two given equations are intersecting lines, parallel lines, or coincident lines.**

1. $5x + 2y = 10$ and $x - 2y = 14$ — 1. ____________

2. $5x - 2y = 10$ and $2y - 5x = 10$ — 2. ____________

3. $4x - 3y = 18$ and $2x + y = 4$ — 3. ____________

4. $2x + 3 = 0$ and $2x + y + 3 = 0$ — 4. ____________

5. $3x - 6 = 0$ and $4y + 8 = 0$ — 5. ____________

6. $2x + 10 = 0$ and $2y + 10 = 0$ — 6. ____________

7. $2y + 8 = 0$ and $2y - 8 = 0$ — 7. ____________

8. $2x = 3y + 6$ and $2x - 3y - 6 = 0$ — 8. ____________

9. $y = 4$ and $y = -3$ — 9. ____________

10. $x = 6$ and $y = 2$ — 10. ____________

NAME ____________ DATE ____________ COURSE ____________

11–15. Determine whether or not each given ordered pair is a solution of the given system.

11. $x = 4y$ and $2x - 9y = 2$
a. $(-2, -8)$ **b.** $(-8, -2)$ **c.** $(0, 0)$

11. **a.** ______
b. ______
c. ______

12. $x - 3y = 2$ and $x - y = 8$
a. $(2, -6)$ **b.** $(11, 3)$ **c.** $(9, -3)$

12. **a.** ______
b. ______
c. ______

13. $2x + 3y = 12$ and $3x = 2y$
a. $(0, 4)$ **b.** $(-2, -3)$ **c.** $(3, 2)$

13. **a.** ______
b. ______
c. ______

14. $x + 2y = 3$ and $3x + 6y = 9$
a. $(3, 0)$ **b.** $\left(-2, 2\frac{1}{2}\right)$ **c.** $\left(0, -\frac{1}{2}\right)$

14. **a.** ______
b. ______
c. ______

15. $2x - 2y = 5$ and $5x - 5y = 4$
a. $\left(3, \frac{1}{2}\right)$ **b.** $(2, 2)$ **c.** $(0, -5)$

15. **a.** ______
b. ______
c. ______

GRAPHICAL SOLUTION 8.2

When the graphs of two linear equations intersect in exactly one point, then the solution set of the system of the two linear equations consists of exactly one ordered pair of real numbers, the coordinates of the point of intersection of the two lines.

To solve a system of two linear equations, graphically:

1. **Graph the two equations on the same coordinate system.**
2. **Read the coordinates of the point of intersection of the two lines.**

The corresponding ordered pair is the solution of the system.

To obtain good results, it is very important that the lines are drawn very carefully.

Drawing aids

1. The drawing pen or pencil should have a **very sharp point.**
2. The line should pass exactly through the plotted points.
3. Use graph paper having squares the same size as or larger than those shown in this book.

Any two points may be selected to graph a line. However, for best results,

Plotting aids

1. **Avoid** points that have **fractions** for coordinates.
2. **Select points** that are **not too close** together.

A small error in joining points close together becomes a larger error for distances beyond the points.

SAMPLE PROBLEM 1

Solve the system $x + 3y = 8$ and $2x - y = 9$ graphically and check.

SOLUTION

$x + 3y = 8$ $2x - y = 9$

x	y
8	0
2	2

x	y
0	−9
2	−5

Note that $-(-9) = 9$.
Note that $2(2) - (-5) = 4 + 5 = 9$.

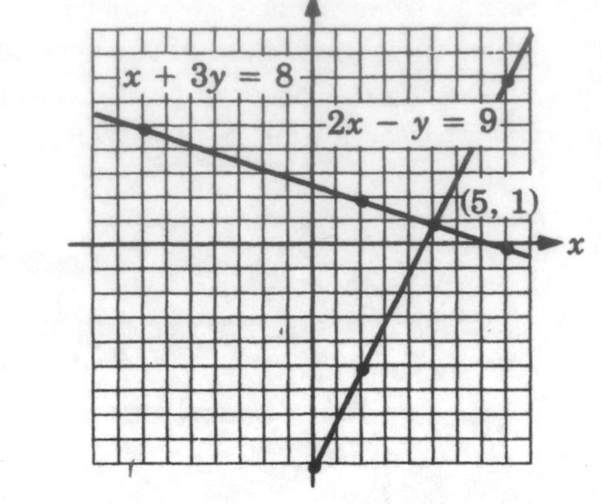

From the graph, the solution is read as (5, 1).

CHECK

For $x + 3y = 8$, $5 + 3(1) = 8$.
For $2x - y = 9$, $2(5) - 1 = 9$.

(Note: Only two points are needed for the graph of each equation. The point of intersection serves as the check point.)

EXERCISES 8.2

Solve each system graphically and check.

1. $x + y = 12$ and $2x - y = 9$

1. ______________

1. Find two points on each line.
2. Graph each line.
3. State the point of intersection as an ordered pair. (____, ____)
4. Check in each equation.

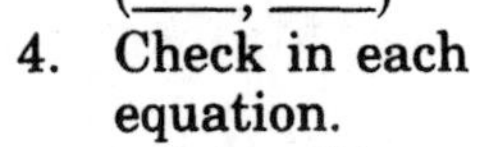

$x + y = 12$ $2x - y = 9$

x	y

x	y

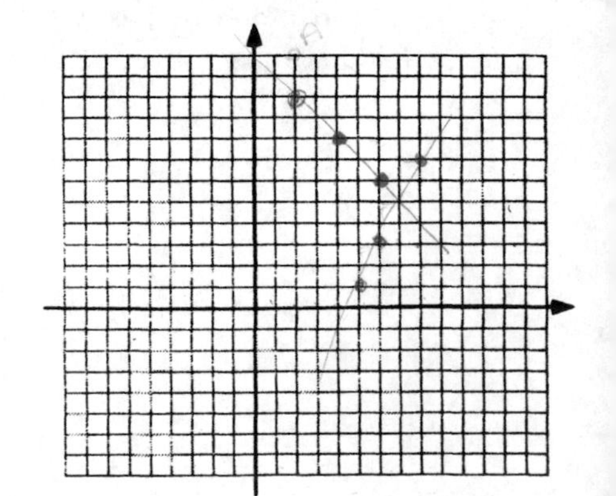

2. $2x + y = 14$ and $x - y = 4$

2. ____________

1. Find two points on each line.
2. Graph each line.
3. State the point of intersection as an ordered pair. (____, ____)
4. Check in each equation.

x	y

x	y

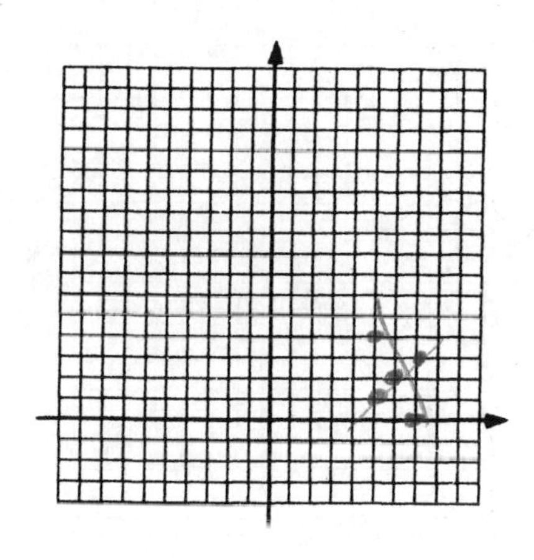

3. $x - 2y = 8$ and $x + 2y = 4$

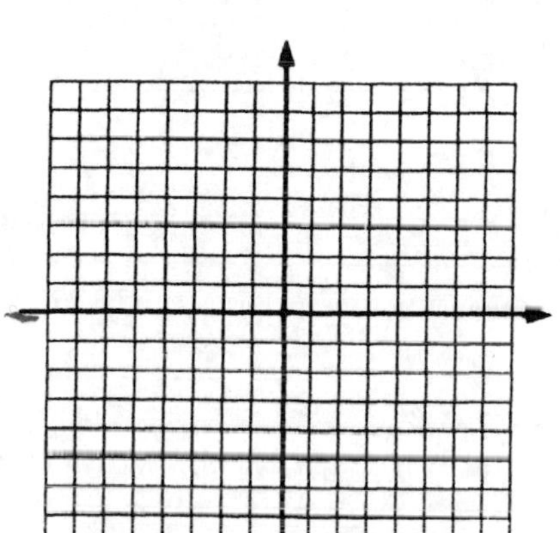

4. $x = 4 - 3y$ and $y = x + 8$

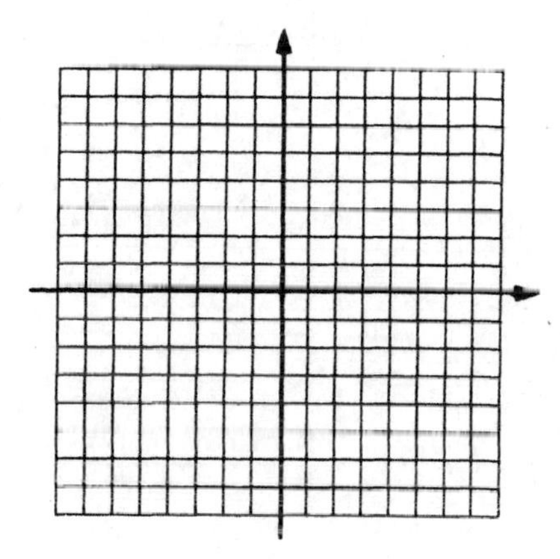

3. ____________

4. ____________

5. $y = 2x + 1$ and $x - 2y = 4$

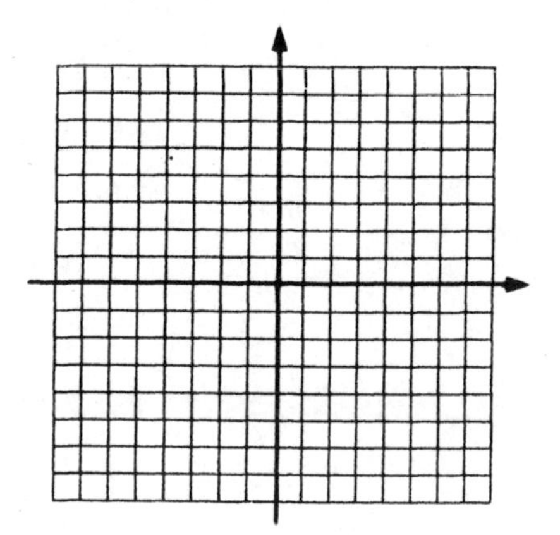

6. $x = 3y - 3$ and $2x - y = 4$

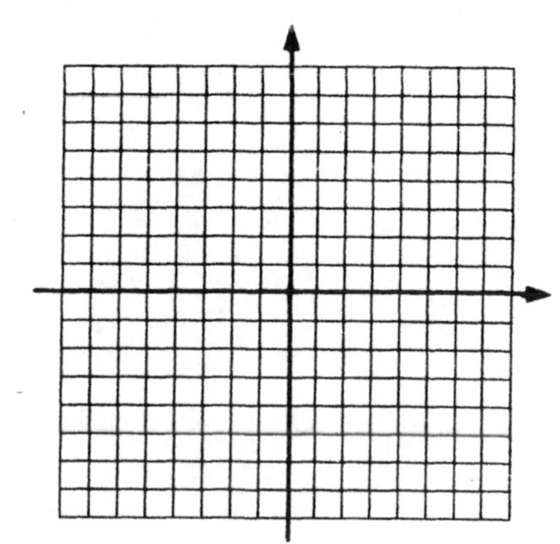

5. ____________

6. ____________

7. $3x + 2y = 6$ and $2x - y = 4$

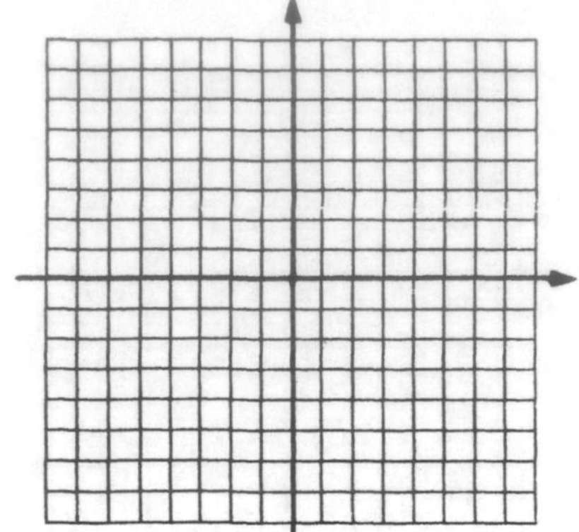

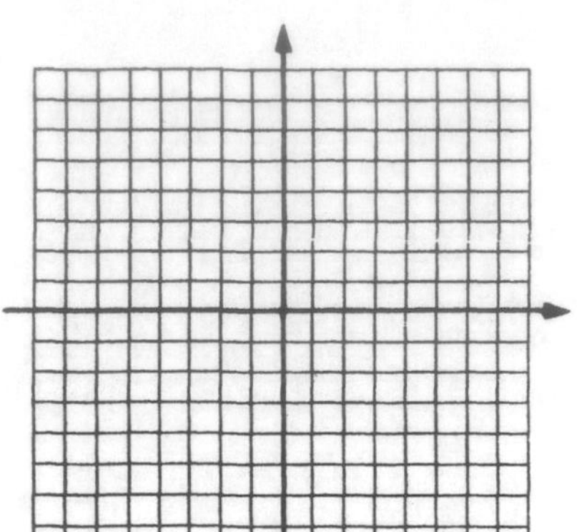

8. $2x + 3y + 12 = 0$ and $x - y + 1 = 0$

7. ______________________

8. ______________________

9. $y = 2x - 3$ and $x = 2y - 3$

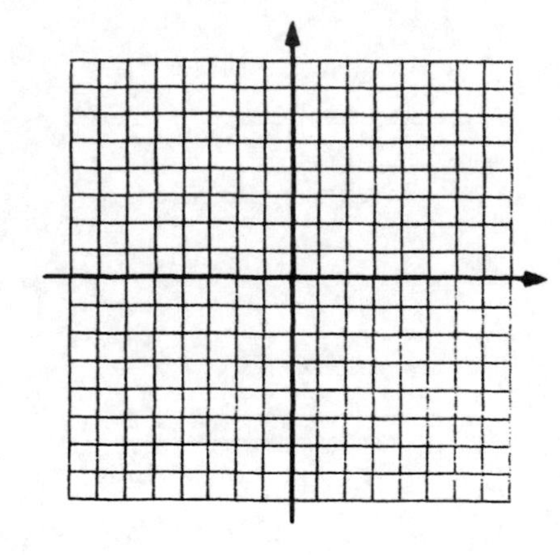

10. $5x + 2y = 10$ and $2y - 5x = 10$

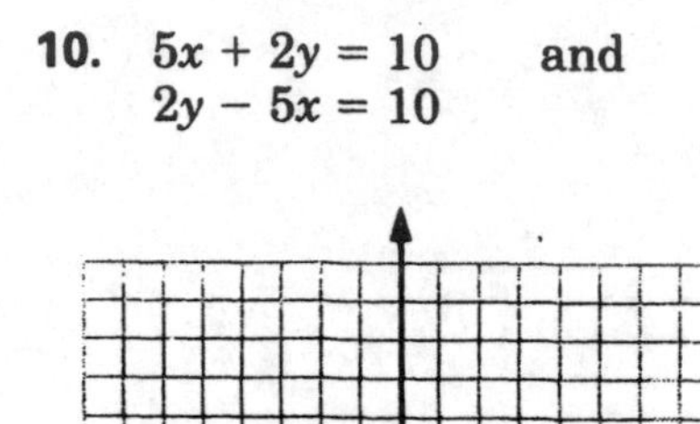

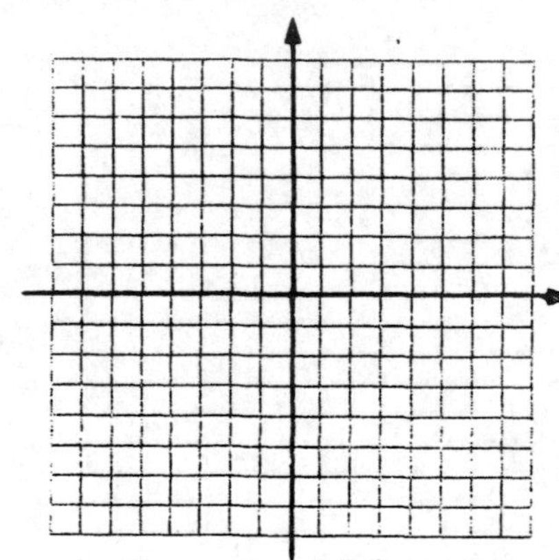

9. ______________________

10. ______________________

EXERCISES 8.2S

Graphically solve each system and check.

1. $x + y = 7$ and $2x - y = 8$

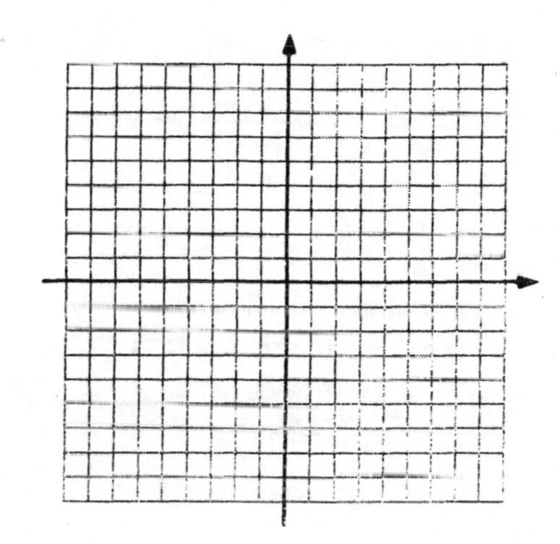

2. $3x + 2y = 6$ and $x - 2y = 10$

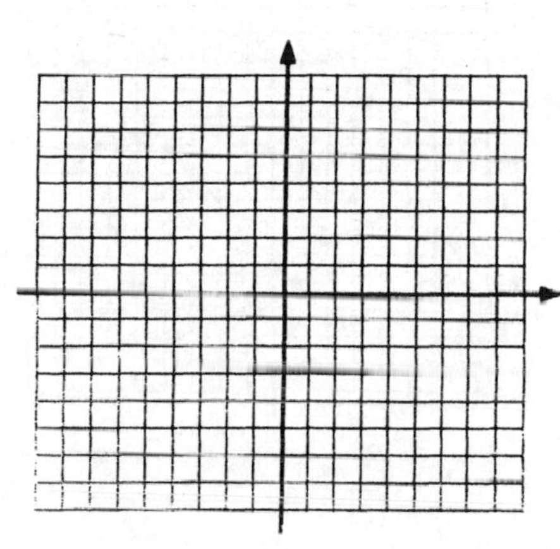

1. ______

2. ______

3. $2x + 3y = 12$ and $3x + y = -3$

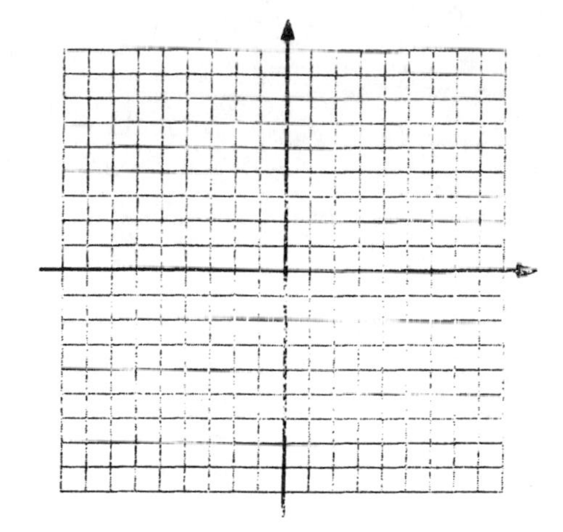

4. $x - 2y = 8$ and $y - 3x = 6$

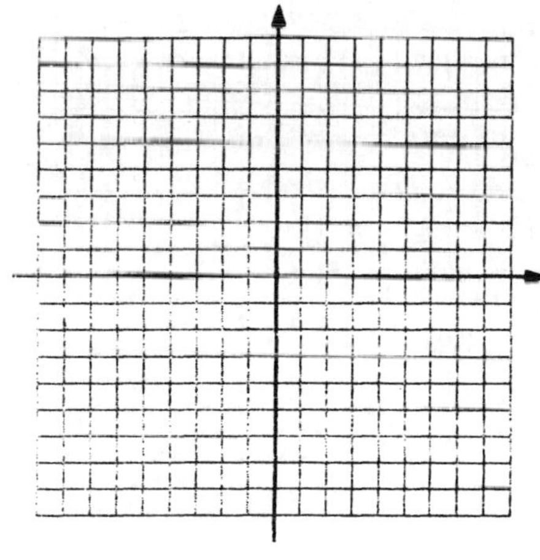

3. ______

4. ______

NAME ______ DATE ______ COURSE ______

5. $3x + y = 9$ and $y = x + 5$

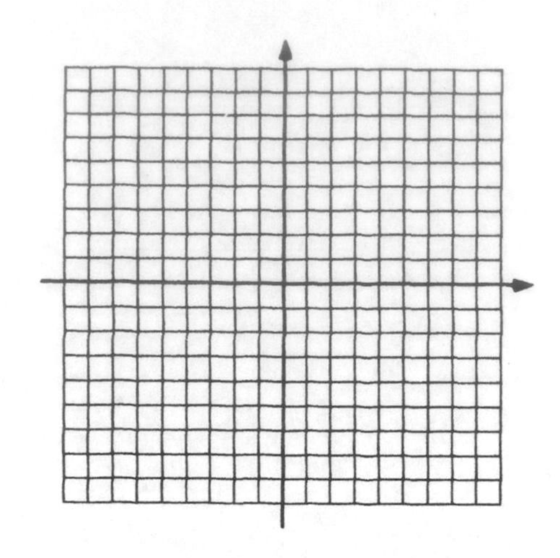

6. $y = 2x - 6$ and $y = 9 - x$

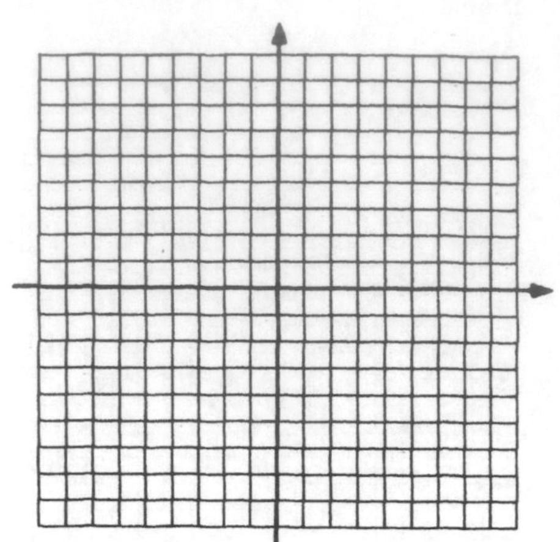

5. ______________

6. ______________

7. $x + 2y - 8 = 0$ and $x - y + 7 = 0$

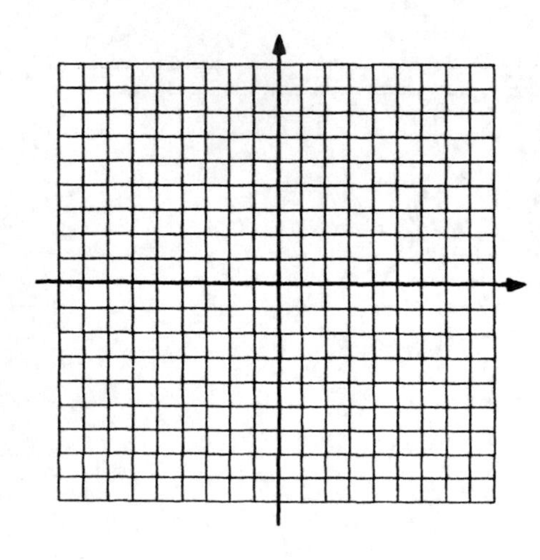

8. $y = 8 - 2x$ and $2x + 3y = 0$

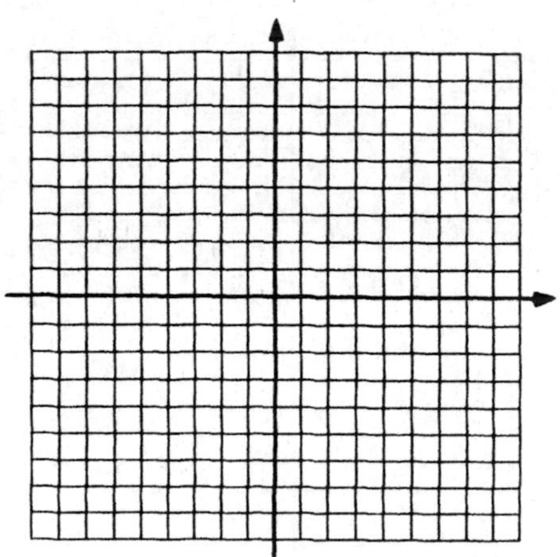

7. ______________

8. ______________

9. $x + y = 0$ and $y = 2x$

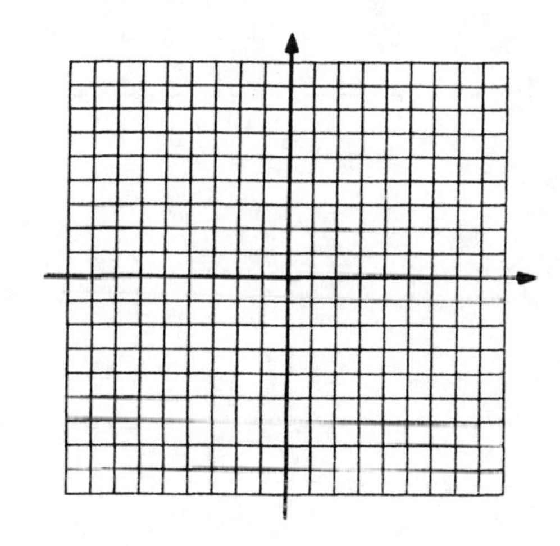

10. $4x + 3y + 12 = 0$ and $x = -6$

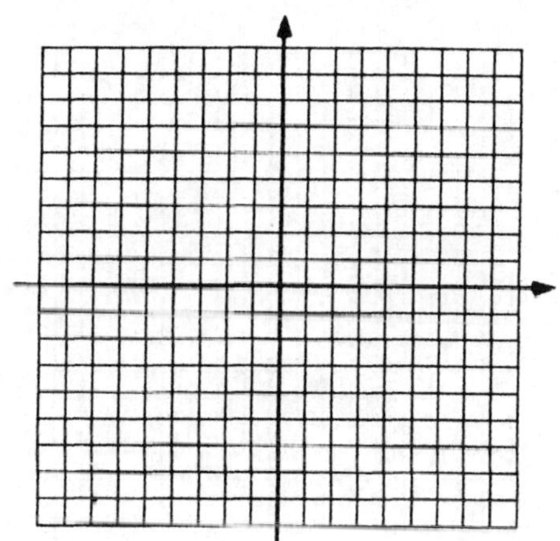

9. ______________________

10. ______________________

11. $4x - y = 8$ and $y = 7 - 2x$

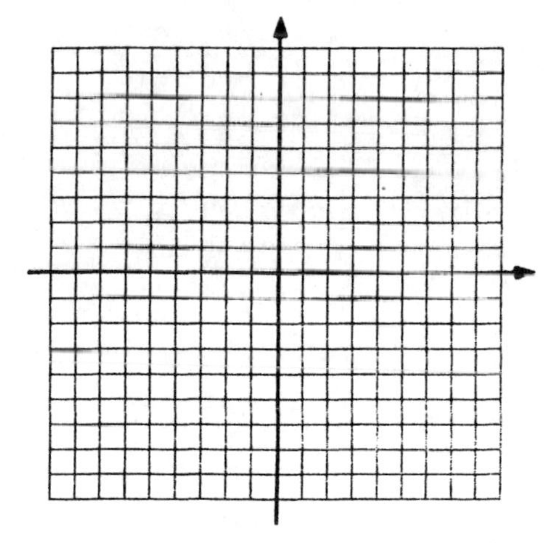

12. $3x + 2y = 6$ and $x + 2y = -4$

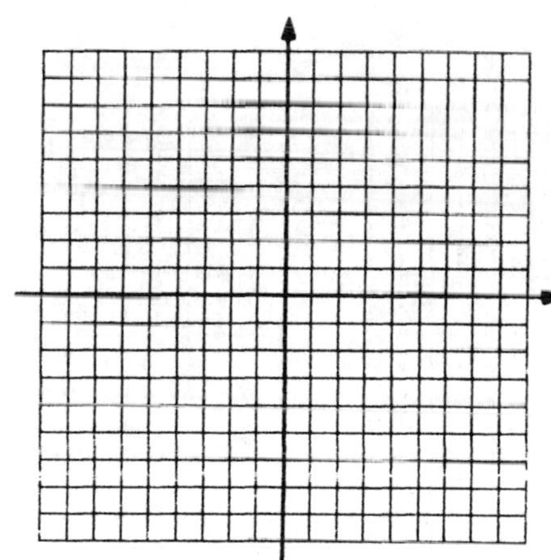

11. ______________________

12. ______________________

13. $x - y = 2$ and $y = 3x + 8$

14. $x + 2y = 0$ and $3x + 2y + 12 = 0$

13. ______________________

14. ______________________

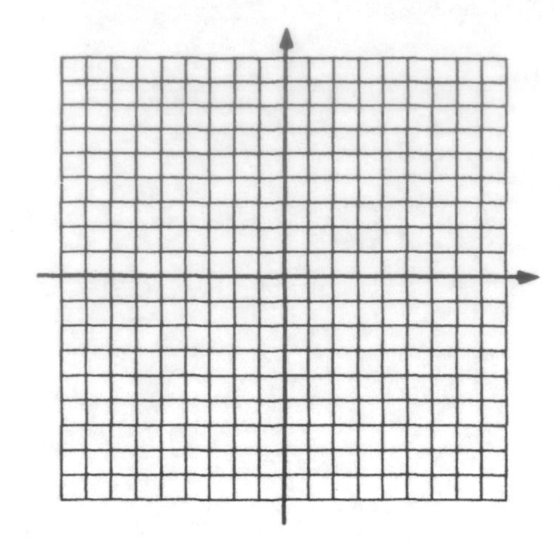

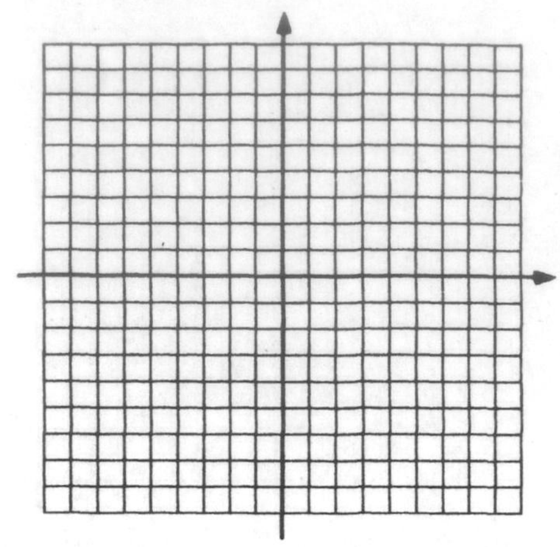

15. $4x + 5y = 200$ and $5y = 4x$
Graph only in Quadrant I.

16. $2x + y = 100$ and $x + 3y = 150$
Graph only in Quadrant I.

15. ______________________

16. ______________________

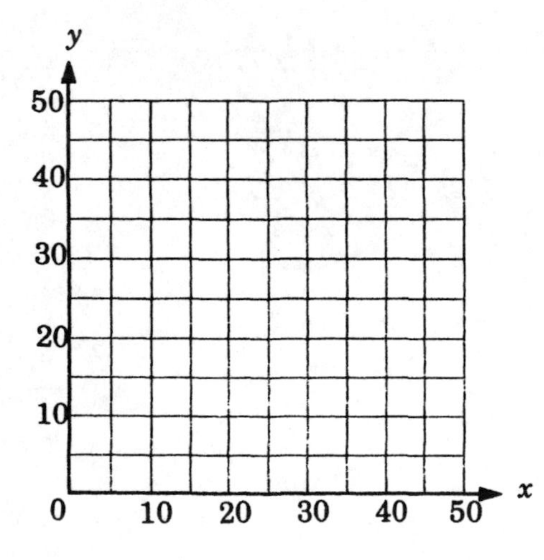

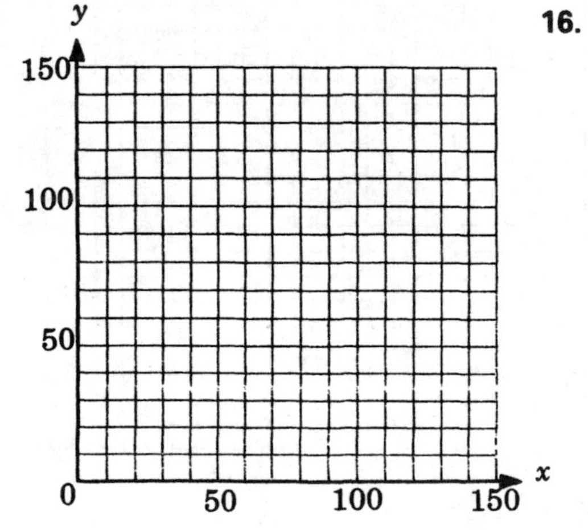

SUBSTITUTION METHOD 8.3

The graphical solution method has certain drawbacks. It requires good drawing techniques and it is time consuming. When the value of x or y is not an integer, this value must often be approximated visually and then verified or adjusted by the checking process. (Compare Problems 11 and 12 in Exercises 8.2 S.)

On the other hand, the graphical method provides us with basic graphing techniques that are used in many applied areas such as science, engineering, and economics.

For our present purposes, the graphical method gives us a visual picture that helps us understand the algebraic solution methods that we will study next.

It is not necessary to graph a linear system to find its solution set. The solution set may also be found by algebraic methods. One such method is called the substitution method. In this method, one of the equations is solved for either x or y, and the resulting expression is substituted into the other equation.

If one equation is solved for y, then the method can be outlined as follows.

SUBSTITUTION METHOD PROCEDURE, SOLVING FOR y

1. **Solve one of the equations for y, obtaining $y = ax + b$.**
2. **In the other equation, replace y by $ax + b$ and solve the resulting equation for x.**
3. **In $y = ax + b$, replace x by the value found in Step 2 and find the value for y.**
4. **State the solution as an ordered pair.**
5. **Check the solution in each equation.**

In the substitution method, the first equation used can be solved for either x or y. The choice depends on the coefficients of x and y in the two equations. If x has a coefficient of 1 in one of the equations, and y does not, then this equation is solved for x. The procedure is as follows.

SUBSTITUTION METHOD PROCEDURE, SOLVING FOR x

1. **Solve one equation for x, obtaining $x = cy + d$.**
2. **In the other equation, replace x by $cy + d$ and solve for y.**
3. **In $x = cy + d$, replace y by the value found in Step 2 and find the value for x.**
4. **State the solution as an ordered pair.**
5. **Check the solution in each equation.**

EXERCISES 8.3

1–6. Solve each of the following systems by first solving for y.

SAMPLE PROBLEM 1

$2x + 3y = 2$ and $3x + y = 10$

SOLUTION

1. Solve one equation for y.	$3x + y = 10$ $y = \mathbf{10 - 3x}$
2. Substitute in the other equation.	$2x + 3y = 2$ $2x + 3(\quad) = 2$ $2x + 3\mathbf{(10 - 3x)} = 2$
3. Solve for x.	$2x + 30 - 9x = 2$ $-7x = -28$ $\mathbf{x = 4}$
4. Replace x using either original equation.	$y = 10 - 3\mathbf{(4)}$
5. Solve for y.	$\mathbf{y = -2}$
6. State the solution as an ordered pair.	The solution is $(4, -2)$.
7. Check the solution in each of the original equations.	$2x + 3y = 2$: $2(4) + 3(-2) = 8 - 6 = 2$ $3x + y = 10$: $3(4) + (-2) = 12 - 2 = 10$

1. $2x + y = 14$ and $3x - 2y = 7$

2. $2x + 3y = 7$ and $y = 5x - 9$

3. $5x + y = 10$ and $3x - 2y - 6 = 0$

4. $5x - 3y - 50 = 0$ and $3x + y = 2$

5. $2x - y = 2$ and $3x + 2y + 25 = 0$

6. $3x - y = 19$ and $2x - 3y = 1$

1. __________

2. __________

3. __________

4. __________

5. __________

6. __________

7–12. Solve each system by the substitution method. First solve for x.

SAMPLE PROBLEM 2

$x - 2y = 1$ and $2x - 3y = 5$

SOLUTION

1. Solve one equation for x. $\quad x - 2y = 1$

   $x = \mathbf{2y + 1}$
2. Replace x in the other equation by the value found in Step (1). $\quad 2x - 3y = 5$

   $2(\mathbf{2y + 1}) - 3y = 5$
3. Solve for y. $\quad 4y + 2 - 3y = 5$

   $\mathbf{y = 3}$
4. In either equation replace y by its value. $\quad x = 2y + 1$ and $y = 3$
5. Solve for x. $\quad x = 2(\mathbf{3}) + 1$

   $\mathbf{x = 7}$
6. State the solution as an ordered pair. $\quad$ The solution is $(7, 3)$.
7. Check in each of the original equations.

   $x - 2y = 1$: $\quad 7 - 2(3) = 7 - 6 = 1$

   $2x - 3y = 5$: $\quad 2(7) - 3(3) = 14 - 9 = 5$

---

7. $2x - 3y = 1$ and
$x - y = 3$

8. $5x - y = 9$ and
$x = y - 3$

9. $2x + 10y = 1$ and
$x = -4y$

10. $x - 4y = 1$ and
$3x + 4y = 19$

11. $2x - 3y = 21$ and
$x + 6y = 3$

12. $3y - 2x = 9$ and
$x = 3y$

7. ____________________

8. ____________________

9. ____________________

10. ____________________

11. ____________________

12. ____________________

13–20. Solve by the substitution method.

13. $4x - 3y = 60$ and $2x + y = 80$ **13.** ______

14. $5x - 2y = 100$ and $x + 2y = 140$ **14.** ______

15. $x = 4y$ and $2x - 9y = 2$ **15.** ______

16. $5x + 2y = 60$ and $2x - y = 15$ **16.** ______

17. $3x + 5y = 35$ and $3y - x = 7$ **17.** ______

18. $90 - y = 4x$ and $40 + 2y = 3x$ **18.** ______

19. $x + y = 300$ and $x - y = 200$ **19.** ______

20. $3x - 2y = 140$ and $x + 2y = 100$ **20.** ______

EXERCISES 8.3 S

Solve each system by the substitution method. Check.

1. $y - 3x = 8$ and $y = 5x$ 1. ______

2. $2x + y = 18$ and $3x + 2y = 29$ 2. ______

3. $x + 4y = 25$ and $2x - 3y = 6$ 3. ______

4. $4y - 3x = 17$ and $x - 2y + 7 = 0$ 4. ______

5. $5x + 2y = 38$ and $3x + y = 24$ 5. ______

6. $y - 2x = 2$ and $3x - 2y = 1$ 6. ______

7. $3x - y = 7$ and $4x + 3y = 18$ 7. ______

NAME ______ DATE ______ COURSE ______

8. $5y - x = 10$ and $3x - 5y + 30 = 0$ 8. ______

9. $x + 3y = 45$ and $x - 4y = 10$ 9. ______

10. $2x + 5y = 8$ and $4x + y + 20 = 0$ 10. ______

11. $x - y = 10$ and $3x + 2y = 70$ 11. ______

12. $2x + y = 2$ and $2y - x = 2$ 12. ______

13. $x - 4y + 1 = 0$ and $6y = 3x + 2$ 13. ______

14. $6x - 5y = 100$ and $x - y = 10$ 14. ______

15. $x + 3y = 3$ and $3x - y = 2$ 15. ______

16. $2x + 2y = 7$ and $5y - x = 4$ 16. ______

ADDITION METHOD 8.4

It can be checked that (2, 5) is a solution of the system
$3x + 2y - 16 = 0$ and $x - y + 3 = 0$.

For $3x + 2y - 16 = 0$, $\quad 3(2) + 2(5) - 16 = 6 + 10 - 16 = 0$.
For $x - y + 3 = 0$, $\quad 2 - 5 + 3 = 0$.
Now consider

$$a(3x + 2y - 16) + b(x - y + 3) = 0$$

where a and b are any real numbers.

No matter what numbers a and b are, (2, 5) will always be a solution of $a(3x + 2y - 16) + b(x - y + 3) = 0$, since $3x + 2y - 16 = 0$ and $x - y + 3 = 0$ for $x = 2$ and $y = 5$.

This means that if each equation of a linear system is multiplied by a constant and the resulting equations are added, then the solution of the system is also a solution of the new equation.

The addition method for solving a linear system is based on this idea. The goal is to choose the constants a and b so that one of the variables is missing in the new equation obtained by addition.

ADDITION METHOD PROCEDURE

1. **Eliminate y.**
   a. **Compare y terms and find a multiplier for each so that the new y terms are addition inverses.**
   b. **Multiply each equation by its multiplier found in Step a.**
   c. **Add the new equations and solve for x.**
2. **Eliminate x.**
   a. **Compare x terms and find a multiplier for each so that the new x terms are addition inverses.**
   b. **Multiply each equation by its multiplier found in Step a.**
   c. **Add the new equations and solve for y.**
3. **State the solution as an ordered pair.**
4. **Check the solution in each of the original equations.**

EXERCISES 8.4

1–4. Solve each system by using the addition method.

SAMPLE PROBLEM 1

$x + 2y = 10$ and $3x - y = 9$

SOLUTION

1. Eliminate y:
   a. Compare y terms and find multipliers, so sum of new y terms is 0.
   $$\begin{aligned} 2y(\mathbf{1}) &= 2y \\ -y(\mathbf{2}) &= -2y \\ \hline &\quad 0 \end{aligned}$$
   b. Multiply equations.
   $\mathbf{1}(x + 2y = 10)$ becomes $x + 2y = 10$
   $\mathbf{2}\,(3x - y = 9)$ becomes $6x - 2y = 18$
   c. Add new equations and solve for x.
   $7x = 28$
   $\mathbf{x = 4}$
2. Eliminate x:
   a. Compare x terms and find multipliers, so sum of new x terms is 0.
   $$\begin{aligned} x(\mathbf{3}) &= 3x \\ 3x(\mathbf{-1}) &= -3x \\ \hline &\quad 0 \end{aligned}$$
   b. Multiply equations.
   $\mathbf{3}(x + 2y = 10)$ becomes $3x + 6y = 30$
   $-\mathbf{1}(3x - y = 9)$ becomes $-3x + y = -9$
   c. Add new equations and solve for y.
   $7y = 21$
   $\mathbf{y = 3}$
3. State the solution as an ordered pair.
   The solution is **(4, 3).**
4. Check the solution in each of the original equations.
   $x + 2y = 10$: $4 + 2(3) = 4 + 6 = 10$
   $3x - y = 9$: $3(4) - 3 = 12 - 3 = 9$

1. $x + 3y = 18$
$2x - y = 8$

2. $x + 6y = 3$
$2x - 3y = 21$

3. $x + y = 14$
$x - y = 10$

4. $3x + 2y = 19$
$x - 2y = 1$

1. ______________________

2. ______________________

3. ______________________

4. ______________________

5–8. Solve by the alternate addition method.

Alternate addition method procedure
After the value of one variable has been found by adding multiples of the two equations, the value of the other variable may be found by using the substitution principle.

SAMPLE PROBLEM 2
$3x + 2y = 11$ and $5x - 3y = 31$.

SOLUTION

1. Eliminate y. Solve for x.

$$\mathbf{3}(3x + 2y = 11) \text{ becomes } 9x + 6y = 33$$
$$\mathbf{2}(5x - 3y = 31) \text{ becomes } 10x - 6y = 62$$
$$19x = 95$$
$$\mathbf{x = 5}$$

2. Substitute the value of x in either of the original equations. Solve for y.

$$3x + 2y = 11$$
$$3(\mathbf{5}) + 2y = 11$$
$$2y = 11 - 15$$
$$2y = -4$$
$$\mathbf{y = -2}$$

3. State the solution as an ordered pair. The solution is **(5, −2)**.

4. Check in each of the original equations.

$3x + 2y = 11$: $3(5) + 2(-2) = 15 - 4 = 11$
$5x - 3y = 31$: $5(5) - 3(-2) = 25 + 6 = 31$

---

5. $2x + y = 13$
$x + 3y = 7$

6. $4x - y = 32$
$x - 5y = 27$

7. $2x + 3y = 31$
$3x - 2y = 1$

8. $5x - 4y = 60$
$4x + 5y = 7$

5. ____________

6. ____________

7. ____________

8. ____________

9–14. Solve by either addition method.

9. $2x - y - 2 = 0$
$3x + 2y + 25 = 0$

10. $3x + y - 2 = 0$
$5x - 3y + 13 = 0$

9. ____________________

10. ____________________

11. $8x + 5y = 550$
$4x + 5y = 350$

12. $7x + 3y = 1740$
$28x + 3y = 4260$

11. ____________________

12. ____________________

13. $9x + 7y = 107$
$7x + 9y = 101$

14. $40x + 30y = 1700$
$25x + 10y = 800$

13. ____________________

14. ____________________

Solve each system by the addition method. Check each solution.

1. $x - y = 18$ and $x + y = 42$ 1. ______

2. $2x + 3y = 23$ and $5x - 3y = 5$ 2. ______

3. $4x - y = 32$ and $4x + 5y = 8$ 3. ______

4. $x + 2y = 12$ and $3x + y + 9 = 0$ 4. ______

5. $2x + 5y = 46$ and $5x + 3y = 58$ 5. ______

6. $3x - 4y = 6$ and $5x - 8y = 14$ 6. ______

7. $3x + 5y = 60$ and $3y - 5x = 36$ 7. ______

NAME ______ DATE ______ COURSE ______

8. $x + 3y = 9$ and $4x - 2y = 1$ 8. ____________

9. $5x + y - 10 = 0$ and $3x - 2y - 6 = 0$ 9. ____________

10. $3x - y - 19 = 0$ and $2x - 3y - 29 = 0$ 10. ____________

11. $3x - 2y = 1$ and $3x + 2y = 3$ 11. ____________

12. $5x + 4y = 6$ and $4x + 5y = 7$ 12. ____________

13. $3x - 7y = 2$ and $7x - 3y = 6$ 13. ____________

14. $2x + 3y = 8$ and $4x + 5y = 9$ 14. ____________

15. $4x + 3y = 1$ and $3x - 4y = 2$ 15. ____________

16. $5x - 5y = 3$ and $20x + 10y = 63$ 16. ____________

ALGEBRAIC METHODS, GENERAL CASE (OPTIONAL) 8.5

There are three types of solution sets for a linear system of the form $Ax + By + C = 0$ and $Dx + Ey + F = 0$.

Methods of solution shown in Sections 8.3 and 8.4.

1. The algebraic solution yields equations of the form $x = a$ and $y = b$. This **solution set consists of exactly one ordered pair,** (a, b), and the graphs of the two equations intersect in exactly one point, the point (a, b). This linear system is said to be **consistent and independent.**
2. The algebraic solution yields an equation that is false for all values of the variables. This **solution set is the empty set, Ø.** The graphs of the two equations are parallel lines, and the linear system is said to be **inconsistent.** (See Sample Problem 1.)
3. The algebraic solution yields an equation that is true for all values of the variables. This **solution set is infinite,** consisting of all solutions of either equation of the system. The graphs of the two equations are coincident lines, and the linear system is said to be **dependent.** (See Sample Problem 2.)

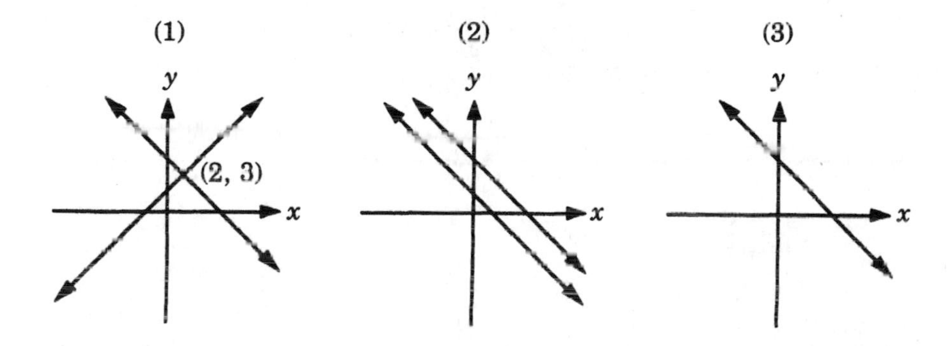

When to use the substitution method

The **substitution method** is convenient to use when either one of the given equations has the form $y = mx + b$ or $x = cy + d$, or when the coefficient of x or y is 1 in one of the given equations. In this latter case, the equation may be easily written in the form $y = mx + b$ or $x = cy + d$.

When to use the addition method

The **addition method** is convenient to use when each of the two given equations has the form $ax + by = c$, and no coefficient of x or y is 1.

EXERCISES 8.5

Find the solution set of each system and state whether the graphs of the two given equations are intersecting lines, parallel lines, or coincident lines. Check each result.

SAMPLE PROBLEM 1

$6x - 3y = 5$ and $y = 2x - 3$

SOLUTION

Use the substitution method: $y = 2x - 3$.

$$\begin{aligned} 6x - 3y &= 5 \\ 6x - 3(2x - 3) &= 5 \\ 6x - 6x + 9 &= 5 \\ 9 &= 5 \end{aligned}$$

(false for all values of x and y)

The solution set is the empty set, ∅.

The graphs are parallel lines.

SAMPLE PROBLEM 2

$2x + 4y = 6$ and $3x + 6y = 9$

SOLUTION

Use the addition method.

$$\begin{aligned} 3(2x + 4y = 6) &\rightarrow \quad 6x + 12y = 18 \\ -2(3x + 6y = 9) &\rightarrow -6x - 12y = -18 \\ & \qquad\qquad\quad 0 = 0 \end{aligned}$$

(true for all values of x and y)

The solution set is $\{(x,y)|2x + 4y = 6\}$.

The solution set can also be written as $\{(x,y)|3x + 6y = 9\}$.

The graphs are coincident lines.

These results can be checked by solving each equation for y and comparing slopes and y-intercepts.

1. $3x - y - 7 = 0$
$x + 3y - 29 = 0$

2. $2x - 4y = 6$
$3x = 6y + 9$

$-3x$

$-3x + 6y = -9$

3. $6x - 3y = 1$
$y = 2x - 5$

4. $x + 4y = 7$
$2x + 8y + 5 = 0$

5. $3x + 4y = 12$
$x - 2y = 4$

6. $x = y - 6$
$y = x + 6$

1. ______________________

2. ______________________

3. ______________________

4. ______________________

5. ______________________

6. ______________________

Solve and check.

1. $5x - 2 = 0$ and $2x - 5 = 0$

1. ______

2. $2x - 5y = 10$ and $6x - 15y = 30$

2. ______

3. $2x - y = 20$ and $4x + 3y = 290$

3. ______

4. $3x - 5y = 100$ and $5x - 3y = 300$

4. ______

5. $10x + 2y - 30 = 0$ and $5x + y - 15 = 0$

5. ______

NAME ______ DATE ______ COURSE ______

6. $4y + 7 = 0$ and $7y + 4 = 0$

6. ______________________

7. $3x - y = 0$ and $3y = x$

7. ______________________

8. $3x + y = 6$ and $x + 3y = 6$

8. ______________________

9. $3x = 6 - y$ and $y = 6 - 3x$

9. ______________________

10. $3x - y = 6$ and $y - 3x = 6$

10. ______________________

APPLICATIONS 8.6

Many word problems are easier to solve when two variables are used to form two equations. Examples of these are shown in this section.

A. MIXTURE PROBLEMS

SAMPLE PROBLEM 1

The cost of printing 400 programs was \$7.50, and the cost of printing 1000 programs was \$10.50. The printer charged a fixed fee for typesetting and then an additional charge for each program. Find the typesetting fee.

SOLUTION

Let $x =$ the typesetting fee.

Let $y =$ the charge per program.

Expressing the cost in cents, $x + 400y = 750$ and $x + 1000y = 1050$.

Multiply the first equation by -1 and add the result to the second equation.

$$\begin{aligned} x + 1000y &= 1050 \\ -x - 400y &= -750 \end{aligned}$$

Add. $$600y = 300$$

Solve for y. $$y = \frac{1}{2}$$

Solve for x. $$x + 400y = 750$$

$$x + 400\left(\frac{1}{2}\right) = 750$$

$$x + 200 = 750$$

$$x = 550$$

The typesetting fee was \$5.50.

SAMPLE PROBLEM 2

Five cans of tomato sauce and two cans of mushrooms cost $1.94. Eight cans of tomato sauce and three cans of mushrooms cost $3.00. Find the cost per can for the tomato sauce and for the mushrooms.

SOLUTION

Let x = the cost in cents per can of tomato sauce.

Let y = the cost in cents per can of mushrooms.

Then, $5x + 2y = 194$ and $8x + 3y = 300$

$-3(5x + 2y = 194)$ becomes $-15x - 6y = -582$

$2(8x + 3y = 300)$ becomes $16x + 6y = 600$

Add. $x = 18$

Substitute in $5x + 2y = 194$ $5(18) + 2y = 194$

Solve for y. $2y = 194 - 90 = 104$

$y = 52$

The cost of one can of tomato sauce was 18¢.

The cost of one can of mushrooms was 52¢.

EXERCISES 8.6A

Solve each of the following by using two variables.

1. Two yards of lace and 3 yd of seam binding cost 96¢. With no change in unit prices, 5 yd of the same lace and 6 yd of the same seam binding cost $2.22. Find the cost per yard of each.

1. ______________

2. 1 qt of oil and 5 gal of gasoline cost $7.05. With no change in unit prices, 2 qt of the same oil and 7 gal of the same gasoline cost $10.38. Find the cost of 1 qt of oil and the cost of 1 gal of gasoline.

2. ______________

3. When 36 lb of copper ore A are combined with 28 lb of copper ore B, then 39 lb of copper are obtained. Fifty pounds of copper ore A combined with 40 lb of copper ore B produce 55 lb of copper. Find the percentage of copper in each ore.

3. ____________________

4. A man received an income of $450 from $5000 invested in stocks and $3000 invested in bonds. His son received $195 from an investment of $2000 in the same stocks and $1500 invested in the same bonds. Find the percentage they received from each investment.

4. ____________________

5. One week when John worked 52 hr and Jill worked 40 hr, their combined income was $420. The next week, John worked 40 hr and Jill worked 25 hr, and their combined income was $300. Find the hourly wage of each.

5. ____________________

B. UNIFORM MOTION PROBLEMS

A uniform motion problem is one in which an object is moving at a constant rate of speed. The product rule is again a key concept. If a car travels at 60 m.p.h. for 2 hr, then the distance traveled is the product $60(2) = 120$ mi.

The uniform motion formula states that the distance, d, traveled is the product of the constant rate, r, and the time of travel, t.

BASIC IDEA

$$d = rt$$

$$\text{distance} = (\text{rate}) \times (\text{time})$$

Basic method

1. **Draw a sketch of the motions.**
2. **Find the values of r (rate) and t (time) for each object.**
3. **Find the value of d (distance) for each object by using the formula, $d = rt$.**
4. **Write an equation by noting how the distances are related.**
5. **Write an equation by noting how the times or rates are related.**
6. **Solve the system and check.**

EXERCISES 8.6B

OPPOSITE DIRECTIONS

SAMPLE PROBLEM 1

Two boats, 81 mi apart, travel toward each other, one traveling 3 m.p.h. faster than the other. They meet in 3 hr. Find the rate of the slower boat.

SOLUTION

Let x = rate of slower boat.

Sketch

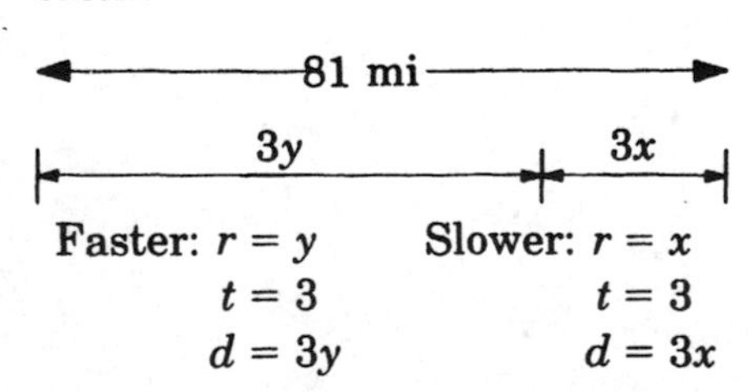

Equations

Sum of distances = 81

$$3x + 3y = 81$$

Rate of faster = 3 + rate of slower

$$y = 3 + x$$

$$3x + 3(x + 3) = 81$$

$$3x + 3x + 9 = 81$$

$$6x = 72$$

$$x = 12 \text{ m.p.h.}$$

SAMPLE PROBLEM 2

Two cars start at the same time from the same place. One travels north at 40 m.p.h. The other travels south at 60 m.p.h. In how many hours are they 300 mi apart?

SOLUTION

Let x = the time of northbound car.

Let y = the time of southbound car.

Sketch

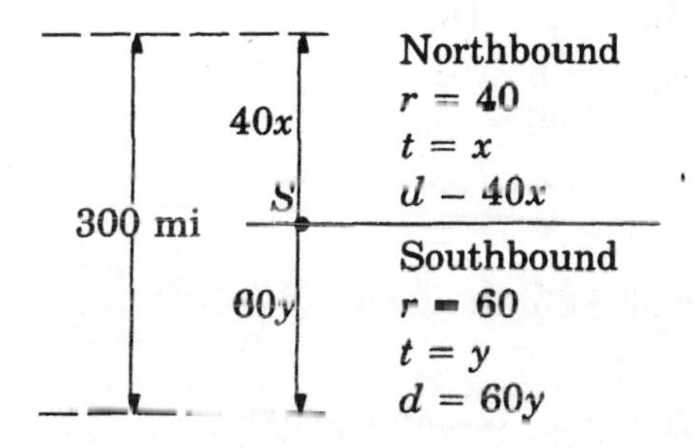

Equations

Sum of distances = 300

$$40x + 60y = 300$$

Time of northbound = time of southbound

$$x = y$$

$$40x + 60x = 300$$
$$100x = 300$$
$$x = 3 \text{ hr}$$

1. At 10:00 A.M., two planes leave the same airport, one traveling east at 460 m.p.h. and the other traveling west at 520 m.p.h. In how many hours are they 1470 mi apart? **1.** ____________

Sketch:

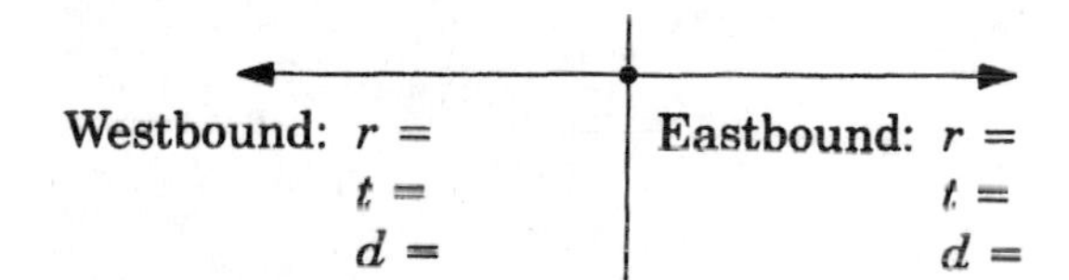

2. A passenger train and a freight train start from depots 550 mi apart and travel in opposite directions toward the same station. The freight train takes 6 hr to reach the station. The passenger train, traveling 50 m.p.h. faster than the freight train, takes 4 hr to reach the station. Find the rate of each train. **2.** ____________

Sketch:

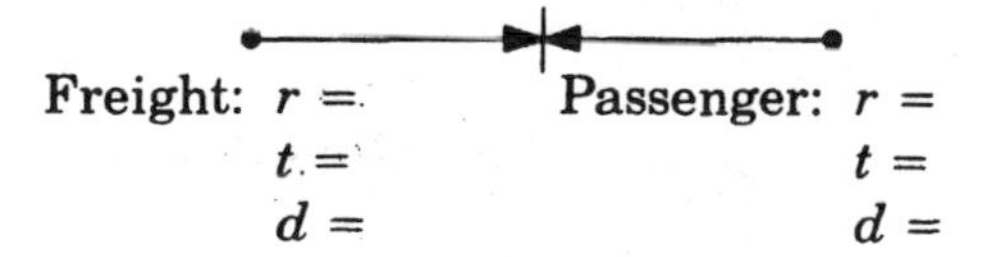

PURSUIT PROBLEMS

SAMPLE PROBLEM 3

A helicopter sets out to overtake a train that is 35 mi ahead. If the rate of the train is 90 m.p.h. and the rate of the helicopter is 160 m.p.h., how long does it take the helicopter to overtake the train?

SOLUTION

Let x = time it takes helicopter.

Let y = time train travels.

Sketch

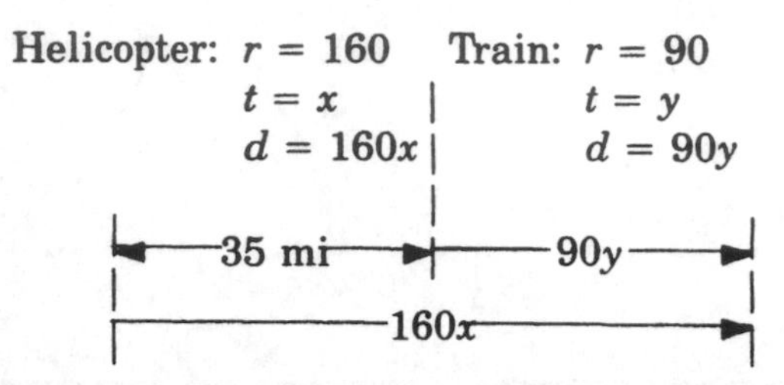

Equations

Distance of train + 35 = distance of helicopter

$$90y + 35 = 160x$$

Time of train = time of helicopter

$$y = x$$

$$160x = 35 + 90x$$

$$70x = 35$$

$$x = \frac{1}{2} \text{ hr} = 30 \text{ min}$$

3. At 8:00 A.M., a man left home and walked in the country at a rate of 3 m.p.h. A sudden thunderstorm arose and at 11:00 A.M. his wife left home and started after him, driving at a rate of 39 m.p.h. At what time did the wife overtake her husband?

3. ________________

Sketch:

Man: $r =$ | Wife: $r =$
$t =$ | $t =$
$d =$ | $d =$

4. A car speeding at 80 m.p.h. is 5 mi beyond a police car when the police car starts in pursuit at 90 m.p.h. How long does it take the police car to overtake the speeding car?

4. ________________

Sketch:

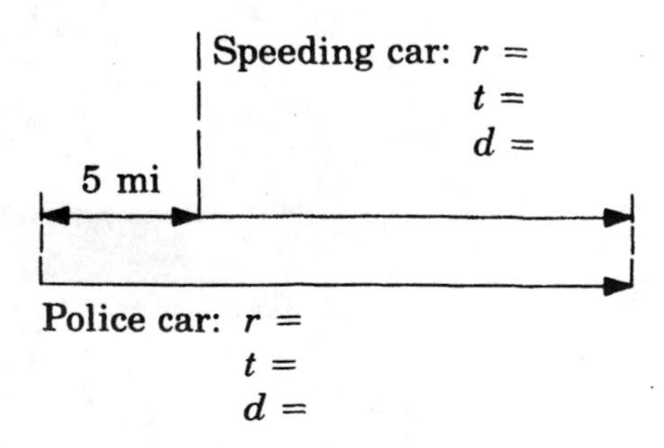

RETURN TRIPS

SAMPLE PROBLEM 4

At 10:00 A.M., a truck left a warehouse to make a delivery to a store. It stayed 1 hr at the store and then began the return trip to the warehouse, arriving back at 4:00 P.M. Because of heavy traffic, the rate returning was 24 m.p.h., although the rate going was 36 m.p.h. Find the distance from the warehouse to the store.

SOLUTION

Let x = time of trip going.

Let y = time of return trip.

Total time of travel = 6 − 1 = 5 hr.

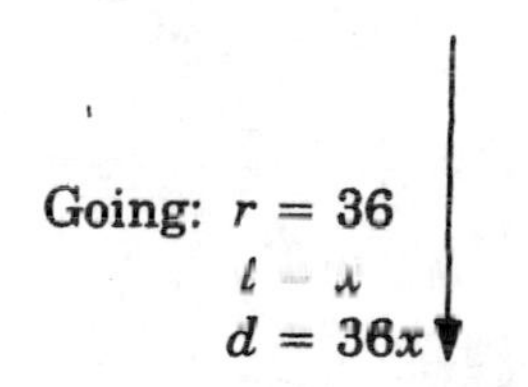

Returning: $r = 24$
$t = y$
$d = 24y$

Equations

For return trips, distance going = distance returning

$$36x = 24y$$

Sum of times = 5

$$x + y = 5$$

$$36x = 24(5 - x)$$

$$36x = 120 - 24x$$

$$60x = 120$$

$$x = 2 \text{ hr}$$

$$d = 36(2) = 72 \text{ mi}$$

5. A plane took 6 hr to fly to Seattle. The return trip took 1 hr longer because a strong wind reduced its speed by 25 m.p.h. Find the rate of the plane going to Seattle.

Returning: $r =$
$t =$
$d =$

Going: $r =$
$t =$
$d =$

5. ______________

6. A plane flew with the wind from Newcity to Oldcity at a rate of 240 m.p.h. On the return trip, flying against the wind, the rate of the plane was 180 m.p.h. The total time of the trip, going and returning, was 7 hr. Find the distance between the two cities.

Going: $r =$
$t =$
$d =$

Returning: $r =$
$t =$
$d =$

6. ______________

CURRENT

Let r be the uniform rate at which an object travels when there is no water current or wind.

Let s be the rate of the current or the wind.

Then $r + s$ = the rate of the object traveling with the current or wind,
and $r - s$ = the rate of the object traveling against the current or wind.

SAMPLE PROBLEM 5

A boat traveling downstream (with a current) goes 36 mi in 2 hr. Traveling upstream (against the same current), the same boat takes 3 hr to go 36 mi. Find the rate of the current.

SOLUTION

Let x = the rate of the boat in still water.

Let y = the rate of the current.

Then $x + y$ = the rate of the boat downstream,
and $x - y$ = the rate of the boat upstream.

Use the formula $rt = d$

Downstream, $2(x + y) = 36$

Upstream, $3(x - y) = 36$

Thus,

$$x + y = 18$$
$$x - y = 12$$
$$2x = 30$$
$$x = 15 \text{ m.p.h.}$$
$$y = 3 \text{ m.p.h., rate of current}$$

7. In 3 hr, an airplane flies 630 mi with a tailwind. Returning against the same wind, the airplane takes $3\frac{1}{2}$ hr to fly 630 mi. Find the speed of the wind.

7. ____________

8. A boat takes 3 hr to travel 42 mi downstream (with a current). The return trip upstream (against the same current) took 7 hr. Find the rate of the current.

8. ____________

Solve by using two variables.

1. A certain amusement park charges a fixed fee for each car and an additional fee for each person in the car. The total cost for 8 cars and 30 persons was $42.50, while the total cost for 10 cars and 40 persons was $55. Find the fixed fee per car and the additional fee per passenger.

1. ______________

2. The mixed price of 9 citrons and 7 fragrant wood-apples is 107; again, the mixed price of 7 citrons and 9 wood-apples is 101. O you arithmetician, tell me quickly the price of a citron and of a wood-apple here, having distinctly separated those prices well. (From the works of the Hindu Mahavira, about A.D. 850.)

2. ______________

3. A certain carpenter receives $8 per hour and his helper receives $5 per hour. Their combined wages for 1 wk was $550. The next week the carpenter worked half as many hours as the week before and his helper worked 10 hr more than the week before. Their combined wages for the second week was $400. Find the number of hours that each worked the first week.

3. ______________

NAME ______________ DATE ______________ COURSE ______________

4. A certain theater charges $2.80 for each adult and $1.20 for each child at all performances. One day, a total of $696 was collected for a matinee performance. At the evening performance, when twice as many adult tickets were sold and half as many children's tickets were sold, a total of $852 was collected. How many adults and how many children attended the matinee performance?

4. ______________________

5. A restaurant bought 80 glasses and 60 coffee mugs for a total cost of $34. Later, because of breakage, 25 glasses and 10 coffee mugs were bought for a total cost of $8. If there was no change in the prices, find the cost of 1 glass and the cost of 1 coffee mug.

5. ______________________

6. Of a total of 500 sheets, some were irregular and were classified as seconds. The irregular sheets were sold for $2.50 each and the other sheets were sold for $4.00 each. A total of $1820 was received for the 500 sheets. How many sheets were seconds?

6. ______________________

7. At the same time, two ships left two different ports 450 mi apart and traveled toward each other on parallel routes until they met. If the rate of one was 40 m.p.h. and the rate of the other was 35 m.p.h., how many hours did each travel?

7. ______________________

8. At 9:00 A.M. a truck left a warehouse and traveled north. At 11:00 A.M. a second truck left the same warehouse and traveled south. They each traveled at a rate of 55 m.p.h. What was the time of day when they were 330 mi apart?

8. ____________________

9. A truck leaves a depot and travels on a certain road at 30 m.p.h. One hour later, a car leaves the same depot and travels on the same road at 60 m.p.h. How long does it take the car to overtake the truck?

9. ____________________

10. At 3:00 P.M. a houseboat on a river is 16 mi upstream from a motorboat. Both are traveling upstream toward a dock, the houseboat at 9 m.p.h. and the motorboat at 25 m.p.h. If they both reach the dock at the same time, how many hours did each travel?

10. ____________________

11. At 10:00 A.M. a messenger left an office, traveled by car to deliver a message at a construction site, and immediately returned to the office after he delivered the message, arriving back at 12:00 noon. His rate going was 60 m.p.h., but his rate returning was 20 m.p.h. because of very heavy traffic. How far from the office was the construction site?

11. ____________________

12. A messenger cycled at the rate of 9 m.p.h. to deliver a message and then returned at the rate of 6 m.p.h. If the total trip took 5 hr, find the distance he cycled at 9 m.p.h.

12. ____________________

13. A boat takes 30 min to go $7\frac{1}{2}$ mi downstream. The return trip takes 45 min. Find the rate of the boat in still water and find the rate of the water current.

13. ____________________

14. An airplane flew 1200 mi in 5 hr, traveling with a tailwind. The return trip, traveling against the same wind, took 6 hr. Find the speed of the wind.

14. ____________________

SAMPLE TEST FOR CHAPTER 8

1–3. Find the slopes and y-intercepts of the graphs of each pair and determine whether the lines intersect, are parallel, or are coincident. (See Section 8.1).

1. $x - 2y = 4$ and $x = 2y + 4$

1. ____________________

2. $3x + 5y = 15$ and $y = 3x + 3$

2. ____________________

3. $x - 3y = 9$ and $3x = 9y + 3$

3. ____________________

4. Determine whether each given ordered pair is a solution of the given system. (See Section 8.1.)

4. $4x + y = 6$ and $3x - y = 15$

a. $(3, -6)$ **b.** $(3, 6)$

4. a. ____________________

b. ____________________

5. Solve graphically and check. (See Section 8.2.)

$2x + 5y = 10$ and $y = 8 + 2x$

5. ____________________

NAME ____________________ DATE ____________________ COURSE ____________________

6–7. Solve by the substitution method and check. (See Section 8.3.)

6. $2x + 5y = 10$
$y = 8 + 2x$

7. $x + 4y = 10$
$5x + 3y + 18 = 0$

6. ____________________

7. ____________________

8. Solve by the addition method and check. (See Section 8.4.)

$5x - 2y = 16$ and $2x + 5y = 18$

8. ____________________

9–11. Solve and identify the graph of each system as intersecting lines, parallel lines, or coincident lines. (See Section 8.5.)

9. $x = 3y - 2$ and $3y = x + 2$

9. ____________________

10. $x + 2y = 8$ and $x = 2y - 8$

10. ____________________

11. $6x - 2y = 9$ and $y - 3x = 6$

11. ____________________

12. A salesman sold some stoves and refrigerators to a buyer from Canada for a total sales of $5050 and to a buyer from Mexico for a total sales of $5350. Each stove was sold at $350 and each refrigerator was sold at $450. The salesman lost his itemized records, but he remembered that the Mexican bought the same number of stoves as the Canadian bought refrigerators and the same number of refrigerators as the Canadian bought stoves. How many stoves and how many refrigerators should he ship to Canada? How many to Mexico? (See Section 8.6.)

12. ____________________

13. On a certain trip an airplane, having a speed in still air of 175 m.p.h., flew for 3 hr with a tailwind. On the return trip traveling against the same wind, the plane had to land at the end of 2 hr, having gone only halfway. Find the speed of the wind. (See Section 8.6.)

13. ____________________

SAMPLE TEST FOR CHAPTERS 7 AND 8

1. For the graph of $3x - 4y = 24$,

   a. state the x-intercept

   b. state the y-intercept

   c. state the slope

1. a. ____________

   b. ____________

   c. ____________

2. Find the slope of the line joining the two given points. Check by graphing the line.

   $A(4, -5)$ and $B(-2, 7)$

2. slope = ____________

3. The equation of line L is $y = 3x - 4$. Find the equation of each specified line by selecting an equation from the column at the left.

   $6x - 2y = 8$ — a. A line parallel to L.
   $3x - y = 9$ — b. A line intersecting L.
   $x = 3y - 4$ — c. A line coincident with L.

3. a. ____________

   b. ____________

   c. ____________

4. a. Find solutions of each equation below by completing the tables.

$2x + y = 14$

x	y
2	
	0

$y = 3x - 6$

x	y
0	
	0

b. Graph the lines given by the equations in Part a.

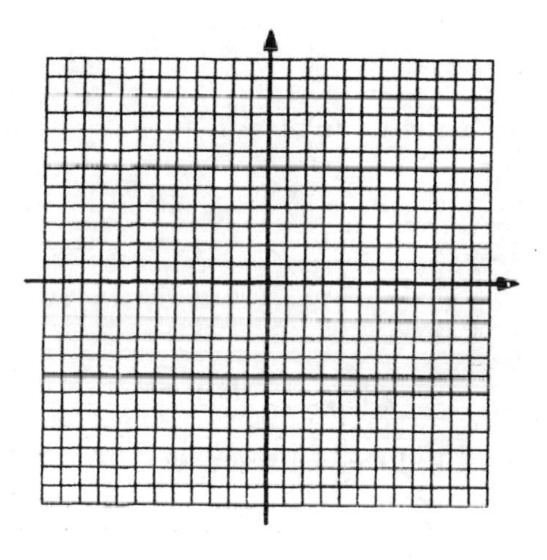

c. State the solution of the system.

d. Check the solution in each equation.

$2x + y = 14$ $\qquad$ $y = 3x - 6$

4. c. ____________

NAME ____________ DATE ____________ COURSE ____________

5. Solve by the substitution method. **5.** ______________

$2x + 3y = 1$ and $x - y = 8$

6. Check Problem 5. **6.** ______________

7. Solve by the addition method. **7.** ______________

$4x + 3y = 8$
$2x + y = 2$

8. Check Problem 7. **8.** ______________

9. Find the solution set. **9.** ______________

$x + y = 10$ and $x + y = 6$

10. Find the solution set. **10.** ______________

$2x + 2y = 20$ and $3x + 3y = 30$

Radicals and Quadratic Equations

ALGEBRA IN ACTION: SPACE

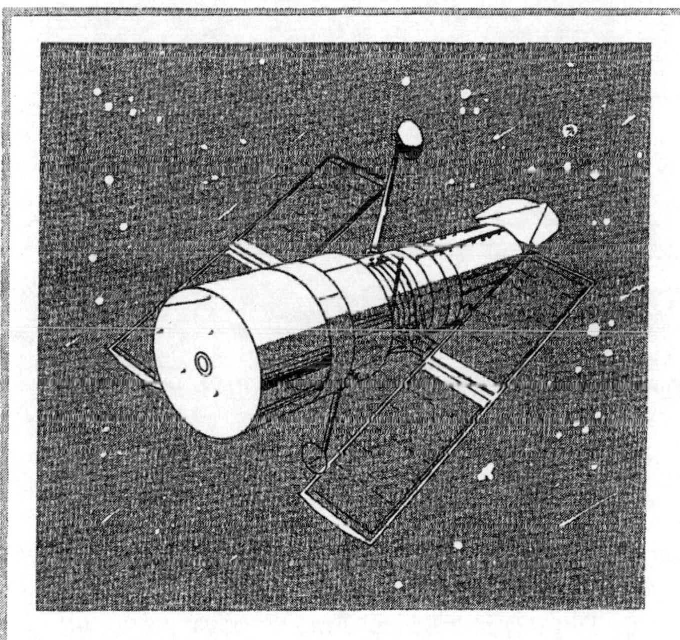

Earth-orbiting satellites help to predict the weather, monitor crop conditions, and allow communication from one part of the globe to another.

Quadratic equations can be used to determine positions of the satellites in their orbits.

Orbit of Satellite

$$\left(\frac{x+750}{5650}\right)^2 + \left(\frac{y}{5600}\right)^2 = 1$$

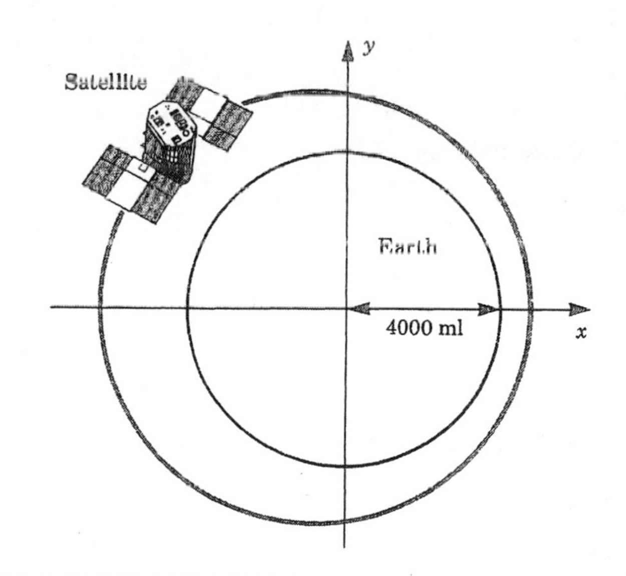

Chapter 9 Objectives

Objectives	Typical Problems
Simplify a square root radical by using a table.	1–4. Simplify. 1. $(\sqrt{6})^2$ 2. $\sqrt{(-6)^2}$ 3. $(-\sqrt{6})^2$ 4. $\sqrt{\frac{49}{64}}$
Approximate a square root radical.	5. Approximate to the nearest hundredth. $10 - \sqrt{30}$
Simplify a square root radical.	6–7. Simplify. 6. $\sqrt{500}$ 7. $\frac{30}{\sqrt{12}}$
Simplify square root radical expressions.	8–10. Simplify. 8. $\sqrt{50} - \sqrt{8}$ 9. $(5 - \sqrt{7})^2$ 10. $\frac{10 + \sqrt{12}}{4}$
Solve a quadratic equation by completing the square.	11. Solve by completing the square. $x^2 - 12x + 16 = 0$
Solve a quadratic equation by using the quadratic formula.	12. Solve by using the quadratic formula. $3x^2 + 2x - 3 = 0$
Solve an application by using a quadratic equation.	13. The distance, D, in feet, needed for a car to come to a stop safely is related to the speed, r, of the car by $20D = r^2 + 22r$ Find r for $D = 70$ ft.

ANSWERS to Typical Problems

1. 6 **2.** 6 **3.** 6 **4.** $\frac{7}{8}$ **5.** 4.52 **6.** $10\sqrt{5}$ **7.** $5\sqrt{3}$ **8.** $3\sqrt{2}$ **9.** $32 - 10\sqrt{7}$

10. $\frac{5 + \sqrt{3}}{2}$ **11.** $\{6 \pm 2\sqrt{5}\}$ **12.** $\left\{\frac{-1 \pm \sqrt{10}}{3}\right\}$ **13.** 28 m.p.h.

In Chapter 4, Section 4.10, we saw how factoring was used to solve certain quadratic equations. Recall that a **quadratic equation** in the variable x is an equation that can be written in the form

$$ax^2 + bx + c = 0 \qquad \text{where} \qquad a \neq 0$$

However, not all polynomials of the form $ax^2 + bx + c$ can be factored over the integers, even when the coefficients a, b, and c are integers. Therefore, we want to develop other methods for solving a quadratic equation.

We noted earlier that $x = \sqrt{2}$ is a solution of $x^2 = 2$, since $(\sqrt{2})^2 = 2$. Also, $\sqrt{5}$ is a solution of $x^2 = 5$, since $(\sqrt{5})^2 = 5$.

We see, then, that some quadratic equations have solutions that involve radicals such as $\sqrt{2}$ and $\sqrt{5}$. This means we must first study radicals in order to develop other methods for solving quadratic equations. After this, we shall see how to solve a quadratic equation by completing the square and by using the quadratic formula. The chapter ends with some applications that can be solved by using quadratic equations.

SQUARE ROOT RADICALS

By solving $x^2 = 25$ by the factoring method, it can be seen that 5 and -5 are the roots of this equation. Since $5^2 = 25$ and since $(-5)^2 = 25$, 5 and -5 are called the **square roots** of 25. Similarly, 4 and -4 are the **square roots** of 16, since $4^2 = 16$ and $(-4)^2 = 16$.

$$x^2 - 25 = 0$$
$$(x + 5)(x - 5) = 0$$
$$x + 5 = 0 \quad \text{or} \quad x - 5 = 0$$
$$x = -5 \quad \text{or} \quad x = 5$$

DEFINITION OF SQUARE ROOT

r is a square root of s if and only if $r^2 = s$.

Each positive real number has two square roots, two real numbers that have the same absolute value but opposite signs. The one and only one square root of 0 is 0. **Negative real numbers do not have real numbers for square roots,** since the product of two positive real numbers is positive and the product of two negative real numbers is also positive.

The **radical sign,** $\sqrt{}$, is used to indicate the positive real square root of a positive real number. Thus, $\sqrt{25} = +5$, the positive square root of 25. The negative square root of a positive real number is indicated by writing a minus sign before the radical. Thus, $-\sqrt{25} = -5$, the negative square root of 25.

The number under the radical sign is called the **radicand.** For example, 25 is the radicand of $\sqrt{25}$.

DEFINITION OF $\sqrt{s}$

If s is any positive real number, then $\sqrt{s}$ names the positive square root of s, and $(\sqrt{s})^2 = s$.

$$\sqrt{s} = r \text{ if and only if } s = r^2$$

The negative square root of s is indicated by $-\sqrt{s}$ and $(-\sqrt{s})^2 = s$.

For most practical purposes, an irrational number is approximated by a rational number expressed in decimal form. For example, using the Table of Squares and Square Roots on the inside cover of this book, $\sqrt{10}$, an irrational number, may be approximated by the rational number 3.162; that is, $\sqrt{10} \approx 3.162$, where $\approx$ is read "equals approximately."

EXERCISES 9.1

1–16. Simplify, if possible.

SAMPLE PROBLEMS 1–8	SOLUTIONS
1. $\sqrt{36}$	1. $\sqrt{36} = 6$
2. $\sqrt{(20)^2}$	2. $\sqrt{(20)^2} = 20$
3. $\sqrt{(-6)^2}$	3. $\sqrt{(-6)^2} = \sqrt{36} = 6$
4. $-\sqrt{9}$	4. $-\sqrt{9} = -3$
5. $\sqrt{-9}$	5. $\sqrt{-9}$ is not a real number.
6. $(\sqrt{3})^2$	6. $(\sqrt{3})^2 = 3$
7. $\sqrt{25}\sqrt{36}$	7. $\sqrt{25}\sqrt{36} = 5(6) = 30$
8. $\sqrt{0}\sqrt{2}$	8. $\sqrt{0}\sqrt{2} = 0 \cdot \sqrt{2} = 0$

1. $\sqrt{64}$

2. $\sqrt{81}$

3. $\sqrt{(11)^2}$

4. $\sqrt{(-10)^2}$

5. $(\sqrt{8})^2$

6. $(-\sqrt{7})^2$

7. $\sqrt{-4}$

8. $-\sqrt{16}$

9. $\sqrt{-16}$

10. $\sqrt{5}\sqrt{0}$

11. $\sqrt{25}\sqrt{4}$

12. $\dfrac{\sqrt{100}}{\sqrt{4}}$

13. $5 + \sqrt{9}$

14. $2 - \sqrt{36}$

15. $\dfrac{3 + \sqrt{49}}{2}$

16. $\dfrac{6 - \sqrt{121}}{3}$

1. ________ **2.** ________

3. ________ **4.** ________

5. ________ **6.** ________

7. ________ **8.** ________

9. ________ **10.** ________

11. ________ **12.** ________

13. ________ **14.** ________

15. ________ **16.** ________

17–22. By using a table or a hand-held calculator, approximate each of the following to the nearest hundredth.

SAMPLE PROBLEM 9

$\sqrt{38}$

SOLUTION

Table method

Locate 38 in Number column.
Find 6.164 in Square Root column.
$\sqrt{38} \approx 6.164$
$\sqrt{38} = 6.16$ to the nearest hundredth.

Calculator method

Enter 38 and press $\sqrt{x}$ key.
Round off display.
$\sqrt{38} = 6.16$ to the nearest hundredth.

SAMPLE PROBLEM 10

$\dfrac{6 - \sqrt{8}}{2}$

SOLUTION

Replace radical by a decimal approximation, correct to the nearest thousandth.

$$\frac{6 - \sqrt{8}}{2} \approx \frac{6 - 2.828}{2}$$

Do the indicated operations.

$$\approx \frac{3.172}{2}$$

$$\approx 1.586$$

Round off to the nearest hundredth.

$$\frac{6 - \sqrt{8}}{2} = 1.59 \text{ to the nearest hundredth.}$$

---

17. $\sqrt{20}$

18. $\sqrt{95}$

19. $2\sqrt{5}$

20. $\dfrac{\sqrt{10}}{2}$

21. $\dfrac{4 - \sqrt{26}}{2}$

22. $\dfrac{6 + 3\sqrt{2}}{3}$

17. ____________________

18. ____________________

19. ____________________

20. ____________________

21. ____________________

22. ____________________

23–28. Solve by using the square root property.

If $A^2 = B^2$, then $A = B$ or $A = -B$

	SAMPLE PROBLEM 11	SAMPLE PROBLEM 12
	$x^2 = 2$	$4x^2 = 5$
	SOLUTION	**SOLUTION**
Form $A^2 = B^2$.	$x^2 = (\sqrt{2})^2$	$(2x)^2 = (\sqrt{5})^2$
State $A = B$ or $A = -B$.	$x = \sqrt{2}$ or $x = -\sqrt{2}$	$2x = \sqrt{5}$ or $2x = -\sqrt{5}$
Solve each equation.		$x = \frac{\sqrt{5}}{2}$ or $x = \frac{-\sqrt{5}}{2}$
State the solution set.	The solution set is $\{\sqrt{2}, -\sqrt{2}\}$.	The solution set is $\left\{\frac{\sqrt{5}}{2}, \frac{-\sqrt{5}}{2}\right\}$.

SAMPLE PROBLEM 13

$0.01x^2 = 42.25$

SOLUTION

Multiply each side by 100 to remove the decimal.

$x^2 = 4225$

Using the table, locate 4225 in the Squares column, beside the number 65.

$x^2 = (65)^2$

$x = 65$ or $x = -65$

The solution set is $\{65, -65\}$.

23. $x^2 = 5$

24. $9x^2 = 40$

23. ______________

24. ______________

25. $0.5x^2 = 4.5$

26. $x^2 = 4900$

25. ______________

26. ______________

27. $0.01x^2 = 0.5625$

28. $100x^2 = 69$

27. ______________

28. ______________

EXERCISES 9.1S

1–16. Simplify, if possible. (See Sample Problems 1–8.)

1. $\sqrt{9}$

2. $-\sqrt{4}$

3. $\sqrt{(13)^2}$

4. $\sqrt{(-15)^2}$

5. $(\sqrt{15})^2$

6. $\sqrt{-36}$

7. $-\sqrt{36}$

8. $(-\sqrt{36})^2$

9. $\sqrt{64}\sqrt{4}$

10. $\dfrac{\sqrt{81}}{\sqrt{9}}$

11. $\sqrt{0}\sqrt{6}$

12. $\sqrt{5}\sqrt{0}\sqrt{7}$

13. $7 - \sqrt{16}$

14. $4 + \sqrt{25}$

15. $\dfrac{4 - \sqrt{36}}{8}$

16. $\dfrac{7 + \sqrt{64}}{3}$

1. __________
2. __________
3. __________
4. __________
5. __________
6. __________
7. __________
8. __________
9. __________
10. __________
11. __________
12. __________
13. __________
14. __________
15. __________
16. __________

NAME __________ DATE __________ COURSE __________

17–22. By using a table or a calculator, approximate each of the following to the nearest hundredth.
(See Sample Problems 9, 10.)

17. $\sqrt{56}$

18. $\sqrt{87}$

19. $3\sqrt{7}$

20. $\dfrac{\sqrt{15}}{3}$

21. $\dfrac{2 + \sqrt{45}}{2}$

22. $\dfrac{15 - 5\sqrt{3}}{10}$

17. ______________

18. ______________

19. ______________

20. ______________

21. ______________

22. ______________

23–28. Solve by factoring. (See Sample Problems 11–13.)

23. $x^2 = 7$

24. $25x^2 = 73$

23. ______________

24. ______________

25. $0.01x^2 = 16$

26. $0.06x^2 = 24$

25. ______________

26. ______________

27. $100x^2 = 529$

28. $10{,}000x^2 = 6724$

27. ______________

28. ______________

PRODUCTS OF RADICALS

Since $\sqrt{4}\sqrt{9} = 2 \cdot 3 = 6$ and $\sqrt{4 \cdot 9} = \sqrt{36} = 6$, it follows that $\sqrt{4}\sqrt{9} = \sqrt{4 \cdot 9}$.

Similarly, $\sqrt{16}\sqrt{25} = 4 \cdot 5 = 20$, and $\sqrt{16 \cdot 25} = \sqrt{400} = 20$, so $\sqrt{16}\sqrt{25} = \sqrt{16 \cdot 25}$.

In general, if two square root radicals have positive radicands, then the product of the radicals is equal to the positive square root of the product of the radicands.

PRODUCT OF RADICALS

$\sqrt{r}\sqrt{s} = \sqrt{rs}$**, where** $r \geq 0$ **and** $s \geq 0$**.**

This property and its symmetric form, $\sqrt{rs} = \sqrt{r}\sqrt{s}$, can be used to simplify certain radicals.

SAMPLE PROBLEM 1

Simplify $\sqrt{5}\sqrt{20}$.

SOLUTION

$\sqrt{5}\sqrt{20} = \sqrt{5 \cdot 20} = \sqrt{100} = 10$

SAMPLE PROBLEM 2

Simplify $\sqrt{3}\sqrt{12}$.

SOLUTION

$\sqrt{3}\sqrt{12} = \sqrt{3 \cdot 12} = \sqrt{36} = 6$

SAMPLE PROBLEM 3

Simplify $\sqrt{3025}$.

SOLUTION

Factor the radicand: $3025 = 5(605) = 5^2(121) = 5^2(11^2)$.

$\sqrt{3025} = \sqrt{5^2(11^2)} = \sqrt{5^2}\sqrt{11^2} = 5 \cdot 11 = 55$

SAMPLE PROBLEM 4

Simplify $\sqrt{784}$.

SOLUTION

$784 = 4(196) = 4^2(49) = 4^2 \cdot 7^2$

$\sqrt{784} = \sqrt{4^2}\sqrt{7^2} = 4 \cdot 7 = 28$

SAMPLE PROBLEM 5

Simplify $\sqrt{20}$.

SOLUTION

$20 = 4 \cdot 5 = 2^2 \cdot 5$

$\sqrt{20} = \sqrt{2^2 \cdot 5} = \sqrt{2^2}\sqrt{5} = 2\sqrt{5}$

Note in the last sample problem that the original radicand is not a product of perfect square factors. However, the radicand can be written as a product of a perfect square and a number that has no perfect square factors. By applying the product property to this form, the radical can then be simplified. Note the following useful axiom:

AXIOM

When r and s are positive, $\sqrt{r^2 s} = r\sqrt{s}$.

DEFINITION

If r and s are positive and if s has no perfect square factors, then $r\sqrt{s}$ is the simplified form of $\sqrt{r^2 s}$.

For example, using $\sqrt{r^2 s} = r\sqrt{s}$, then $\sqrt{9^2 \cdot 3} = 9\sqrt{3}$. $9\sqrt{3}$ is the simplified form of $\sqrt{9^2 \cdot 3}$.

SAMPLE PROBLEM 6

Simplify $\sqrt{108}$.

SOLUTION

Factor 108 in the form $r^2 s$.	$108 = 4(27) = 4(9)(3) = 36(3) = 6^2(3)$
Rewrite radical as $\sqrt{r^2 s}$.	$\sqrt{108} = \sqrt{6^2(3)}$
Use $\sqrt{r^2 s} = \sqrt{r^2}\sqrt{s}$.	$= \sqrt{6^2}\sqrt{3}$
Use $\sqrt{r^2 s} = r\sqrt{s}$.	$= 6\sqrt{3}$

SAMPLE PROBLEM 7

Simplify $\sqrt{10}\sqrt{80}$.

SOLUTION

Use $\sqrt{r}\sqrt{s} = \sqrt{rs}$.	$\sqrt{10}\sqrt{80} = \sqrt{800}$
Factor radicand as $r^2 s$.	$800 = 400(2) = (20)^2(2)$
Use $\sqrt{r^2 s} = r\sqrt{s}$.	$\sqrt{800} = \sqrt{(20)^2(2)} = 20\sqrt{2}$

SAMPLE PROBLEM 8
Simplify $\sqrt{6}\sqrt{75}$.

SOLUTION

Use $\sqrt{a}\sqrt{b} = \sqrt{ab}$.	$\sqrt{6}\sqrt{75} = \sqrt{6(75)}$
Factor radicand to find perfect square factors.	$= \sqrt{2(3)(3)(25)}$
	$= \sqrt{25}\sqrt{9}\sqrt{2}$
Use $\sqrt{r^2 s} = r\sqrt{s}$.	$= 5(3)\sqrt{2}$
	$= 15\sqrt{2}$

EXERCISES 9.2

Simplify.

1. $\sqrt{36}\sqrt{5}$

2. $\sqrt{9}\sqrt{6}$

1. ______________
2. ______________

3. $\sqrt{7}\sqrt{25}$

4. $\sqrt{3}\sqrt{49}$

3. ______________
4. ______________

5. $\sqrt{100}\sqrt{40}$

6. $\sqrt{7}\sqrt{28}$

5. ______________
6. ______________

7. $\sqrt{48}\sqrt{3}$

8. $\sqrt{2}\sqrt{50}$

7. ______________
8. ______________

9. $\sqrt{196}$

10. $\sqrt{1296}$

9. ______

10. ______

11. $\sqrt{7225}$

12. $\sqrt{12}$

11. ______

12. ______

13. $\sqrt{45}$

14. $\sqrt{600}$

13. ______

14. ______

15. $\sqrt{98}$

16. $\sqrt{8}$

15. ______

16. ______

17. $\sqrt{160}$

18. $\sqrt{30}\sqrt{6}$

17. ______

18. ______

19. $\sqrt{18}\sqrt{50}$

20. $\sqrt{2}\sqrt{6}\sqrt{15}$

19. ______

20. ______

EXERCISES 9.2 S

Simplify.

1. $\sqrt{49}\sqrt{17}$

2. $\sqrt{64}\sqrt{3}$

1. ______

2. ______

3. $\sqrt{10}\sqrt{81}$

4. $\sqrt{2}\sqrt{16}$

3. ______

4. ______

5. $\sqrt{4}\sqrt{20}$

6. $\sqrt{2}\sqrt{18}$

5. ______

6. ______

7. $\sqrt{6}\sqrt{24}$

8. $\sqrt{5}\sqrt{45}$

7. ______

8. ______

9. $\sqrt{324}$

10. $\sqrt{1936}$

9. ______

10. ______

NAME ______ DATE ______ COURSE ______

11. $\sqrt{9801}$

12. $\sqrt{28}$

11. ______

12. ______

13. $\sqrt{54}$

14. $\sqrt{75}$

13. ______

14. ______

15. $\sqrt{180}$

16. $\sqrt{27}$

15. ______

16. ______

17. $\sqrt{192}$

18. $\sqrt{7}\sqrt{14}$

17. ______

18. ______

19. $\sqrt{640}\sqrt{50}$

20. $\sqrt{3}\sqrt{15}\sqrt{35}$

19. ______

20. ______

QUOTIENTS OF RADICALS

Since $\frac{\sqrt{100}}{\sqrt{4}} = \frac{10}{2} = 5$ and since $\sqrt{\frac{100}{4}} = \sqrt{25} = 5$, it follows that $\frac{\sqrt{100}}{\sqrt{4}} = \sqrt{\frac{100}{4}}$ and $\sqrt{\frac{100}{4}} = \frac{\sqrt{100}}{\sqrt{4}}$.

Similarly, $\frac{\sqrt{49}}{\sqrt{64}} = \frac{7}{8}$ and $\sqrt{\frac{49}{64}} = \frac{7}{8}$.

Thus $\frac{\sqrt{49}}{\sqrt{64}} = \sqrt{\frac{49}{64}}$ and $\sqrt{\frac{49}{64}} = \frac{\sqrt{49}}{\sqrt{64}}$.

This property is true in general for radicals having positive radicands.

QUOTIENT OF RADICALS

For r and s positive, $\frac{\sqrt{r}}{\sqrt{s}} = \sqrt{\frac{r}{s}}$ and $\sqrt{\frac{r}{s}} = \frac{\sqrt{r}}{\sqrt{s}}$.

An expression is not in simplified form if it contains a radical in a denominator or if a fraction occurs in a radicand.

To simplify an expression having the form $\sqrt{\frac{r}{s}}$ where r and s are natural numbers:

1. Multiply r and s by the same number so the new denominator is a perfect square. (Use the smallest number possible.)
2. Use the quotient of radicals property.
3. Simplify the denominator.
4. Simplify the numerator and the fraction, if possible.

SAMPLE PROBLEM 1

Simplify $\sqrt{\frac{1}{5}}$.

SOLUTION

Multiply numerator and denominator by 5 to obtain the perfect square 25 in the denominator.

$$\sqrt{\frac{1}{5}} = \sqrt{\frac{1(5)}{5(5)}} = \sqrt{\frac{5}{25}}$$

Apply the quotient of radicals property.

$$= \frac{\sqrt{5}}{\sqrt{25}}$$

Simplify the denominator.

$$= \frac{\sqrt{5}}{5}$$

SAMPLE PROBLEM 2

Simplify $\sqrt{\frac{3}{8}}$.

SOLUTION

While multiplying the denominator by 8 would produce the perfect square 64, a smaller perfect square 16 can be obtained by multiplying the denominator by 2.

Finding the smallest perfect square denominator simplifies the work that follows.

$$\sqrt{\frac{3}{8}} = \sqrt{\frac{3(2)}{8(2)}} = \sqrt{\frac{6}{16}} = \frac{\sqrt{6}}{\sqrt{16}} = \frac{\sqrt{6}}{4}$$

SAMPLE PROBLEM 3

Simplify $\frac{1}{\sqrt{20}}$.

SOLUTION (Alternate Method)

1. Simplify radical. $\frac{1}{\sqrt{20}} = \frac{1}{2\sqrt{5}}$
2. Multiply numerator and denominator by resulting radical. $= \frac{1\sqrt{5}}{2\sqrt{5}\sqrt{5}}$
3. Simplify. $= \frac{\sqrt{5}}{2 \cdot 5}$

$= \frac{\sqrt{5}}{10}$

SAMPLE PROBLEM 4

Simplify $\frac{2}{\sqrt{14}}$.

SOLUTION (Alternate Method)

1. $\sqrt{14}$ is simplified.
2. $\frac{2}{\sqrt{14}} = \frac{2\sqrt{14}}{\sqrt{14}\sqrt{14}}$

$= \frac{2\sqrt{14}}{14}$

$= \frac{\sqrt{14}}{7}$

Note that a resulting fraction should always be simplified, whenever possible.

SAMPLE PROBLEM 5

Simplify $\frac{\sqrt{3}}{\sqrt{15}}$.

SOLUTION

$$\frac{\sqrt{3}}{\sqrt{15}} = \sqrt{\frac{3}{15}} = \sqrt{\frac{1}{5}} \text{ (Simplify a fraction whenever possible.)}$$

$$= \sqrt{\frac{5}{25}} = \frac{\sqrt{5}}{\sqrt{25}} = \frac{\sqrt{5}}{5}$$

EXERCISES 9.3

Simplify.

1. $\sqrt{\frac{1}{2}}$

2. $\sqrt{\frac{2}{7}}$

3. $\sqrt{\frac{1}{27}}$

4. $\sqrt{\frac{4}{75}}$

5. $\frac{1}{\sqrt{6}}$

6. $\frac{2}{\sqrt{28}}$

7. $\frac{1}{\sqrt{24}}$

8. $\frac{\sqrt{2}}{\sqrt{28}}$

1. ____________________
2. ____________________
3. ____________________
4. ____________________
5. ____________________
6. ____________________
7. ____________________
8. ____________________

9. $\frac{\sqrt{162}}{\sqrt{6}}$

10. $\sqrt{\frac{81}{100}}$

9. ______________________

10. ______________________

11. $\sqrt{0.56}$

12. $\sqrt{0.012}$

11. ______________________

12. ______________________

13. $\frac{\sqrt{5}}{\sqrt{10}}$

14. $\frac{3\sqrt{7}}{\sqrt{6}}$

13. ______________________

14. ______________________

15. $\sqrt{\frac{75}{4}}$

16. $\sqrt{0.01}$

15. ______________________

16. ______________________

EXERCISES 9.3 S

Simplify.

1. $\sqrt{\frac{1}{3}}$

2. $\sqrt{\frac{5}{6}}$

1. ______

2. ______

3. $\sqrt{\frac{1}{50}}$

4. $\sqrt{\frac{9}{32}}$

3. ______

4. ______

5. $\frac{2}{\sqrt{10}}$

6. $\frac{7}{\sqrt{14}}$

5. ______

6. ______

7. $\frac{1}{\sqrt{56}}$

8. $\frac{\sqrt{6}}{\sqrt{108}}$

7. ______

8. ______

NAME ______ DATE ______ COURSE ______

9. $\sqrt{\frac{49}{1000}}$

10. $\sqrt{0.042}$

9. ____________________

10. ____________________

11. $\frac{\sqrt{5}}{\sqrt{8}}$

12. $\sqrt{2\frac{1}{4}}$

11. ____________________

12. ____________________

13. $\frac{\sqrt{12}}{\sqrt{18}}$

14. $\frac{5\sqrt{2}}{\sqrt{20}}$

13. ____________________

14. ____________________

15. $\sqrt{\frac{32}{9}}$

16. $\frac{1}{\sqrt{5}}$

15. ____________________

16. ____________________

RADICALS: SUMS AND DIFFERENCES 9.4

Square root radicals having the same radicand are called **like radicals.**

As examples, $3\sqrt{5}$ and $4\sqrt{5}$ are like radicals, and $\sqrt{7}$ and $2\sqrt{7}$ are like radicals.

On the other hand, $4\sqrt{3}$ and $4\sqrt{5}$ are unlike radicals, since the radicands are different.

Like radicals can be combined by using the distributive axiom for real numbers. Some unlike radicals may become like radicals after simplification, and then these can be combined.

SAMPLE PROBLEM 1

Simplify $\sqrt{12} + \sqrt{75}$

SOLUTION

Simplify each radical. $\sqrt{12} = \sqrt{4}\sqrt{3} = 2\sqrt{3}$ and $\sqrt{75} = \sqrt{25}\sqrt{3} = 5\sqrt{3}$

$\sqrt{12} + \sqrt{75} = 2\sqrt{3} + 5\sqrt{3}$

Use the distributive axiom. $= (2 + 5)\sqrt{3}$

Simplify. $= 7\sqrt{3}$

SAMPLE PROBLEM 2

Simplify $\sqrt{36} + \sqrt{24}$

SOLUTION

Simplify each radicand. $\sqrt{36} = 6$ and $\sqrt{24} = \sqrt{4}\sqrt{6} = 2\sqrt{6}$

Rewrite sum. $\sqrt{36} + \sqrt{24} = 6 + 2\sqrt{6}$

No further simplification is possible.

SAMPLE PROBLEM 3

Simplify $\sqrt{45} - \sqrt{20}$

SOLUTION

Simplify radicals. $\sqrt{45} - \sqrt{20} = \sqrt{9}\sqrt{5} - \sqrt{4}\sqrt{5}$

$= 3\sqrt{5} - 2\sqrt{5}$

Use distributive axiom. $= (3 - 2)\sqrt{5}$

Simplify. $= 1\sqrt{5}$

$= \sqrt{5}$

SAMPLE PROBLEM 4

Simplify $\sqrt{63} + \sqrt{147}$

SOLUTION

Simplify radicals.

$$\sqrt{63} + \sqrt{147} = \sqrt{9}\sqrt{7} + \sqrt{49}\sqrt{3} = 3\sqrt{7} + 7\sqrt{3}$$

No further simplification is possible.

The forms $\frac{r + \sqrt{s}}{t}$ and $\frac{r - \sqrt{s}}{t}$ occur in the general solution of the quadratic equation. Therefore, it is useful to learn how to simplify these forms.

SAMPLE PROBLEM 5

Simplify $\frac{6 + \sqrt{45}}{3}$

SOLUTION

First, simplify the radical: $\sqrt{45} = \sqrt{9}\sqrt{5} = 3\sqrt{5}$.

Then, simplify the resulting fraction, if possible, by finding a **common factor** of the denominator and **each** term of the numerator.

NOTE TO STUDENT: This factoring step is very important. A common error is to write $\frac{6 + 3\sqrt{5}}{3}$ as $2 + 3\sqrt{5}$. Factoring first will help to avoid this error.

1. Simplify radical. $\frac{6 + \sqrt{45}}{3} = \frac{6 + 3\sqrt{5}}{3}$
2. Factor numerator and denominator. $= \frac{3(2 + \sqrt{5})}{3}$
3. Line out common **factors.** $= \frac{2 + \sqrt{5}}{1}$
4. Simplify. $= 2 + \sqrt{5}$

SAMPLE PROBLEM 6

Simplify $\frac{28 - \sqrt{800}}{12}$

SOLUTION

1. Simplify radical. $\frac{28 - \sqrt{800}}{12} = \frac{28 - \sqrt{400}\sqrt{2}}{12}$

   $= \frac{28 - 20\sqrt{2}}{12}$
2. Factor numerator and denominator. $= \frac{4(7 - 5\sqrt{2})}{4(3)}$
3. Simplify. $= \frac{7 - 5\sqrt{2}}{3}$

EXERCISES 9.4

Simplify.

1. $\sqrt{45} + \sqrt{80}$

2. $\sqrt{98} - \sqrt{32}$

3. $\sqrt{90} - \sqrt{40}$

4. $\sqrt{72} + \sqrt{75}$

5. $\sqrt{49} + \sqrt{28}$

6. $\sqrt{6} - \sqrt{36}$

7. $\dfrac{15 + \sqrt{50}}{10}$

8. $\dfrac{10 - \sqrt{12}}{5}$

9. $\dfrac{4 - \sqrt{56}}{4}$

10. $\dfrac{7 + \sqrt{81}}{24}$

11. $\dfrac{12 - \sqrt{8}}{6}$

12. $\dfrac{6 - \sqrt{20}}{4}$

1. ____________________

2. ____________________

3. ____________________

4. ____________________

5. ____________________

6. ____________________

7. ____________________

8. ____________________

9. ____________________

10. ____________________

11. ____________________

12. ____________________

13–16. Evaluate the given expressions for the given value of x.

SAMPLE PROBLEM 7

a. x^2 b. $10x - 18$ for $x = 5 - \sqrt{7}$

SOLUTION

a. $x^2 = (5 - \sqrt{7})^2 = 5^2 - 2(5)\sqrt{7} + (\sqrt{7})^2$
$= 25 - 10\sqrt{7} + 7 = \mathbf{32 - 10\sqrt{7}}$

b. $10x - 18 = 10(5 - \sqrt{7}) - 18$
$= 50 - 10\sqrt{7} - 18 = \mathbf{32 - 10\sqrt{7}}$

Note that $x^2 = 10x - 18$ for $x = 5 - \sqrt{7}$.
Therefore, $5 - \sqrt{7}$ is a solution of $x^2 = 10x - 18$.

13. For $x = 3 - \sqrt{6}$
a. x^2 **b.** $6x - 3$

13. a. ______________

b. ______________

14. For $x = 4 + \sqrt{5}$
a. x^2 **b.** $8x - 11$

14. a. ______________

b. ______________

15. For $x = -2 + \sqrt{7}$
a. x^2 **b.** $4x - 3$

15. a. ______________

b. ______________

16. For $x = -2 - \sqrt{3}$
a. x^2 **b.** $4x + 1$

16. a. ______________

b. ______________

1–12. Simplify.

1. $\sqrt{75} + \sqrt{12}$

2. $\sqrt{54} - \sqrt{24}$

3. $\sqrt{40} - \sqrt{50}$

4. $\sqrt{12} + \sqrt{32}$

5. $\sqrt{108} - \sqrt{81}$

6. $5 + \sqrt{20}$

7. $\dfrac{2 + \sqrt{20}}{4}$

8. $\dfrac{6 - \sqrt{54}}{15}$

9. $\dfrac{14 + \sqrt{60}}{2}$

10. $\dfrac{3 - \sqrt{121}}{16}$

11. $\dfrac{2 - \sqrt{12}}{4}$

12. $\dfrac{28 - \sqrt{300}}{14}$

1. ______________

2. ______________

3. ______________

4. ______________

5. ______________

6. ______________

7. ______________

8. ______________

9. ______________

10. ______________

11. ______________

12. ______________

NAME ______________ DATE ______________ COURSE ______________

13–18. Evaluate the given expressions for the given value of x.

13. For $x = 3 + \sqrt{2}$

a. x^2

b. $6x - 7$

13. a. ______________

b. ______________

14. For $x = 2 + \sqrt{5}$

a. x^2

b. $4x + 1$

14. a. ______________

b. ______________

15. For $x = 5 - \sqrt{10}$

a. x^2

b. $10x - 15$

15. a. ______________

b. ______________

16. For $x = 2 - \sqrt{7}$

a. x^2

b. $4x + 3$

16. a. ______________

b. ______________

17. For $x = -4 + \sqrt{6}$

a. x^2

b. $-8x - 10$

17. a. ______________

b. ______________

18. For $x = -3 - \sqrt{15}$

a. x^2

b. $6 - 6x$

18. a. ______________

b. ______________

19. Show that $4 + \sqrt{7}$ and $4 - \sqrt{7}$ are solutions of $x^2 = 8x - 9$.

19. ______________

20. Show that $-2 + \sqrt{6}$ and $-2 - \sqrt{6}$ are solutions of $x^2 + 4x - 2 = 0$.

20. ______________

9.5 COMPLETING THE SQUARE

By the factoring method, it can be seen that if $x^2 = 2$, then $x = \sqrt{2}$ or $x = -\sqrt{2}$. Similarly, if $x^2 = 5$, then $x = \sqrt{5}$ or $x = -\sqrt{5}$. This leads to the following general statement.

SQUARE ROOT PROPERTY

For r, any positive real number, and for X, any polynomial,

$$\text{if } X^2 = r, \text{ then } X = \sqrt{r} \text{ or } X = -\sqrt{r}$$

To solve a general quadratic equation by this method, it is necessary to rewrite the given equation in the form $(rx + s)^2 = t$. This involves a process called "completing the square."

In Chapter 4, the perfect square trinomial formula was studied. This formula is used in solving a quadratic equation by completing the square.

PERFECT SQUARE TRINOMIAL FORMULA

$$X^2 + 2AX + A^2 = (X + A)^2$$

COMPLETING THE SQUARE SOLUTION METHOD

To solve $X^2 + 2AX + C = 0$:

1. **Rewrite the equation as $X^2 + 2AX = -C$.**
2. **Add A^2 (the square of one half the coefficient of X) to each side to form the equation $X^2 + 2AX + A^2 = A^2 - C$.**
3. **Rewrite the left side as the square of a binomial and simplify the right side.**
4. **Solve the resulting equation by using the square root property illustrated in Sample Problems 1 and 2.**

EXERCISES 9.5

Solve by completing the square, when necessary.

SAMPLE PROBLEM 1

$(x - 5)^2 = 3$

SOLUTION

Use the square root property.	$x - 5 = \sqrt{3}$	or	$x - 5 = -\sqrt{3}$	**$x - 5$ equals the positive or the negative square root of 3.**
Solve each equation.	$x = 5 + \sqrt{3}$	or	$x = 5 - \sqrt{3}$	

SAMPLE PROBLEM 2

$(3x + 2)^2 = 7$

SOLUTION

1. Use the square root property.
2. Solve each equation.

$$3x + 2 = \sqrt{7} \quad \text{or} \quad 3x + 2 = -\sqrt{7}$$
$$3x = -2 + \sqrt{7} \qquad 3x = -2 - \sqrt{7}$$
$$x = \frac{-2 + \sqrt{7}}{3} \qquad x = \frac{-2 - \sqrt{7}}{3}$$

$3x + 2$ equals the positive or the negative square root of 7.

1. $(x - 4)^2 = 5$

2. $(x + 2)^2 = 3$

1. ______________

2. ______________

3. $(x - 8)^2 = 12$

4. $(2x + 3)^2 = 10$

3. ______________

4. ______________

5. $(5x - 1)^2 = 20$

6. $(x + 1)^2 = 2$

5. ______________

6. ______________

SAMPLE PROBLEM 3

$x^2 + 6x + 7 = 0$

SOLUTION

1. Variables on the left side, constant on the right side. $\quad x^2 + 6x = -7$
2. Add $\left(\frac{\text{coefficient of } x}{2}\right)^2$ to each side. $\quad x^2 + 6x + 3^2 = -7 + 3^2 \qquad \frac{6}{2} = 3$
3. Write left side as the square of a binomial. Simplify right side. $\quad (x + 3)^2 = 2 \qquad x^2 + 6x + 9 = (x + 3)^2$
4. Take square roots of each side. Solve for x.

$$x + 3 = +\sqrt{2} \quad \text{or} \quad x + 3 = -\sqrt{2}$$
$$x = -3 + \sqrt{2} \qquad x = -3 - \sqrt{2}$$

We also write $x = -3 \pm \sqrt{2}$ for the two solutions.

SAMPLE PROBLEM 4

$x^2 - 12x - 14 = 0$

SOLUTION

1. Rewrite as $X^2 + 2AX = -C$. $\quad x^2 - 12x = 14 \qquad \frac{-12}{2} = -6$
2. Add A^2 to each side. $\quad x^2 - 12x + (-6)^2 = 14 + 36$
3. Write left side as square of a binomial. Simplify right side. $\quad (x - 6)^2 = 50 \qquad x^2 - 12x + 36 = (x - 6)^2$
4. Solve, using the square root property of this section.

$$x - 6 = +\sqrt{50} \quad \text{or} \quad x - 6 = -\sqrt{50}$$
$$x = 6 + \sqrt{50} \qquad x = 6 - \sqrt{50}$$
$$x = 6 + 5\sqrt{2} \qquad x = 6 - 5\sqrt{2}$$

We also write $x = 6 \pm 5\sqrt{2}$.

Note that the radical is left in simplified form.

7. $x^2 - 6x + 3 = 0$ **8.** $x^2 - 8x - 34 = 0$

9. $x^2 + 4x = 6$ **10.** $x^2 - 10x = -18$

11. $x^2 - 2x - 7 = 0$ **12.** $x^2 + 12x + 24 = 0$

7. ______

8. ______

9. ______

10. ______

11. ______

12. ______

SAMPLE PROBLEM 5

$2x^2 = 12x - 13$

SOLUTION

1. Divide each term by the coefficient of x^2. $\quad x^2 = 6x - \frac{13}{2}$
2. Write variable terms on left side, constant term on right side. $\quad x^2 - 6x = \frac{-13}{2}$
3. Add $\left(\frac{\text{coefficient of } x}{2}\right)^2$ to each side. $\quad x^2 - 6x + \left(\frac{-6}{2}\right)^2 = \frac{-13}{2} + 9$
4. Simplify each side. $\quad x^2 - 6x + 9 = \frac{5}{2}$
5. Write left side as the square of a binomial. $\quad (x-3)^2 = \frac{5}{2}$
6. Take square root of each side. $\quad x - 3 = \pm\sqrt{\frac{5}{2}}$
7. Simplify radical. $\quad x - 3 = \pm\frac{\sqrt{5}\sqrt{2}}{\sqrt{2}\sqrt{2}} = \frac{\pm\sqrt{10}}{2}$
8. Solve for x. $\quad x = 3 \pm \frac{\sqrt{10}}{2}$ and $x = \mathbf{\frac{6 \pm \sqrt{10}}{2}}$

Check for $x = \frac{6+\sqrt{10}}{2}$.

$$2x^2 = 2\left(\frac{6+\sqrt{10}}{2}\right)^2 = 2\left(\frac{36 + 12\sqrt{10} + 10}{4}\right) = \frac{46 + 12\sqrt{10}}{2} = \mathbf{23 + 6\sqrt{10}}$$

$$12x - 13 = 12\left(\frac{6+\sqrt{10}}{2}\right) - 13 = 6(6 + \sqrt{10}) - 13 = \mathbf{23 + 6\sqrt{10}}$$

13. $3x^2 = 12x - 11$

14. $5x^2 + 10x + 3 = 0$

15. $4x^2 - 24x + 9 = 0$

16. $2x^2 = 16x - 31$

17. $4x^2 = 1 - 12x$

18. $3x^2 = 2x + 21$

19. $16x^2 + 25 = 40x$

20. $20x^2 + 20x = 80$

13. ______________

14. ______________

15. ______________

16. ______________

17. ______________

18. ______________

19. ______________

20. ______________

Solve by completing the square.

1. $(x + 7)^2 = 6$

2. $(x - 5)^2 = 7$

1. ______________

2. ______________

3. $(x + 9)^2 = 18$

4. $(3x - 4)^2 = 14$

3. ______________

4. ______________

5. $(6x + 1)^2 = 32$

6. $x^2 - 4x - 1 = 0$

5. ______________

6. ______________

7. $x^2 + 10x + 15 = 0$

8. $x^2 + 12x + 8 = 0$

7. ______________

8. ______________

NAME ______________ DATE ______________ COURSE ______________

9. $x^2 = 2x + 11$

10. $x^2 = 6x + 36$

9. ____________

10. ____________

11. $x^2 + 14x + 31 = 0$

12. $x^2 - 20x + 50 = 0$

11. ____________

12. ____________

13. $x^2 = 50 - 4x$

14. $x^2 = 4 - 2x$

13. ____________

14. ____________

15. $x^2 = 10x + 15$

16. $x^2 = 12x + 36$

15. ____________

16. ____________

17. $2x^2 - 2x = 3$

18. $4x^2 + 12x - 7 = 0$

17. ____________

18. ____________

19. $9x^2 + 36x + 29 = 0$

20. $4x^2 - 24x + 33 = 0$

19. ____________

20. ____________

QUADRATIC FORMULA 9.6

There is a formula that can be used to solve a quadratic equation by applying the method of completing the square to the general quadratic equation $ax^2 + bx + c = 0$, where $a \neq 0$.

The quadratic formula method of solution is most useful when the quadratic equation is not easily factored and when completing the square involves fractions.

Solving $ax^2 + bx + c = 0$,

1. Divide each term by a. $\quad x^2 + \frac{b}{a}x + \frac{c}{a} = 0$

2. Subtract $\frac{c}{a}$ from each side. $\quad x^2 + \frac{b}{a}x = \frac{-c}{a}$

3. Complete the square. $\quad x^2 + \frac{b}{a}x + \left(\frac{b}{2a}\right)^2 = \frac{-c}{a} + \frac{b^2}{4a^2}$ $\qquad \frac{-c(4a)}{a(4a)} = \frac{-4ac}{4a^2}$

4. Form the square. Simplify the right side. $\quad \left(x + \frac{b}{2a}\right)^2 = \frac{b^2 - 4ac}{4a^2}$

5. If $b^2 - 4ac \geq 0$, take the square root of each side. $\quad x + \frac{b}{2a} = \frac{\pm\sqrt{b^2-4ac}}{2a}$

6. Solve for x. $\quad x = \frac{-b \pm \sqrt{b^2-4ac}}{2a}$

Since a, b, and c were chosen completely arbitrarily, except $a \neq 0$ and $b^2 - 4ac \geq 0$, the set of equations

$$\left\{x = \frac{-b + \sqrt{b^2-4ac}}{2a},\ x = \frac{-b - \sqrt{b^2-4ac}}{2a}\right\}$$

can be used as a formula for the solution of a quadratic equation.

QUADRATIC FORMULA

If $ax^2 + bx + c = 0$, where $a \neq 0$, then

$$x = \frac{-b + \sqrt{b^2-4ac}}{2a} \quad \text{or} \quad x = \frac{-b - \sqrt{b^2-4ac}}{2a}$$

The quadratic formula is often condensed and written as follows:

$$x = \frac{-b \pm \sqrt{b^2-4ac}}{2a}$$

TO SOLVE A QUADRATIC EQUATION BY USING THE QUADRATIC FORMULA

1. **Write the equation in standard form.**
2. **Determine the coefficients a, b, and c.**
3. **In the formula replace a, b, and c by their numerical values.**
4. **Simplify the radicand.**
5. **Simplify the radical.**
6. **Simplify the fraction.**

SAMPLE PROBLEM 1

Solve by using the quadratic formula.

$3x^2 + 6x + 2 = 0$

SOLUTION

1. Write as $ax^2 + bx + c = 0$. $\quad 3x^2 + 6x + 2 = 0$
2. State the values of a, b, and c. $\quad a = 3,\ b = 6,\ c = 2$
3. Substitute into the quadratic formula. $\quad x = \dfrac{-b \pm \sqrt{b^2-4ac}}{2a}$
4. Simplify $b^2 - 4ac$ $\quad x = \dfrac{-6 \pm \sqrt{36 - 4(3)(2)}}{2(3)}$

$$= \frac{-6 \pm \sqrt{36 - 24}}{6}$$

5. Simplify the radical. $\quad = \dfrac{-6 \pm \sqrt{12}}{6}$
6. Simplify the fraction. $\quad = \dfrac{-6 \pm 2\sqrt{3}}{6} = \dfrac{\not{2}(-3 \pm \sqrt{3})}{\not{2}(3)}$

$$= \frac{-3 \pm \sqrt{3}}{3}$$

Note that the lined out factor, 2, is a factor of each term of the numerator.

SAMPLE PROBLEM 2

Solve by using the quadratic formula.

$5x^2 = 3x + 2$

SOLUTION

1. $5x^2 - 3x - 2 = 0$
2. $a = 5$, $\mathbf{b = -3}$, $c = -2$
3. $x = \dfrac{-b \pm \sqrt{b^2 - 4ac}}{2a}$

   $x = \dfrac{\mathbf{-(-3)} \pm \sqrt{9 - 4(5)(-2)}}{2(5)}$
4. $= \dfrac{\mathbf{+3} \pm \sqrt{\mathbf{9 - (-40)}}}{10}$ Note 1
5. $= \dfrac{3 \pm \sqrt{9 \mathbf{+ 40}}}{10}$

   $= \dfrac{3 \pm \sqrt{49}}{10}$
6. $x = \dfrac{3 + 7}{10}$ or $x = \dfrac{3 - 7}{10}$ Note 2

   $\mathbf{x = 1}$ $\qquad x = \mathbf{\dfrac{-2}{5}}$

SAMPLE PROBLEM 3

Solve by using the quadratic formula.

$x^2 + 23 = 10x$

SOLUTION

1. $x^2 - 10x + 23 = 0$
2. $a = 1$, $\mathbf{b = -10}$, $c = 23$
3. $x = \dfrac{-b \pm \sqrt{b^2 - 4ac}}{2a}$

   $x = \dfrac{\mathbf{-(-10)} \pm \sqrt{100 - 4(1)(23)}}{2(1)}$ Note 3
4. $= \dfrac{\mathbf{10} \pm \sqrt{100 - 92}}{2}$
5. $= \dfrac{10 \pm \sqrt{8}}{2}$
6. $= \dfrac{10 \pm 2\sqrt{2}}{2}$

   $= \dfrac{2(5 \pm \sqrt{2})}{2}$

   $= \mathbf{5 \pm \sqrt{2}}$

Note 1. When the constant term, c, is negative, it is important to be very careful when evaluating $b^2 - 4ac$, the quantity under the radical sign. This is a common source of errors.

Note 2. When the solutions are rational, they are always left in their simplified rational form.

Note 3. It is important to note that when b is negative, then $-b$ in the formula is positive; in this case, $-(-10) = +10$.

EXERCISES 9.6

Solve by using the quadratic formula.

1. $3x^2 + 5x + 1 = 0$

2. $x^2 = 3x + 1$

3. $x^2 = 2(3x - 2)$

4. $3x^2 = 2 - 2x$

1. ______________

2. ______________

3. ______________

4. ______________

5. $5x^2 + 9x = 18$

6. $x^2 - 80x + 1200 = 0$

5. ______________________

6. ______________________

7. $x^2 + x = 1$

8. $x^2 + 4x = 2$

7. ______________________

8. ______________________

9. $x(x - 2) = 1$

10. $16x^2 - 40x + 25 = 0$

9. ______________________

10. ______________________

11. $4x(x - 1) = 35$

12. $2x^2 = x + 1$

11. ______________________

12. ______________________

13. $x^2 + 2bx + c = 0$ (solve for x)

13. ______________________

14. $2x^2 - 5xy + 2y^2 = 0$ (solve for x)

14. ______________________

EXERCISES 9.6 S

Solve by using the quadratic formula.

1. $3x^2 + 9x + 5 = 0$

2. $9x^2 = 12x + 1$

1. ______

2. ______

3. $2x^2 = 4 - 5x$

4. $7x = 2x^2 + 3$

3. ______

4. ______

5. $10x^2 = 11x + 18$

6. $x^2 + 170x - 6000 = 0$

5. ______

6. ______

7. $x^2 = x + 1$

8. $x^2 - 10x = 3$

7. ______

8. ______

NAME ______ DATE ______ COURSE ______

9. $x^2 = 12(x + 1)$

10. $36x^2 + 84x + 49 = 0$

9. ____________

10. ____________

11. $3x(3x - 2) = 1$

12. $x^2 = 20(x + 240)$

11. ____________

12. ____________

13. $x^2 - 2rx - s = 0$ (solve for x)

13. ____________

14. $y^2 + 2xy - 15x^2 = 0$ (solve for y)

14. ____________

APPLICATIONS 9.7

We already saw some applications of quadratic equations in Chapter 4, Section 4.11. More applications are shown in this chapter.

GUIDELINES FOR SOLVING QUADRATIC EQUATIONS

1. **First, simplify the equation by dividing each side by the GCMF, the greatest common monomial factor of each term.**
2. **Use the factoring method when the factors can be found easily.**
3. **Solve by completing the square when the coefficient of the square term is 1 and the coefficient of the variable is even.**
4. **Use the quadratic formula for all other cases.**

SAMPLE PROBLEM 1

The weight, W, in pounds, of an astronaut who is x thousand mi above the surface of the earth is given by

$$W(x + 4)^2 = 2304$$

where 4000 mi is used as the radius of the earth.

On the surface of the earth, $x = 0$ and $W(16) = 2304$ and $W = 144$ lb.

How far above the surface of the earth is this astronaut when his weight is 24 lb?

SOLUTION

$24(x + 4)^2 = 2304$

$(x + 4)^2 = 96$ Simplify by dividing each side by 24.

$x + 4 = \sqrt{96}$ or $x + 4 = -\sqrt{96}$

$x = -4 + \sqrt{96}$ or $x = -4 - \sqrt{96}$

Approximating,

$x = -4 + 9.80$ or $x = -4 - 9.80$

$x = 5.80$ Reject, since distance is not negative.

The astronaut is approximately 5800 mi above the surface of the earth.

SAMPLE PROBLEM 2

The orbit of an artificial satellite of the earth is given by

$$\left(\frac{x + 750}{5650}\right)^2 + \left(\frac{y}{5600}\right)^2 = 1$$

where x and y are measured in miles.

The center of the earth is at the origin, and the radius of the earth is taken as 4000 mi.

Find the position (coordinates) of the satellite for

a. $y = 0$ b. $x = -750$ c. $y = 3360$

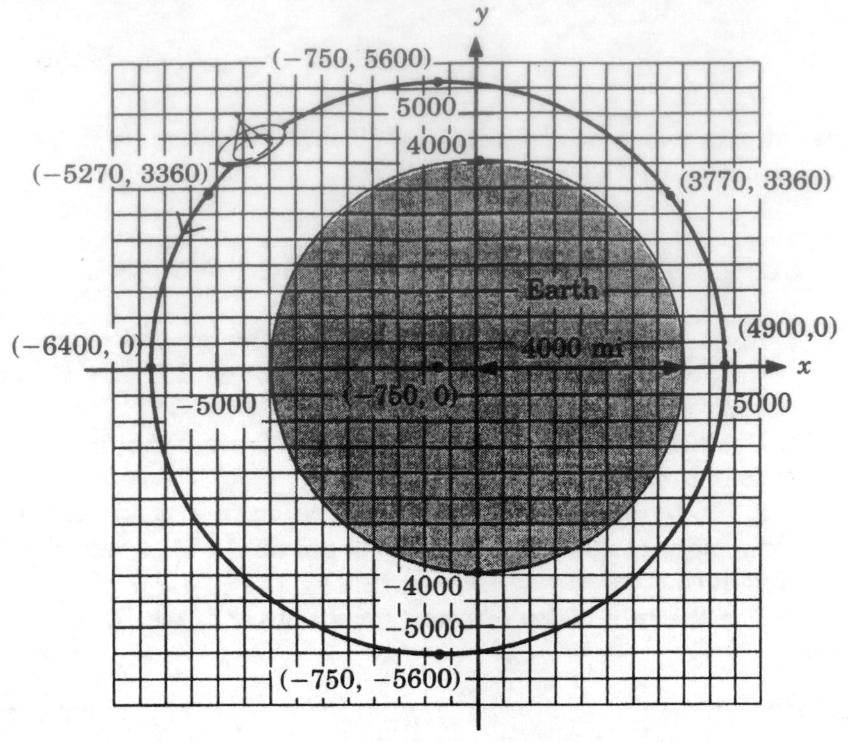

Orbit of Satellite

SOLUTION

a. $(x + 750)^2 = (5650)^2$

$x + 750 = \pm 5650$

$x = -750 \pm 5650$

$x = -6400$ or $x = 4900$

Both answers are accepted.

The satellite is at $(-6400, 0)$ or $(4900, 0)$.

b. $0 + \left(\frac{y}{5600}\right)^2 = 1$

$y^2 = (5600)^2$

$y = \pm 5600$

The satellite is at $(-750, 5600)$ or $(-750, -5600)$.

c. $\left(\frac{x + 750}{5650}\right)^2 + \left(\frac{3360}{5600}\right)^2 \doteq 1$

$\left(\frac{x + 750}{5650}\right)^2 + (0.6)^2 = 1$

$(x + 750)^2 = (5650)^2(0.64)$

$x + 750 = \pm 4520$

$x = 3770$ or $x = -5270$

The satellite is at $(3770, 3360)$ or $(-5270, 3360)$.

SAMPLE PROBLEM 3

The trajectory (path) of a research rocket is given by the equation

$$25y = 20x - 2x^2$$

where x and y are measured in thousands of miles.

Note when $x = 0$, then $y = 0$. Also, when $x = 10$, then $y = 0$.

a. Find y for $x = 5$. b. Find x for $y = 1$.

SOLUTION

a. $25y = 20(5) - 2(5^2)$

$25y = 100 - 50 = 50$

$y = 2$ When the horizontal distance is 5000 mi, the elevation is 2000 mi. This is the maximum elevation.

b. $y = 1$

$25(1) = 20x - 2x^2$

$2x^2 - 20x + 25 = 0$ **Since the coefficient of $x^2 \neq 1$, the quadratic formula is used.**

$$x = \frac{-(-20) \pm \sqrt{(-20)^2 - 4(2)(25)}}{2(2)}$$

$$x = \frac{20 \pm \sqrt{200}}{4}$$

$$x = \frac{20 \pm 10\sqrt{2}}{4}$$

$$x = \frac{10 \pm 5\sqrt{2}}{2}$$

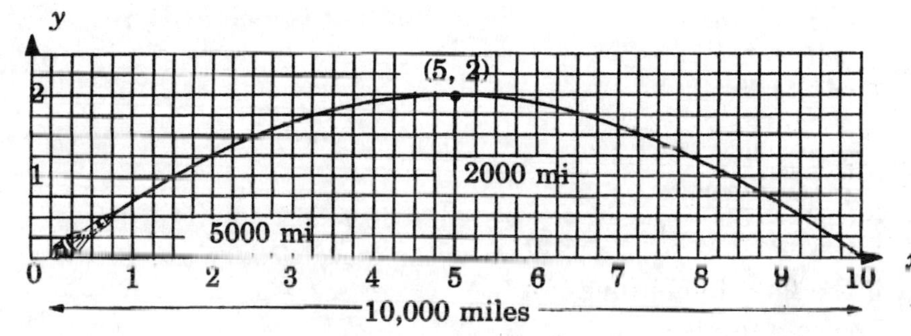

$25y = 20x - 2x^2$

Trajectory of a Research Rocket

Approximating,

$$x = \frac{10 \pm 7.07}{2}$$

$x = 8.5$ or $x = 1.5$

When the elevation is 1000 mi, the horizontal distance is approximately 1500 mi or 8500 mi.

SAMPLE PROBLEM 4

On a reforestation project, it took one boy scout troop 2 hr longer to plant 100 trees than it took another troop. When both troops worked together, they planted 100 trees in 5 hr. How long did it take each troop working alone to plant 100 trees?

SOLUTION

Let x = time of the faster troop and $x + 2$ = time of the slower troop.

Then, $\frac{1}{x}$ = the rate of the faster troop and $\frac{1}{x+2}$ = the rate of the slower troop.

Use (work = rate × time) and (work of A + work of B = total work).

$$\frac{5}{x} + \frac{5}{x+2} = 1$$

$$5(x + 2) + 5x = x(x + 2)$$

$$10x + 10 = x^2 + 2x$$

$$x^2 - 8x = 10$$

$$x^2 - 8x + 16 = 26$$

We solve by completing the square because the coefficient of $x^2 = 1$ and the coefficient of x is even.

$$(x - 4)^2 = 26$$

$x - 4 = \sqrt{26}$ or $x - 4 = -\sqrt{26}$

$x = 4 + \sqrt{26}$ $\qquad$ $x = 4 - \sqrt{26}$

$x \approx 4 + 5.099$ $\qquad$ Reject, since it is a negative result.

$x \approx 9.1$ hr

$x + 2 \approx 11.1$ hr

SAMPLE PROBLEM 5

A plane flies 300 mi with a tailwind of 10 m.p.h. and returns against a wind of 20 m.p.h. What is the speed of the plane in still air if the total flying time is 4 hr?

SOLUTION

Let x represent the speed of the plane in still air.

Then $x + 10$ = speed with the tailwind

and $x - 20$ = speed against the wind.

$$\frac{300}{x+10} + \frac{300}{x-20} = 4$$

Time going + time returning = total time

Time = $\frac{\text{Distance}}{\text{Rate}}$

$$300(x - 20) + 300(x + 10) = 4(x + 10)(x - 20)$$

$$75(x - 20) + 75(x + 10) = (x + 10)(x - 20)$$

Divide both sides by 4.

$$150x - 750 = x^2 - 10x - 200$$

$$x^2 - 160x + 550 = 0$$

$$x^2 - 160x + (-80)^2 = -550 + 6400$$

$$(x - 80)^2 = 5850$$

$$x - 80 = \pm\sqrt{5850}$$

$$= \pm 76.5 \text{ to the nearest tenth}$$

Therefore $x = 80 + 76.5$ or $x = 80 - 76.5$

$x = 156.5$ or $x = 3.5$

Since speed is always a positive number, $x \neq 3.5$, because $x - 20$ would then be a negative number. Therefore, the speed of the plane in still air is 156.5 m.p.h., correct to the nearest tenth.

EXERCISES 9.7S

1. In reference to Sample Problem 1, how far is the astronaut from the center of the earth when his weight is 12 lb?

1. ____________________

2. The Kilauea volcano on the island of Hawaii hurled lava 1500 ft high. How many miles away could this be seen?

$$\text{Use } 50h = 33d^2$$

where h is in feet and d is in miles.
Find d for $h = 1500$.

2. ____________________

3. In reference to Sample Problem 2, find y for $x = 3770$.

3. ____________________

4. In reference to Sample Problem 2, find x for $y = 4480$.

4. ____________________

5. The trajectory of a research rocket is given by

$$12y = 12x - x^2$$

where x and y are measured in thousands of miles.
a. Find x for $y = 0$.
b. Find y for $x = 6$. (This y is the maximum height.)
c. Find x for $y = 2$.

5. a. ____________________

b. ____________________

c. ____________________

NAME ____________________ DATE ____________________ COURSE ____________________

6. On a soil conservation project, a crew takes 15 hr longer digging holes by hand than a machine takes to do the same amount of work. When the crew and the machine work together, the same amount of work can be completed in 4 hr. How long would it have taken the crew working alone to do the job?

6. ______________________

7. A new machine can produce a certain number of articles in 4 hr less time than an old machine. After the new machine had been working for 3 hr, it broke down, and the old machine completed the job in 5 more hr. How long would it have taken the new machine alone to do the whole job?

7. ______________________

8. It takes 48 min longer to cut a certain amount of lumber by using a hand saw than it does by using a power saw. When both the hand saw and the power saw are used, the same amount of lumber can be cut in 7 min. How long does it take to cut this lumber using the hand saw alone?

8. ______________________

9. In order to finish a job as soon as possible, a contractor rents a spray outfit that can paint a building in 9 hr less time than the spray outfit he owns. Using both outfits, he has the building painted in 20 hr. How long would it have taken if he had used only his own spray outfit?

9. ______________________

10. A projectile is fired vertically upward with an initial speed of 800 ft per second. The equation

$$s = 800t - 16t^2$$

gives the distance s in feet above the ground that the projectile reaches t seconds after it is fired. When is the projectile 9600 ft above the ground? Explain the two answers.

10. ______________________

11. A bomb is released by an airplane 2064 ft above the earth with an initial speed of 640 ft per second vertically downward. The equation $s = 640t + 16t^2$ gives the distance, s, in feet below the point where the bomb was dropped, t sec after it was dropped. How long does it take the bomb to reach the earth?

11. ______________

12. A certain manufacturing company uses the profit equation

$$P = 80x - 2x^2$$

that relates the profit, P, in dollars that can be made by selling x units of an article that the company manufactures. Find how many units should be sold to obtain a profit of the following amounts.

a. $750 b. $800

12. a. ________ b. ________

13. A man rows a boat 24 mi downstream and then immediately returns to his starting point. If the entire trip took $4\frac{1}{2}$ hr and if his speed in still water was 12 m.p.h., find the rate of the current.

13. ______________

14. An airplane flew 1140 mi with a tailwind of 20 m.p.h. and then flew an additional 870 mi with a tailwind of 30 m.p.h. If the total time of the flight was $3\frac{1}{2}$ hr, find the speed of the plane in still air.

14. ______________

15. Two light sources having illuminations of 90 and 40 candlepower (cp), respectively, are placed 5 ft apart. Find the distance on the line joining the sources so that an object placed in this location will be equally illuminated by both sources.

15. ______________

Formula: Intensity $= \dfrac{\text{candlepower}}{(\text{distance})^2}$

Equation: $\dfrac{90}{x^2} = \dfrac{40}{(5 - x)^2}$

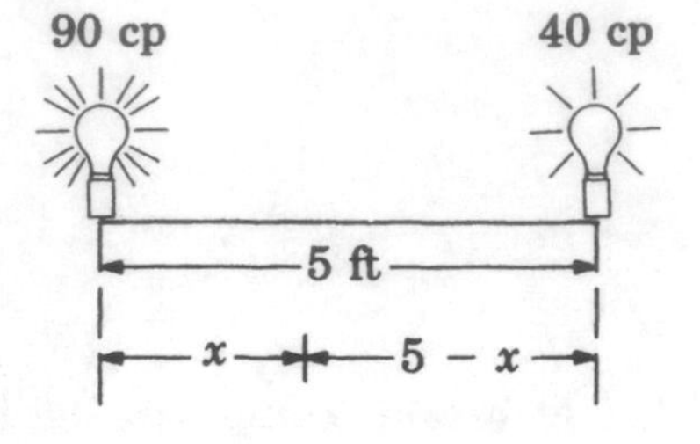

16. Repeat Exercise 15 for illuminations of 75 and 60 cp placed 8 ft apart.

16.

SAMPLE TEST FOR CHAPTER 9

1–2. Simplify. (See Section 9.1.)

1. $\frac{2\sqrt{81}}{3}$ **2.** $\frac{(\sqrt{6})^2 + \sqrt{100}}{5}$

1. ______

2. ______

3–4. By using a table or a calculator, approximate each of the following to the nearest hundredth. (See Section 9.1.)

3. $5\sqrt{68}$ **4.** $\frac{3 - \sqrt{15}}{4}$

3. ______

4. ______

5–6. Solve by factoring. (See Section 9.1.)

5. $9x^2 = 35$ **6.** $0.01x^2 = 0.2916$

5. ______

6. ______

7–8. Simplify. (See Section 9.2.)

7. $\sqrt{18}\sqrt{98}$ **8.** $\sqrt{192}$

7. ______

8. ______

9–10. Simplify. (See Section 9.3.)

9. $\frac{5}{\sqrt{150}}$ **10.** $\sqrt{\frac{7}{12}}$

9. ______

10. ______

NAME ______ DATE ______ COURSE ______

11–12. Simplify. (See Section 9.4.)

11. $\sqrt{150} + \sqrt{96}$

12. $\dfrac{12 - \sqrt{40}}{4}$

11. ____________________

12. ____________________

13–14. Solve by completing the square. (See Section 9.5.)

13. $x^2 + 8x + 4 = 0$

14. $3x^2 - 4x - 8 = 0$

13. ____________________

14. ____________________

15–16. Solve by using the quadratic formula. (See Section 9.6.)

15. $9x^2 = 2(6x + 1)$

16. $2x^2 + 2x - 1 = 0$

15. ____________________

16. ____________________

17. A glass window in a building is in the shape of a rectangle surmounted by a triangle. The height of the rectangle is 3 ft longer than the width of the rectangle. The base and altitude of the triangle are both equal to the width of the rectangle. Find the dimensions of the rectangle so that the window will admit 72 sq ft of light. (See Section 9.7.)

17. 

18. A motorboat takes 45 min longer to travel 36 mi upstream than it takes to travel the same distance downstream. If the boat can travel 20 m.p.h. in still water, find the rate of the current. (See Section 9.7.)

18. 

Answers

CHAPTER 1

EXERCISES 1.1A

1. $7y$ **2.** $s+11$ **3.** $x+3$ **4.** $5-x$ **5.** $x-5$ **6.** $\frac{n}{4}$ **7.** $\frac{x}{8}$ **8.** $\frac{4}{n}$ **9.** $q+6$ **10.** $6-q$
11. $6-q$ **12.** $\frac{6}{q}$ **13.** yz **14.** $9y$ **15.** $3 \cdot 25$, or $3(25)$, or $(3)(25)$ **16.** $8-r$ **17.** $\frac{7}{z}$ **18.** $x+9$
19. $9x$ **20.** $t-2$ **21.** $t-z$ **22.** xy **23.** $2 \cdot 3$, or $2(3)$, or $(2)(3)$ **24.** $2x$ **25.** $\frac{n}{2}$ **26.** $\frac{2}{n}$
27. $y-x$ **28.** $x-y$ **29.** $y+z$ **30.** $\frac{r}{s}$ **31.** $2y$ **32.** $\frac{y}{2}$ **33.** $\frac{t}{3}$ **34.** $\frac{n}{4}$ **35.** $\frac{x}{5}$

EXERCISES 1.1B

1. y^2 **2.** n^2 **3.** 4^3 **4.** n^3 **5.** 6^3 **6.** t^3 **7.** t^2 **8.** x^3 **9.** 10^2 or 100 **10.** 9^2 or 81
11. 9^3 or 729 **12.** y^3 **13.** y^4 **14.** 10^4 or 10,000 **15.** 8^5 **16.** x^5 **17.** 4^5 **18.** 9^7 **19.** n^6
20. y^8 **21.** 7^6 **22.** n^6 **23.** t^4 **24.** x^2

EXERCISES 1.1C

1. k **2.** n **3.** z **4.** y **5.** b **6.** t **7.** n **8.** 0 **9.** 0 **10.** Undefined **11.** x
12. 0 **13.** x **14.** $x+1$ **15.** Undefined **16.** x **17.** $\frac{1}{x}$ **18.** 0 **19.** Undefined **20.** 0

EXERCISES 1.1 S

1. $n+8$ **2.** $8n$ **3.** $t-12$ **4.** $12-t$ **5.** $\frac{x}{7}$ **6.** $\frac{7}{x}$ **7.** ab **8.** $a+b$ **9.** $\frac{a}{b}$ **10.** $b-a$
11. n **12.** $x-4$ **13.** $\frac{t}{4}$ **14.** $\frac{4}{t}$ **15.** $y+9$ **16.** $9-y$ **17.** $2t$ **18.** $3r$ **19.** $\frac{z}{2}$ **20.** $\frac{n}{3}$
21. b^2 **22.** z^3 **23.** 7^3, or 343 **24.** 18^2, or 324 **25.** 5^3, or 125 **26.** r^4 **27.** s^3 **28.** t^5
29. x **30.** y **31.** t **32.** $k+20$ **33.** z **34.** x **35.** $\frac{1}{c}$ **36.** xy **37.** $x+y$ **38.** 0
39. 0 **40.** Undefined **41.** 0 **42.** k **43.** $x+7$ **44.** $7y$ **45.** $10-z$ **46.** $\frac{x}{6}$ **47.** x^4
48. x^5 **49.** y^6 **50.** t^n

EXERCISES 1.2A

1. {0, 1, 2, 3, 4} **2.** {7, 8, 9} **3.** {7, 8, 9, 10, 11, . . .} **4.** {1, 2, 3, 4} **5.** {6, 7, 8, 9} **6.** {6, 7, 8, 9, 10, . . .}
7. {1, 3, 5, 9, 15, 45} **8.** {1, 3, 5, 9} **9.** {0, 2, 4, 6, 8} **10.** {2, 4, 6, 8, 10, . . .} **11.** {0, 2, 4, 6, 8, . . .} **12.** W
13. D **14.** ∅ **15.** {0} **16.** N **17.** ∅ **18.** {1} **19.** W **20.** ∅

EXERCISES 1.2B

1. {1, 2, 4, 5, 10, 20} **2.** {20, 40, 60, 80, 100, . . .} **3.** {1, 3, 7, 21} **4.** {8, 16, 24, 32, 40, . . .} **5.** {5, 7}
6. ∅, or { } **7.** {30, 60, 90, 120, . . .} **8.** {1, 2, 3, 5, 6, 10, 15, 30} **9.** {9, 18, 27, 36, 45, . . .}
10. N, or {1, 2, 3, 4, 5, . . .}

EXERCISES 1.2 S

1. {1, 2, 3, 4, 5, 6} **2.** {0, 1, 2, 3, 4, 5, 6} **3.** {5, 6, 7, 8, 9} **4.** {5, 6, 7, 8, 9, . . .} **5.** {1, 2, 3, 4, 5, 6}
6. {1, 2, 3, 4, 5, 6, 10, 12, 15, 20, 30, 60} **7.** {0, 3, 6, 9} **8.** {3, 6, 9, 12, 15, . . .} **9.** N, or {1, 2, 3, 4, 5, . . .}
10. {1, 2, 3, 4, 5, 6, 7, 8, 9} **11.** {45, 90, 135, 180, 225, . . .} **12.** {1, 3, 5, 9, 15, 45} **13.** {30, 60, 90, 120, 150, . . .}
14. {1, 2, 3, 5, 6, 10, 15, 30} **15.** {2, 4, 7, 14} **16.** ∅, or { } **17.** {1, 2, 3, 6, 7, 14, 21, 42}
18. {10, 20, 30, 40, 50, . . .} **19.** W, or {0, 1, 2, 3, 4, . . .} **20.** { } or ∅

EXERCISES 1.3A

1. $\{12, 14, 16, 18\}$ **2.** $\{21, 23, 25, 27, 29\}$ **3.** $\{10, 20, 30, 40, \ldots\}$ **4.** $\{5, 15, 25, 35, \ldots\}$ **5.** $\{2, 6, 10, 30\}$
6. $\{1, 3, 5, 15\}$ **7.** $\{1\}$ **8.** $\varnothing$ **9.** $\varnothing$ **10.** N

EXERCISES 1.3B

1. $\{23, 29\}$ **2.** $\{21, 22, 24, 25, 26, 27, 28\}$ **3.** $\{2\}$ **4.** $\{3, 5, 7, 11, 13, \ldots\}$ **5.** $\{5, 7\}$ **6.** $\{35\}$
7. $\{6, 22, 33, 66\}$ **8.** $\{2, 3, 11\}$ **9.** $\{1\}$ **10.** $\varnothing$

EXERCISES 1.3 S

1. $\{31, 37, 41, 43, 47\}$ **2.** $\{36, 38, 39, 40, 42, 44\}$ **3.** $\{22, 44, 66, 88, 110, \ldots\}$ **4.** $\{11, 33, 55, 77, 99, \ldots\}$
5. $\{2, 7, 11\}$ **6.** $\{14, 22, 77, 154\}$ **7.** $\{1, 5, 7, 35\}$ **8.** $\{2, 10, 14, 70\}$ **9.** $\varnothing$, or $\{\ \}$ **10.** $\{2\}$
11. $\{1, 2, 3, 6, 9, 18, 27, 54\}$ **12.** $\{2, 3\}$ **13.** $\{6, 9, 18, 27, 54\}$ **14.** $\{2, 6, 18, 54\}$ **15.** $\{1, 3, 9, 27\}$ **16.** N
17. $\varnothing$ **18.** $\{1, 5, 13, 65\}$ **19.** $\{2, 5, 13\}$ **20.** $\{10, 26, 65, 130\}$ **21.** 1(35), 5(7) **22.** 1(28), 2(14), 4(7)
23. 1(12), 2(6), 3(4) **24.** 1(36), 2(18), 3(12), 4(9), 6(6) **25.** 1(45), 3(15), 5(9)
26. 1(60), 2(30), 3(20), 4(15), 5(12), 6(10) **27.** 1(90), 2(45), 3(30), 5(18), 6(15), 9(10) **28.** 1(75), 3(25), 5(15)
29. 1(42), 2(21), 3(14), 6(7) **30.** 1(30), 2(15), 3(10), 5(6)

EXERCISES 1.4A

1. 4 **2.** 17 **3.** 8 **4.** 7 **5.** 1 **6.** 15 **7.** 50 **8.** 100 **9.** 6 **10.** 10 **11.** 5 **12.** 3
13. 36 **14.** 13 **15.** 400 **16.** 8 **17.** 30 **18.** 144 **19.** 150 **20.** 6 **21.** 0 **22.** 0
23. 41 **24.** 10

EXERCISES 1.4B

1. $3(x+5)$ **2.** $3x+5$ **3.** $\frac{8-x}{4}$ **4.** $\frac{x}{4}-8$ **5.** $(5x)^2$ **6.** $5x^2$ **7.** $x-(y-8)$ **8.** $(x-y)-8$

9. $y-2(x+3)$ **10.** $(y-2)(x+3)$ **11.** $9(x+3)$ **12.** $\frac{x-3}{2}$ **13.** $\frac{y+2}{5}$ **14.** $4(y-6)$

15. $3\left(\frac{y}{8}\right)$ or $\frac{3y}{8}$ **16.** $5x+4$ **17.** $5(x+4)$ **18.** $6y-1$ **19.** $6(y-1)$ **20.** $s-(t+6)$ **21.** $x+(8-y)$

22. $x+\frac{3}{x}$ **23.** $6-2r$ **24.** $9z+5$ **25.** $x(y-3)$ **26.** $\frac{z+4}{9}$ **27.** $t(t+8)$ **28.** $(7z)^2$

29. $(x+y+z)^3$ **30.** n^2+3^2 or n^2+9 **31.** $\frac{5}{n+3}$ **32.** $q+7q$ or $8q$ **33.** $\frac{4t}{9}$ **34.** $\frac{t^3}{9}$ **35.** $\frac{n-6}{2}$

EXERCISES 1.4C

1. 30 **2.** 2 **3.** 36 **4.** 360 **5.** 14 **6.** 12 **7.** 56 **8.** 4 **9.** 36 **10.** 12 **11.** 231
12. 56 **13.** 6 **14.** 28 **15.** 900 **16.** 30

EXERCISES 1.4 S

1. 30 **2.** 48 **3.** 100 **4.** 1000 **5.** 24 **6.** 4 **7.** 10 **8.** 0 **9.** 3 **10.** 15 **11.** 75
12. 49 **13.** 16 **14.** 6 **15.** 5 **16.** 15 **17.** 3 **18.** 20 **19.** 10 **20.** 28 **21.** x^3-3x

22. $5z+7$ **23.** $\frac{8y}{3}$ **24.** $9(9-s)$ **25.** $x+(x-6)$ **26.** $y-(x+6)$ **27.** $12-(6-x)$ **28.** $2(t+3)$

29. $n-(n+4)$ **30.** $2(xy)$, or $2xy$ **31.** $\frac{xy}{2}$ **32.** $5y^2$ **33.** $6x^3$ **34.** x^2-1 **35.** y^3+1

36. x^2-y^2 **37.** $(x-y)^2$ **38.** x^2+5^2 or x^2+25 **39.** y^3+2^3 or y^3+8 **40.** $x-(y-z)$ **41.** $x+(y-z)$

42. $x(x-3)$ **43.** $x+3(x-3)$ **44.** $\frac{5(y+3)}{2}$ **45.** $\frac{x+y}{2(x-y)}$ **46.** 6 **47.** 48 **48.** 81 **49.** 21

50. 80 **51.** 400 **52.** 3 **53.** 12 **54.** 11 **55.** 3 **56.** 6 **57.** 2 **58.** 18 **59.** 36
60. 8

EXERCISES 1.5A

1. 37 **2.** 80 **3.** 100 **4.** 1 **5.** 77 **6.** 39 **7.** 196 **8.** 120 **9.** 2 **10.** 45 **11.** 17
12. 65 **13.** 10 **14.** 54 **15.** 7 **16.** 590 **17.** 40 **18.** 16 **19.** 0 **20.** 46

EXERCISES 1.5B

1. 2. **2.** 286 **3.** 20 **4.** 240 **5.** 27 **6.** 11 **7.** 24 **8.** 21 **9.** 21 **10.** 37 **11.** 26
12. 1 **13.** 175 **14.** 11,000 **15.** 35 **16.** 175

EXERCISES 1.5 S

1. 4 **2.** 52 **3.** 20 **4.** 11 **5.** 23 **6.** 6 **7.** 0 **8.** 32 **9.** 20 **10.** 63 **11.** 31 **12.** 20 **13.** 5 **14.** 18 **15.** 45 **16.** 12 **17.** 165 **18.** 280 **19.** 98 **20.** 28 **21.** 100 **22.** 143 **23.** 1 **24.** 5 **25.** 5 **26.** 1

EXERCISES 1.6 S

1. 9 **2.** 90 **3.** \$12 **4.** \$300 **5.** 40% **6.** 300% **7.** 50% **8.** 12% **9.** 54 meters **10.** 500 ft **11.** 60° C **12.** 10° C **13.** 0 **14.** 1 **15.** 4 **16.** 7 **17.** 1275 **18.** 92 **19.** 50% **20.** 84% **21a.** 28 **21b.** 49 **22.** 84 **23a.** 12 **23b.** 8 **23c.** 18 **23d.** 24 **24.** \$13,080 **25a.** 10% **25b.** 14% **26.** 730 **27.** 88¢ **28.** 159 **29.** 120 **30.** 5 cu yd **31a.** \$4738 **31b.** \$7788 **32.** \$20,000 **33.** \$188,600 **34a.** 78 ft **34b.** 180 ft **35.** \$428 **36.** 26 **37.** 164 **38.** 138 **39a.** 68 **39b.** 95 **39c.** 212 **40.** 6 ohms

SAMPLE TEST FOR CHAPTER 1

1. $8x$ **2.** $n+6$ **3.** $y-4$ **4.** $\frac{x}{3}$ **5.** t^2 **6.** cd **7.** $c+d$ **8.** $d-c$ **9.** $\frac{c}{d}$ **10.** $\frac{b}{3}$ **11.** x **12.** x **13.** x **14.** 0 **15.** y **16.** {0, 1, 2, 3} **17.** {0, 6, 12, 18, . . .} **18.** {1, 2, 4, 5, 10, 20} **19.** {1, 2, 3, 6, 7, 14, 21, 42} **20.** N, or {1, 2, 3, . . .} **21.** {9, 18, 27, 36, . . .} **22.** $\varnothing$, or { } **23.** {101, 103, 105, 107, 109} **24.** {2, 4, 6, 8, 12, 16, 24, 48} **25.** {41, 43, 47} **26.** {4, 10, 20} **27.** $7(n+8)$ **28.** $3x-5$ **29.** $(y-8)^2$ **30.** $y-(x+2)^3$ **31.** $\frac{2x}{x+2}$ **32.** 29 **33.** 4 **34.** 45 **35.** 9 **36.** 3 **37.** 20 **38.** 16 **39.** 2 **40.** 279 **41.** 40 **42.** 1296 **43.** 60 **44.** 48 **45.** 50 **46.** 9

CHAPTER 2

EXERCISES 2.1A

1. A withdrawal of \$50 **2.** A loss of 8 yd **3.** An altitude 200 ft below sea level **4.** A velocity of 45 m.p.h. in the south direction **5.** No gain or loss in weight **6.** A decrease of 75¢ in the price of a stock **7.** The year 1250 B.C. **8.** A force of 20 lb upward **9.** No rotation **10.** A current of 8 amp for a discharging battery

EXERCISES 2.1B

1. $A(-11)$, $B(-3)$, $C(2)$, $D(6)$ **2.** $A(-6)$, $B(-8)$, $C(0)$, $D(-3)$ **3.** $A(5)$, $B(0)$, $C(11)$, $D(7)$

4. $A\left(\frac{1}{2}\right)$, $B\left(1\frac{1}{4}\right)$, $C\left(\frac{-3}{4}\right)$, $D\left(-2\frac{1}{2}\right)$ **5.** $A(0)$, $B\left(1\frac{1}{8}\right)$, $C\left(\frac{-3}{8}\right)$, $D\left(-1\frac{1}{8}\right)$

6–10.

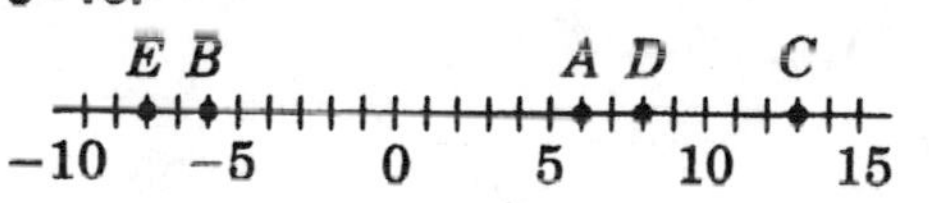

11–15.

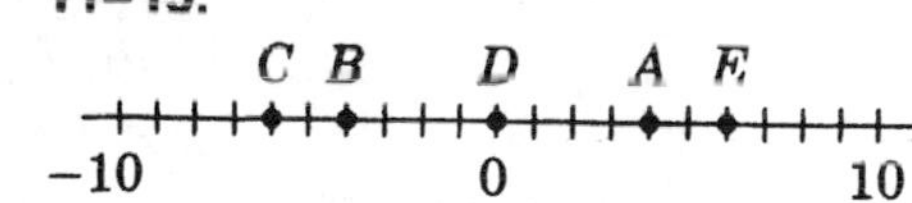

16. $A(8)$, $B(3)$, $C(-2)$, $D(-8)$ **17.** $A(-6)$, $B(14)$, $C(-18)$, $D(6)$

18–22.

+
C(14)
10
5
D(3)
0
B(0)
A(−3)
−5
E(−7)

23–27.

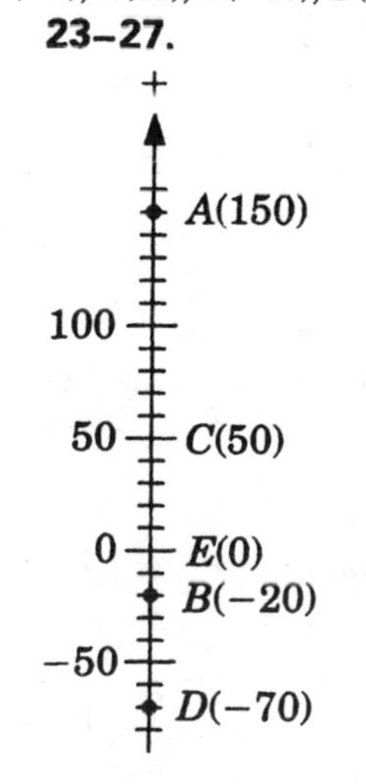

28–32.

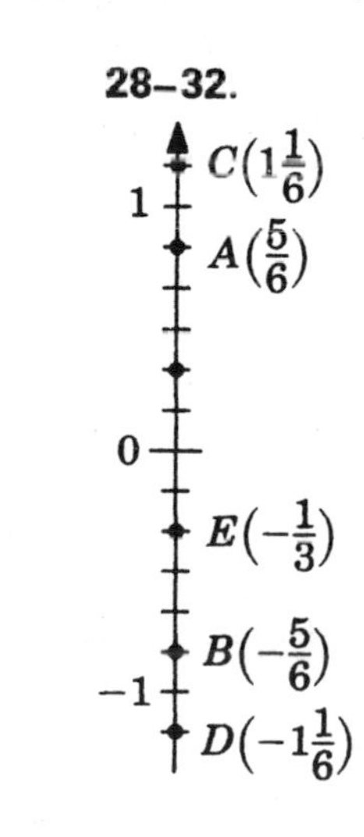

EXERCISES 2.1C

1. 6 **2.** 7 **3.** 0 **4.** 4 **5.** 8 **6.** 9 **7.** 18 **8.** 3 **9.** 7 **10.** 3 **11.** 7 **12.** 4
13. 12 **14.** 14 **15.** 20 **16.** 0 **17.** −5 **18.** 0 **19.** 3 **20.** 8 **21.** 1 **22.** 4
23. $|+20|$ **24.** $|0|$ **25.** $|-20|$ **26.** $|x|$ **27.** Yes **28.** Yes **29.** No **30.** Yes **31.** Yes
32. Yes

EXERCISES 2.1D

1. $+4 < +7$, +4 is less than +7, point +4 is left of point +7
2. $+8 > +3$, +8 is greater than +3, point +8 is right of point +3
3. $-4 > -7$, −4 is greater than −7, point −4 is right of point −7
4. $-8 < -3$, −8 is less than −3, point −8 is left of point −3 **5.** $-2 < +5$, −2 is less than +5, point −2 is left of point +5
6. $+2 > -5$, +2 is greater than −5, point +2 is right of point −5
7. $0 > -4$, 0 is greater than −4, point 0 is right of point −4 **8.** $|-6| = 6$, $|-6|$ equals 6, point $|-6|$ is the same as point 6
9. $6 > -3$, 6 is greater than −3, point 6 is right of point −3 **10.** $-6 < 0$, −6 is less than 0, point −6 is left of point 0
11. $x + 7 > 12$ **12.** $5y < 30$ **13.** $n + t < 0$ **14.** $x + 4 > 10$ **15.** $y - 5 < 8$ **16.** $|x| > 0$
17. $2(t + 6) > 0$ **18.** $ab < 0$ **19.** $y < 3x + 9$ **20.** $3x + 9 > y$

EXERCISES 2.1 S

1. 30° longitude east of Greenwich **2.** A temperature 20° below zero **3.** Zero latitude (on the equator)
4. A loss of 5 yd **5.** No increase or decrease **6.** A loss of \$100 **7.** A deceleration of 15 m.p.h.
8. The distance of an image 24 cm in back of a mirror **9.** A loss of 32 points **10.** A score of 17 points below average
11. $A(6), B(-8), C(16), D(-16)$ **12.** $A(9), B(-17), C(-22), D(2)$ **13.** $A\left(\frac{3}{5}\right), B\left(-1\frac{4}{5}\right), C\left(\frac{-3}{5}\right), D\left(1\frac{3}{5}\right)$
14. $A(20), B(-40), C(35), D(-15)$ **15.** $A(-40), B(20), C(-90), D(90)$ **16.** $A\left(\frac{1}{3}\right), B\left(-\frac{1}{2}\right), C\left(-1\frac{1}{6}\right), D\left(1\frac{2}{3}\right)$

17–21.

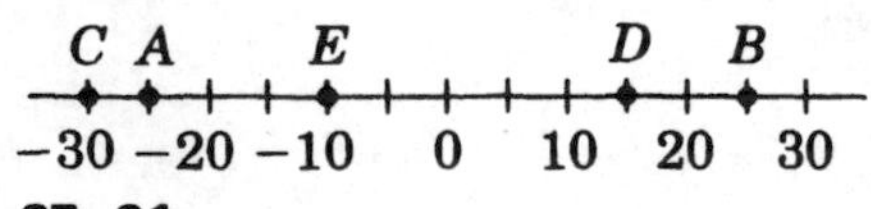

22–26.

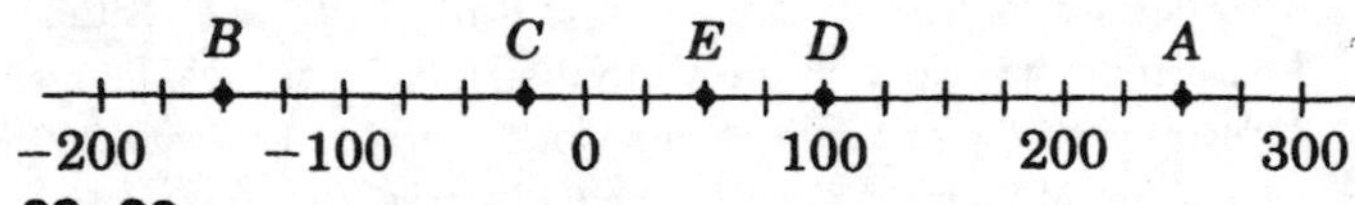

27–31.

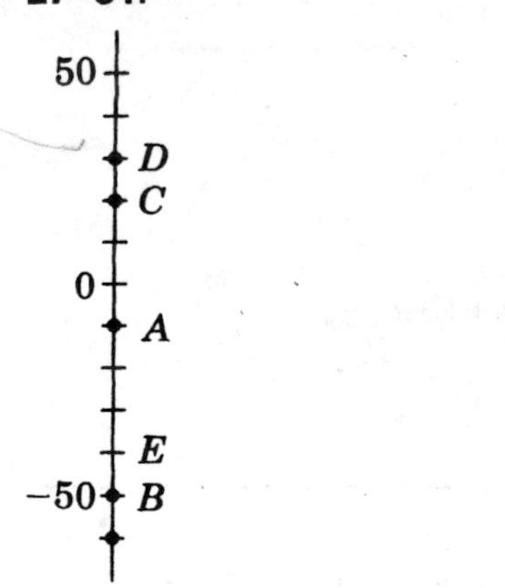

32–36.

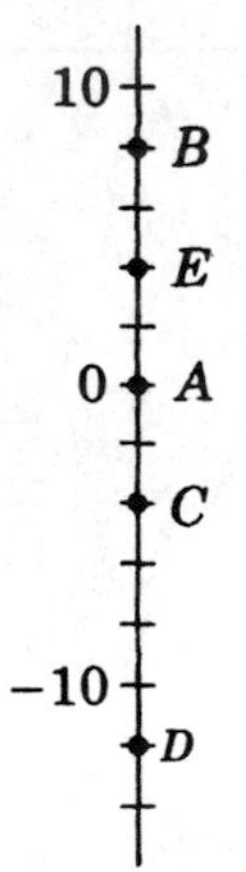

37. 9 **38.** 20 **39.** 17 **40.** 7 **41.** 13 **42.** 7 **43.** 16 **44.** 20 **45.** 60 **46.** 40
47. 5 **48.** 8 **49.** 7 **50.** 1 **51.** $2 > -3$, 2 is greater than −3, point 2 is right of point −3
52. $4 > 0$, 4 is greater than 0, point 4 is right of point 0 **53.** $0 < 5$, 0 is less than 5, point 0 is left of point 5
54. $7 > 5$, 7 is greater than 5, point 7 is right of point 5 **55.** $-12 < -4$, −12 is less than −4, point −12 is left of point −4
56. $|-3| = +3$, $|-3|$ equals +3, point $|-3|$ is the same as point +3
57. $-12 < 0$, −12 is less than 0, point −12 is left of point 0
58. $|-7| > -7$, $|-7|$ is greater than −7, point $|-7|$ is right of point −7
59. $-8 < 4$, −8 is less than 4, point −8 is left of point 4 **60.** $-4 > -9$, −4 is greater than −9, point −4 is right of point −9

61. $6t < 24$ **62.** $x + y > xy$ **63.** $4y < 0$ **64.** $|ab| > 0$ **65.** $x - 7 > y$ **66.** $x + 8 < y$

EXERCISES 2.2A

1. +9 **2.** +2 **3.** −9 **4.** +3 **5.** −4 **6.** −4 **7.** 0 **8.** −6 **9.** +10 **10.** −15
11. −9 **12.** +4 **13.** −4 **14.** 0 **15.** 0

EXERCISES 2.2B

1. 8 **2.** 11 **3.** −10 **4.** −94 **5.** +24 **6.** −7 **7.** −14 **8.** −26 **9.** 70 **10.** −100
11. 1 **12.** −3 **13.** −7 **14.** −5 **15.** 7 **16.** −5 **17.** −5 **18.** 3 **19.** 0 **20.** 0
21. −5 **22.** −4 **23.** −19 **24.** −100 **25.** 75 **26.** 75 **27.** −10 **28.** −9 **29.** −8
30. −8 **31.** 0 **32.** 0 **33.** $\frac{-3}{4}$ **34.** $-8\frac{2}{5}$ **35.** 0 **36.** 0 **37.** −7 **38.** 12 **39.** 9
40. −41 **41.** −3 **42.** −10 **43.** −25 **44.** −5 **45.** −1 **46.** −50 **47.** −4 **48.** 30
49. 20 **50.** −20

EXERCISES 2.2 S

1. +80 **2.** +75 **3.** −10 **4.** −42 **5.** −150 **6.** +24 **7.** +9 **8.** −5 **9.** −3 **10.** +7
11. −4 **12.** −17 **13.** −1 **14.** −21 **15.** 0 **16.** 13 **17.** 1 **18.** −1 **19.** −13 **20.** 1
21. −1 **22.** 0 **23.** 1 **24.** −1 **25.** 0 **26.** 12 **27.** 100 **28.** −100 **29.** −1 **30.** −5
31. 0 **32.** 0 **33.** −15 **34.** 50 **35.** 32 **36.** −32 **37.** −100 **38.** −100 **39.** −30
40. 80

EXERCISES 2.3A

1. 6 **2.** 2 **3.** −3 **4.** −4 **5.** −7 **6.** 5 **7.** 8 **8.** −2 **9.** −9 **10.** 9 **11.** 7
12. 4 **13.** 9 **14.** 8 **15.** 6 **16.** 12 **17.** 4 **18.** 6 **19.** 8 **20.** 6 **21.** 4 **22.** −6
23. −7 **24.** 10 **25.** 5 **26.** −5 **27.** −3 **28.** 2 **29.** −9 **30.** 8

EXERCISES 2.3B

1. −6 **2.** −1 **3.** 0 **4.** 1 **5.** 6 **6.** 10 **7.** 10 **8.** 0 **9.** 2 **10.** −2 **11.** 7
12. 9 **13.** −9 **14.** −8 **15.** 6 **16.** 2 **17.** 1 **18.** −3 **19.** −8 **20.** −9 **21.** −7
22. 6 **23.** 5 **24.** 7 **25.** 1

EXERCISES 2.3C

1. 2 **2.** 21 **3.** −2 **4.** 2 **5.** 10 **6.** −4 **7.** −7 **8.** −10 **9.** 10 **10.** 10 **11.** 2
12. −6 **13.** −2 **14.** 14 **15.** −12 **16.** 0 **17.** 0 **18.** −7 **19.** 6 **20.** 0 **21.** −12
22. 0 **23.** 30 **24.** 30 **25.** −30 **26.** −35 **27.** 17 **28.** −17 **29.** 55 **30.** −92 **31.** 8
32. 7 **33.** 20 **34.** 15 **35.** 15 **36.** 12 **37.** 10 **38.** 6 **39.** 8 **40.** 9 **41.** 4.5
42. 4.1 **43.** 3.4 **44.** 2.4 **45.** −0.4(P.M. tide lower) **46.** 0.9 **47.** −1.5(P.M. tide lower than A.M.)
48. −0.4(P.M. tide lower) **49.** 0.3 **50.** 1.2 **51.** 14 P.M.(2 P.M.) **52.** 20 P.M.(8 P.M.) **53.** 9 A.M.
54. 7 A.M. **55.** 4 A.M. **56.** 13 P.M.(1 P.M.) **57.** 21 P.M.(9 P.M.) **58.** 2 A.M. **59.** 11 A.M.
60. 16 P.M.(4 P.M.) **61.** 9 A.M. **62.** 18 P.M.(6 P.M.) **63.** 20 P.M.(8 P.M.) **64.** 2 A.M. **65.** 5 A.M.
66. 19 P.M.(7 P.M.) **67.** −6 west **68.** 6 east **69.** −8 west **70.** 7 east **71.** −5 west **72.** 8 east
73. −5 west **74.** −9 west **75.** −4 west **76.** 4 east **77.** 7 east **78.** 146 west **79.** 19 east
80. −34 west **81.** −47 west **82.** 38 east **83.** −11 west **84.** −25 west **85.** 38 east **86.** 6 east
87. 15 north **88.** −23 south **89.** −63 south **90.** 32 north **91.** −21 south **92.** 9 north
93. −56 south **94.** −14 south **95.** 11 north **96.** −73 south

EXERCISES 2.3 S

1. −10 **2.** 7 **3.** 0 **4.** −10 **5.** 10 **6.** −4 **7.** −3 **8.** 2 **9.** 0 **10.** −9 **11.** 5
12. 5 **13.** −16 **14.** 1 **15.** 2 **16.** 10 **17.** 6 **18.** −6 **19.** 10 **20.** −2 **21.** 25
22. −55 **23.** 55 **24.** −25 **25.** −12 **26.** −140 **27.** 135 **28.** 56 **29.** −25 **30.** 78
31. −100 **32.** −30 **33.** 0 **34.** 80 **35.** −7 **36.** 25 **37.** −30 **38.** −70 **39.** −2
40. −1 **41.** 27 **42.** 47 **43.** $13\frac{3}{4}$ **44.** $4\frac{1}{4}$ **45.** 25,028 up **46.** −9140 down **47.** 74 up
48. −163 down **49.** −19,757 down **50.** 3662 up

EXERCISES 2.4A

1. −10 **2.** 10 **3.** −10 **4.** 10 **5.** 27 **6.** −40 **7.** 4 **8.** −4 **9.** 1 **10.** 36 **11.** −56
12. −100 **13.** 9 **14.** 100 **15.** 1000 **16.** −1000 **17.** −64 **18.** −9 **19.** −1000
20. −10,000 **21.** −30 **22.** −100 **23.** −42 **24.** −18 **25.** 81 **26.** 3600 **27.** 100 **28.** 400
29. 0 **30.** 0 **31.** −6 **32.** −10 **33.** −9 **34.** −12 **35.** −8 **36.** −15 **37.** 7 **38.** 0
39. 0 **40.** 0

EXERCISES 2.4B

1. 3 **2.** 5 **3.** −5 **4.** 3 **5.** −2 **6.** −10 **7.** 5 **8.** −3 **9.** −4 **10.** −2 **11.** −6 **12.** −100 **13.** 10 **14.** 1 **15.** −1 **16.** −1 **17.** 1 **18.** 0 **19.** 0 **20.** −25 **21.** Undefined **22.** Undefined **23.** 1 **24.** Undefined **25.** 0 **26.** −20 **27.** −4 **28.** 100 **29.** 50 **30.** −30

EXERCISES 2.4C

1. $\frac{1}{4}$ **2.** 5 **3.** $\frac{-1}{3}$ **4.** −7 **5.** $\frac{5}{3}$ **6.** $\frac{-9}{4}$ **7.** $\frac{-2}{11}$ **8.** −1 **9.** $\frac{1}{6}$ **10.** $\frac{-1}{6}$ **11.** 6 **12.** −6 **13.** $\frac{3}{2}$ **14.** $\frac{-3}{2}$ **15.** 1 **16.** −1 **17.** 3 **18.** −8 **19.** −7 **20.** 5 **21.** 1 **22.** 1 **23.** $\frac{-4}{5}$ **24.** $\frac{2}{3}$ **25.** $\frac{7}{10}$ **26.** −2 **27.** $\frac{-5}{8}$ **28.** $\frac{7}{20}$ **29.** $\frac{1}{15}$ **30.** $\frac{-15}{2}$ **31.** $\frac{-15}{2}$ **32.** $\frac{35}{2}$ **33.** $\frac{3}{4}$ **34.** $\frac{-5}{4}$ **35.** $\frac{-1}{2}$ **36.** 2 **37.** −8 **38.** $\frac{-6}{25}$ **39.** $\frac{8}{7}$ **40.** 1

EXERCISES 2.4 S

1. 0 **2.** 0 **3.** 108 **4.** −1000 **5.** 100 **6.** 0 **7.** 0 **8.** 0 **9.** 0 **10.** −15 **11.** −25 **12.** 10 **13.** −4600 **14.** 58,000 **15.** 36 **16.** 64 **17.** 49 **18.** 81 **19.** −600 **20.** −3000 **21.** −150 **22.** −280 **23.** 7200 **24.** 144 **25.** −18 **26.** 36 **27.** 0 **28.** 0 **29.** −64 **30.** 216 **31.** −8 **32.** −7 **33.** −15 **34.** −20 **35.** 1 **36.** −29 **37.** 25 **38.** 0 **39.** 80 **40.** 400 **41.** −32 **42.** −13 **43.** −80 **44.** −4 **45.** 0 **46.** 14 **47.** 7 **48.** −1000 **49.** 1000 **50.** 100 **51.** −9 **52.** −1 **53.** 0 **54.** 1 **55.** −19 **56.** −4 **57.** −27 **58.** 125 **59.** 16 **60.** −625 **61.** Undefined **62.** Undefined **63.** 1 **64.** −7 **65.** Undefined **66.** 0 **67.** +1 **68.** −1 **69.** 0 **70.** −1.5 **71.** −1 **72.** −3 **73.** −4 **74.** +1 **75.** −1 **76.** −2 **77.** 240 CCW **78.** −4000 CW **79.** −200,000 CW **80.** 360 CCW **81.** +150,000; repel **82.** −40,000; attract **83.** −120,000; attract **84.** +60,000; repel

EXERCISES 2.5A

1. 5 **2.** 0 **3.** 0 **4.** 0 **5.** 0 **6.** −5 **7.** −1 **8.** −1 **9.** −4 **10.** 6 **11.** −27 **12.** 625 **13.** 20 **14.** 13 **15.** 6 **16.** 0 **17.** −5 **18.** −9 **19.** 5 **20.** 11 **21.** −3 **22.** −120 **23.** −25 **24.** −900 **25.** 65 **26.** 60 **27.** −1500 **28.** −30 **29.** −20 **30.** −324

EXERCISES 2.5B

1. −50 **2.** 56 **3.** −15 **4.** 27 **5.** −4 **6.** −13 **7.** 52 **8.** −34 **9.** 68 **10.** −70 **11.** 36 **12.** 64 **13.** 20 **14.** 96 **15.** 55 **16.** −36 **17.** 16 **18.** −4 **19.** 6 **20.** −5 **21.** −30 **22.** −16 **23.** −20 **24.** −30 **25.** 50 **26.** −4 **27.** −36 **28.** −10 **29.** 48 **30.** −24 **31.** −21 **32.** 10 **33.** 60 **34.** −120

EXERCISES 2.5 S

1. 32 **2.** 4 **3.** −8 **4.** 2 **5.** −2 **6.** 1 **7.** −3 **8.** −14 **9.** 7 **10.** −7 **11.** −24 **12.** 6 **13.** 200 **14.** −200 **15.** −64 **16.** 64 **17.** 16 **18.** −1125 **19.** −7 **20.** 12 **21.** −4 **22.** −1 **23.** 28 **24.** −20 **25.** 105 **26.** −127 **27.** −4 **28.** 2 **29.** 4 **30.** 0 **31.** −2 **32.** 63 **33.** −6 **34.** 23 **35.** −3 **36.** −1250 **37.** $\frac{1}{2}$ **38.** −6

SAMPLE TEST FOR CHAPTER 2

1. 14 **2.** 11 **3.** $y - 6 > 5$ **4.** $y + 9 < 0$ **5.** $+4 < +8$, +4 is less than +8 **6.** $+4 > -8$, +4 is greater than −8 **7.** $-4 > -8$, −4 is greater than −8 **8.** $-8 < -4$, −8 is less than −4 **9.** −57 **10.** −13 **11.** 13 **12.** −12 **13.** −25 **14.** 0 **15.** −20 **16.** −25 **17.** −12 **18.** 0 **19.** −45 **20.** 0 **21.** −1000 **22.** 150 **23.** −621 **24.** 35 **25.** 0 **26.** −17 **27.** 6 **28.** −9 **29.** −13 **30.** 0 **31.** 1 **32.** −8 **33.** −9 **34.** Undefined **35.** −3800 **36.** −4 **37.** 44 **38.** −48 **39.** −26 **40.** 0 **41.** −6 **42.** 15

SAMPLE TEST FOR CHAPTERS 1 AND 2

1. $5y$ **2.** $x + 7$ **3.** $\frac{x}{4}$ **4.** $y - 8$ **5.** t^2 **6.** r^3 **7.** $6(x + 3)$ **8.** $n^2 - 4$ **9.** $\frac{y + 9}{3}$ **10.** $(2x)^3$

11. $\{1, 2, 3, 4, 5\}$ **12.** $\{2, 4, 6, 8\}$ **13.** $\{15, 30, 45, \ldots\}$ **14.** $\{15, 45, 75, \ldots\}$ **15.** $\{1, 3, 9, 11, 33, 99\}$
16. $\{3, 11\}$ **17.** $\varnothing$ **18.** $\{9, 33, 99\}$ **19.** 20 **20.** −36 **21.** −60 **22.** −158 **23.** −45 **24.** 60
25. −17 **26.** 4 **27.** 4 **28.** 14 **29.** 6 **30.** 1 **31.** −4 **32.** 30

CHAPTER 3

EXERCISES 3.1A

1. $x = 5$ **2.** $x = -3$ **3.** $2x - 7 = 9$ **4.** $3x + 1 = 16$ **5.** $6 - x = 10$ **6.** $\frac{x}{4} + 5 = 0$ **7.** $x + 5 = 7 - x$
8. $4x - 13 = x + 2$ **9.** $x + y = 10$ **10.** $2x - 3y = 6$ **11.** $-x + 4 = 4 - x$ **12.** $-x = -1(x)$
13. $ab + ac = a(b + c)$ **14.** $a(b - c) = ab - ac$ **15.** $y - x = -1(x - y)$ **16.** $\frac{x}{x} = 1$ **17.** $-1(a - b) = b - a$
18. $\frac{x}{1} = x$ **19.** cd **20.** $c + d$ **21.** $7x$ **22.** $x + 5$ **23.** $2(y + 5)$ **24.** $xy + 2$ **25.** $x + y + 3$
26. $3(x + 6)$ **27.** $3x + (5x + 4)$ **28.** $8xy$ **29.** $a(b + c)$ **30.** $4y + (2y - 9)$

EXERCISES 3.1B

1. $(4 + 6) + 7 = 17$ **2.** $3 + (2 + 6) = 11$ **3.** $3 + (9 + 5) = 17$ **4.** $(25 + 75) + 87 = 187$ **5.** $(3 \cdot 7)4 = 84$
6. $(2 \cdot 50)8 = 800$ **7.** $21(4 \cdot 5) = 420$ **8.** $57(25 \cdot 4) = 5700$ **9.** $45 + (55 + 87) = 187$ **10.** $96(8 \cdot 125) = 96{,}000$
11. $x + y + 2$ **12.** $4cd$ **13.** $5rs$ **14.** $7(x + 6)$ **15.** $5(x + y + 2)$ **16.** $x + y + 3$ **17.** $6xy$
18. $15pq$ **19.** $2(x + y)$ **20.** $xy + z + 5$ **21.** $5x + 7x + 2$ **22.** $8y + 4y + 6$ **23.** $8x - 10y + 6$
24. $8y - 9y + 5$ **25.** $2x + 4x + y + 3y$ **26.** $9x - 5x - 2y + 6y$ **27.** $3x - x - 3$ **28.** $6t - 2t - 1$
29. $-a + b + c$ **30.** $a - b + c$ **31.** $-a - b + c$ **32.** $-x + 4$ **33.** $-y + 6$ **34.** $x - 5$ **35.** $x + y + 5$
36. $a + b - 7$ **37.** $x - y$ **38.** $2t + 4t - 5$ **39.** $-r + 3r + 8t - 6t$ **40.** 0

EXERCISES 3.1 S

1. $2x + 1 = 15$ **2.** $3y - 12 = 0$ **3.** $6y - 25 = y$ **4.** $3x + 7 = 2x$ **5.** $a(b + c) = ab + ac$
6. $ab - ac = a(b - c)$ **7.** $-1(x - y) = y - x$ **8.** $(5 - 2)x = 5x - 2x$ **9.** $8x + 8y = 8(x + y)$ **10.** $t - 5 = -1(5 - t)$
11. $cd + 4$ **12.** $8(xy)$ **13.** $(c + d) + 4$ or $c + d + 4$ **14.** $4(c - d)$ **15.** $5(a + b)$ **16.** $3(x + y)$
17. $(3x)y$ or $3xy$ **18.** $st + 7$ **19.** $rs + t$ **20.** $b(a + 4)$ **21.** $(15 + 10) + 5 = 30$ **22.** $(8 + 7) + 4 = 19$
23. $68 + (35 + 65) = 168$ **24.** $59 + (49 + 51) = 159$ **25.** $(3 \cdot 4)5 = 60$ **26.** $(4 \cdot 25)9 = 900$
27. $47(2 \cdot 50) = 4700$ **28.** $92(4 \cdot 25) = 9200$ **29.** $(8 \cdot 125)25 = 25{,}000$ **30.** $(43 + 57) + 39 = 139$ **31.** $2xy$
32. $b + c + d$ **33.** $x + y + 5$ **34.** xyz **35.** $n(m + 1)$ **36.** $7kn$ **37.** $5mn$ **38.** $15xy$
39. $9(x + y + 4)$ **40.** $6(a + b)$ **41.** $6x + 2x + 9$ **42.** $7y - 3y + 4$ **43.** $2x - 5y + 10$ **44.** $-x + y - 1$
45. $4x + 5x - 8$ **46.** $8x - 3x - 3$ **47.** $5x - 2x + 3y - 6y$ **48.** $6a - 5a + 6b - 7b$

EXERCISES 3.2A

1. $7x$ **2.** $3y$ **3.** $2x$ **4.** y **5.** 0 **6.** $-6z$ **7.** $2y$ **8.** $10x$ **9.** y **10.** $8z$ **11.** $2x + 5$
12. $x - 2$ **13.** $y - 6$ **14.** $-x + 5$ **15.** $4x + 10$ **16.** $x^2 - x - 6$ **17.** $3y^2 - 7y + 2$ **18.** $5x + 3y$
19. $s + t$ **20.** y **21.** $9xy$ **22.** $-5xy$ **23.** $x^2 - xy + y^2$ **24.** $9xy + xz$

EXERCISES 3.2B

1. $2x + 8$ **2.** $5x - 15$ **3.** $-3x - 18$ **4.** $-4x + 4$ **5.** $12x + 18$ **6.** $21x + 28$ **7.** $12y - 20$
8. $y^2 - 4y$ **9.** $-6x - 7$ **10.** $-18r - 9$ **11.** $-8t + 16$ **12.** $-4y + 5$ **13.** $-6y + 9$ **14.** $-5t - 2$
15. $5x + 5y - 15$ **16.** $8x - 12y + 8$ **17.** $3x + 15y + 21$ **18.** $50x - 60y - 80$ **19.** $-3x - 12y + 3$
20. $-14x + 35y - 42$ **21.** $-x - y - 1$ **22.** $-x + y - 1$ **23.** $-2x + y + 2$ **24.** $-4x - 9y + 8$
25. $6x^3 - 4x^2 - 2x$ **26.** $-x^3 - 3x^2 - 4x$ **27.** $2x^2 - 8x$ **28.** $3y^2 + 15y$ **29.** $-4t^2 - 28t$ **30.** $5n^2 - 5n$

EXERCISES 3.2C

1. $3x + 4$ **2.** $-2x + 4$ **3.** $3y - 5$ **4.** $-6t + 7$ **5.** $-x + 36$ **6.** $2y - 12$ **7.** $-3z + 5$ **8.** $8n - 2$
9. $x + 5$ **10.** $-6x + 18$ **11.** $-2x + 7y$ **12.** $y^2 - 3y + 6$ **13.** $9a + b$ **14.** y **15.** $6r + 8s$
16. $x - y$ **17.** $2x$ **18.** $-3r - 3s + 6t$ **19.** $-7y + 4$ **20.** x

EXERCISES 3.2 S

1. $6x$ **2.** $6y$ **3.** $-3y$ **4.** $-x$ **5.** 0 **6.** $7n$ **7.** $2x$ **8.** x **9.** 0 **10.** $10k$ **11.** $8x - 2$
12. $2y + 5$ **13.** −2 **14.** $2x - 2$ **15.** $-5t + 6$ **16.** $y^2 + y - 6$ **17.** $5k^2 - 3k - 8$ **18.** $-2x + 5y$
19. $2s$ **20.** $3m + 9n$ **21.** xy **22.** $-5xy$ **23.** $r^2 - 8rs - s^2$ **24.** $ab - ac$ **25.** $-3xyz$ **26.** $x^2 + 2x$

27. $2y + xy$ **28.** $2t - 6$ **29.** $2x$ **30.** $5 - 2y$ **31.** $3y + 18$ **32.** $-4x - 28$ **33.** $7x - 63$
34. $-6x + 48$ **35.** $2x^2 - 10x$ **36.** $-12y + 3y^2$ **37.** $-5x - 8$ **38.** $-6y + 9$ **39.** $10x - 15y + 30$
40. $-32x - 40y + 16$ **41.** $-r + s + t$ **42.** $-r + 2s - 1$ **43.** $2x^3 - 2x^2 - 2x$ **44.** $-8y - 4y^2 + 4y^3$
45. $-a^2 - ab - ac$ **46.** $a^3b - a^2b^2 - ab^3$ **47.** $x^3y - 5x^2y^2 - 6xy^3$ **48.** $-2x^2y - 14xy^2 + 2xy$
49. $4x^3 - 8x^2 - 4x$ **50.** $-6t^3 - 8t^2 + 10t$ **51.** $4x - 10$ **52.** $4y - 8$ **53.** $-5x - 2$ **54.** $2t - 7$
55. $x + 6$ **56.** $2y - 4$ **57.** $-4n - 3$ **58.** $x - 3$ **59.** $x - 6$ **60.** $5x + 6y$ **61.** $x^2 + 3x - 6$
62. $x + 5y$ **63.** $2r - 19s$ **64.** x **65.** $2c$ **66.** $x - 9$ **67.** $-2x + 2y - 2z$ **68.** $4x - y + 12$
69. $x - 3$ **70.** $8x - 7y - 9$

EXERCISES 3.3A

1. 12, addition **2.** 4, subtraction **3.** 3, division **4.** 36, multiplication **5.** −5, subtraction
6. −2, addition **7.** −16, division (or multiplication) **8.** 50, multiplication **9.** 0, subtraction
10. 4, multiplication by −1 **11.** −6, division **12.** −12, multiplication **13.** −13, subtraction
14. −2, addition **15.** 0, addition **16.** 4, division **17.** 32, multiplication **18.** 0, division
19. −50, multiplication **20.** −400, division

EXERCISES 3.3B

1. 4 **2.** −7 **3.** 45 **4.** −6 **5.** 9 **6.** 13 **7.** 12 **8.** −4 **9.** −2 **10.** 3 **11.** 4
12. 18 **13.** 5 **14.** −3 **15.** −8 **16.** −13 **17.** 7 **18.** 8 **19.** −10 **20.** 5

EXERCISES 3.3 S

1. 7, subtraction **2.** 15, addition **3.** 75, multiplication **4.** 6, division **5.** −5, addition **6.** −7, division
7. −4, subtraction **8.** −80, multiplication **9.** 24, multiplication **10.** −1, subtraction **11.** 0, subtraction
12. 8, division **13.** 0, division **14.** 30, division (or multiplication) **15.** −2, division
16. −200, multiplication **17.** 2, division **18.** −10, subtraction **19.** 0, addition **20.** −6, multiplication
21. 6 **22.** 10 **23.** 8 **24.** 7 **25.** −15 **26.** 5 **27.** 9 **28.** 0 **29.** −35 **30.** −1
31. 6 **32.** −9 **33.** 11 **34.** −5 **35.** 0 **36.** 21

EXERCISES 3.4A

1. 3 **2.** 7 **3.** −6 **4.** 2 **5.** −3 **6.** 9 **7.** 7 **8.** 8 **9.** −2 **10.** 13 **11.** 13
12. 2

EXERCISES 3.4B

1. ∅ **2.** All x **3.** All x **4.** ∅ **5.** ∅ **6.** ∅ **7.** All x **8.** All x **9.** All x **10.** ∅
11. ∅ **12.** 3 **13.** All x **14.** 0 **15.** −2 **16.** 0 **17.** ∅ **18.** All x **19.** 0 **20.** ∅
21. 3 **22.** All x

EXERCISES 3.4 S

1. −14 **2.** −4 **3.** 5 **4.** −6 **5.** ∅ **6.** −9 **7.** 10 **8.** −30 **9.** 11 **10.** −7
11. All x **12.** 0 **13.** −5 **14.** 12 **15.** ∅ **16.** All x

EXERCISES 3.5A

1. 18 cm **2.** 10 in **3.** 20 **4.** 83 **5.** 85 **6.** 3 hr **7.** 18K **8.** $4 **9.** 2 in **10.** 12 oz
11a. 3 ft **11b.** 9 sq ft **12.** 180 sq ft

EXERCISES 3.5B

1. $x = y - 4$ **2.** $x = \frac{y}{3}$ **3.** $y = x + 5$ **4.** $y = 2x$ **5.** $y = 6 - x$ **6.** $x = y + 6$ **7.** $y = x - 8$

8. $x = \frac{8 - y}{2}$ **9.** $P = S - C$ **10.** $n = \frac{72}{r}$ **11.** $B = N + C$ **12.** $H = AT$ **13.** $t = \frac{d}{r}$ **14.** $b = \frac{A}{h}$

15. $c = P - a - b$ **16.** $r = \frac{I}{Pt}$ **17.** $L = 2S - A$ **18.** $L = \frac{P - 2W}{2}$ **19.** $B = \frac{pA}{100}$ **20.** $h = \frac{2A}{b}$

21. $y = 10 - 2x$ **22.** $x = 3y + 6$ **23.** $x = \frac{10 - y}{2}$ **24.** $y = \frac{12 - 2x}{3}$ **25.** $y = x - 1$ **26.** $x = y - 1$

27. $x = \frac{5y + 20}{4}$ **28.** $y = \frac{4x - 20}{5}$ **29.** $y = \frac{-3x - 12}{4}$ **30.** $y = \frac{-ax - c}{b}$ **31.** $c = 3A - a - b$

32. $c = \frac{V}{ab}$ **33.** $C = Dn + S$ **34.** $L = \frac{R + 10}{8}$ **35.** $n = \frac{C - 3}{17}$ **36.** $n = \frac{d + L - a}{d}$ **37.** $a = \frac{mg - T}{m}$

38. $C = P - 3D$ **39.** $N = \frac{F - 386}{6}$ **40.** $p = \frac{200I}{(n + 1)L}$

EXERCISES 3.5 S

1. $200 **2.** $24,800 **3.** 46 hr **4.** 7 **5.** 4 **6.** 500 ml **7a.** $200 **7b.** $16 **8a.** $36 **8b.** 1200 sq ft **9a.** 5 cu yd **9b.** 6 in **10a.** 29 **10b.** 6705 **11a.** $45 **11b.** 3 **12a.** 10 **12b.** 5 **12c.** 40 **13.** $y = 7 - 3x$ **14.** $y = 2x - 6$ **15.** $x = \frac{-3y - 12}{2}$ **16.** $x = \frac{2y + 10}{5}$ **17.** $y = mx + b$ **18.** $I = \frac{E}{R}$ **19.** $W = \frac{P - 2L}{2}$ **20.** $b = \frac{2A}{h} - a$ **21.** $d = \frac{L - a}{n - 1}$ **22.** $R = \frac{PV}{T}$ **23.** $C = S - P$ **24.** $c = \frac{L - a}{at}$ **25.** $A = \frac{2L}{cdv^2}$ **26.** $M = \frac{QC}{100}$ **27.** $c = \frac{K - aK}{a^2}$ **28.** $r = \frac{12Mn - 12C}{Cn}$ **29.** $t = \frac{A - P}{Pr}$ **30.** $f = \frac{ab}{a + b}$

EXERCISES 3.6A

1. $x + 5 = 5(x - 3); x = 5$ **2.** $4x + 1 = 2x - 3; x = -2$ **3.** $4(x - 4) = 5(x - 6); x = 14$ **4.** $10x - 2(x - 1) = 58; x = 7$ **5.** $5 + 4x = 17 - 2x; x = 2$

EXERCISES 3.6B

1. 150 **2.** $1.50 **3.** 34 **4.** $480 **5.** $1.44 **6.** $40 **7.** $4.00 **8.** $2.70 **9** $1600 **10.** $4250 **11.** $275 **12.** $450 **13.** 160 **14.** 800 **15.** 25 **16.** 1200 **17.** $300 **18.** 7000 **19.** 25 **20.** 35 **21.** $800 **22.** $20,000 **23.** $1750 **24.** $7150 **25.** 25% **26.** 8% **27.** 20% **28.** 4% **29.** 40% **30.** 75% **31.** 55% **32.** 90% **33.** 15% **34.** 38% **35.** 20% **36.** 6.65%

EXERCISES 3.6 S

1. $3(7 - x) + 12 = 8x; x = 3$ **2.** $\frac{7 + x}{2} = x + 1; x = 5$ **3.** $3x + 5 = 5x - 3; x = 4$ **4.** $4 = x + 2(3x - 5); x = 2$ **5.** $6(x - 3) = 5(x - 5); x = -7$ **6.** -10 **7.** 36 **8.** 21 **9.** 600 **10.** 32 **11.** 150% **12.** 14% **13.** $180 **14.** 250 **15.** 100% **16.** 70% **17.** $2.16 **18.** $24 **19.** $1200 **20.** 40%

EXERCISES 3.7A

1. $30 - 5$ **2.** $x - 5$ **3.** $20 - x$ **4.** $25 + 12$ **5.** $x + 12$ **6.** $x + 24$ **7.** $40 - 7$ **8.** $40 - x$ **9.** $x - 7$ **10.** $8 + 10$ **11.** $y + 10$ **12.** $x + 8$ **13.** $x - 3$ **14.** $2x + 6$

15.

Age 3 yr ago	Age 7 yr from now
$x - 3$	$x + 7$
$x - 8$	$x + 2$
$4x - 3$	$4x + 7$

16.

Age 10 yr ago	Age 50 yr from now
$x - 10$	$x + 50$
$x + 10$	$x + 70$
$x - 40$	$x + 20$

17.

Age now	Age 6 yr ago	Age 9 yr from now
14	8	23
$14 - x$	$8 - x$	$23 - x$
$14 + y$	$8 + y$	$23 + y$

18.

Age now	Age 8 yr ago	Age 4 yr from now
x	$x - 8$	$x + 4$
$2x$	$2x - 8$	$2x + 4$
$x + 5$	$x - 3$	$x + 9$

EXERCISES 3.7B

1. $x + 20 = 3(x - 8)$; 22 yr **2.** $\frac{x}{2} = x - 9$; 18 yr **3.** $(x - 2) - 5 + (x - 5) = (x + 5) - 5$; 12 yr **4.** $(x + 25) + (x + 2) + 25 = 100$; 26 yr and 24 yr **5.** $3x + 14 = 2(x + 14)$; 42 yr

EXERCISES 3.7C

1. $2x + 2(x + 9) = 50$; 8 in, 17 in **2.** $2x + 2(4x) = 150$; 15 meters, 60 meters **3.** $x + (x + 10) + 3x = 30$; 4 ft, 14 ft, 12 ft **4.** $x + x + 2x - 5 = 35$; 10 cm, 10 cm, 15 cm **5.** $2x + 2(x + 10) = 180$; 40 ft, 50 ft **6.** $x + (x + 6) + (x + 3) = 36$; 9 cm, 12 cm, 15 cm **7.** 2 in **8.** 4 meters **9.** 15 in **10.** 2 ft

EXERCISES 3.7 S

1. $x - 14 = \frac{x + 8}{3}$; 25 yr **2.** $x + 20 = 2x$, present age is 20; $20 + y = 3(20)$, 40 yr **3.** $x + 14 = 3x$; 7 yr **4.** $x + (x - 3) + 2x = 29$; 8 yr **5.** $x + 54 + 5 = 4(x + 5)$; 13 yr **6.** $2x + 2(2x) = 54$; 9 meters, 18 meters **7.** $2x + 2(x + 12) = 60$; 9 ft, 21 ft **8.** $2x + 2(2x - 3) = 30$; 6 in, 9 in **9.** 15 ft by 30 ft **10.** $x + (3x - 8) + (2x + 1) = 35$; 7 in, 13 in, 15 in **11.** $x + 2x + 2x = 160$; 32 cm, 64 cm, 64 cm **12.** $x + (2x - 30) + x + 30 = 360$; 90 ft, 120 ft, 150 ft **13.** $(8 + x) + (9 + x) + (10 + x) = 2(8 + 9 + 10)$; 9 dm

14. $6(4 + x) = 42$; 3 ft **15.** $12 = \frac{6(8 - x)}{2}$; 4 meters **16.** 20 cm **17.** 2 ft
18. $2(48 + 2x) + 2(36 + 2x) = 176$; 1 in **19.** $2(5 + 2x) + 2(8 + 2x) = 42$; 2 meters
20. $\frac{x(x - 2)}{2} = \frac{x(x - 8)}{2} + 60$; 20 in, 12 in

SAMPLE TEST FOR CHAPTER 3

1. $rs + 6$ **2.** $9(x + 3)$ **3.** $x + y + 4$ **4.** $9cx$ **5.** $-5x - 2$ **6.** $2x^2 - 3x - 5$ **7.** $x^2 - 2xy - 8y^2$
8. $x^2y + 4xy^2$ **9.** $10 - 4x$ **10.** $4x - 4$ **11.** $6x^2 - 11x + 4$ **12.** $6x^2 - 11xy - 2y^2$ **13.** 100 **14.** -4
15. -8 **16.** -225 **17.** 5 **18.** 7 **19.** 6 **20.** -5 **21.** All x **22.** $\varnothing$ **23.** $c = 2s - a - b$
24. $y = \frac{9x - 18}{2}$ **25.** 9 ft **26.** -8 **27.** 30% **28.** \$320 **29.** 70 and 74 **30.** 10 ft by 14 ft

CHAPTER 4

EXERCISES 4.1A

1. $20x^4$ **2.** $-42y^6$ **3.** $40y^5$ **4.** $100x^7$ **5.** $-36x^6$ **6.** $49x^4$ **7.** $64x^6$ **8.** $36y^6$ **9.** $81y^4$
10. x^3y^3 **11.** $-24x^4y^4$ **12.** $8x^3$ **13.** $-27x^6$ **14.** $-25x^2y^2$ **15.** $25x^2y^2$ **16.** $-64x^6$ **17.** $-4x^6$
18. $75x^5$ **19.** $80x^5$ **20.** $56x^7$

EXERCISES 4.1B

1. $3x^2 - 5x + 8$ **2.** $x^3 - x^2 + x - 1$ **3.** $-x^4 + 14x^2 - 49$ **4.** $5x^4 + 7x^3 + 6x^2 + 4x$ **5.** $x^2 + 2x - y^2 - 4y - 3$
6. $-8x^3y + xy^2 + y^4$ **7.** $-4x^2 + 8x + 9y^2 - 6y - 10$ **8.** $-x^3 - x - y^3 - y + 1$ **9.** $-x^2 + 6x + 36y^2 - 72y + 5$
10. $2x^3y + 4x^2y^2 - 7xy^3$

EXERCISES 4.1C

1. $8x^2 - 9x + 1$ **2.** $7x - 3y + 3$ **3.** $-5x + 9$ **4.** $2y - 12$ **5.** $10x - 2$ **6.** $x^3 + 1$ **7.** $y - 7$
8. $-8x - 5y + 8z$ **9.** $y^4 - y^2 + 4$ **10.** $x^3 - 3x^2y + 3xy^2 - y^3$

EXERCISES 4.1D

1. $5x^2 + 4x + 3$ **2.** $x^2 - 3x - 4$ **3.** $-x - 9y + 5$ **4.** $-10x + 4y + 15$ **5.** $x^4 - 28x^2 + 8x + 35$
6. $-x^2 - 14x - 24$ **7.** $2y - 2z$ **8.** $5x^2y - 2x^2y^2 - 10xy^2$ **9.** $-12y^2 + 14$ **10.** $x^2 + x - 2xy - y + y^2$

EXERCISES 4.1 S

1. $21x^6$ **2.** $-16y^6$ **3.** $30t^8$ **4.** $-16x^3y^3$ **5.** $81x^4$ **6.** $100x^4$ **7.** $25y^6$ **8.** $144y^6$ **9.** $30r^4s^4$
10. $-64t^6$ **11.** $-216t^6$ **12.** $-8x^3y^3$ **13.** $3x^2 - 2x + 5$ **14.** $x^3 - 6x + 1$ **15.** $-x^4 + 10x^2 - 25$
16. $x^5 - 2x^3 + 4x - 10$ **17.** $x^4 - 4x^3 - x + 4$ **18.** $x^2 + 4xy + y^2 + 6y$ **19.** $-x^3 + 5x - 2$ **20.** $x^4 + x^2y^2 + y^4$
21. $6x^2 + x - 9$ **22.** $-5y^2 - 2y$ **23.** $x^3 + 3x^2 + 3x + 1$ **24.** $5x - 15$ **25.** $y^3 - 8$ **26.** $2y - 3$
27. $-x + y - 9$ **28.** $-x^2 + 2y^2$ **29.** $x^3 + 27y^3$ **30.** $x^4 + 4$ **31.** $3x - 3y - 25$ **32.** $11x^2 + 10xy - y^2$
33. $x^3 + 343$ **34.** $x^2 - xy - 7y^2$ **35.** $4xy - 2y^2$ **36.** $x^2y^2 - x^2 - 36y^2 + 81$ **37.** $1 - xy - xz - yz + tx$
38. $10 - 2x - 2y$ **39.** $x^3 - 9x^2 + 27x - 27$ **40.** $x + y - z - 7$

EXERCISES 4.2A

1. $14x^3 - 6x^2 - 8x$ **2.** $-20x^2 + 8xy - 28x$ **3.** $5x^4 - 10x^3y + 5x^2y^2$ **4.** $12y^4 - 18xy^3 - 6x^2y^2$
5. $6x^3y - 24x^2y^2 + 18xy^3$ **6.** $7a^3 - 35a^2b + 7a$ **7.** $-9b^4 + 9b^3 + 54b$ **8.** $32x^3y^2 - 40x^2y^3 - 48x^2y^2$
9. $-60a^4b^2 + 60a^3b^2 - 15a^2b^2$ **10.** $5x^4 + 150x^3 - 425x^2$

EXERCISES 4.2B

1. $6x^2 + 23x + 20$ **2.** $20x^2 - 19x + 3$ **3.** $12y^2 - 16y - 35$ **4.** $12y^2 + 16y - 35$ **5.** $12y^2 - 44y + 35$
6. $12y^2 + 44y + 35$ **7.** $15x^2 + 14xy - 8y^2$ **8.** $15x^2 + 2xy - 8y^2$ **9.** $15x^2 - 14xy - 8y^2$ **10.** $15x^2 - 2xy - 8y^2$
11. $4x^4 + 3x^2 - 1$ **12.** $36y^4 - 25y^2 + 4$ **13.** $xy + 2x + 3y + 6$ **14.** $xy - 5x - 4y + 20$ **15.** $xy - x + 7y - 7$
16. $xy + 4x - 6y - 24$ **17.** $x^3 - x^2 - x + 1$ **18.** $y^3 - 6y^2 + 5y - 30$ **19.** $x^4 - 5x^3 + x - 5$
20. $y^5 + 4y^3 - 2y^2 - 8$

EXERCISES 4.2C

1. $x^3 + 7x^2 + 17x + 15$ **2.** $x^3 - 2x^2y - xy^2 + 2y^3$ **3.** $x^3 - 216$ **4.** $y^3 + 6y^2 + 12y + 8$ **5.** $x^4 - 1$
6. $x^2 + 2xy + 2xz + y^2 + 2yz + z^2$ **7.** $x^4 - 5x^2 + 4$ **8.** $y^4 - 26y^2 + 25$ **9.** $y^4 + y^3 - 3y^2 - y + 2$
10. $4x^2 + y^2 + 9z^2 + 4xy - 12xz - 6yz$

EXERCISES 4.2 S

1. $-48r^3s + 48r^2s^2 - 12rs$ **2.** $-15st^3 + 75st^2 - 60st$ **3.** $-r^4 + r^3 - r^2 + r$ **4.** $-c^4d - c^3d + c^2d^2 + c^2d^3$ **5.** $8x^6 + 4x^5 - 8x^4 - 4x^3$ **6.** $8y^3 - 8y^4 - 8y^5 - 8y^6$ **7.** $10y^4 - 6y^3 + 2y^2$ **8.** $-xy + x^2y - xy^2 + x^2y^2$ **9.** $7s^3t - 7s^2t - 7st^2 + 7st^3$ **10.** $5a^2b^2c^3 + 10ab^3c^3 - 10ab^2c^4 + 25ab^2c^3$ **11.** $3x^2 + 2x - 8$ **12.** $3x^2 - 2x - 8$ **13.** $49x^2 - 4y^2$ **14.** $49x^2 - 28xy + 4y^2$ **15.** $25x^2 - 4$ **16.** $x^3 - x^2y + xy^2 - y^3$ **17.** $2x^2 - 3xy - 2y^2$ **18.** $-25x^2 + 40xy - 16y^2$ **19.** $x^2 + 4xy + 4y^2$ **20.** $49x^2 - 84xy + 36y^2$ **21.** $a^4 - b^4$ **22.** $x^4 - 16$ **23.** $x^6 - 1$ **24.** $64x^6 - 729$ **25.** $4x^2 + 24x - 7xy - 42y$ **26.** $x^3 - 8x^2 - 9x + 72$ **27.** $2y^3 - 7y^2 + 7y - 2$ **28.** $-x^3 - 9x^2y - 19xy^2 + 4y^3$ **29.** $x^3 + 125$ **30.** $y^3 - 12y^2 + 48y - 64$ **31.** $r^2 + 4rs + 4s^2 - 25$ **32.** $x^4 + 4$ **33.** $4y^4 - 20y^3 + 2y^2 - 11y + 5$ **34.** $x^4 - 3x^2y + x^2 - 6y - 2$ **35.** $x^2 + y^2 + z^2 - 2xy - 2xz + 2yz$ **36.** $9a^2 + 25b^2 + c^2 - 30ab + 6ac - 10bc$

EXERCISES 4.3

1. $6x^2(2x + 1)$ **2.** $6y^3(3y - 2)$ **3.** $5xy(2 + 3xy)$ **4.** $7xy(2y - 3x)$ **5.** $8x(2x^2 + 1)$ **6.** $9x^2(2x^2 + 1)$ **7.** $8x^2y^2(2x - 5y)$ **8.** $9x^2y^3(6x + y^2)$ **9.** $15at^2(t^2 + 5)$ **10.** $6c^2s(4s - 7c)$ **11.** $7(x + y + z)$ **12.** $4(x - y + 1)$ **13.** $15a(x - 3y - 1)$ **14.** $4a^2(x^2 + 2ax - 1)$ **15.** $9y^2(y^2 + y - 3)$ **16.** $50y(3y^2 - 4y - 5)$ **17.** $35t^2(2t^2 - 3t + 4)$ **18.** $abc(ab + bc + ac)$ **19.** $r^2s^2t^2(r^2 - s^2 - t^2)$ **20.** $24xyz(x^2 + 2y^2 - 3z^2)$ **21.** $-3x^2(x^2 + 4)$ **22.** $-2y(y^2 - 2y - 3)$ **23.** $-9x^3(5x^2 - 1)$ **24.** $-1(t^2 + 5t - 10)$ **25.** $-4y^2(y^3 + 3y + 1)$ **26.** $-n^2(n^2 - 2n - 5)$ **27.** $-5t^4(t^2 - 4t + 1)$ **28.** $-6x^3(2x^3 + 3)$ **29.** $-n^2(n^4 - n^2 + 1)$ **30.** $a^2x^2(a^2x^2 + ax + 1)$

EXERCISES 4.3 S

1. $10xy(4x - 5y)$ **2.** $25x^3(x - 4)$ **3.** $4x^4(x^2 + 16)$ **4.** $-15y^2(x^2 + y^2)$ **5.** $6x^2(7x + 1)$ **6.** $-8xy^2(5x - 4y)$ **7.** $21(x - 2y + z)$ **8.** $x^2y^2(x^2 - y^2 - 1)$ **9.** $12(4r - 7s + 2)$ **10.** $6(x^2 - 3x + 5)$ **11.** $4x(x^2 + 4x - 1)$ **12.** $5xy(2x^2 - xy - 2y^2)$ **13.** $-8y^2(3y^2 - 5y + 4)$ **14.** $rst(r + s + t)$ **15.** $-3a^2b^2c^2(a^2 + 2b^2 - c^2)$

EXERCISES 4.4

1. $x^2 - 9$ **2.** $x^2 - 25$ **3.** $y^2 - 4$ **4.** $y^2 - 16$ **5.** $25x^2 - 1$ **6.** $16x^2 - 49$ **7.** $x^2 - y^2$ **8.** $s^2 - r^2$ **9.** $36r^2 - s^2$ **10.** $x^4 - 36$ **11.** $x^4 - 64y^4$ **12.** $9y^4 - 100$ **13.** $b^4 - a^4$ **14.** $49a^2 - 4b^2$ **15.** $1 - x^2y^2$ **16.** $(x + 5)(x - 5)$ **17.** $(x + 4)(x - 4)$ **18.** $(y + 6)(y - 6)$ **19.** $(y + 7)(y - 7)$ **20.** $(8x + 1)(8x - 1)$ **21.** $(3y + 1)(3y - 1)$ **22.** $(9x + 5)(9x - 5)$ **23.** $(10y + 7)(10y - 7)$ **24.** $(5x + 8y)(5x - 8y)$ **25.** $(2x + 3y)(2x - 3y)$ **26.** $(4x^2 + 5y)(4x^2 - 5y)$ **27.** $(6y + x^2)(6y - x^2)$ **28.** $(xy^2 + z)(xy^2 - z)$ **29.** $(20ab + c)(20ab - c)$ **30.** $(x^3 + y)(x^3 - y)$ **31.** $(9x^3 - 10y^2)(9x^3 + 10y^2)$ **32.** $5x^2 - 80$ **33.** $x^3 - 9x$ **34.** $3x^3 - 12x$ **35.** $4y^3 - 4y$ **36.** $y^4 - 49y^2$ **37.** $-25t^4 + t^2$ **38.** $5(x + 4)(x - 4)$ **39.** $x(x + 3)(x - 3)$ **40.** $7x(x + 1)(x - 1)$ **41.** $x^2(2x + 1)(2x - 1)$ **42.** $2y^2(y + 10)(y - 10)$ **43.** $xy(xy + 1)(xy - 1)$ **44.** $9(x^2 + 9)$ **45.** $25x^2(x^2 + 4)$

EXERCISES 4.4 S

1. $x^2 - 36y^2$ **2.** $4a^2 - 49b^2$ **3.** $16 - x^4$ **4.** $r^2s^2 - 1$ **5.** $x^2y^2 - z^2$ **6.** $9x^6 - 64$ **7.** $81r^4 - s^4$ **8.** $y^2 - x^2$ **9.** $b^2y^2 - a^2x^2$ **10.** $y^4 - x^4$ **11.** $(x + 1)(x - 1)$ **12.** $(4x + y)(4x - y)$ **13.** $(6x + 7)(6x - 7)$ **14.** $(8y + 5)(8y - 5)$ **15.** $(x + 10y)(x - 10y)$ **16.** $(3xy + 2)(3xy - 2)$ **17.** $(9 + k)(9 - k)$ **18.** $(xy + z^2)(xy - z^2)$ **19.** $(30a^2 + 1)(30a^2 - 1)$ **20.** $(a^3 + bc)(a^3 - bc)$ **21.** $(2 + 9y^3)(2 - 9y^3)$ **22.** $(c^3 + a^2)(c^3 - a^2)$ **23.** $(11x^3 + y^2)(11x^3 - y^2)$ **24.** $(y^3 + 12x^2)(y^3 - 12x^2)$ **25.** $(x^3 + 13y^3)(x^3 - 13y^3)$ **26.** $(1 - 10t)(1 + 10t)$ **27.** $(x^2 - 7)(x^2 + 7)$ **28.** $(y^2 - 8)(y^2 + 8)$ **29.** $(x^2 - 3y^2)(x^2 + 3y^2)$ **30.** $(5a^2 - 4b^2)(5a^2 + 4b^2)$ **31.** $3(5x + 1)(5x - 1)$ **32.** $4y(y - 2)(y + 2)$ **33.** $t^2(t + 6)(t - 6)$ **34.** $x^2y^2(x + y)(x - y)$ **35.** $16x^2(2x + 1)(2x - 1)$ **36.** $16x^3(4x - 1)$ **37.** $16x^2(4x^2 + 1)$ **38.** $9(y + 3)(y - 3)$ **39.** $9y(y - 9)$ **40.** $9(y^2 + 9)$

EXERCISES 4.5

1. $(x + 2)(x + 5)$ **2.** $(y + 2)(y + 4)$ **3.** $(x - 1)(x - 2)$ **4.** $(t - 3)(t - 6)$ **5.** Prime **6.** $(n - 1)(n - 6)$ **7.** $(x + 5)(x - 2)$ **8.** $(x - 10)(x + 1)$ **9.** $(y - 6)(y + 2)$ **10.** $(x + 4y)(x - 3y)$ **11.** $(r + 7t)(r - 2t)$ **12.** Prime **13.** $(x + 7)(x + 2)$ **14.** $(x - 2)(x - 7)$ **15.** $(x + 2)(x - 7)$ **16.** $(x - 2)(x + 7)$ **17.** $(x - 1)(x - 14)$ **18.** $(x + 1)(x - 14)$ **19.** $(x - 1)(x + 14)$ **20.** $(x + 6)(x + 5)$ **21.** $(x - 4)(x - 1)$ **22.** $(x - 3)(x - 2)$ **23.** $(x - 7)(x + 3)$ **24.** $(x - 3)(x + 7)$ **25.** $(y - 4)(y + 5)$ **26.** $(y - 5)(y + 4)$ **27.** $(y - 15)(y + 1)$ **28.** Prime **29.** $(x + 13)(x + 2)$ **30.** $(x + 13)(x - 2)$ **31.** $(x - 13)(x + 2)$ **32.** $(t - 13)(t - 2)$ **33.** $(x + 6y)(x - 2y)$ **34.** $(x - 6y)(x + 2y)$ **35.** $(r - 12s)(r + s)$ **36.** $(s - 12t)(s - t)$ **37.** $(u - 5v)^2$ **38.** $(x + 6y)^2$

EXERCISES 4.5 S

1. $(x+4)(x+2)$ **2.** $(x-5)(x-6)$ **3.** $(x-6)(x+4)$ **4.** $(x+4)(x-2)$ **5.** Prime **6.** $(y-5)(y+1)$
7. $(y+5)(y-1)$ **8.** $(y-1)(y-4)$ **9.** Prime **10.** $(x+2y)(x+9y)$ **11.** $(x+8y)(x-3y)$
12. $(x-7y)(x+5y)$ **13.** Prime **14.** $(u-2v)(u+v)$ **15.** Prime **16.** $(u+8v)(u-7v)$
17. $(r-9s)(r+6s)$ **18.** $(x^2-3y^2)(x^2+y^2)$ **19.** $(x^2-5)(x^2-10)$ **20.** $(x^2-8)^2$ **21.** $(x+6)(x+8)$
22. $(x-6)(x-8)$ **23.** $(x+8)(x-6)$ **24.** $(x-8)(x+6)$ **25.** $(y+4)^2$ **26.** $(y-9)^2$ **27.** $(x^2+1)^2$
28. $(x^2-6)^2$ **29.** $(x^2-2)(x^2+1)$ **30.** Prime **31.** $(x-y)(x-9y)$ **32.** $(x-3y)^2$ **33.** $(x^2+5y)^2$
34. $(x^2+25y)(x^2+y)$ **35.** $2(x+2)(x+3)$ **36.** $x(x-3)(x-7)$ **37.** $5x(x-5)(x+2)$ **38.** $2x(x-7)^2$
39. $6y(y+6)(y-4)$ **40.** $100y^2(y-1)^2$ **41.** $16x^2(x^2+x-1)$ **42.** $16x^2(x-1)^2$

EXERCISES 4.6

1. $x^2+8x+16$ **2.** x^2-6x+9 **3.** $4x^2-20x+25$ **4.** $4x^2-20x+25$ **5.** $25x^2-20x+4$
6. $x^2-14xy+49y^2$ **7.** $x^2-2xy+y^2$ **8.** x^4+12x^2+36 **9.** $16y^6+8y^3+1$ **10.** $x^2y^2+16xy+64$
11. $(x+4)^2$ **12.** $(y+1)^2$ **13.** $(y-2)^2$ **14.** $(y-3)^2$ **15.** Not a perfect square **16.** Not a perfect square
17. $(2x+5)^2$ **18.** $(2x-5)^2$ **19.** Not a perfect square **20.** Not a perfect square **21.** Not a perfect square
22. $(x-8y)^2$ **23.** $(x+9y)^2$ **24.** Not a perfect square **25.** $(x-10y)^2$ **26.** $(3x+1)^2$ **27.** $(4y-1)^2$
28. $(5t^2-1)^2$ **29.** $(7x+6y)^2$ **30.** Not a perfect square **31.** $120x$; $(x+60)^2$ **32.** $14xy$; $(7x-y)^2$
33. $4y^2$; $(2y^2-1)^2$ **34.** $144xy$; $(8x+9y)^2$ **35.** $20xy$; $(10xy+1)^2$ **36.** $60x^2y^2$; $(6x^2-5y^2)^2$ **37.** $2abc$; $(ab+c)^2$
38. $4t^2$; $(t^2-2)^2$ **39.** $22xy$; $(x-11y)^2$ **40.** $24x$; $(x+12)^2$

EXERCISES 4.6 S

1. $x^2+10x+25$ **2.** $y^2-12y+36$ **3.** $100x^2-60xy+9y^2$ **4.** $x^2y^2+8xy+16$ **5.** $64x^4-16x^2+1$
6. $25x^2+10xy+y^2$ **7.** $81a^2-72ab+16b^2$ **8.** $81a^2-72ab+16b^2$ **9.** $81a^2+72ab+16b^2$
10. $16a^2-72ab+81b^2$ **11.** $81x^4-18x^2+1$ **12.** y^4-20y^2+100 **13.** $x^4+2x^2y^2+y^4$ **14.** $x^4-2x^2y^2+y^4$
15. $x^4-2x^2y^2+y^4$ **16.** $36x^4-60x^2y^2+25y^4$ **17.** x^4+16x^2+64 **18.** x^4-4x^2+4 **19.** $16t^4-8t^2+1$
20. x^6+2x^3+1 **21.** $(3x-7y)^2$ **22.** $(x^2-5)^2$ **23.** $(y^2+2)^2$ **24.** $(5x^2+6y^2)^2$ **25.** Not a perfect square
26. Not a perfect square **27.** $(5x+1)^2$ **28.** $(7y-1)^2$ **29.** Not a perfect square **30.** $(6x-1)^2$
31. $(10t^2+1)^2$ **32.** $(8t^2-1)^2$ **33.** $(2x-y)^2$ **34.** Not a perfect square **35.** $(3x^2-1)^2$ **36.** $(x^2+1)^2$
37. Not a perfect square **38.** $(10c^2-9)^2$ **39.** $(x^3+y^3)^2$ **40.** Not a perfect square

EXERCISES 4.7

1. $3x+4$ **2.** $2x+1$ **3.** $y-9$ **4.** $2y-3$ **5.** $2x-3$ **6.** $x+2$ **7.** $x-10$ **8.** $5x-3$
9. $2t+3$ **10.** $6t-7$ **11.** $(6x-7)(x+1)$ **12.** $(6x+7)(x-1)$ **13.** $(6x-7)(x-1)$ **14.** $(6x+7)(x+1)$
15. $(3y+1)(3y+2)$ **16.** $(3y-1)(3y+2)$ **17.** $(3y+1)(3y-2)$ **18.** $(3y-1)(3y-2)$ **19.** $(5x+1)(x+2)$
20. $(x-7)(3x-1)$ **21.** $(x-3)(2x+1)$ **22.** $(x+2)(7x-1)$ **23.** $(y-1)(3y+5)$ **24.** $(y+1)(2y-5)$
25. $(t-1)(7t+5)$ **26.** Prime **27.** $(x+y)(11x-2y)$ **28.** $(3x-y)(11x+y)$ **29.** $(x-2y)(2x-3y)$
30. $(x-2y)(2x+3y)$ **31.** $(y+2)(5y-4)$ **32.** $(y-2)(5y+4)$ **33.** $(2x-5)(3x-1)$ **34.** Prime
35. $(2x-3)(2x-1)$ **36.** $(3x-5)(3x+1)$ **37.** $(4x-1)(x-4)$ **38.** $(9x-5)(x+1)$ **39.** $(5x+1)(x-1)$
40. Prime

EXERCISES 4.7 S

1. $3r-2$ **2.** $6r-7$ **3.** $13x+10y$ **4.** $2x-y$ **5.** $5x-2y$ **6.** $2x+3y$ **7.** $x-4y$ **8.** $3x-2$
9. $3x+2$ **10.** $3x-2$ **11.** $(4s-5t)(2s+t)$ **12.** $(8s+3t)(5s-4t)$ **13.** $(12x-y)(x-8y)$
14. $(4x-25y)(5x+4y)$ **15.** $(2a+7b)(5a-2b)$ **16.** $(2x+3y)(15x-7y)$ **17.** $(7a-4b)(3a-2b)$
18. $(7a-4b)(3a+2b)$ **19.** $(3y+1)(5y-2)$ **20.** $(3y-1)(5y+2)$ **21.** $(3x-y)(5x-2y)$ **22.** $(x-y)(15x-2y)$
23. $(x+2)(15x-1)$ **24.** Prime **25.** $(2t-1)(3t+2)$ **26.** $(3a+5)(9a+1)$ **27.** $(2b+1)(5b-2)$
28. $(x-5)(8x+3)$ **29.** $(2x+5)(4x-3)$ **30.** $(3x-2y)(5x-4y)$ **31.** Prime **32.** $(3r+4s)(6r-7s)$
33. $(3r-4s)(6r+7s)$ **34.** $(2r+7s)(9r-4s)$ **35.** $(2x-9)(5x+1)$ **36.** $(2x-1)(5x+9)$ **37.** $(2x-3)(5x-3)$
38. $(10x-1)(x+9)$ **39.** $(10x-9)(x-1)$ **40.** $(10x+3)(x-3)$

EXERCISES 4.8

1. e **2.** b **3.** f **4.** h **5.** a **6.** g **7.** d **8.** c **9.** $(3x-5)(3x+5)$ **10.** $x(9x-25)$
11. $9(x^2+4)$ **12.** Prime **13.** $(x-6)^2$ **14.** $(3x-5)^2$ **15.** $(x-9)(x+4)$ **16.** $(9x+1)(x+2)$
17. $9(x^2+2x+3)$ **18.** $(7x-1)(7x+1)$ **19.** $(x+8)(x-4)$ **20.** $(5x+3)(x-1)$ **21.** $(x^2-10)(x^2+10)$
22. $x^3(x-100)$ **23.** $(x^2+10)^2$ **24.** $x^2(x^2-x+1)$ **25.** $(x^2-y)(x^2+2y)$ **26.** $(x^2+5)(x^2+4)$

EXERCISES 4.8 S

1. e **2.** g **3.** f **4.** h **5.** b **6.** a **7.** d **8.** c **9.** $(x + 6)(x - 6)$ **10.** $(x^2 + 6)(x^2 - 6)$
11. $x^2(x^2 + 36)$ **12.** $(x - 7)^2$ **13.** $(x^2 - 7)^2$ **14.** $(x - 2)(x + 8)$ **15.** $(x^2 - 2)(x^2 + 8)$ **16.** $(7x - 3)(x + 1)$
17. $(7x^2 - 3)(x^2 + 1)$ **18.** $(x^2 - 8y)(x^2 + 8y)$ **19.** $x^2(x^2 + 64)$ **20.** $(6x^2 + 1)^2$ **21.** $(2x - 5)(2x + 5)$
22. $4x(x - 25)$ **23.** $27y^2(y - 3)$ **24.** $(2x - 7)^2$ **25.** $(t - 6)(t + 5)$ **26.** $3x(x^2 - x - 10)$ **27.** $(3x + 2)(x - 5)$
28. $(x^2 - 6y)(x^2 + 6y)$ **29.** $(x^2 - 2)(x^2 - 6)$ **30.** $(5x^2 + 2)^2$ **31.** $xy(x^2 + 4y^2)$ **32.** $(5n^2 + 1)(n^2 + 8)$
33. $(x^2 + 4)(x^2 - 3)$ **34.** $(3y^2 - 7)^2$ **35.** $(5x^2 - 2)(x^2 + 1)$ **36.** Prime **37.** $4(x^2 + 25)$ **38.** Prime
39. $16(y^2 + 4)$ **40.** Prime

EXERCISES 4.9

1. $5x(x - 2)(x - 5)$ **2.** $(2x - 3y)(2x + 3y)(4x^2 + 9y^2)$ **3.** $4(5x - y)(5x + y)$ **4.** $4x(25x - 1)$ **5.** $16y^2(y^2 + 4)$
6. $6x^2(x - 7)(x + 3)$ **7.** $9x^2(x - 4)^2$ **8.** $25(x^2 - 2x + 4)$ **9.** $(4x - 5y)(4x + 5y)(16x^2 + 25y^2)$ **10.** $15(x + 2)^2$
11. $12(x - 1)(x + 6)$ **12.** $36x^2(x - 2)(x + 2)$ **13.** $21x(x^2 - x - 1)$ **14.** $x(x - 1)(3x - 5)$ **15.** $3(2x - 1)(7x + 2)$
16. $4(6x - 7)(3x + 1)$ **17.** $(x - 1)(x + 1)(x - 3)(x + 3)$ **18.** $4(2x + 1)(2x - 1)(x + 2)(x - 2)$
19. $25(5x - 1)(5x + 1)(x^2 + 1)$ **20.** $(6x - 1)(6x + 1)(2x^2 + 5)$

EXERCISES 4.9 S

1. $9(x - 2)(x + 1)$ **2.** $9y(x + 2)(x - 1)$ **3.** $x(x - 6)(3x + 1)$ **4.** $4xy(6x - 5y)(6x + 5y)$
5. $4(3x - y)(3x + y)(9x^2 + y^2)$ **6.** $9t(3t^2 + 4t + 7)$ **7.** $x(7 - x)(7 + x)$ **8.** $y^4(y - 2)(y + 2)(y^2 + 4)$
9. $5(x - 5)(x + 2)$ **10.** $(2y - 7)^2$ **11.** $2(x - 2)(3x - 4)$ **12.** $2c^2(5c - 1)^2$ **13.** $y(y - 1)(y + 1)(y^2 + 1)$
14. $6(x - 2)(x - 7)$ **15.** $-3x(x^2 + 9)$ **16.** $5a(a - 2)(a + 2)(a^2 + 16)$ **17.** $b^2(3b - 4)(2b + 3)$
18. $x^2(x - 4)(x + 4)$ **19.** $(x - 2)(x + 2)(x - 3)(x + 3)$ **20.** $4(a^2 - 2b^2)(a^2 + 2b^2)$ **21.** $2(xy - 2)(xy + 2)(x^2y^2 + 4)$
22. $t^2(t - 2)(t + 2)(t^2 + 9)$ **23.** $4x^2(x - 6)^2$ **24.** $(x - 5)(x + 5)(x^2 + 1)$ **25.** $4(2u^2 - 3v^2)(2u^2 + 3v^2)$
26. $-x^2(x - 5)(x + 1)$ **27.** $3x(x - 7)(x + 2)$ **28.** $4(y - 6)^2$ **29.** $(2n - 1)(2n + 1)(4n^2 + 1)$
30. $(x - 3)(x + 3)(x^2 + 4)$ **31.** $x^2(2x - 5)(x - 1)$ **32.** $(x - 3y)(x + 3y)(x^2 + 9y^2)$ **33.** $x^2(x - 8)(x + 8)$
34. $(x - 3)^2(x + 3)^2$ **35.** $4(y - 1)(y + 1)(y^2 + 10)$ **36.** $10xy(x + 2y)^2$

EXERCISES 4.10

1. $\{4, -2\}$ **2.** $\{-7, 2\}$ **3.** $\{5, -5\}$ **4.** $\{7, -7\}$ **5.** $\{0, 6\}$ **6.** $\{6\}$ **7.** $\{4\}$ **8.** $\{-6, 5\}$ **9.** $\{5, -5\}$
10. $\{-2, -3\}$ **11.** $\{0, 3\}$ **12.** $\{2\}$ **13.** $\{0, 4\}$ **14.** $\{-1, -4\}$ **15.** $\{10, -10\}$ **16.** $\{0, -10\}$
17. $\{2, -1\}$ **18.** $\{2, 5\}$ **19.** $\{0, 16\}$ **20.** $\{0\}$

EXERCISES 4.10 S

1. $\{-1, -5\}$ **2.** $\{6, -3\}$ **3.** $\{3, -3\}$ **4.** $\{8, -8\}$ **5.** $\{0, -7\}$ **6.** $\{4, -2\}$ **7.** $\{-7\}$ **8.** $\{2, -2\}$
9. $\{8, -6\}$ **10.** $\{1, -1\}$ **11.** $\{3, 5\}$ **12.** $\{0, 9\}$ **13.** $\{-1\}$ **14.** $\{6, -7\}$ **15.** $\{0, 4\}$ **16.** $\{4, 7\}$
17. $\{8, -5\}$ **18.** $\{1, 5\}$ **19.** $\{0, -4\}$ **20.** $\{0, 12\}$

EXERCISES 4.11

1. 5 meters **2.** 7 in **3.** 20 ft **4.** 50 meters **5.** 12 ft **6.** 12 ft **7.** 15 ft **8.** 8 ft **9.** 2 ft
10. 20 meters by 40 meters **11.** 1 ft **12.** 3 cm **13.** 6 **14.** 2 moles **15.** 2 mg **16.** 5 hr
17. 2 amp or 20 amp **18a.** 78 ft **18b.** 50 m.p.h. **19.** 30 cm

SAMPLE TEST FOR CHAPTER 4

1a. $x^4 - x^2 + 18x - 81$ **1b.** $-2x^3y^2 + 5x^2y^3 - 10xy^4$ **2.** $x^2 - 3$ **3.** $5x - 7$ **4.** $x^4 - 25x^2 + 12x + 60$
5. $9x^2 + y^2$ **6.** $-14x^5 + 2x^3 - 6x^2$ **7.** $20x^2 - 7x - 6$ **8.** $18xy - 3x - 42y + 7$ **9.** $36x^2 - 84x + 49$
10. $12x^3 - 8x^2y - 21xy^2 - 5y^3$ **11.** $3ax^2(6x + 1)$ **12.** $(y + 9)(y - 3)$ **13.** $(x - 1)(2x - 1)$ **14.** $(y^2 + 8)^2$
15. $(x + 2)(x - 2)(x^2 + 4)$ **16.** $3(x - 7)(x + 5)$ **17.** $x^2(x - 5)(2x + 3)$ **18.** $(y - 2)^2(y + 2)^2$ **19.** $\{-7, 10\}$
20. $\{-6, 6\}$

CHAPTER 5

EXERCISES 5.1A

1. Yes **2.** No **3.** No **4.** Yes **5.** No **6.** Yes **7.** Yes, if $k \neq 0$ **8.** No **9.** 5 **10.** 6
11. 3 **12.** $\frac{1}{2}$ **13.** 4 **14.** 6 **15.** -5 **16.** 0 **17.** $\frac{5}{4}$ **18.** $\varnothing$ **19.** All x but 5
20. All x but 0

EXERCISES 5.1B

1. $\frac{2}{3}$ **2.** $\frac{7}{8}$ **3.** $\frac{1}{9}$ **4.** $\frac{1}{25}$ **5.** $\frac{3}{5}$ **6.** $\frac{1}{8}$ **7.** $\frac{2}{3x}$ **8.** $\frac{4}{9xy}$ **9.** $4x$ **10.** $\frac{2x}{3y}$ **11.** $\frac{2x}{3}$
12. $\frac{1}{5y}$ **13.** $\frac{3a}{5b}$ **14.** $\frac{5s^3}{3r}$ **15.** $\frac{5c}{4b}$ **16.** $\frac{3xyz}{4}$ **17.** $\frac{rt^2}{3}$ **18.** $\frac{3a^2}{2t^3}$ **19.** $\frac{x^2-1}{x^2+1}$ **20.** $\frac{x}{x-5}$
21. $\frac{x+1}{x+2}$ **22.** $\frac{4x}{x+1}$ **23.** $\frac{x+7}{x+4}$ **24.** $\frac{2x+1}{x+2}$ **25.** $\frac{x-7}{x+7}$ **26.** $\frac{2y+3}{2y-3}$ **27.** $\frac{1}{3x}$ **28.** y^2+4

EXERCISES 5.1C

1. $\frac{-1}{3x}$ **2.** $\frac{-2x}{9y}$ **3.** $\frac{-1}{3}$ **4.** -1 **5.** $\frac{-2}{y}$ **6.** -1 **7.** $\frac{-3}{x}$ **8.** -1 **9.** $\frac{-x}{x+1}$
10. $\frac{-x-6}{x}$

EXERCISES 5.1 S

1. $\frac{3}{4}$ **2.** $\frac{1}{4}$ **3.** $\frac{9}{16}$ **4.** $\frac{1}{4}$ **5.** $\frac{4}{5}$ **6.** $\frac{2}{125}$ **7.** $\frac{2x}{3y}$ **8.** $\frac{2x}{3}$ **9.** $\frac{1}{5y}$ **10.** $\frac{4}{x-2}$
11. $\frac{x^2+9}{2(x^2-9)}$ **12.** $\frac{x+6}{x}$ **13.** $\frac{x-2}{x-3}$ **14.** $\frac{x+3}{x}$ **15.** $\frac{x-5}{x-11}$ **16.** $\frac{x+3}{x-3}$ **17.** $\frac{y+6}{y-6}$
18. $\frac{5x+2y}{5x-2y}$ **19.** $x+2$ **20.** $\frac{1}{2(x^2-8)}$ **21.** $\frac{-2}{5x}$ **22.** $\frac{-2}{3x}$ **23.** $\frac{-3y^2}{5x^2}$ **24.** -1 **25.** -1
26. $\frac{-1}{x}$ **27.** $\frac{-5}{x}$ **28.** $\frac{-x}{x+2}$ **29.** $\frac{1-y}{1+y}$ **30.** $\frac{-y-5}{y-5}$

EXERCISES 5.2A

1. 0.125 **2.** 0.1875 **3.** 0.08 **4.** $0.\overline{6}$ **5.** $0.\overline{5}$ **6.** $0.\overline{72}$ **7.** $0.\overline{142857}$ **8.** 1.75 **9.** 7.5
10. $8.\overline{3}$ **11.** 1.03125 **12.** $3.\overline{142857}$ **13.** $\frac{2}{3}$ **14.** $\frac{25}{99}$ **15.** $\frac{13}{45}$ **16.** 6 **17.** $\frac{499}{99}$ **18.** $\frac{26}{11}$
19. $\frac{1373}{333}$ **20.** $\frac{631}{165}$

EXERCISES 5.2B

1. 11 gal **2.** 480 mi **3.** \$9 **4.** 20 **5.** 12 **6.** \$12.50 **7.** \$84 **8.** \$300 **9.** \$7.50
10. 60¢ **11.** \$18 **12.** \$180 **13.** 20 oz **14.** 6 **15.** 40 **16.** 36 oz **17.** 100 grams
18. $1\frac{3}{4}$ oz **19.** 12 cases **20.** $1\frac{2}{3}$ yd **21.** $7\frac{1}{2}$ meters **22.** 2 **23.** 9 yd **24.** \$2.50
25a. 10 ears **b.** \$1.32 **c.** $16\frac{1}{2}$¢ **d.** No **26a.** 20 lb **b.** \$29 **c.** \$2.90 **d.** Yes **27.** 8 lb
28. 45 **29.** 125 **30.** $19\frac{1}{5}$ lb **31.** 540 cal **32.** $7\frac{1}{2}$ oz

EXERCISES 5.2 S

1. 0.625 **2.** 4.25 **3.** $0.1\overline{6}$ **4.** $0.\overline{5}$ **5.** $2.\overline{27}$ **6.** $0.41\overline{6}$ **7.** $\frac{7}{9}$ **8.** $\frac{7}{8}$ **9.** $\frac{5}{33}$ **10.** $\frac{21}{4}$
11. $\frac{701}{99}$ **12.** $\frac{5}{37}$ **13.** 450 cal **14.** 7.5 oz **15.** 40 min **16.** 210 cal **17.** \$16 **18.** 25 bulbs
19. 20 oz **20.** 725 mg **21.** $7\frac{1}{2}$ lb **22.** 150 **23.** \$1.72 **24.** No

EXERCISES 5.3A

1. $\frac{1}{5}$ **2.** $\frac{-2}{7}$ **3.** 12 **4.** $\frac{-2}{13}$ **5.** $\frac{-5}{2}$ **6.** 0 **7.** $\frac{2x^2}{7}$ **8.** $\frac{4y}{3}$ **9.** $2x$ **10.** $\frac{1}{7a}$
11. $\frac{3}{t+3}$ **12.** $\frac{8}{2x+3y}$ **13.** $\frac{x^2}{3(x-7)}$ **14.** $\frac{2(x+1)}{x^2(x+2)}$ **15.** -1 **16.** $\frac{t^2-4}{t^2-1}$ **17.** -1 **18.** 1
19. 1 **20.** $\frac{-3}{x}$

EXERCISES 5.3B

1. $\frac{2y}{5x}$ **2.** $\frac{4x}{3}$ **3.** $\frac{t^3}{3}$ **4.** $\frac{3}{50bc^4}$ **5.** $\frac{(x+y)^2}{4(x-y)^2}$ **6.** 1 **7.** $(5a+6b)^2$ **8.** $\frac{-2}{3x}$ **9.** $\frac{a-1}{a+1}$
10. $\frac{x(x+2y)}{5y(x-2y)}$ **11.** $5(x+5)^2$ **12.** $\frac{-1}{x(x-2)(x+2)}$

EXERCISES 5.3 S

1. $\frac{8}{15}$ **2.** $\frac{-7}{33}$ **3.** 27 **4.** 2 **5.** $\frac{-9}{2}$ **6.** 1 **7.** $\frac{y^2}{4}$ **8.** 1 **9.** $4x$ **10.** $x - y$
11. $\frac{1}{2(x + 8)}$ **12.** $\frac{7(x - 10)}{9}$ **13.** $\frac{4(x + 3y)}{3(x - 2y)}$ **14.** $\frac{y(x - 1)}{x(y + 1)}$ **15.** $\frac{b - 3}{b + 2}$ **16.** $\frac{3r + s}{4(r - 3s)}$ **17.** -1
18. $\frac{x - 7}{x + 7}$ **19.** 1 **20.** $\frac{-x}{4}$ **21.** $\frac{y}{3}$ **22.** $5x$ **23.** $\frac{5s^3}{6}$ **24.** $\frac{1}{25t^2}$ **25.** $\frac{2(a + 2b)}{a - 2b}$ **26.** -1
27. $\frac{1}{(7r - 2s)^2}$ **28.** $\frac{2(y - 6)}{y(y + 6)}$ **29.** $\frac{2(t + 5)}{5(t - 5)}$ **30.** $\frac{b^3}{a^3}$

EXERCISES 5.4

1. 15 **2.** 12 **3.** 60 **4.** 40 **5.** $10x$ **6.** $-x$ **7.** -4 **8.** $-6x^2$ **9.** -1 **10.** -4
11. $x^2 + 6x + 8$ **12.** $y^2 - 16$ **13.** $3x + 9$ **14.** $3x - 9$ **15.** $-x^2 + x$ **16.** $2y + 8$ **17.** $5x$ **18.** $12t$
19. $x^2 - 3x + 2$ **20.** $5x - 25$ **21.** $3t^2 + 3t$ **22.** $5y^2 + 20y$ **23.** $2t^2$ **24.** $8x^2 - 4x - 4$ **25.** $10x - 40$
26. $x^2 + 2x + 1$ **27.** $x^2 - 16$ **28.** $x^2 + 3x - 4$ **29.** $5x^2 + 26x + 5$ **30.** $x^3 + 16x^2 + 64x$

EXERCISES 5.4 S

1. 28 **2.** 20 **3.** 48 **4.** 35 **5.** 125 **6.** 8 **7.** 16 **8.** $63x^2$ **9.** 10 **10.** $8y$ **11.** $-y$
12. -2 **13.** -1 **14.** $5x + 30$ **15.** $3x + 12$ **16.** $2x + 14$ **17.** $40x + 80$ **18.** $x^2 - 10x + 25$
19. $x^2 - 9$ **20.** $x^2 - 4x + 4$ **21.** $4x^2 + x - 3$ **22.** $5x^2 - 90x + 405$ **23.** $t^2 - 16$ **24.** $6t^2 + 12t$
25. $5x^3$ **26.** $3x^3 - 3x$ **27.** $y^2 - 1$ **28.** $y^2 - 9$ **29.** $t^3 - 16t$ **30.** $5t^3 - 20t$

EXERCISES 5.5A

1. $\frac{1}{2}$ **2.** $\frac{4}{5}$ **3.** $\frac{1}{4}$ **4.** $\frac{2}{3}$ **5.** $\frac{4}{x}$ **6.** $\frac{1}{3y}$ **7.** $\frac{x + 2}{x - 2}$ **8.** 1 **9.** 2 **10.** $\frac{3}{2}$
11. $\frac{2(x - 2)}{x + 1}$ **12.** $\frac{4}{x - 2}$ **13.** 1 **14.** -1 **15.** $x + 2$ **16.** $\frac{1}{2}$ **17.** $\frac{3}{x}$ **18.** $\frac{x - 5}{5}$ **19.** $\frac{x + 2}{x + 5}$
20. $\frac{1}{2}$

EXERCISES 5.5B

1. 36 **2.** 150 **3.** 24 **4.** 36 **5.** 210 **6.** 144 **7.** 100 **8.** 42 **9.** 375 **10.** 432
11. 144 **12.** 12 **13.** 16 **14.** 50 **15.** 144 **16.** 36 **17.** 50 **18.** 168 **19.** 180 **20.** 150
21. $5x^2$ **22.** $12y^3$ **23.** $225x^3y^2$ **24.** $30x^3$ **25.** $12(x + y)^2$ **26.** $30(x + 5)$ **27.** $2x(x + 2)$
28. $4y(y - 4)$ **29.** $3y(y - 2)$ **30.** $4x^2(x + 2)$ **31.** $x(x - 2)(x + 2)$ **32.** $y(y + 3)(y - 3)$ **33.** $(y - 5)(y + 5)^2$
34. $(6x + 7)(6x - 7)^2$ **35.** $(x + 2)(x + 3)(x + 4)$ **36.** $(y - 4)(y - 5)(y - 6)$ **37.** $(x + 4)(x - 4)$
38. $(x + 5)^2(x - 5)^2$ **39.** $6(x + 3)$ **40.** $20(x - 5)$

EXERCISES 5.5C

1. $\frac{4}{9}$ **2.** $\frac{2}{3}$ **3.** $\frac{1}{6}$ **4.** $\frac{-1}{14}$ **5.** $\frac{41}{24}$ **6.** $\frac{1}{40}$ **7.** $\frac{1}{6}$ **8.** $\frac{1}{10}$ **9.** $\frac{5}{3}$ **10.** $\frac{7}{5}$ **11.** $\frac{x}{10}$
12. $\frac{x}{6}$ **13.** $\frac{5x + 6}{3x^2}$ **14.** $\frac{21 - 2y}{9y^2}$ **15.** $\frac{x + 1}{15x}$ **16.** $\frac{1}{20}$ **17.** $\frac{1}{y^2}$ **18.** $\frac{2x^2 - 1}{2x}$ **19.** $\frac{x + y}{xy}$
20. $\frac{3y - 2x}{36xy^2}$ **21.** $\frac{9}{2(x + 3)}$ **22.** $\frac{4}{15(y - 3)}$ **23.** $\frac{1}{7x}$ **24.** $\frac{4}{(y + 5)(y - 5)}$ **25.** $\frac{2}{y(y + 2)}$ **26.** $\frac{1}{2x}$
27. $\frac{1}{6}$ **28.** $\frac{2}{x^2(x + 4)}$ **29.** $\frac{12t}{(t - 3)(t + 3)}$ **30.** $\frac{2y^2 - 4}{(y + 1)(y - 1)}$ **31.** $\frac{7x - 6}{(x + 6)(x - 6)^2}$ **32.** $\frac{-8}{(x - 4)(x + 4)^2}$
33. $\frac{5x + 8}{x^2 - 49}$ **34.** $\frac{-y}{(y + 3)^2}$ **35.** $\frac{1}{(y + 1)(y + 2)}$ **36.** $\frac{1}{(2t + 1)(3t + 1)}$ **37.** $\frac{2}{(x + 1)(x + 3)}$ **38.** $\frac{-1}{(2t + 1)(3t - 1)}$
39. $\frac{2x - 7}{(x - 2)(x - 3)}$ **40.** $\frac{1}{(y - 1)(y - 2)}$ **41.** $\frac{8x^2 + 9x - 10}{12x^3}$ **42.** 0 **43.** $\frac{1}{t + 1}$ **44.** $\frac{4 - 3y}{4y(y - 4)}$
45. $\frac{2}{x + 3}$ **46.** $\frac{32t}{(t - 2)(t + 2)^2}$ **47.** $\frac{25x - 25}{x(x - 5)(x + 5)}$ **48.** $\frac{36}{(x - 3)^2(x + 3)^2}$ **49.** $\frac{1}{(x + 1)(x + 2)}$ **50.** 0

EXERCISES 5.5 S

1. $\frac{1}{3}$ **2.** $\frac{-1}{12}$ **3.** $\frac{1}{10}$ **4.** $\frac{1}{3x}$ **5.** $\frac{x + 5}{x - 5}$ **6.** 1 **7.** $\frac{6}{y - 6}$ **8.** $\frac{2}{x - 10}$ **9.** $\frac{2y + 4}{y}$
10. $\frac{8x - 15}{10x^2}$ **11.** $\frac{-1}{24x^2}$ **12.** $\frac{2x}{x^2 - 25}$ **13.** $\frac{-2x + 15}{(x - 6)^2}$ **14.** $\frac{50}{(x - 5)(x + 5)}$ **15.** $\frac{60x}{(5x + 3)(5x - 3)^2}$

16. $\frac{y+10}{5(y+5)}$ **17.** $\frac{1}{3y}$ **18.** $\frac{4}{4k^2-1}$ **19.** $\frac{n-6}{2n-6}$ **20.** $\frac{-8}{(2x+1)(2x-1)(2x-1)}$ **21.** $\frac{x^2+2x-1}{x^2-1}$
22. $\frac{25x^2+x+5}{5x^2}$ **23.** $\frac{3x+5y-xy}{15x^2y^2}$ **24.** $\frac{x^2-4x+16}{4x(x-4)}$ **25.** $\frac{-1}{3x}$ **26.** $\frac{4x^2}{(x-1)^2(x+1)^2}$ **27.** $\frac{-6x^2}{x^2-9}$
28. $\frac{2x^2-5x+4}{(x-2)^2}$ **29.** $\frac{x(x^3+4)}{2(x+4)(x^2+4)}$ **30.** $\frac{4}{(x-2)(x+3)}$

EXERCISES 5.6A

1. 2 **2.** $\frac{11}{8}$ **3.** $\frac{x-1}{x^2}$ **4.** $2(x-5)$, or $2x-10$ **5.** $\frac{5}{12}$ **6.** $\frac{10x^2}{x+5}$ **7.** $\frac{3x}{x+3}$ **8.** $\frac{1}{4}$
9. $\frac{t^2+5t}{t^2+25}$ **10.** $\frac{x+1}{x-1}$

EXERCISES 5.6B

1. $\frac{1}{2}$ **2.** 6 **3.** 48 **4.** 5 **5.** $2=2$ **6.** $10=10$ **7.** $\frac{15}{4}=\frac{15}{4}$ **8.** $4=4$ **9.** $-1=-1$
10. $\frac{12}{5}=\frac{12}{5}$

EXERCISES 5.6 S

1. $\frac{2x}{27y}$ **2.** $\frac{y-x}{2y}$ **3.** $7x+49$ **4.** $\frac{x-1}{x+1}$ **5.** $\frac{5}{4}$ **6.** $\frac{32}{y-4}$ **7.** $\frac{x^2+25}{x^2-25}$ **8.** $\frac{2xy}{x+y}$
9. $\frac{t}{7-t}$ or $\frac{-t}{t-7}$ **10.** $\frac{x-4}{x+4}$ **11.** 2916 **12.** $\frac{99}{7}$ **13.** $\frac{31}{16}$ **14.** 7 **15.** $-2=-2$ **16.** $\frac{1}{2}=\frac{1}{2}$
17. $\frac{15}{14}=\frac{15}{14}$ **18.** $-1=-1$ **19.** $\frac{-4}{75}=\frac{-4}{75}$ **20.** $\frac{-3}{8}=\frac{-3}{8}$

EXERCISES 5.7

1. $x-9+\frac{28}{x+2}$ **2.** $y-1$ **3.** $3x-1+\frac{-5}{2x-1}$ **4.** $4x-3$ **5.** $2y^2+2y-3+\frac{-2}{y+1}$ **6.** x^2+4x+6
7. $y^2-5y+25$ **8.** $2x-3+\frac{18}{2x+3}$ **9.** x^2-4 **10.** $y^3+2y^2-6y-12$

EXERCISES 5.7 S

1. $x+1+\frac{-11}{x+3}$ **2.** $y-5$ **3.** $2x+\frac{-9}{4x+3}$ **4.** $x-4$ **5.** $4y^2+3y-1+\frac{-8}{2y-1}$ **6.** x^2+3x-6
7. y^2+3y+9 **8.** $5x-4y+\frac{32y^2}{5x+4y}$ **9.** x^2-4x+3 **10.** $y^3-3y^2-5y+15$

SAMPLE TEST FOR CHAPTER 5

1. $x+2$ **2.** $\frac{-3x}{2y}$ **3.** $\frac{2x+7}{2x}$ **4.** $\frac{-x-2}{4}$ **5.** 100 qt **6.** 3.8 liters **7.** $\frac{2x}{x+3}$ **8.** $\frac{x(x+4)}{3x+2}$
9. $6t^3+36t^2$ **10.** x^2+x-12 **11.** $\frac{x^2-10}{(x+3)^2}$ **12.** $\frac{-1}{2x-1}$ **13.** $\frac{x-1}{x+1}$ **14.** $4x^2-20x+25$

CHAPTER 6

EXERCISES 6.1

1. 9 **2.** 8 **3.** 24 **4.** 30 **5.** 3 **6.** 5 **7.** 9 **8.** 8 **9.** 2 **10.** 2 **11.** 320
12. 3500 **13.** 3 **14.** 40 **15.** 6 **16.** 45 **17.** 40 **18.** 360 **19.** 48 **20.** 4000

EXERCISES 6.1 S

1. 14 **2.** 10 **3.** 40 **4.** 72 **5.** -2 **6.** 10 **7.** $\frac{1}{2}$ **8.** 3 **9.** 4 **10.** 4 **11.** 10
12. 50 **13.** 4 **14.** 35 **15.** 30 **16.** 120 **17.** 15,000 **18.** 2000 **19.** 40 **20.** 8

EXERCISES 6.2A

1. 100 **2.** $25x$ **3.** 0.05(2000) **4.** 0.065(5000) **5.** $0.05d$ **6.** $0.05(x+500)$ **7.** 5(15) **8.** $15y$
9. $15(20-x)$ **10.** $5c$ **11.** 0.05(40) **12.** $0.05x$ **13.** $0.05(x+15)$ **14.** $0.05(50-x)$ **15.** 40(3.10)
16. $3.10h$ **17.** $40x$ **18.** 0.6(20) **19.** $0.6y$ **20.** $0.6(x+20)$

EXERCISES 6.2B

1. $5 + 3$ **2.** $x + 3$ **3.** $12 - 3$ **4.** $12 - x$ **5.** $2000 + 3000$ **6.** $d + 2000$ **7.** $5000 - 3000$
8. $5000 - x$ **9.** $20 + 40$ **10.** $x + 40$ **11.** $75 - 50$ **12.** $x - 50$

EXERCISES 6.2C

1. 15 lb **2.** $30,000 **3.** 30 grams **4.** 2.5 liters **5.** 320 **6.** 3 qt **7.** 8 lb **8.** $40,000
9. 40 gal **10.** 10 oz **11.** 12 lb **12.** 20 lb

EXERCISES 6.2 S

1. 45 lb **2.** 40 qt **3.** 300 gal **4.** $4000 bonds, $8000 stocks **5.** 160 oz **6.** 60 cc **7.** $40,000
8. 500 lb **9.** 4 kg **10.** 30,000

EXERCISES 6.3A

1. 7 **2.** -4 **3.** 1 **4.** 4 **5.** 2 **6.** 6 **7.** 4 **8.** 20 **9.** 2 **10.** 8

EXERCISES 6.3B

1. 5 **2.** 1 **3.** 4 **4.** 6 **5.** -3 **6.** 4 **7.** 3 **8.** -2 **9.** 6 **10.** 8 **11.** -2 **12.** 2
13. $\frac{1}{3}$ **14.** $\frac{1}{2}$

EXERCISES 6.3C

1. $\varnothing$ **2.** All x, $x \neq 0$ **3.** All t, $t \neq 1$ **4.** $\varnothing$ **5.** $\varnothing$ **6.** $\{0\}$ **7.** All x, $x \neq 6$ **8.** $\varnothing$ **9.** $\varnothing$
10. $\varnothing$

EXERCISES 6.3 S

1. 5 **2.** -5 **3.** 8 **4.** 6 **5.** 10 **6.** 9 **7.** -2 **8.** 4 **9.** -3 **10.** 3 **11.** $\frac{1}{5}$
12. $\frac{1}{2}$ **13.** $\varnothing$ **14.** All x, $x \neq -5$ **15.** All x, $x \neq 0$, $x \neq 2$ **16.** $\varnothing$ **17.** 7 **18.** -1 **19.** -8
20. -5 **21.** 6 **22.** 2 **23.** -1 **24.** $\frac{2}{5}$

EXERCISES 6.4

1. 12 min **2.** 10 hr **3.** 24 min **4.** 12 hr **5.** 15 days, 30 days **6.** $26\frac{2}{3}$ min, 80 min **7.** 8 min
8. 3 hr **9.** 3 hr **10.** 18 min **11.** 21 hr **12.** 40 days **13.** 5 min **14.** 42 min **15.** 8 days

EXERCISES 6.4 S

1. 15 min **2.** 50 min **3.** 22 hr, 44 hr, 66 hr **4.** $4\frac{1}{2}$ hr **5.** 30 hr **6.** 8 days

SAMPLE TEST FOR CHAPTER 6

1. 12 **2.** 50 **3.** 20 liters **4.** 30 lb **5.** 20 **6.** $\varnothing$ **7.** All x, $x \neq 4$, $x \neq -4$ **8.** 2 **9.** 5
10. 10 **11.** 36 min, 180 min **12.** 6 hr

SAMPLE TEST FOR CHAPTERS 5 AND 6

1. $\frac{-x}{x+4}$ **2.** $\frac{2(n-2)}{n+3}$ **3.** $x + 6$ **4.** $x^2 + 4x - 2$ **5.** $\frac{5}{2}$ **6.** $\frac{-4}{75} = \frac{-4}{75}$ **7.** 3 **8.** $\frac{-1}{10} = \frac{-1}{10}$
9. All x, $x \neq 4$ **10.** $\varnothing$

CHAPTER 7

EXERCISES 7.1A

1. Yes **2.** No **3.** Yes **4.** Yes **5.** Yes **6.** No **7.** No **8.** Yes **9.** Yes **10.** No
11. Yes **12.** Yes **13.** No **14.** Yes **15.** Yes **16.** No **17.** No **18.** No **19.** Yes
20. Yes **21.** No **22.** Yes **23.** Yes **24.** Yes **25.** Yes **26.** Yes **27.** Yes **28.** No
29. Yes **30.** Yes

EXERCISES 7.1B

1. (7, 5) **2.** (14, −2) **3.** (2, −3) **4.** (−4, −9) **5.** (1, 6) **6.** (−2, 13) **7.** (−1, −8) **8.** (2, −2) **9.** (0, −2) **10.** (5, 0)

EXERCISES 7.1C

1–8.

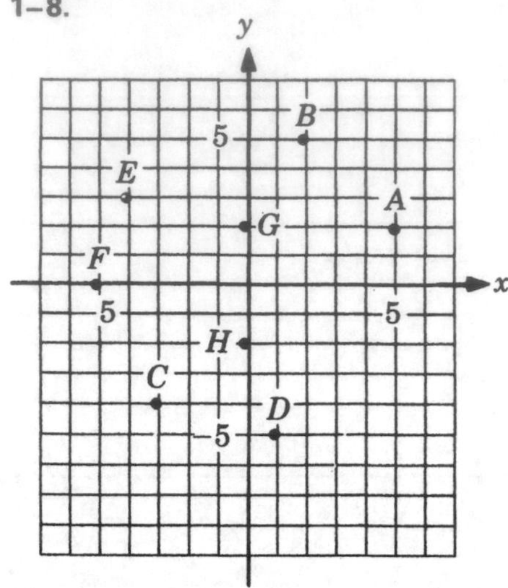

9. $A(5, 3)$
10. $B(-4, 4)$
11. $C(-3, 0)$
12. $D(0, -2)$
13. $E(-5, -4)$
14. $F(4, -2)$

15.

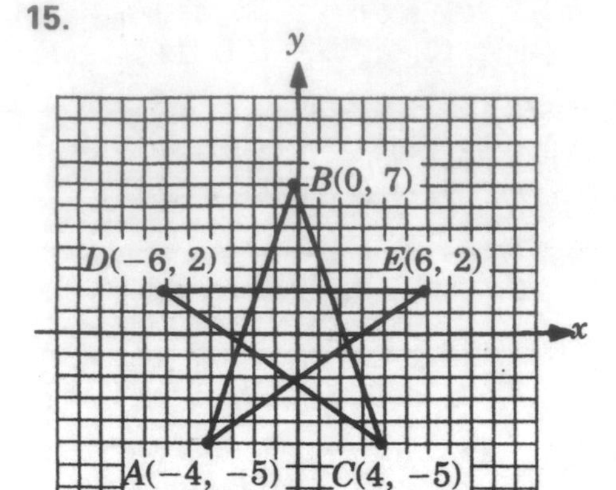

16.

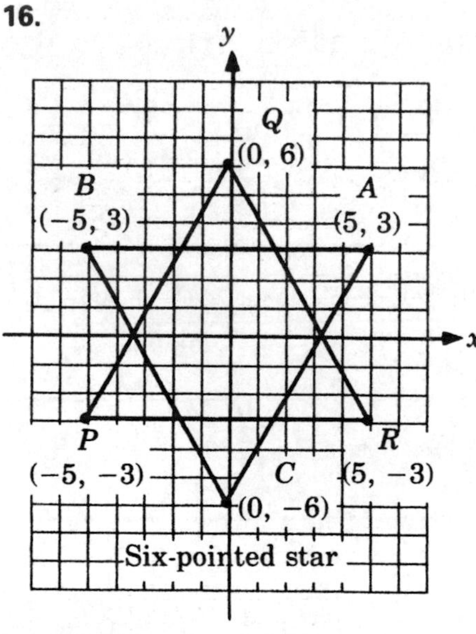

17. (4, 2) **18.** (−2, 3) **19.** (3, −1) **20.** (−3, −4) **21.** (3, 5) **22.** (−3, −5) **23.** (−3, 5) **24.** (5, −3) **25.** (−5, 0) **26.** (0, 3)

EXERCISES 7.1 S

1. Yes **2.** No **3.** Yes **4.** No **5.** Yes **6.** Yes **7.** Yes **8.** No **9.** No **10.** Yes **11.** Yes **12.** No **13.** Yes **14.** No **15.** No **16.** Yes **17.** No **18.** Yes **19.** No **20.** Yes **21.** No **22.** Yes **23.** Yes **24.** Yes **25.** Yes **26.** Yes **27.** Yes **28.** No **29.** Yes **30.** Yes **31.** No **32.** Yes **33.** Yes **34.** Yes **35.** Yes **36.** Yes **37.** Yes **38.** No **39.** No **40.** Yes **41.** (0, −3) **42.** (20, −2) **43.** (4, 1) **44.** (3, 3) **45.** (5, 24) **46.** $\left(\frac{-4}{3}, 5\right)$ **47.** (1, −2) **48.** $\left(\frac{-5}{2}, 0\right)$ **49.** (0, −2) **50.** $\left(\frac{11}{2}, 1\right)$

51–58.

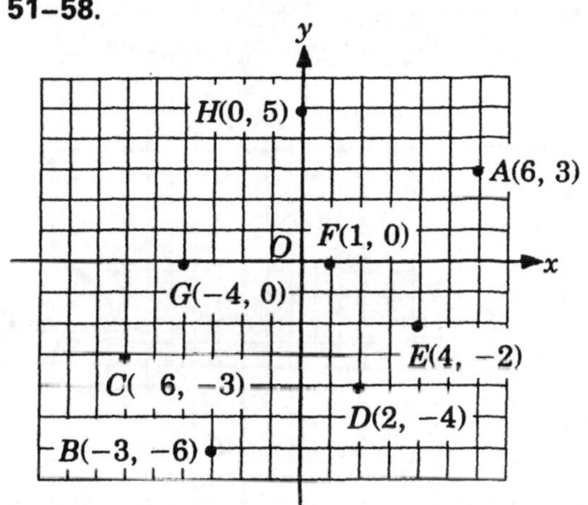

59. $P(0, 4)$
60. $Q(0, 0)$
61. $R(4, -3)$
62. $S(-5, 0)$
63. $T(0, -4)$
64. $U(-3, -1)$

65. (2, 5) **66.** (−4, −2) **67.** (3, −1) **68.** (−2, 3) **69.** (−4, 5), (0, 3), (4, 1), (6, 0), (8, −1); straight line
70. (−1, −6), (0, −4), (2, 0), (3, 2), (5, 6); straight line

EXERCISES 7.2A

1. $y = x + 4$

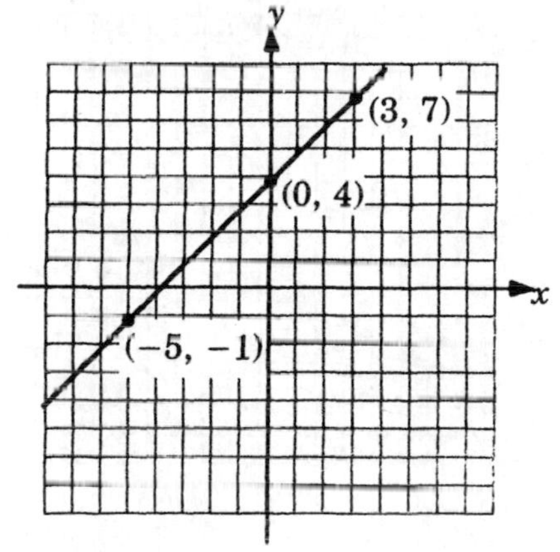

2. $y = 2x + 3$

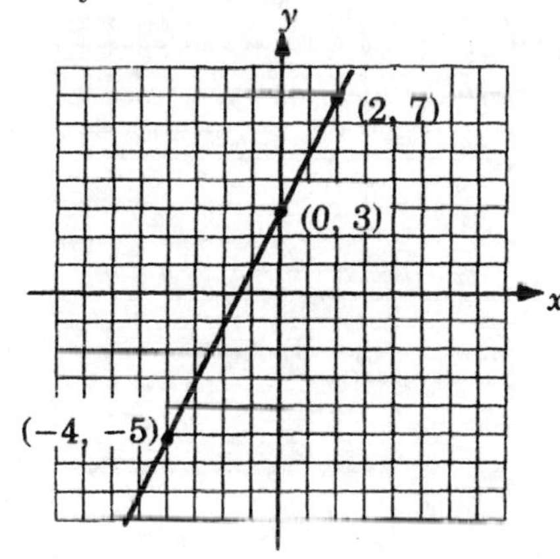

3. $y = 5 - x$

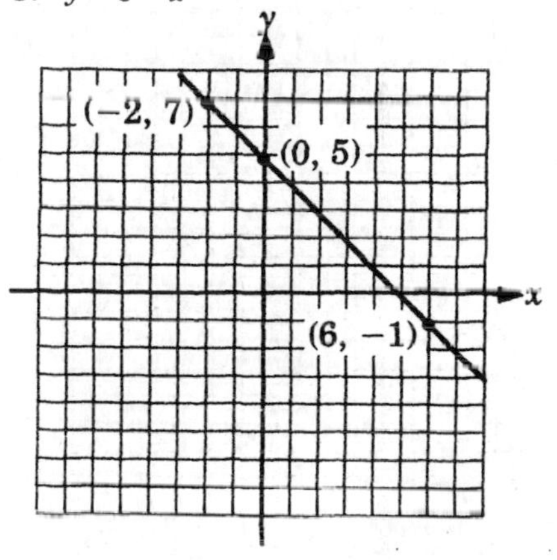

4. $y = 3x - 6$

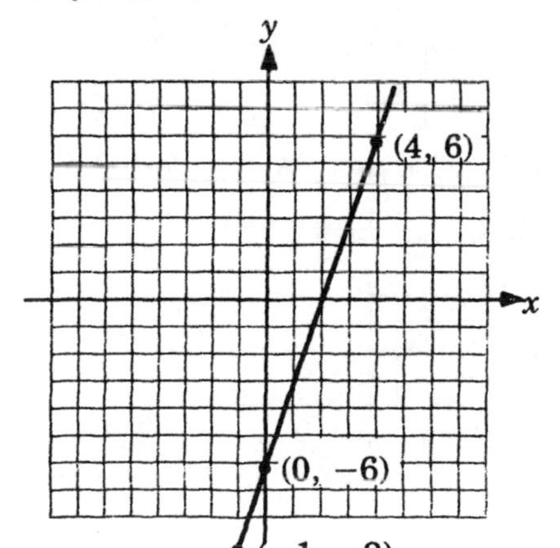

5. $y = 4 - 2x$

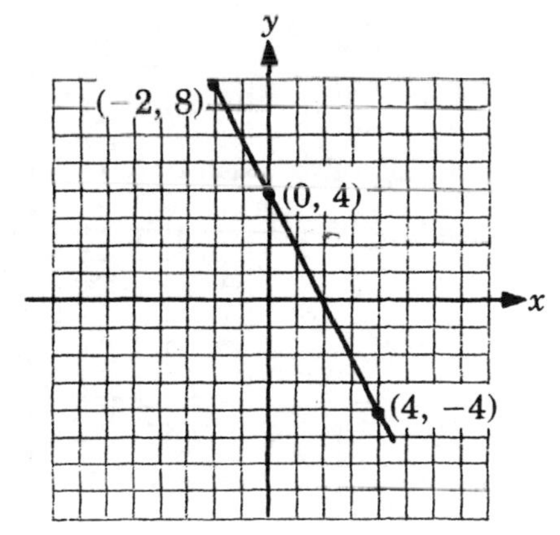

6. $y = \frac{-x}{3}$

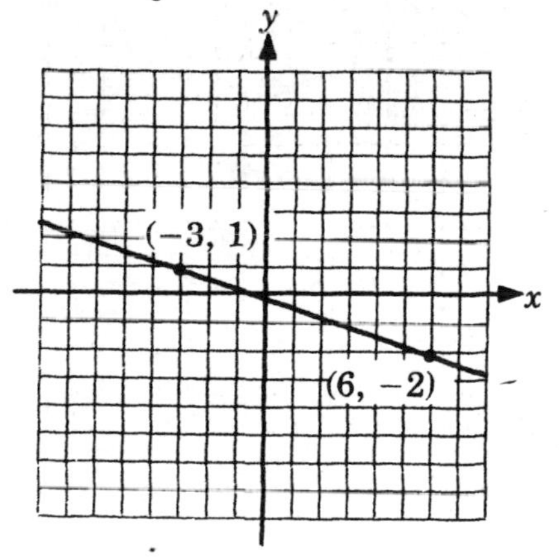

EXERCISES 7.2B

1. x-intercept = 2; y-intercept = 4

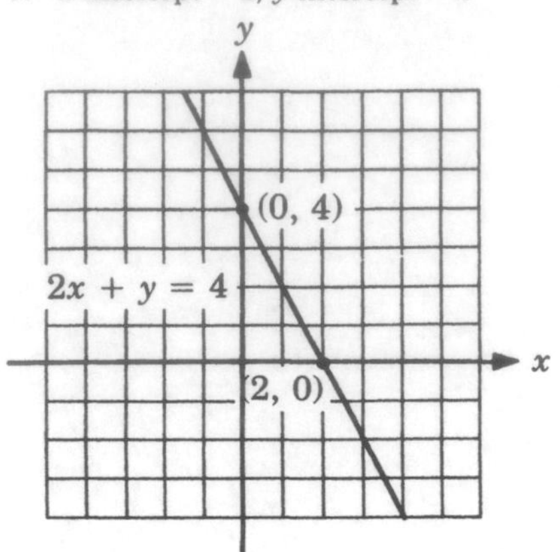

2. x-intercept = −3; y-intercept = 9

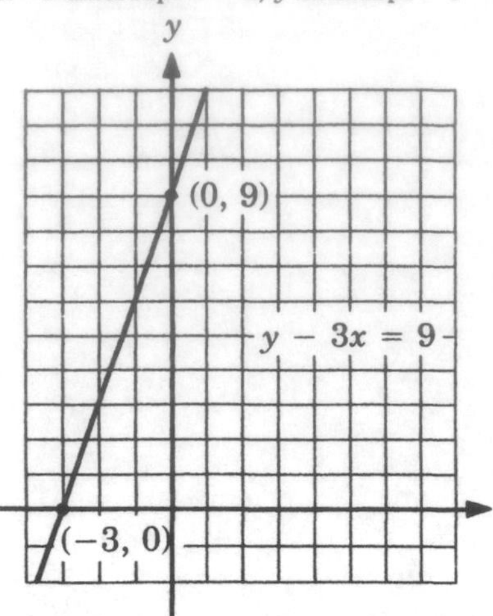

3. x-intercept = 5; y-intercept = −5

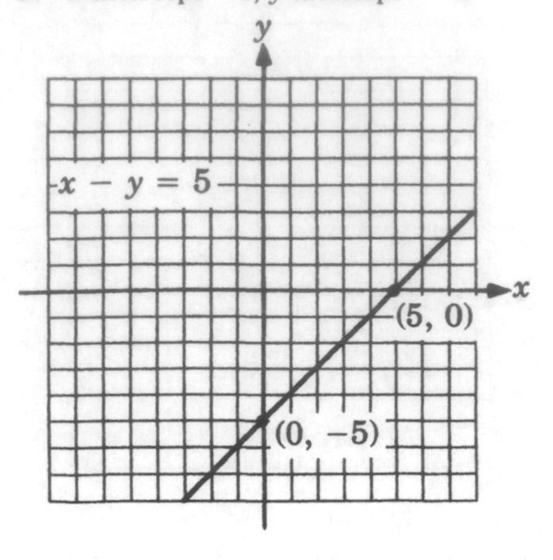

4. x-intercept = 2; y-intercept = 5

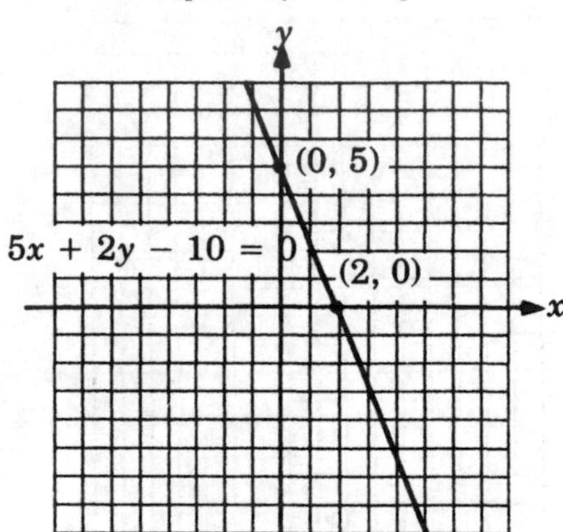

5. x-intercept = −4; y-intercept = −3

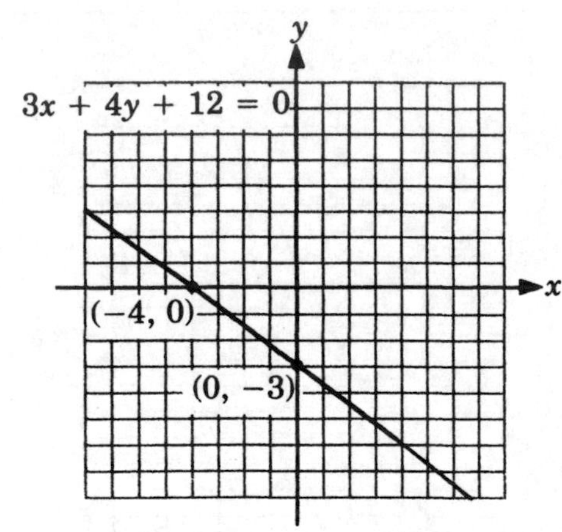

6. x-intercept = 5; y-intercept = −2

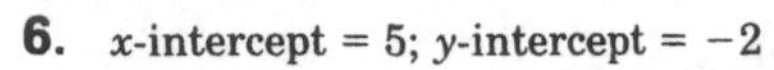
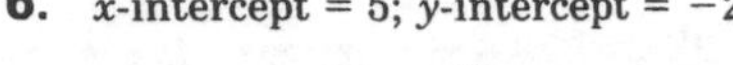
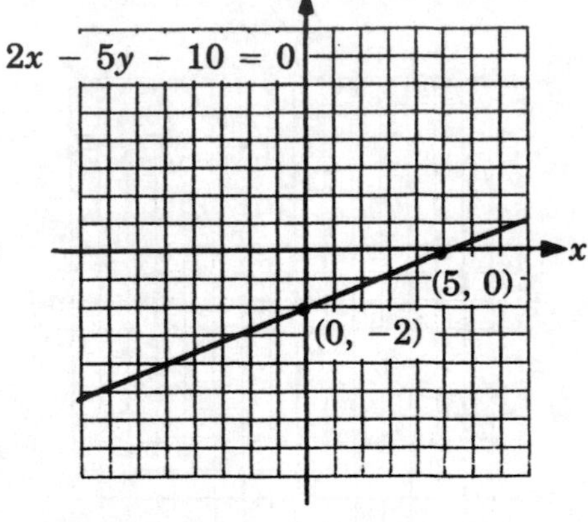

7. x-intercept = −8; y-intercept = 2

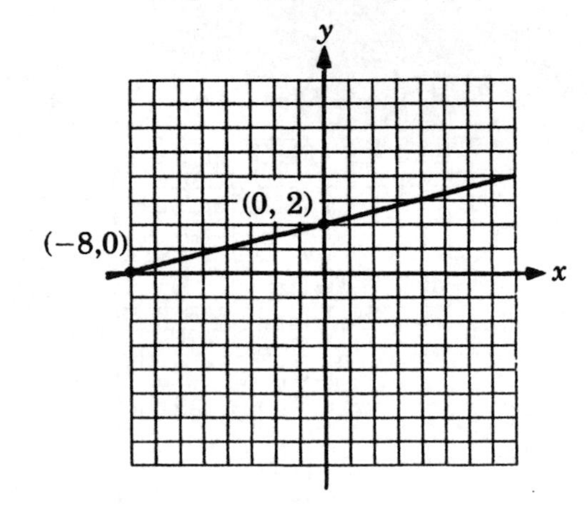

8. x-intercept = −4; y-intercept = −4

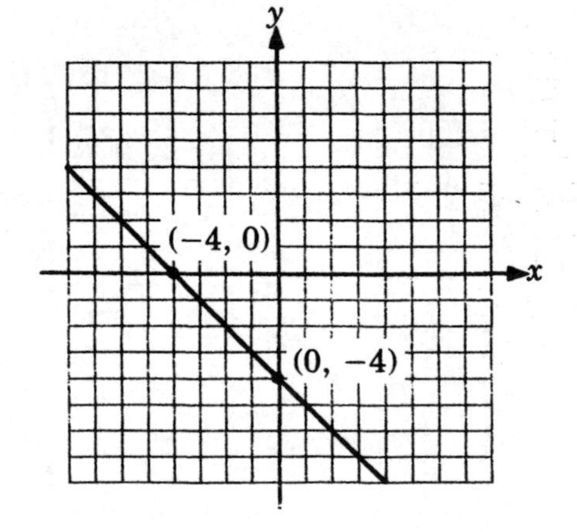

9. $y = 4 - x$; x-intercept = 4, y-intercept = 4

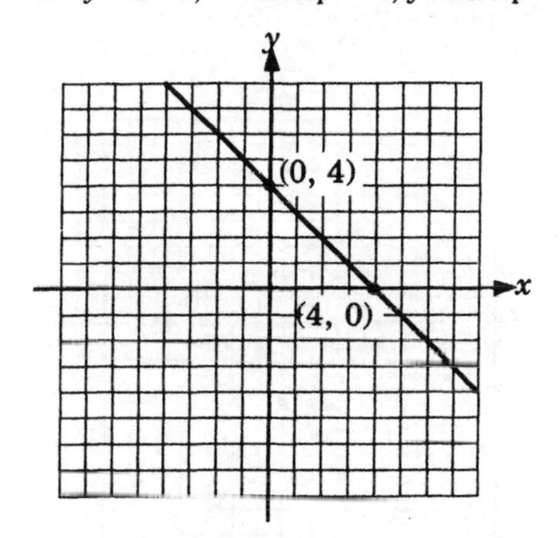

EXERCISES 7.2C

1. x-intercept = 2; no y-intercept

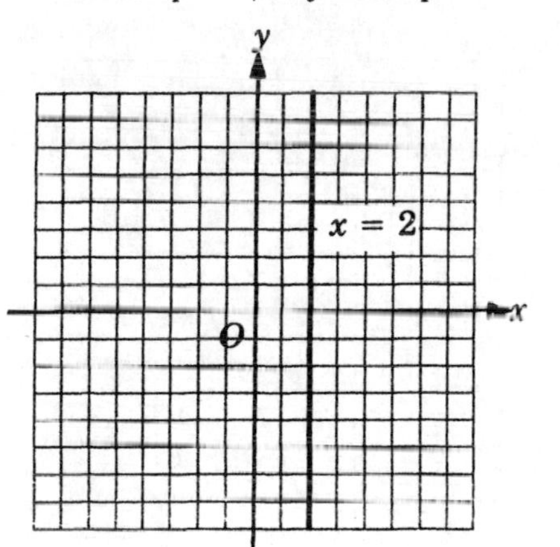

2. x-intercept = −4; no y-intercept

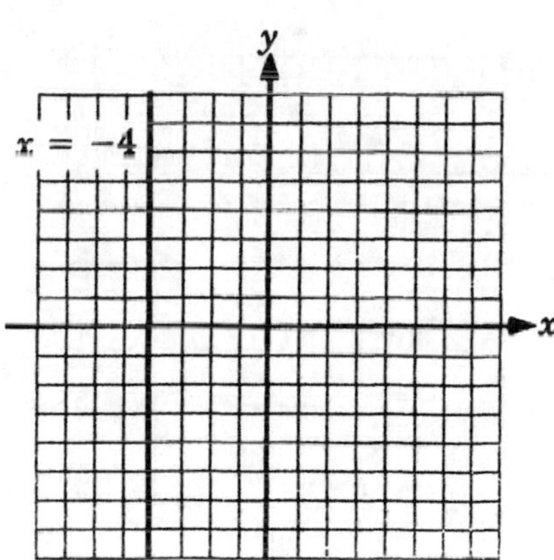

3. no x-intercept; y-intercept = −3

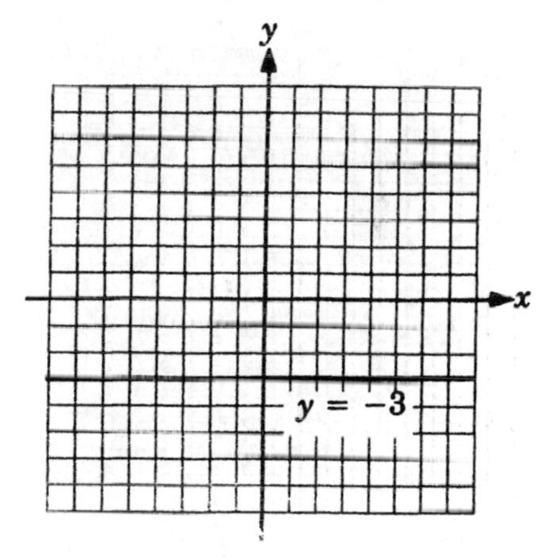

4. no x-intercept, y-intercept = 5

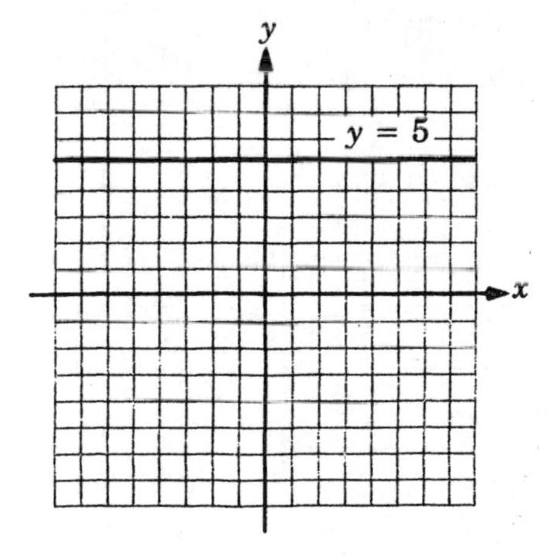

EXERCISES 7.2 S

1. $y = x$

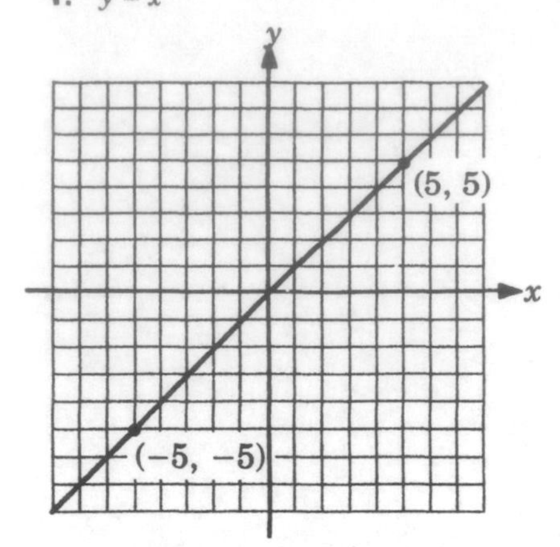

2. $y = x + 5$

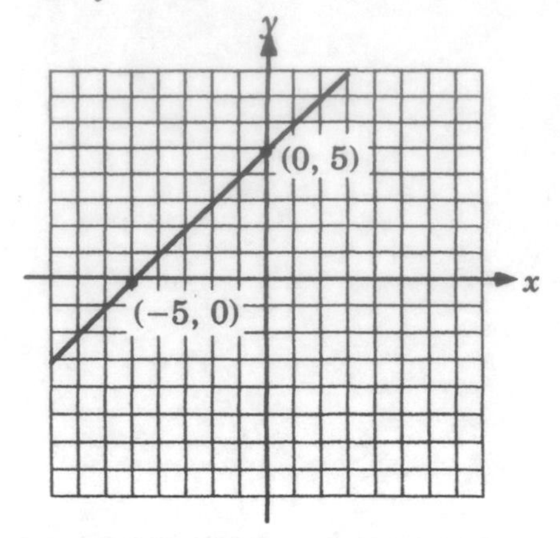

3. $y = x - 3$

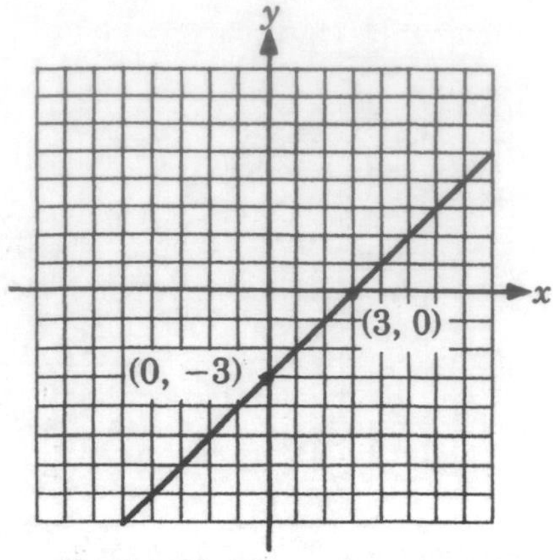

4. $y = 6 - x$

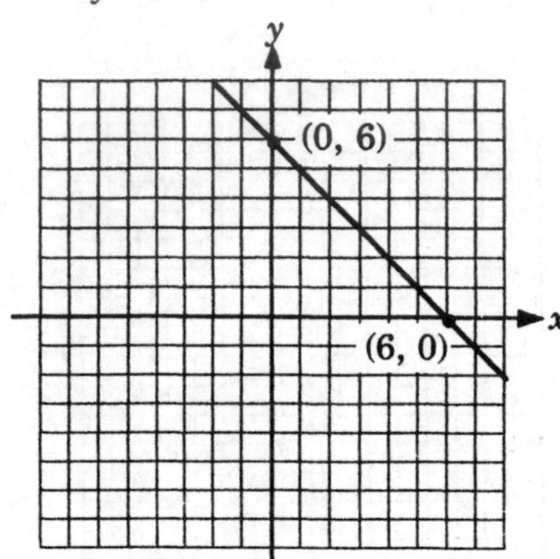

5. $y = -x - 8$

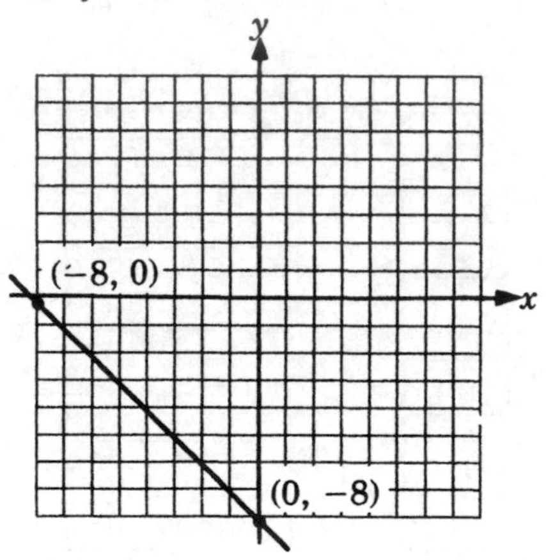

6. $y = 2x - 6$

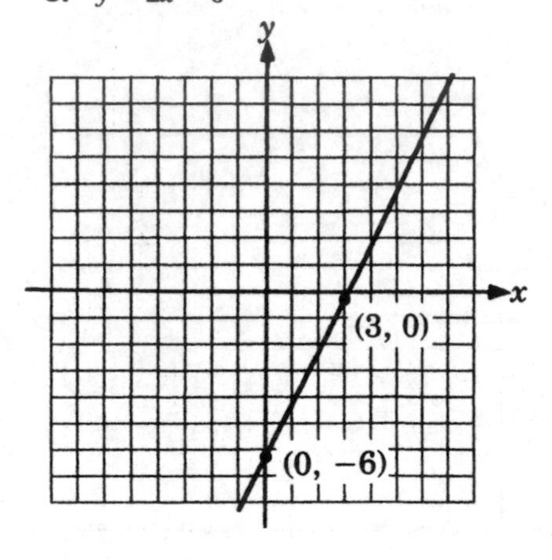

7. $y = 3x + 12$

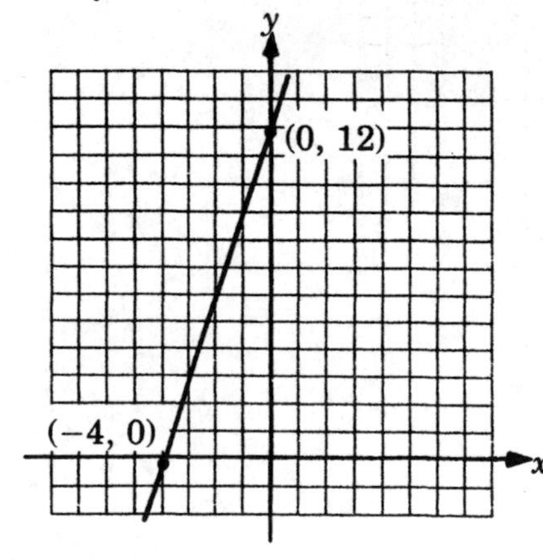

8. $y = 8 - 4x$

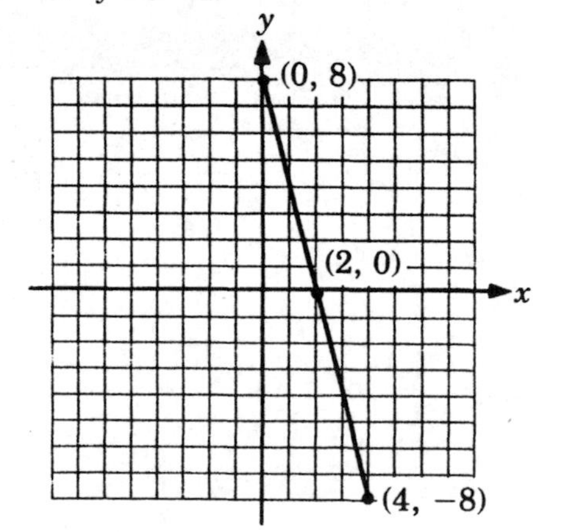

9. $y = 6 - 3x$

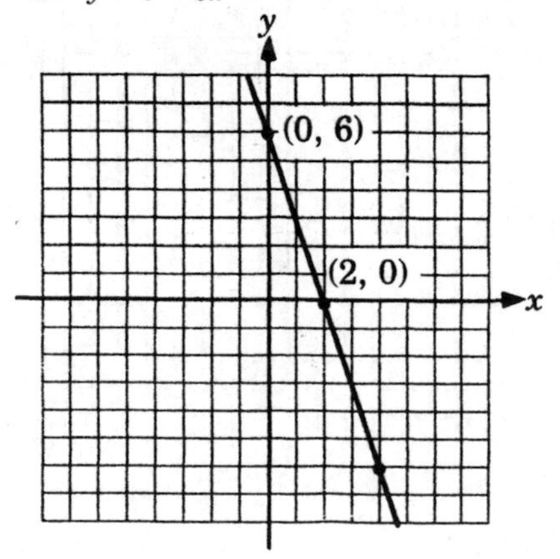

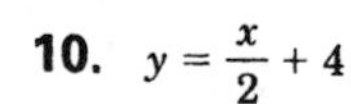

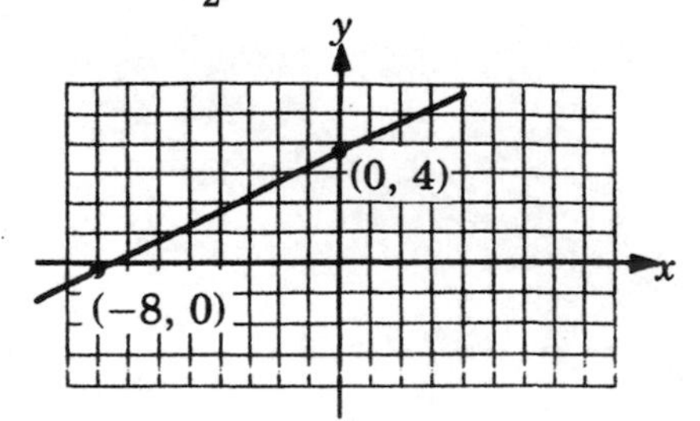

11. $x = -5$; x-intercept $= -5$; no y-intercept

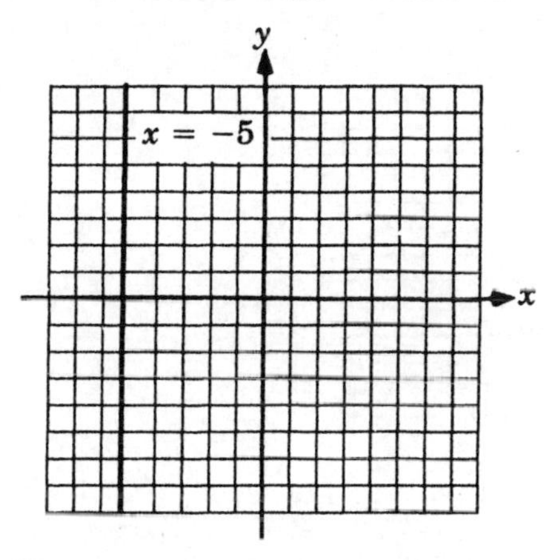

12. $y = 4$; no x-intercept; y-intercept $= 4$

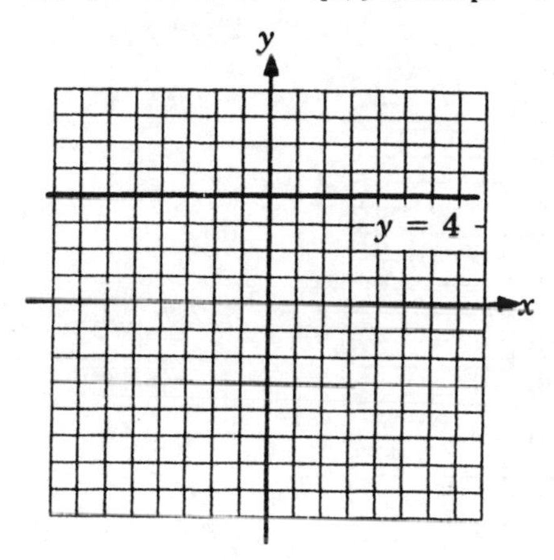

13. x-intercept $= 4$; no y-intercept

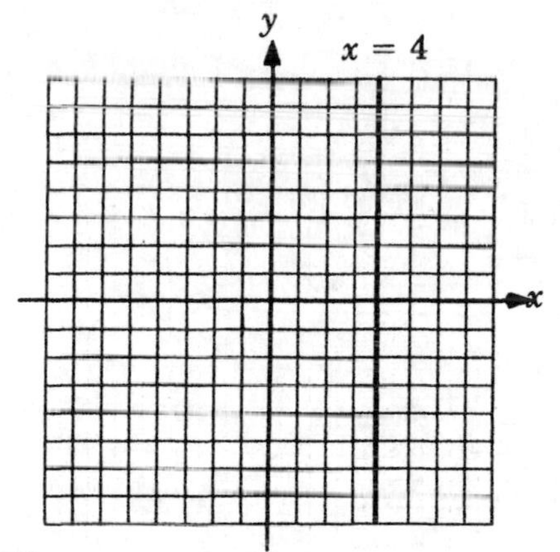

14. no x-intercept; y-intercept $= -1$

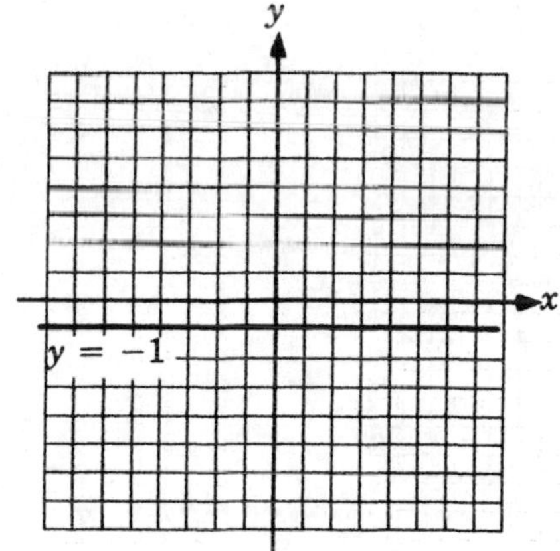

15. $3x - y = 6$; x-intercept $= 2$; y-intercept $= -6$

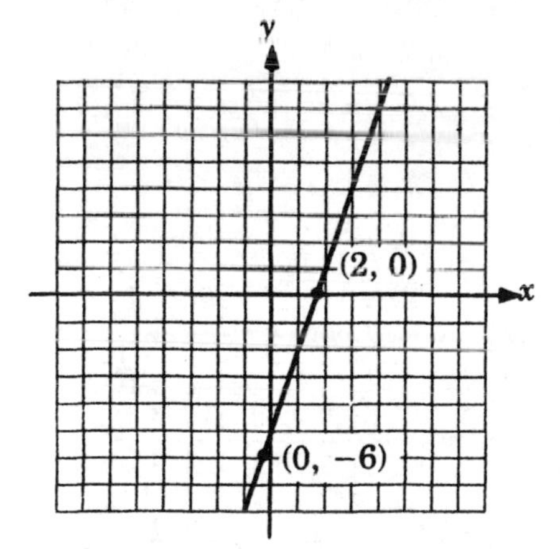

16. $5x - 2y + 10 = 0$; x-intercept $= -2$; y-intercept $= 5$

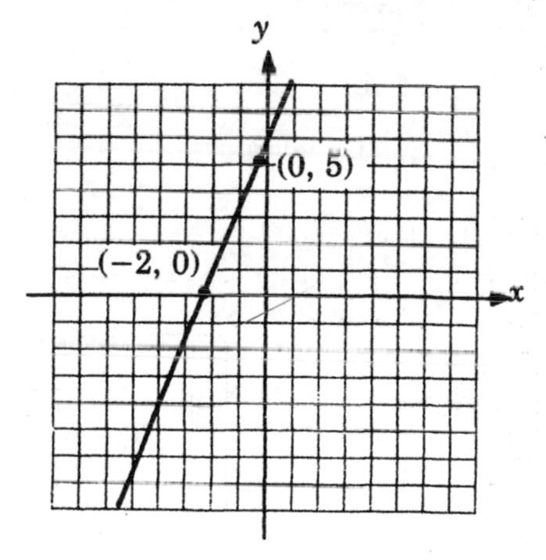

17. $5x - 6y - 30 = 0$; x-intercept $= 6$; y-intercept $= -5$

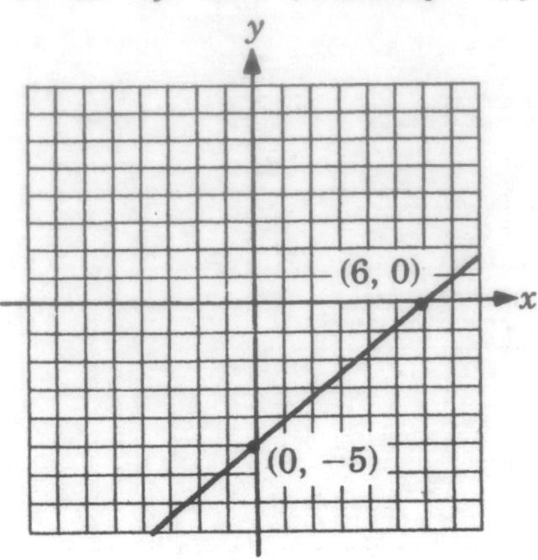

18. $2x + 7y + 14 = 0$; x-intercept $= -7$; y-intercept $= -2$

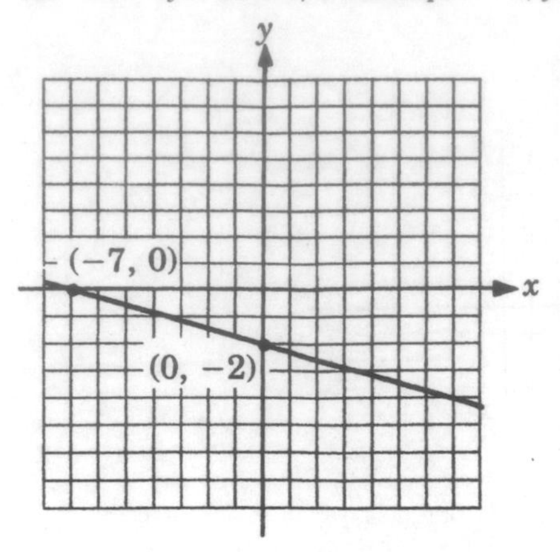

19. $x + 2y = 6$; x-intercept $= 6$; y-intercept $= 3$

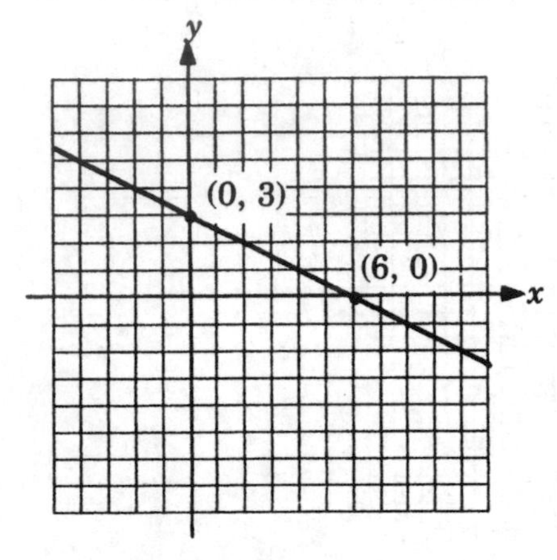

20. $x - 2y = 0$; x-intercept $= 0$; y-intercept $= 0$

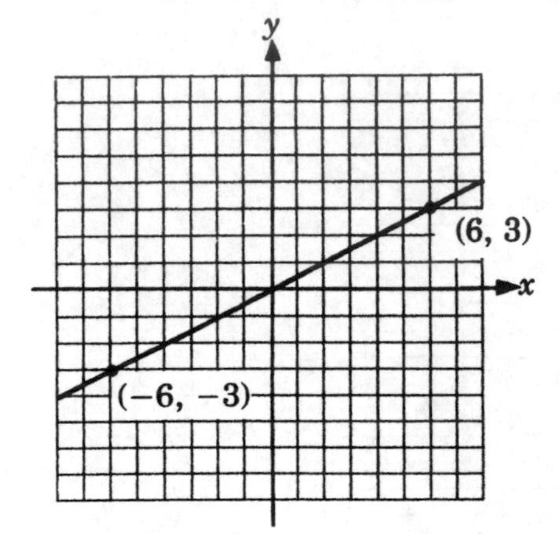

21. $y = \frac{-x}{2}$; x-intercept $= 0$; y-intercept $= 0$

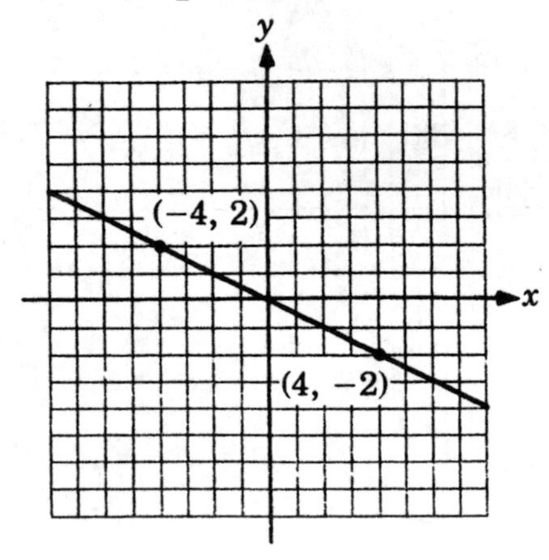

22. $C = \frac{5(F - 32)}{9}$

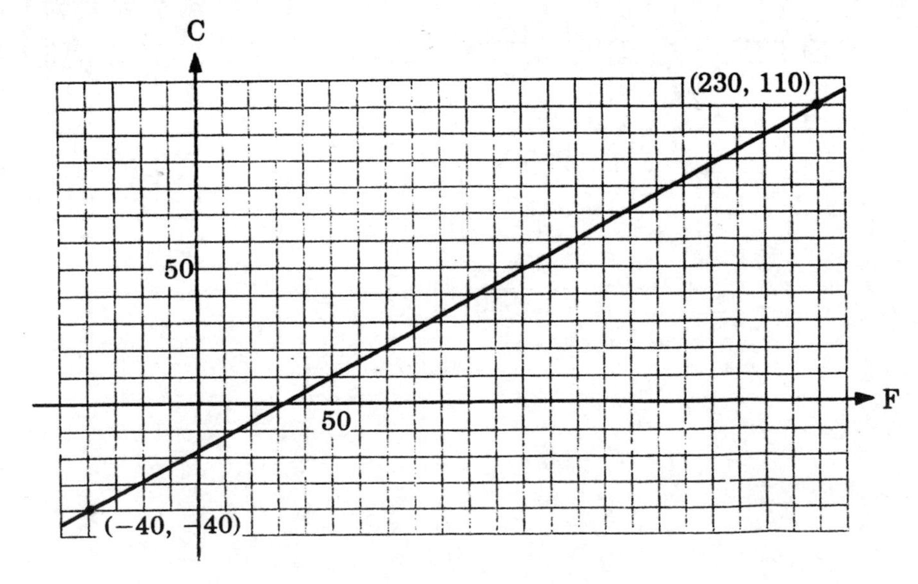

23. $y = 12 - \frac{3x}{4}$

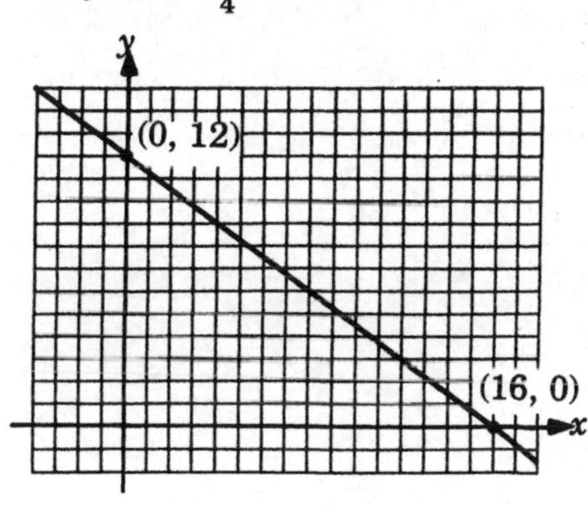

24.

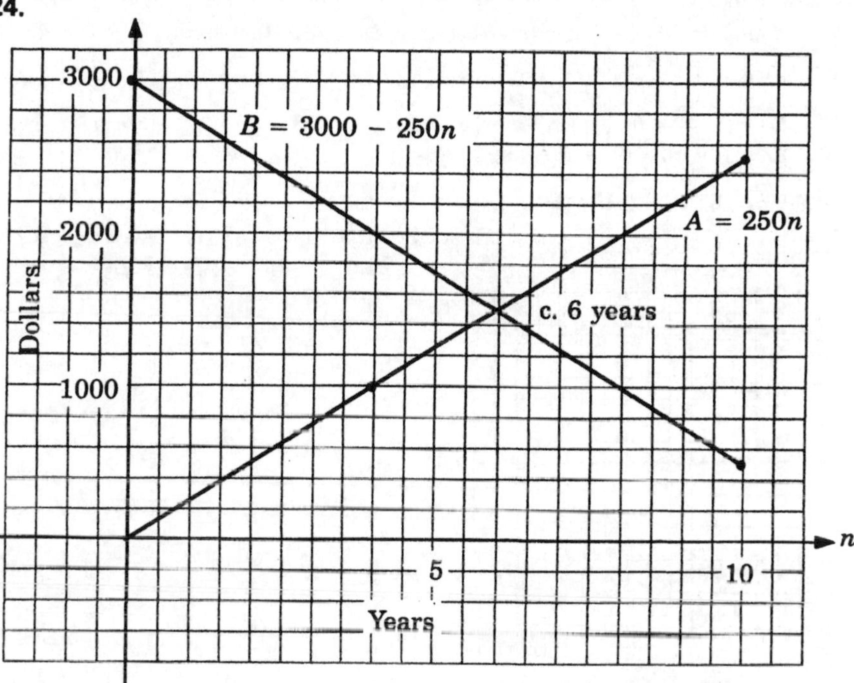

EXERCISES 7.3A

1. 2 **2.** −4 **3.** $\frac{1}{3}$ **4.** $\frac{-3}{2}$ **5.** $\frac{5}{4}$ **6.** 0 **7.** $\frac{-2}{5}$ **8.** Undefined **9.** Undefined **10.** 0

EXERCISES 7.3B

1. $m = 2, b = 6$ **2.** $m = -2, b = 6$ **3.** $m = 2, b = 0$ **4.** $m = 0, b = 2$ **5.** Undefined **6.** $m = \frac{1}{2}, b = 4$
7. $m = \frac{-2}{3}, b = 2$ **8.** $m = \frac{2}{3}, b = 2$ **9.** $m = \frac{-2}{3}, b = -2$ **10.** $m = \frac{2}{3}, b = 4$

EXERCISES 7.3C

1. 200 ft **2.** 1000 ft **3.** 90 ft **4.** 30 ft **5.** $\frac{1}{4}$ **6.** 1 **7.** $\frac{8}{5}$ **8.** 20 ft

EXERCISES 7.3 S

1. 3 **2.** −2 **3.** $\frac{7}{6}$ **4.** $\frac{-5}{3}$ **5.** $\frac{-2}{3}$ **6.** $\frac{1}{6}$ **7.** Undefined **8.** 0 **9.** 92.4 ft **10.** 237.6 ft
11. a and b **12a.** $4\frac{1}{2}$ ft **12b.** 18 ft **13.** $m = 3, b = -6$ **14.** $m = -2, b = 8$ **15.** $m = 1, b = -6$
16. $m = -2, b = 12$ **17.** $m = 0, b = \frac{-5}{2}$ **18.** $m = \frac{1}{2}, b = 4$ **19.** $m = 2, b = -4$ **20.** $m = 1, b = \frac{5}{2}$
21. $m = \frac{-1}{2}, b = 0$ **22.** $m = \frac{2}{3}, b = 0$

EXERCISES 7.4A

1. Lines parallel, $m = 1$ **2.** Lines intersect at (0, 4) **3.** Lines intersect at (2, 0) **4.** Lines parallel, $m = -2$
5. Lines intersect at (6, 2) **6.** Lines parallel, $m = \frac{1}{3}$

EXERCISES 7.4B

1. $x + y - 5 = 0$ **2.** $x - 2y - 8 = 0$ **3.** $x - 3y = 0$ **4.** $4x + 5y = 0$ **5.** $2x - 5y - 10 = 0$
6. $5x + 2y - 10 = 0$ **7.** $2x + 3y - 6 = 0$ **8.** $3x - 4y - 6 = 0$ **9.** $3x - 2y = 0$ **10.** $4x + 3y = 0$

EXERCISES 7.4 S

1. Parallel, $m = -1$ **2.** Intersect at (0, 6) **3.** Intersect at (4, 2) **4.** Parallel, $m = 3$ **5.** Intersect at (−3, 2)
6. Parallel, $m = \frac{1}{3}$ **7.** Intersect at (3, −2) **8.** Parallel, $m = \frac{-1}{4}$ **9.** Intersect at (5, 4) **10.** $3x - 4y + 20 = 0$
11. $5x + y - 20 = 0$ **12.** $x - 3y + 13 = 0$ **13.** $5x + 2y - 13 = 0$ **14.** $2x + 5y - 10 = 0$ **15.** $3x - 2y - 12 = 0$
16. $5x + 2y - 10 = 0$ **17.** $3x - 4y - 12 = 0$ **18.** $x + y - 7 = 0$

SAMPLE TEST FOR CHAPTER 7

1–5.

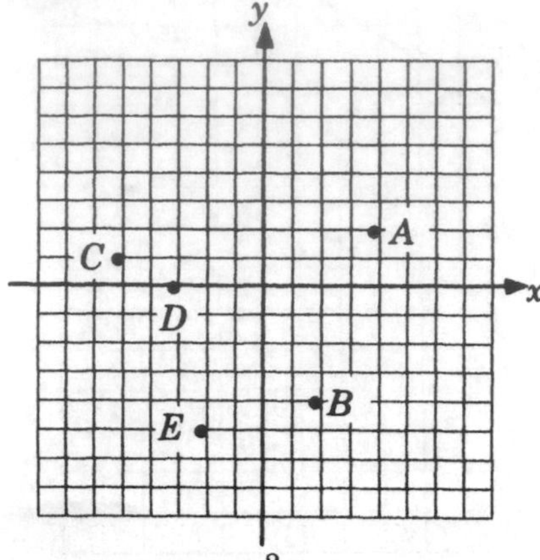

6. $P(4, -2)$ **7.** $Q(0, 3)$ **8.** $R(0, 0)$ **9.** $S(-5, -4)$ **10.** $T(5, 5)$
11. (5, −6) **12.** (4, 2)

13. x-intercept $= \frac{2}{3}$, y-intercept $= -2$ **14.** x-intercept = 8, y-intercept = −2 **15.** x-intercept = −2, y-intercept = −10

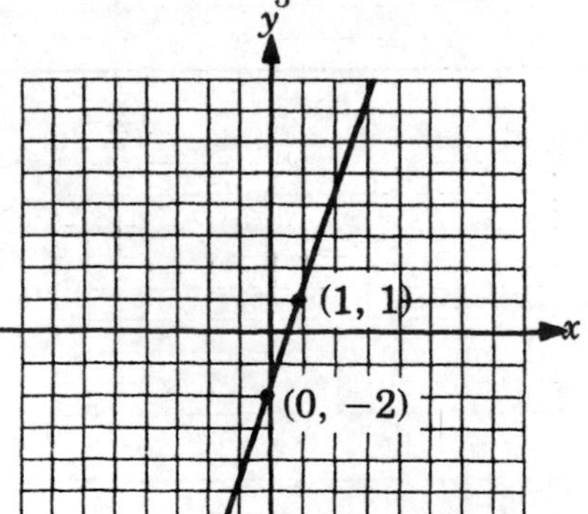

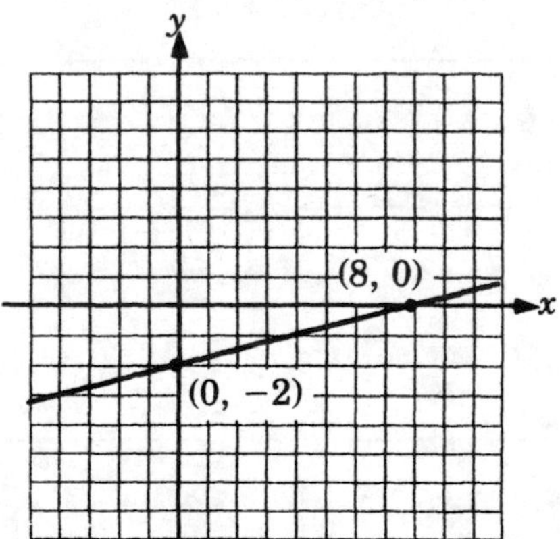

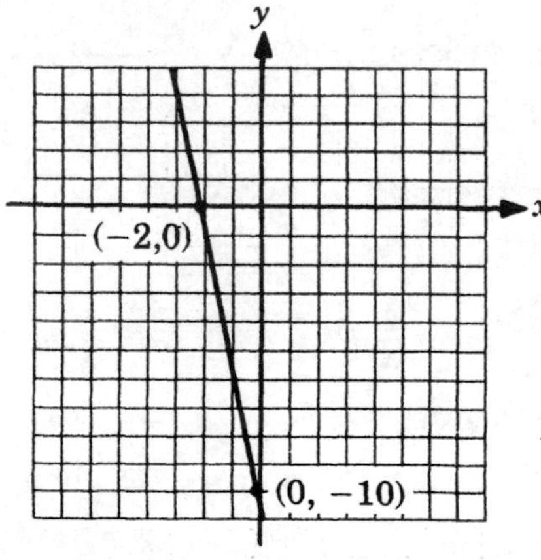

16. $m = \frac{-1}{2}$ **17.** $m = \frac{1}{3}$ **18.** $m = 2$, $b = 6$ **19.** $m = \frac{-1}{2}$, $b = 4$ **20.** Intersect at (3, −3)
21. Parallel, $m = 3$ **22.** $3x - 5y - 20 = 0$ **23.** $2x + y - 9 = 0$ **24.** $4x + y - 13 = 0$

CHAPTER 8

EXERCISES 8.1A

1. Intersecting **2.** Coincident **3.** Parallel **4.** Intersecting **5.** Coincident **6.** Parallel

EXERCISES 8.1B

1. Yes, no, no **2.** No, yes, no **3.** No, no, yes **4.** No, no, yes **5.** No, yes, no **6a.** No **6b.** No
6c. No **7a.** Yes **7b.** Yes **7c.** Yes **8a.** No **8b.** No **8c.** Yes **9a.** Yes **9b.** Yes
9c. Yes **10a.** No **10b.** No **10c.** No

EXERCISES 8.1 S

1. Intersecting **2.** Parallel **3.** Intersecting **4.** Intersecting **5.** Intersecting **6.** Intersecting
7. Parallel **8.** Coincident **9.** Parallel **10.** Intersecting **11.** No, yes, no **12.** No, yes, no
13. No, no, no **14.** Yes, yes, no **15.** No, no, no

EXERCISES 8.2

1. (7, 5)

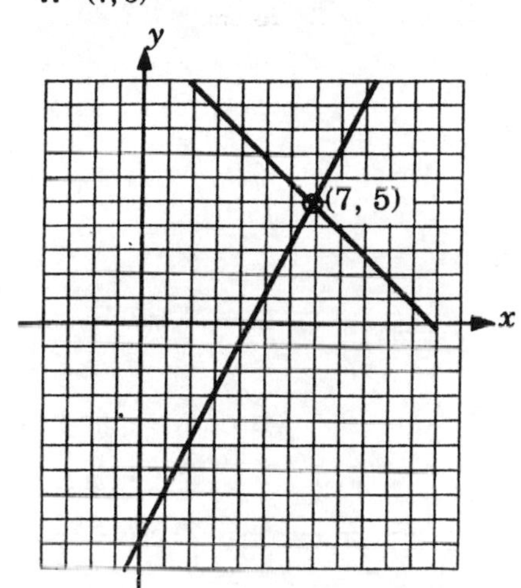

2. (6, 2)

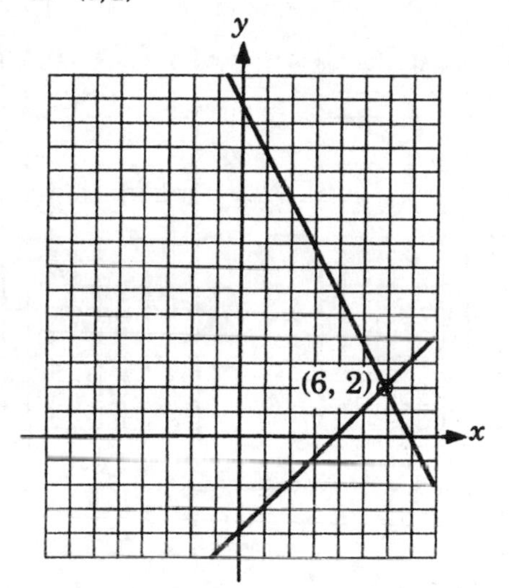

3. (6, −1)

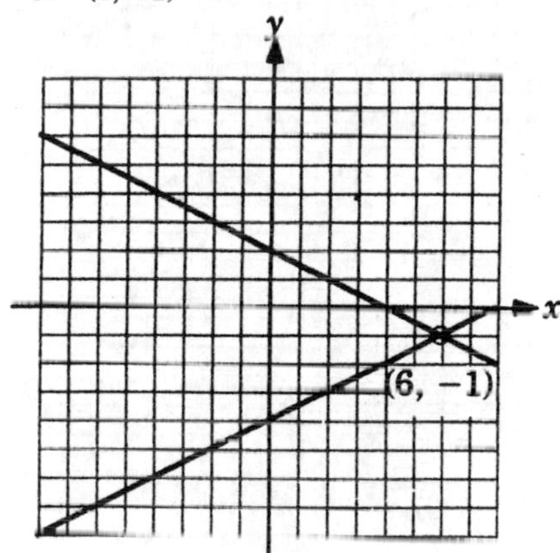

4. (−5, 3)

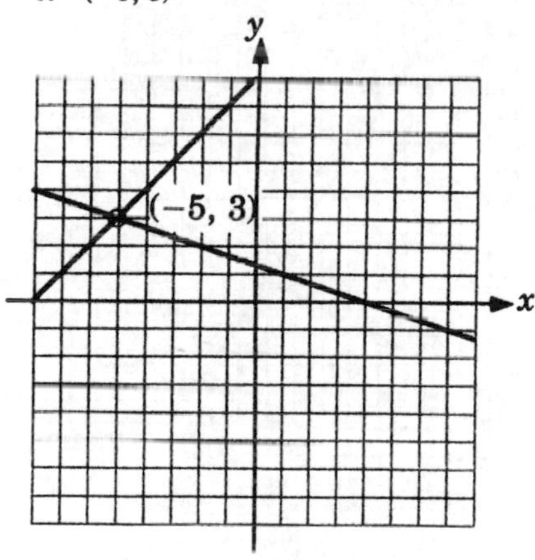

5. (−2, −3)

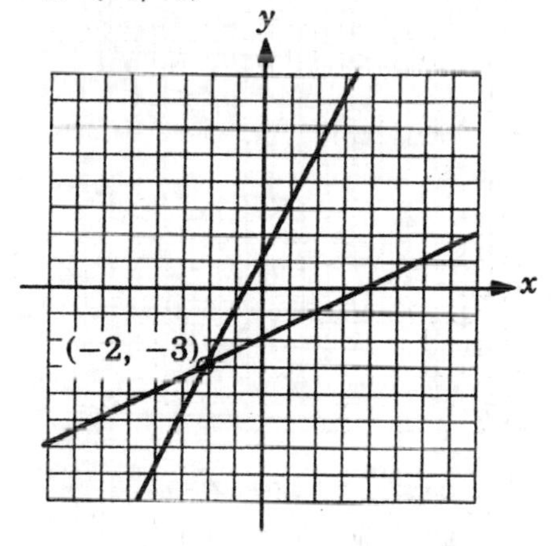

6. (3, 2)

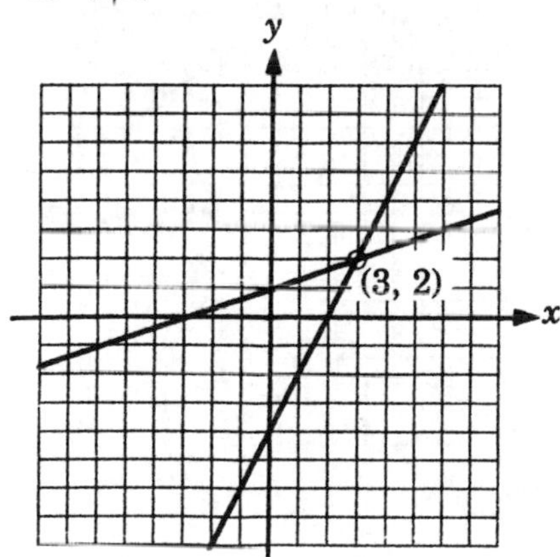

7. (2, 0)

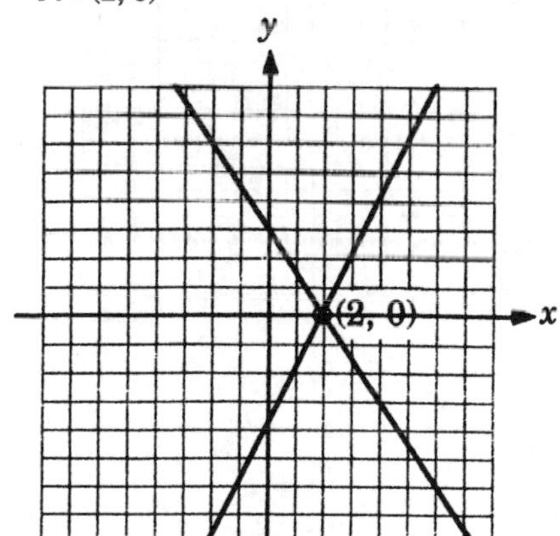

8. (−3, −2)

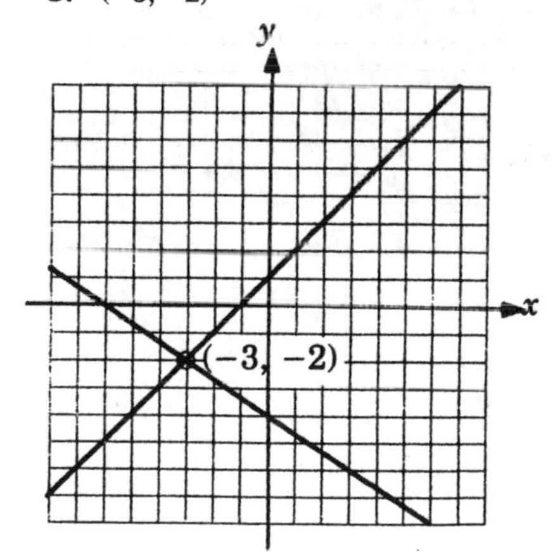

9. (3, 3)

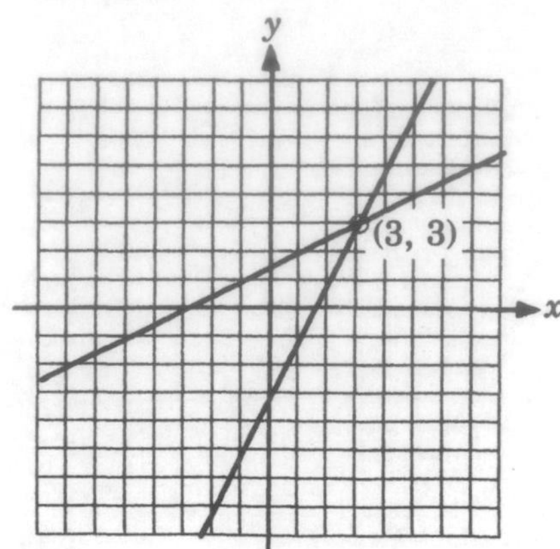

10. (0, 5)

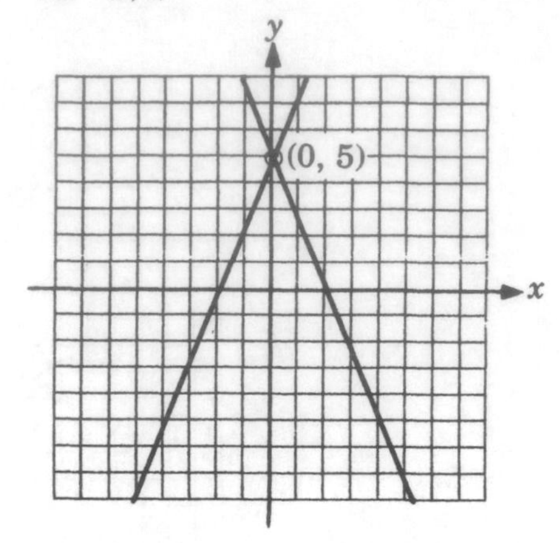

EXERCISES 8.2 S

1. (5, 2)

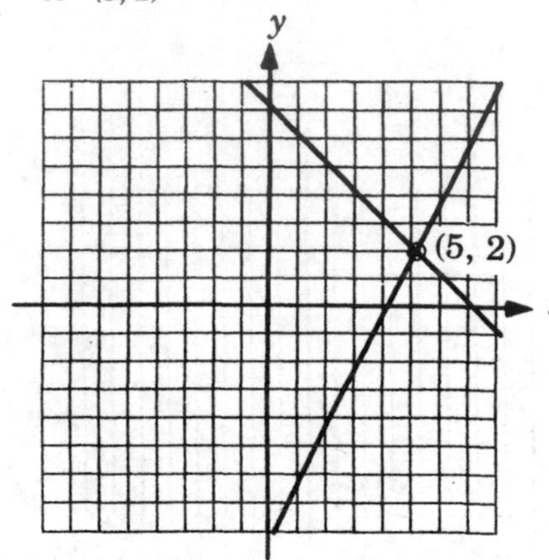

2. (4, −3)

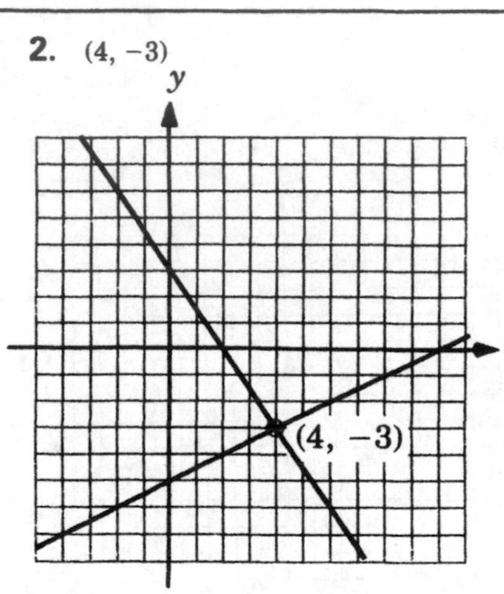

3. (−3, 6)

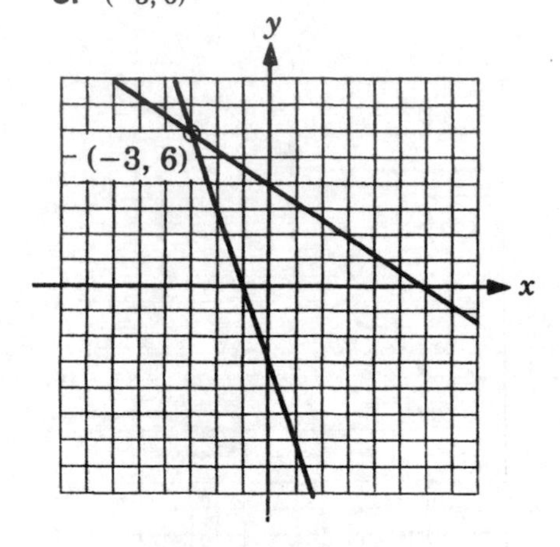

4. (−4, −6)

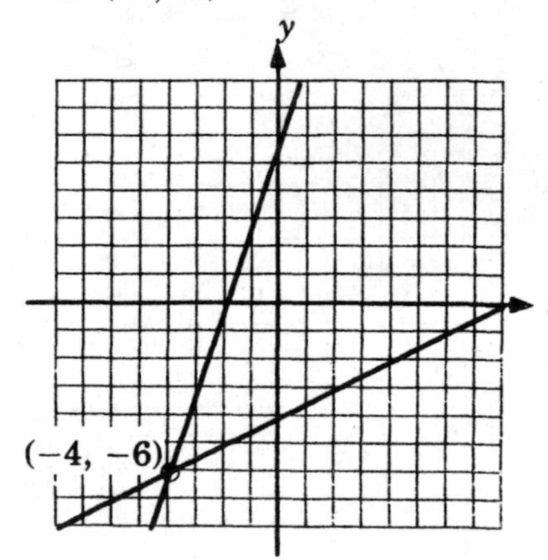

5. (1, 6)

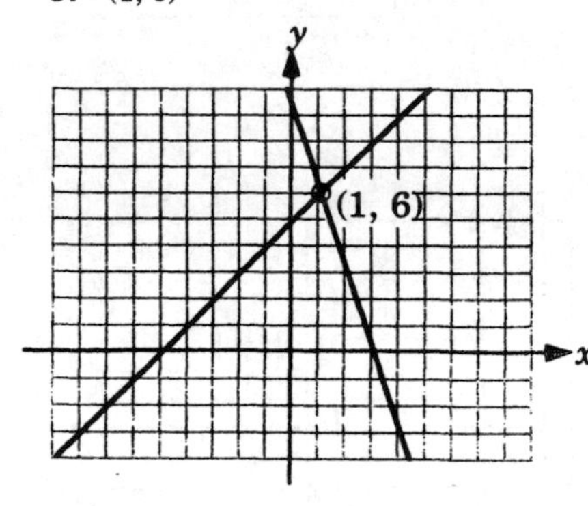

6. (5, 4)

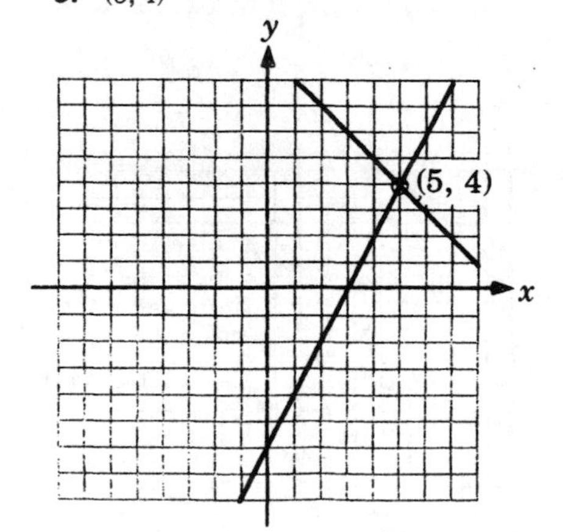

7. $(-2, 5)$

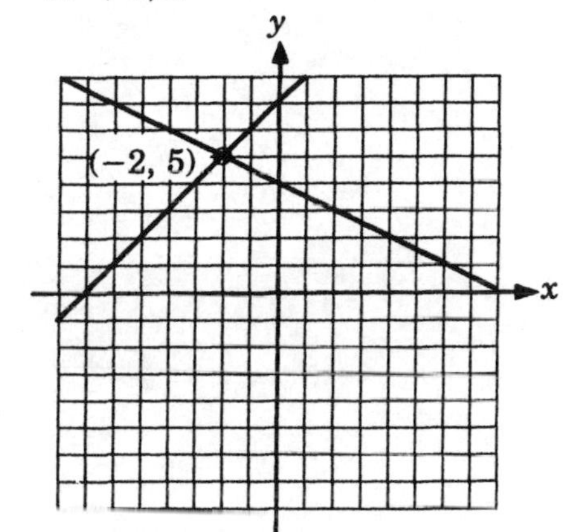

8. $(6, -4)$

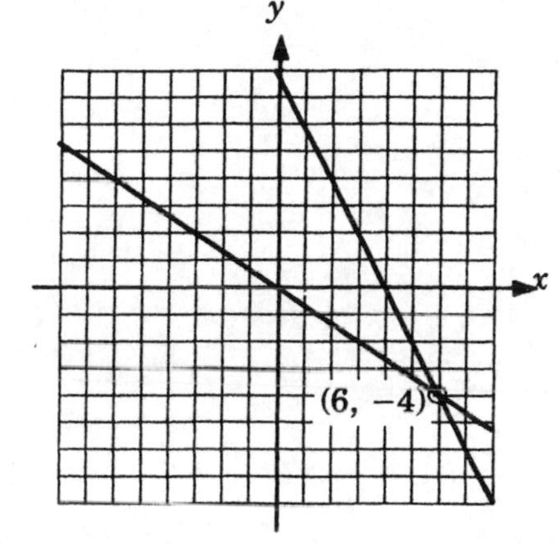

9. $(0, 0)$

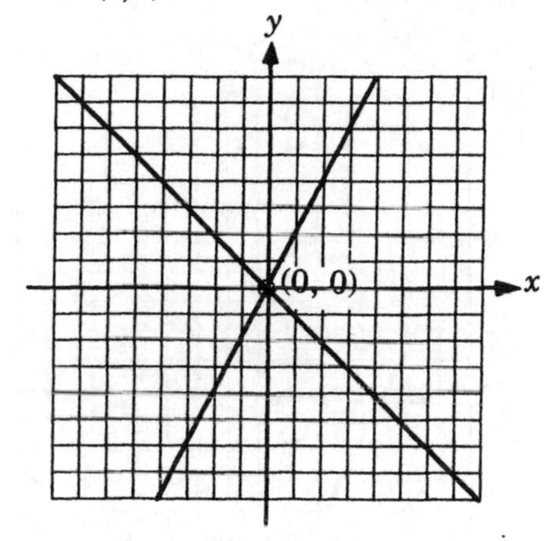

10. $(-6, 4)$

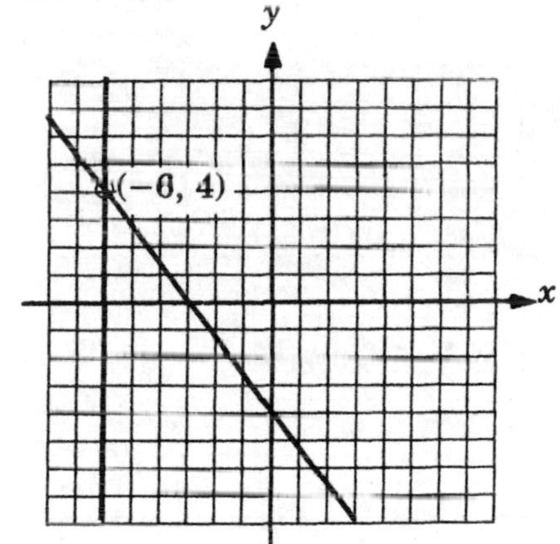

11. $\left(2\frac{1}{2}, 2\right)$

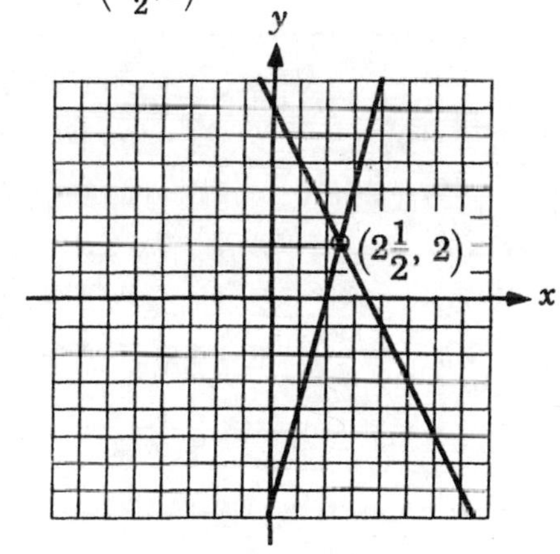

12. $\left(5, -4\frac{1}{2}\right)$

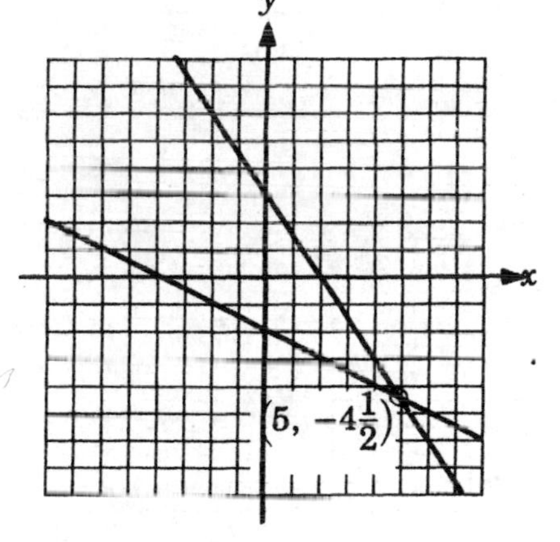

13. $(-5, -7)$

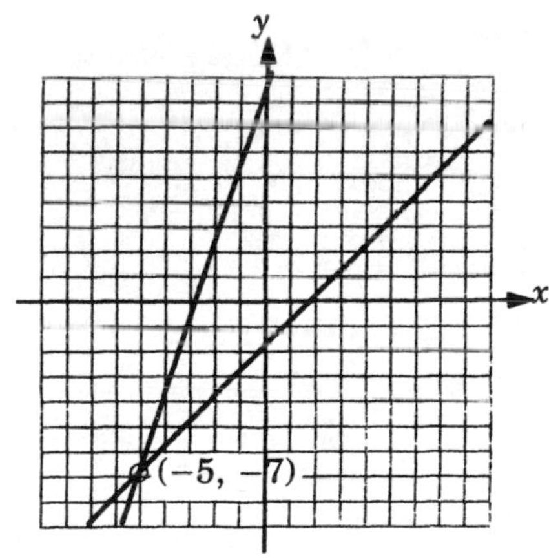

14. $(-6, 3)$

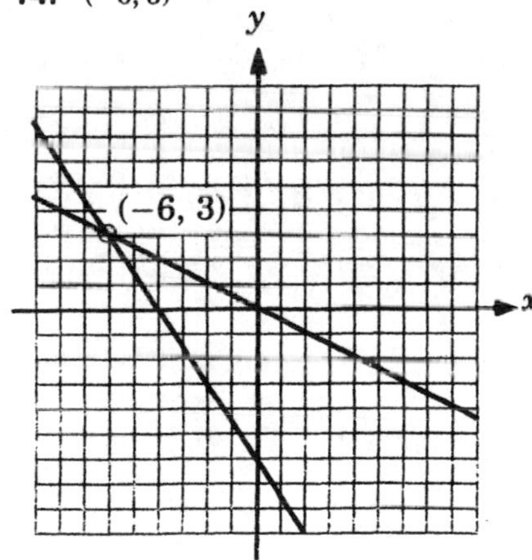

15. $(25, 20)$

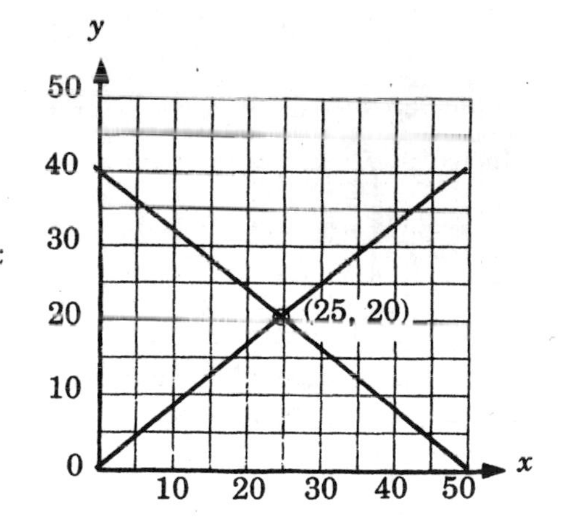

16. (30, 40)

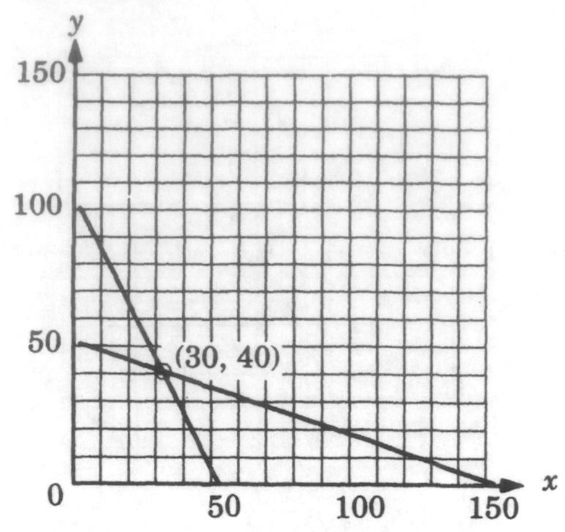

EXERCISES 8.3

1. (5, 4) **2.** (2, 1) **3.** (2, 0) **4.** (4, −10) **5.** (−3, −8) **6.** (8, 5) **7.** (8, 5) **8.** (3, 6)
9. $\left(-2, \frac{1}{2}\right)$ **10.** (5, 1) **11.** (9, −1) **12.** (−9, −3) **13.** (30, 20) **14.** (40, 50) **15.** (−8, −2)
16. (10, 5) **17.** (5, 4) **18.** (20, 10) **19.** (250, 50) **20.** (60, 20)

EXERCISES 8.3 S

1. (4, 20) **2.** (7, 4) **3.** (9, 4) **4.** (−3, 2) **5.** (10, −6) **6.** (−5, −8) **7.** (3, 2) **8.** (−10, 0)
9. (30, 5) **10.** (−6, 4) **11.** (18, 8) **12.** $\left(\frac{2}{5}, \frac{6}{5}\right)$ or (0.4, 1.2) **13.** $\left(\frac{-1}{3}, \frac{1}{6}\right)$ **14.** (50, 40)
15. (0.9, 0.7) **16.** $\left(\frac{9}{4}, \frac{5}{4}\right)$

EXERCISES 8.4

1. (6, 4) **2.** (9, −1) **3.** (12, 2) **4.** (5, 2) **5.** $\left(\frac{32}{5}, \frac{1}{5}\right)$ **6.** (7, −4) **7.** (5, 7) **8.** (8, −5)
9. (−3, −8) **10.** $\left(\frac{-1}{2}, \frac{7}{2}\right)$ **11.** (50, 30) **12.** (120, 300) **13.** (8, 5) **14.** (20, 30)

EXERCISES 8.4 S

1. (30, 12) **2.** (4, 5) **3.** (7, −4) **4.** (−6, 9) **5.** (8, 6) **6.** (−2, −3) **7.** (0, 12) **8.** $\left(\frac{3}{2}, \frac{5}{2}\right)$
9. (2, 0) **10.** (4, −7) **11.** $\left(\frac{2}{3}, \frac{1}{2}\right)$ **12.** $\left(\frac{2}{9}, \frac{11}{9}\right)$ **13.** (0.9, 0.1) **14.** $\left(\frac{-13}{2}, 7\right)$ **15.** $\left(\frac{2}{5}, \frac{-1}{5}\right)$
16. (2.3, 1.7)

EXERCISES 8.5

1. (5, 8), intersecting **2.** $\{(x, y) \mid x = 2y + 3\}$, coincident **3.** ∅, parallel **4.** ∅, parallel **5.** (4, 0), intersecting
6. $\{(x, y) \mid y = x + 6\}$, coincident

EXERCISES 8.5 S

1. ∅, parallel **2.** $\{(x, y) \mid 2x - 5y = 10\}$, coincident **3.** (35, 50), intersecting **4.** (75, 25), intersecting
5. $\{(x, y) \mid y = 15 - 5x\}$, coincident **6.** ∅, parallel **7.** (0, 0), intersecting **8.** $\left(\frac{3}{2}, \frac{3}{2}\right)$, intersecting
9. $\{(x, y) \mid y = 6 - 3x\}$, coincident **10.** ∅, parallel

EXERCISES 8.6A

1. Let x = price of lace per yard
Let y = price of seam binding per yard
$2x + 3y = 96$, $5x + 6y = 222$; $x = 30¢$, $y = 12¢$

2. Let x = price of oil per quart
Let y = price of gasoline per gallon
$x + 5y = 705$, $2x + 7y = 1038$; $x = 85¢$, $y = \$1.24$

3. Let a = percentage of copper in ore A
Let b = percentage of copper in ore B
$\frac{36a}{100} + \frac{28b}{100} = 39$, $\frac{50a}{100} + \frac{40b}{100} = 55$; $a = 50\%$, $b = 75\%$

4. Let x = percentage for stocks
Let y = percentage for bonds
$\frac{5000x}{100} + \frac{3000y}{100} = 450$, $\frac{2000x}{100} + \frac{1500y}{100} = 195$;
$x = 6\%$, $y = 5\%$

5. Let x = hourly wage of John
Let y = hourly wage of Jill
$52x + 40y = 420$, $40x + 25y = 300$; $x = \$5$, $y = \$4$

EXERCISES 8.6B

1. $520x + 460x = 1470$; $1\frac{1}{2}$ hr **2.** $6x + 4(x + 50) = 550$; 35 m.p.h. (freight), 85 m.p.h. (passenger)

3. $3(x + 3) = 39x$; $x = \frac{1}{4}$, 11:00 + 0:15 = 11:15 A.M. **4.** $90x = 80x + 5$; $x = \frac{1}{2}$ hr or 30 min

5. $6x = 7(x - 25)$; 175 m.p.h. **6.** $240x = 180(7 - x)$; $x = 3$ hr, $3(240) = 720$ mi

7. $3(a + w) = 630$; $\frac{7}{2}(a - w) = 630$
$w = 15$ m.p.h. (speed of wind)

8. $3(x + y) = 42$; $7(x - y) = 42$
$y = 4$ m.p.h. (rate of current)

EXERCISES 8.6 S

1. Let x = fixed fee per car
Let y = additional fee per passenger
$8x + 30y = 42.5$, $10x + 40y = 55$
$x = \$2.50$ (car), $y = 75¢$ (passenger)

2. Let x = price of each citron
Let y = price of each wood-apple
$9x + 7y = 107$, $7x + 9y = 101$;
$x = 8$ (citron), $y = 5$ (wood-apples)

3. Let x = number of hours carpenter worked first week
Let y = number of hours helper worked first week
$8x + 5y = 550$, $4x + 5(y + 10) = 400$
$x = 50$ hr, $y = 30$ hr

4. Let x = number of adults
Let y = number of children
$280x + 120y = 69{,}600$, $280(2x) + 120\left(\frac{y}{2}\right) = 85{,}200$;
$x = 120$ adults, $y = 300$ children

5. $80x + 60y = 3400$, $25x + 10y = 800$
$x = 20¢$ (each glass), $y = 30¢$ (each mug)

6. $2.5x + 4y = 1820$, $x + y = 500$
$x = 120$ sheets (seconds)

7. $40x + 35x = 450$; 6 hr **8.** $55x + 55(x + 2) = 330$; 1:00 P.M. **9.** $60x = 30(x + 1)$; 1 hr **10.** $25x = 9x + 16$; 1 hr

11. $60x = 20(2 - x)$; $x = \frac{1}{2}$ hr, $d = 30$ mi **12.** $9x = 6(5 - x)$; $x = 2$ hr, $d = 18$ mi

13. Let x = rate of boat; let y = rate of current
$\frac{1}{2}(x + y) = \frac{15}{2}$, $\frac{3}{4}(x - y) = \frac{15}{2}$
$x = 12\frac{1}{2}$ m.p.h., $y = 2\frac{1}{2}$ m.p.h.

14. $5(x + y) = 1200$; $6(x - y) = 1200$
$y = 20$ m.p.h., speed of wind

SAMPLE TEST FOR CHAPTER 8

1. Coincident **2.** Intersect **3.** Parallel **4a.** Yes **4b.** No

5. $\left(\frac{-5}{2}, 3\right)$

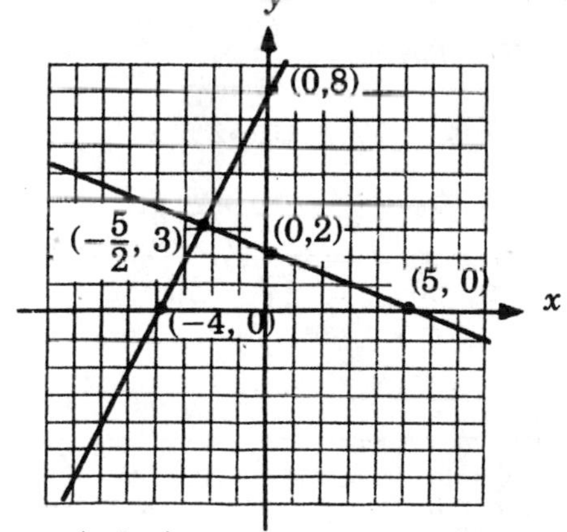

6. $\left(\frac{-5}{2}, 3\right)$ **7.** $(-6, 4)$ **8.** $(4, 2)$ **9.** $\{(x, y)|3y = x + 2\}$; coincident lines **10.** $(0, 4)$, intersecting lines

11. $\varnothing$, parallel lines **12.** 5 stoves and 8 refrigerators to Mexico, 8 stoves and 5 refrigerators to Canada

13. 25 m.p.h.

SAMPLE TEST FOR CHAPTERS 7 AND 8

1a. 8 **1b.** −6 **1c.** $\frac{3}{4}$ **2.** Slope = −2
3a. $3x - y = 9$ **3b.** $x = 3y - 4$ **3c.** $6x - 2y = 8$ **4a.** $2x + y = 14$: (2, 10) and (7, 0); $y = 3x - 6$: (0, −6) and (2, 0)
4b.

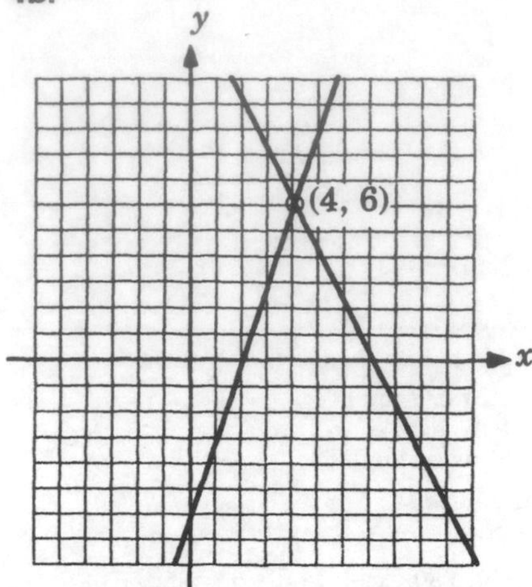

4c. (4, 6) **4d.** 14 = 14 and 6 = 6 **5.** (5, −3) **6.** 8 = 8 and 1 = 1 **7.** (−1, 4) **8.** 8 = 8 and 2 = 2
9. ∅ **10.** $\{(x, y) | 2x + 2y = 20\}$

CHAPTER 9

EXERCISES 9.1

1. 8 **2.** 9 **3.** 11 **4.** 10 **5.** 8 **6.** 7 **7.** Not real **8.** −4 **9.** Not real **10.** 0
11. 10 **12.** 5 **13.** 8 **14.** −4 **15.** 5 **16.** $\frac{-5}{3}$ **17.** 4.47 **18.** 9.75 **19.** 4.47 **20.** 1.58
21. −0.55 **22.** 3.41 **23.** $\{\sqrt{5}, -\sqrt{5}\}$ **24.** $\left\{\frac{2\sqrt{10}}{3}, \frac{-2\sqrt{10}}{3}\right\}$ **25.** {3, −3} **26.** {70, −70}
27. {7.5, −7.5} **28.** $\left\{\frac{\sqrt{69}}{10}, \frac{-\sqrt{69}}{10}\right\}$

EXERCISES 9.1 S

1. 3 **2.** −2 **3.** 13 **4.** 15 **5.** 15 **6.** Not real **7.** −6 **8.** 36 **9.** 16 **10.** 3 **11.** 0
12. 0 **13.** 3 **14.** 9 **15.** $\frac{-1}{4}$ **16.** 5 **17.** 7.48 **18.** 9.33 **19.** 7.94 **20.** 1.29 **21.** 4.35
22. 0.63 **23.** $\{\sqrt{7}, -\sqrt{7}\}$ **24.** $\left\{\frac{\sqrt{73}}{5}, \frac{-\sqrt{73}}{5}\right\}$ **25.** {40, −40} **26.** {20, −20} **27.** {2.3, −2.3}
28. {0.82, −0.82}

EXERCISES 9.2

1. $6\sqrt{5}$ **2.** $3\sqrt{6}$ **3.** $5\sqrt{7}$ **4.** $7\sqrt{3}$ **5.** $20\sqrt{10}$ **6.** 14 **7.** 12 **8.** 10 **9.** 14 **10.** 36
11. 85 **12.** $2\sqrt{3}$ **13.** $3\sqrt{5}$ **14.** $10\sqrt{6}$ **15.** $7\sqrt{2}$ **16.** $2\sqrt{2}$ **17.** $4\sqrt{10}$ **18.** $6\sqrt{5}$ **19.** 30
20. $6\sqrt{5}$

EXERCISES 9.2 S

1. $7\sqrt{17}$ **2.** $8\sqrt{3}$ **3.** $9\sqrt{10}$ **4.** $4\sqrt{2}$ **5.** $4\sqrt{5}$ **6.** 6 **7.** 12 **8.** 15 **9.** 18 **10.** 44
11. 99 **12.** $2\sqrt{7}$ **13.** $3\sqrt{6}$ **14.** $5\sqrt{3}$ **15.** $6\sqrt{5}$ **16.** $3\sqrt{3}$ **17.** $8\sqrt{3}$ **18.** $7\sqrt{2}$ **19.** $80\sqrt{5}$
20. $15\sqrt{7}$

EXERCISES 9.3

1. $\frac{\sqrt{2}}{2}$ **2.** $\frac{\sqrt{14}}{7}$ **3.** $\frac{\sqrt{3}}{9}$ **4.** $\frac{2\sqrt{3}}{15}$ **5.** $\frac{\sqrt{6}}{6}$ **6.** $\frac{\sqrt{7}}{7}$ **7.** $\frac{\sqrt{6}}{12}$ **8.** $\frac{\sqrt{14}}{14}$ **9.** $3\sqrt{3}$
10. $\frac{9}{10}$ **11.** $\frac{\sqrt{14}}{5}$ **12.** $\frac{\sqrt{30}}{50}$ **13.** $\frac{\sqrt{2}}{2}$ **14.** $\frac{\sqrt{42}}{2}$ **15.** $\frac{5\sqrt{3}}{2}$ **16.** $0.1 = \frac{1}{10}$

EXERCISES 9.3 S

1. $\frac{\sqrt{3}}{3}$ **2.** $\frac{\sqrt{30}}{6}$ **3.** $\frac{\sqrt{2}}{10}$ **4.** $\frac{3\sqrt{2}}{8}$ **5.** $\frac{\sqrt{10}}{5}$ **6.** $\frac{\sqrt{14}}{2}$ **7.** $\frac{\sqrt{14}}{28}$ **8.** $\frac{\sqrt{2}}{6}$ **9.** $\frac{7\sqrt{10}}{100}$
10. $\frac{\sqrt{105}}{50}$ **11.** $\frac{\sqrt{10}}{4}$ **12.** $\frac{3}{2}$ **13.** $\frac{\sqrt{6}}{3}$ **14.** $\frac{\sqrt{10}}{2}$ **15.** $\frac{4\sqrt{2}}{3}$ **16.** $\frac{\sqrt{5}}{5}$

EXERCISES 9.4

1. $7\sqrt{5}$ **2.** $3\sqrt{2}$ **3.** $\sqrt{10}$ **4.** $6\sqrt{2} + 5\sqrt{3}$ **5.** $7 + 2\sqrt{7}$ **6.** $\sqrt{6} - 6$ **7.** $\frac{3 + \sqrt{2}}{2}$ **8.** $\frac{10 - 2\sqrt{3}}{5}$
9. $\frac{2 - \sqrt{14}}{2}$ **10.** $\frac{2}{3}$ **11.** $\frac{6 - \sqrt{2}}{3}$ **12.** $\frac{3 - \sqrt{5}}{2}$ **13a,b.** $15 - 6\sqrt{6}$ **14a,b.** $21 + 8\sqrt{5}$
15a. $11 - 4\sqrt{7}$ **15b.** $-11 + 4\sqrt{7}$ **16a.** $7 + 4\sqrt{3}$ **16b.** $-7 - 4\sqrt{3}$

EXERCISES 9.4 S

1. $7\sqrt{3}$ **2.** $\sqrt{6}$ **3.** $2\sqrt{10} - 5\sqrt{2}$ **4.** $2\sqrt{3} + 4\sqrt{2}$ **5.** $6\sqrt{3} - 9$ **6.** $5 + 2\sqrt{5}$ **7.** $\frac{1 + \sqrt{5}}{2}$
8. $\frac{2 - \sqrt{6}}{5}$ **9.** $7 + \sqrt{15}$ **10.** $-\frac{1}{2}$ **11.** $\frac{1 - \sqrt{3}}{2}$ **12.** $\frac{14 - 5\sqrt{3}}{7}$ **13a,b.** $11 + 6\sqrt{2}$
14a,b. $9 + 4\sqrt{5}$ **15a,b.** $35 - 10\sqrt{10}$ **16a,b.** $11 - 4\sqrt{7}$ **17a,b.** $22 - 8\sqrt{6}$ **18a,b.** $24 + 6\sqrt{15}$
19. $23 + 8\sqrt{7} = 23 + 8\sqrt{7}$ and $23 - 8\sqrt{7} = 23 - 8\sqrt{7}$ **20.** $10 - 4\sqrt{6} + (-10 + 4\sqrt{6}) = 0$ and $10 + 4\sqrt{6} + (-10 - 4\sqrt{6}) = 0$

EXERCISES 9.5

1. $\{4 + \sqrt{5}, 4 - \sqrt{5}\}$ **2.** $\{-2 + \sqrt{3}, -2 - \sqrt{3}\}$ **3.** $\{8 + 2\sqrt{3}, 8 - 2\sqrt{3}\}$ **4.** $\left\{\frac{-3 + \sqrt{10}}{2}, \frac{-3 - \sqrt{10}}{2}\right\}$
5. $\left\{\frac{1 + 2\sqrt{5}}{5}, \frac{1 - 2\sqrt{5}}{5}\right\}$ **6.** $\{-1 + \sqrt{2}, -1 - \sqrt{2}\}$ **7.** $\{3 + \sqrt{6}, 3 - \sqrt{6}\}$ **8.** $\{4 + 5\sqrt{2}, 4 - 5\sqrt{2}\}$
9. $\{-2 + \sqrt{10}, -2 - \sqrt{10}\}$ **10.** $\{5 + \sqrt{7}, 5 - \sqrt{7}\}$ **11.** $\{1 + 2\sqrt{2}, 1 - 2\sqrt{2}\}$ **12.** $\{-6 + 2\sqrt{3}, -6 - 2\sqrt{3}\}$
13. $\left\{\frac{6 \pm \sqrt{3}}{3}\right\}$ **14.** $\left\{\frac{-5 \pm \sqrt{10}}{5}\right\}$ **15.** $\left\{\frac{6 \pm 3\sqrt{3}}{2}\right\}$ **16.** $\left\{\frac{8 \pm \sqrt{2}}{2}\right\}$ **17.** $\left\{\frac{-3 \pm \sqrt{10}}{2}\right\}$ **18.** $\left\{3, \frac{-7}{3}\right\}$
19. $\left\{\frac{5}{4}\right\}$ **20.** $\left\{\frac{-1 \pm \sqrt{17}}{2}\right\}$

EXERCISES 9.5 S

1. $\{-7 + \sqrt{6}, -7 - \sqrt{6}\}$ **2.** $\{5 + \sqrt{7}, 5 - \sqrt{7}\}$ **3.** $\{-9 + 3\sqrt{2}, -9 - 3\sqrt{2}\}$ **4.** $\left\{\frac{4 + \sqrt{14}}{3}, \frac{4 - \sqrt{14}}{3}\right\}$
5. $\left\{\frac{-1 + 4\sqrt{2}}{6}, \frac{-1 - 4\sqrt{2}}{6}\right\}$ **6.** $\{2 + \sqrt{5}, 2 - \sqrt{5}\}$ **7.** $\{-5 + \sqrt{10}, -5 - \sqrt{10}\}$ **8.** $\{-6 + 2\sqrt{7}, -6 - 2\sqrt{7}\}$
9. $\{1 \pm 2\sqrt{3}\}$ **10.** $\{3 \pm 3\sqrt{5}\}$ **11.** $\{-7 \pm 3\sqrt{2}\}$ **12.** $\{10 \pm 5\sqrt{3}\}$ **13.** $\{-3 \pm 3\sqrt{6}\}$ **14.** $\{-1 \pm \sqrt{5}\}$
15. $\{5 \pm 2\sqrt{10}\}$ **16.** $\{6 \pm 6\sqrt{2}\}$ **17.** $\left\{\frac{1 \pm \sqrt{7}}{2}\right\}$ **18.** $\left\{\frac{1}{2}, \frac{-7}{2}\right\}$ **19.** $\left\{\frac{-6 \pm \sqrt{7}}{3}\right\}$ **20.** $\left\{\frac{6 \pm \sqrt{3}}{2}\right\}$

EXERCISES 9.6

1. $\frac{-5 \pm \sqrt{13}}{6}$ **2.** $\frac{3 \pm \sqrt{13}}{2}$ **3.** $3 \pm \sqrt{5}$ **4.** $\frac{-1 \pm \sqrt{7}}{3}$ **5.** $-3, \frac{6}{5}$ **6.** $60, 20$ **7.** $\frac{-1 \pm \sqrt{5}}{2}$
8. $-2 \pm \sqrt{6}$ **9.** $1 \pm \sqrt{2}$ **10.** $\frac{5}{4}$ **11.** $\frac{7}{2}, \frac{-5}{2}$ **12.** $\frac{-1}{2}, 1$ **13.** $-b \pm \sqrt{b^2 - c}$ **14.** $x = 2y$ or $x = \frac{1}{2}y$

EXERCISES 9.6 S

1. $\frac{-9 \pm \sqrt{21}}{6}$ **2.** $\frac{2 \pm \sqrt{5}}{3}$ **3.** $\frac{-5 \pm \sqrt{57}}{4}$ **4.** $3, \frac{1}{2}$ **5.** $2, \frac{-9}{10}$ **6.** $30, -200$ **7.** $\frac{1 \pm \sqrt{5}}{2}$
8. $5 \pm 2\sqrt{7}$ **9.** $6 \pm 4\sqrt{3}$ **10.** $\frac{-7}{6}$ **11.** $\frac{1 \pm \sqrt{2}}{3}$ **12.** $80, -60$ **13.** $r \pm \sqrt{r^2 + s}$
14. $y = -5x$ or $y = 3x$

EXERCISES 9.7 S

1. 9.86 thousand mi **2.** 47.7 mi **3.** 3360, −3360 **4.** 2640, −4140 **5a.** 0, 12 **5b.** 3
5c. $6 \pm 2\sqrt{3}$; approx. 9.5 and 2.5 **6.** 20 hr **7.** 6 hr **8.** 56 min **9.** 45 hr
10. 20 sec (going up), 30 sec (coming down) **11.** 3 sec **12a.** 15 or 25 **12b.** 20 **13.** 4 m.p.h.
14. 550 m.p.h. **15.** 3 ft **16.** $40 - 16\sqrt{5} = 4.2$ ft, approx.

SAMPLE TEST FOR CHAPTER 9

1. 6 **2.** $\frac{16}{5}$ **3.** 41.23 **4.** -0.22 **5.** $\left\{\frac{\sqrt{35}}{3}, \frac{-\sqrt{35}}{3}\right\}$ **6.** $\{5.4, -5.4\}$ **7.** 42 **8.** $8\sqrt{3}$ **9.** $\frac{\sqrt{6}}{6}$

10. $\frac{\sqrt{21}}{6}$ **11.** $9\sqrt{6}$ **12.** $\frac{6-\sqrt{10}}{2}$ **13.** $\{-4+2\sqrt{3}, -4-2\sqrt{3}\}$ **14.** $\left\{\frac{2+2\sqrt{7}}{3}, \frac{2-2\sqrt{7}}{3}\right\}$

15. $\left\{\frac{2+\sqrt{6}}{3}, \frac{2-\sqrt{6}}{3}\right\}$ **16.** $\left\{\frac{-1+\sqrt{3}}{2}, \frac{-1-\sqrt{3}}{2}\right\}$ **17.** $x(x+3)+\frac{x^2}{2}=72$; 6 ft by 9 ft

18. $\frac{36}{20-x}=\frac{36}{20+x}+\frac{3}{4}$; 4 m.p.h

Index

39th Edition

Warman's®

Antiques & Collectibles Price Guide

Edited by

Ellen T. Schroy

Published by

700 East State Street • Iola, WI 54990-0001
715-445-2214 • 888-457-2873

Our toll-free number to place an order or obtain
a free catalog is (800) 258-0929.

Library of Congress Catalog Number: 1076-1985

ISBN: 0-87349-990-5

Designed by Kay Sanders
Edited by Kristine Manty

Printed in the United States of America

On the front cover, from top left: Brooch/pin, Arts and Crafts, silver, enamel, opal, fresh-water pearl, c. 1900, shield-shaped plaque, stylized branchlike motif bezel-set with six small circular blue-green opal cabochons across top, enclosing large oval bezel-set opal with predominately red play-of-color, on a green plique à jour enameled ground, fresh-water pearl drop, reverse marked "DEPOSÉ," Fr import mark stamped on C-catch, attributed to Heinrich Levinger, Pforzhein (Germany), 1-1/4" w x 1-1/4" l, $1,200; carnival glass punch bowl, Northwood Grape and Cable, horehound/marigold, mid-size, part of a set with eight cups, $2,100; Gustav Stickley humpback rocking chair, Arts and Crafts, 1902, oak, original finish, small mark, seat cover replaced, 38" h, 28" w, $1,600-$1,800.
On the back cover: Roseville pottery, Gardenia vase, bulbous, golden tan, 684-8", $350-$400.